Turfgrass Science and Management

SECOND EDITION

Turfgrass Science and Management

SECOND EDITION

Robert D. Emmons

Delmar Publishers' Online Services
To access Delmar on the World Wide Web, point your browser to:
http://www.delmar.com/delmar.html
To access through Gopher: gopher://gopher.delmar.com
(Delmar Online is part of "thomson.com", an Internet site with information on more than 30 publishers of the International Thomson Publishing organization.)
For information on our products and services:
email: info@delmar.com
or call 800-347-7707

Delmar Publishers™
I(T)P® An International Thomson Publishing Company

Albany • Bonn • Boston • Cincinnati • Detroit • London • Madrid • Melbourne • Mexico City
New York • Pacific Grove • Paris • San Francisco • Singapore • Tokyo • Toronto • Washington

NOTICE TO THE READER

Publisher does not warrant or guarantee any of the products described herein or perform any independent analysis in connection with any of the product information contained herein. Publisher does not assume, and expressly disclaims, any obligation to obtain and include information other than that provided to it by the manufacturer.

The reader is expressly warned to consider and adopt all safety precautions that might be indicated by the activities described herein and to avoid all potential hazards. By following the instructions contained herein, the reader willingly assumes all risks in connection with such instructions.

The publisher makes no representations or warranties of any kind, including but not limited to, the warranties of fitness for particular purpose or merchantability, nor are any such representations implied with respect to the material set forth herein, and the publisher takes no responsibility with respect to such material. The publisher shall not be liable for any special, consequential or exemplary damages resulting, in whole or in part, from the readers' use of, or reliance upon, this material.

Cover photo courtesy of Albany Country Club. Photo by Michael Dzaman.
Cover Design: Wendy Troeger

Delmar Staff
Publisher: Tim O'Leary
Administrative Editor: Cathy Esperti
Developmental Editor: Carolyn Ellis
Senior Project Editor: Andrea Edwards Myers
Production Manager: Wendy Troeger

Printed in the United States of America

For more information, contact:

Delmar Publishers
3 Columbia Circle, Box 15015
Albany, NY 12212-5015

International Thomson Publishing
Berkshire House 168-173
High Holborn
London, WC1V7AA
England

Thomas Nelson Australia
102 Dodds Street
South Melbourne, 3205
Victoria, Australia

Nelson Canada
1120 Birchmont Road
Scarborough, Ontario
Canada M1K 5G4

International Thomson Editores
Campos Eliseos 385, Piso 7
Col Polanco
11560 Mexico D F Mexico

International Thomson Publishing GmbH
Königswinterer Strasse 418
53227 Bonn
Germany

International Thomson Publishing Asia
221 Henderson Road
#05-10 Henderson Building
Singapore 0315

International Thomson Publishing Japan
Hirakawacho Kyowa Building, 3F
2-2-1 Hirakawacho
Chiyoda-ku, Tokyo 102
Japan

5 6 7 8 9 10 XXX 01 00 99 98 97

Library of Congress Cataloging-in-Publication Data

Emmons, Robert D., 1945–
Turfgrass science and management / Robert D. Emmons. — 2nd ed.
p. cm.
Includes bibliographical references and index.
ISBN 0-8273-6598-5 (textbook)
1. Turfgrasses. 2. Turf management. 3. Turfgrasses industry.
I. Title.
SB433.E44 1995
635.9'642—dc20 94-30692
CIP

Contents

Preface

THE INDUSTRY

The turfgrass industry in the United States and Canada is a multibillion-dollar-a-year business. It is one of the fastest-growing segments of the horticulture industry.

One important factor in the growth of the turfgrass industry is the amount of change that has occurred in the American lifestyle in the last several decades. The technological revolution has speeded up the pace of life so that a peculiar contrast exists: On one side, individuals face increased pressures from responsibilities to family and career; on the other side, there is more free time available (as a result of laborsaving devices and services). Both of these situations have led to a demand for "green space." Both private homes and public buildings show the increased emphasis on the use of plant materials to create an attractive, soothing background in which to conduct everyday activities. In parks, recreational facilities, and golf courses, leisure time may be enjoyed in healthful, attractive surroundings.

The establishment and maintenance of these "green spaces" requires a large support system to provide necessary services. As a result, there are many career opportunities for students to consider. The turfgrass industry provides jobs not only in the areas of installation and maintenance of turf but also in the manufacture of equipment, sales and service of equipment, horticultural chemicals, horticultural supplies, sod farming, seed production, garden centers, and other retail outlets. Each of these areas needs men and women who are turf specialists. Personnel for direct installation and/or maintenance of turf areas are needed for golf courses, national, state, and local parks, recreational areas, playing fields, and lawn care businesses, as well as schools, industrial parks, hospitals, apartments, condominiums, government buildings, and other institutions.

DESIGN OF THE TEXT

This text is intended both for people who are presently working as turfgrass managers and for students—the turfgrass managers of the future. It discusses the establishment and maintenance practices used by successful turfgrass managers. It also deals with the relevant scientific theory and practical management skills.

This book begins with an overview of the industry, showing its relative size and the opportunities available to people knowledgeable in turf practices. The author describes the general responsibilities of a turf manager and discusses the educational background desirable. The following chapters thoroughly cover specific aspects of turfgrass science, beginning with the classification of turfgrasses and descriptions of plant characteristics. The chapters on warm season and cool season grasses describe specific grass varieties for the major climatic regions of the United States. Tables summarize the characteristics of the species presented to assist the turf worker in selecting suitable turfgrasses for actual installation. The use of both cool season and warm season grasses in the transition zones between the major regions is also investigated.

Chapters 5 to 8 thoroughly discuss the role of soil in successful turf installation and maintenance. Soil criteria are established to meet the needs of various species. The student learns how to evaluate the condition of the soil, how to test it for various characteristics, including nutrients, and then how to modify the soil to achieve the desired medium for turf growth.

Chapter 9 utilizes the previous material by showing how an evaluation of factors results in the selection of turfgrass species for a specific site. Methods of establishment are discussed. Once growth has been initiated, there are a number of practices required to promote a healthy turf area. Chapters 10 to 18 cover the topics of fertilization and fertilizer programs; mowing techniques and equipment; irrigation needs and systems; pesticide types and application; weeds and their control with herbicides; insect pests and their control; diseases common to turfgrasses and their control; integrated pest management programs; tolerance of grass species to shade; the effects of compaction on turf growth; and the problems associated with thatch.

Following these thorough discussions of the problems which can afflict turfgrasses, the student is given clear guidelines on how to evaluate an unsatisfactory turf and determine the steps required to correct the problem.

Chapter 20 discusses the special requirements of managing golf course turf. In addition to establishment and maintenance practices, the author discusses the special needs and expectations of golfers versus the costs of creating and maintaining high-quality turf.

Chapter 21 provides guidelines for maintenance of lawns, athletic fields, parks, cemeteries, sod farms, and other sites.

An introduction to business management practices for the turf manager is provided in Chapter 22. Recommendations for supervising personnel are included, as well as other topics such as recordkeeping and budget preparation.

The final chapter presents typical math calculations required of turf workers. These examples show students how to approach each type of practical problem by setting up the known information and solving for the unknown quantity or quantities.

FEATURES OF THE TEXT

- Thorough, well-explained presentation of turf principles.
- Blend of turf science and practical applications.
- Warm season and cool season grasses are described, as well as the use of both types in the transition zones.
- Recommended turf practices are always related to the benefits to be obtained.
- The chapter on golf course management provides information on specialized needs of turf species used on golf courses and the methods used to satisfy these needs.
- The chapter on business management provides both general information on business practices and more specialized guidelines of use to the turf manager.
- Mathematical problems relating to turf science are presented in solved form; additional practice problems are given.
- A glossary provides definitions of special terms relating to turfgrass science and management.
- The appendices include tables of customary-to-metric and metric-to-customary conversions, a sprayer calibration procedure, spreader calibration procedures, and a chart covering the identification of turfgrasses by vegetative characteristics.

The information in this text has been derived from numerous resource materials. Many valuable contributions have resulted from discussions with turfgrass managers and turfgrass extension specialists and visits to research facilities. The author's own experiences as a golf course superintendent, extension specialist, consultant, and teacher are drawn upon.

An Instructor's Guide provides the answers to the chapter-end review questions, plus additional questions and answers for the instructor's use in developing tests.

AUTHOR'S ACKNOWLEDGMENTS

I an indebted to the many people who have contributed to the preparation of this text.

To the following reviewers I express my gratitude for their in-depth critiques and recommendations.

Robert Boufford, Loudonville, Ohio
Dr. William Knoop, Dallas, Texas
Gary Parkert, Buckley, Washington

Appreciation is also expressed to photographers James R. Bates and Ted Bruetsch for sharing their expertise and preparing photographs of various turf conditions.

I would especially like to thank Ralph Smalley, Malcolm Johnson, Art Wick, Jack Murray, Stan Frederiksen, and Bill Knoop, who inspired me to pursue a career in turfgrass education and whose contribution to this book were invaluable. My editor, Carolyn Ellis, has been wonderfully supportive, helpful, and patient. Finally, I am eternally grateful to my wife Holly for her numerous contributions. Her support throughout this lengthy project was the reason I finished it rather than slipping away to Tahiti. I would also like to thank my son Lee for his patience on those occasions when I had to work on the book revision rather than take him fishing or playing ball with him.

ABOUT THE AUTHOR

Robert D. Emmons is a professor in the Plant Science department of the State University of New York at Cobleskill. He is responsible for the turfgrass management and golf course management programs at the school and he teaches an integrated pest management course. Mr. Emmons has had over 5,000 students in his classes.

Before joining the state university system, Mr. Emmons worked as a golf course superintendent for several years and as a turfgrass extension specialist. In addition to his present teaching responsibilities, Mr. Emmons works very closely with the New York State Turfgrass Association. He has been awarded that organization's Citation of Merit for outstanding contributions to the turfgrass industry. He has also received a Certificate of Merit from the United States Department of Agriculture. Mr. Emmons is a regular contributor to the *New York Times* and *Horticulture Magazine* .

Acknowledgments

The following individuals helped in the preparation of a number of photographs.

Sharlotte Albert
Steven Bailey
Wendy Bemis
Betty Choy
Mariano dela Cruz
Carol Grudzien
Kirk D. Humphrey
Princess R. James
Crystal Joseph
Kwesi Folson
Michael F. Ritchie
Deborah J. Stevens
Brenda Watson
Carol Wehnau

The following companies and associations provided technical assistance and/or illustrative material for this text.

Agricultural Products/3M
Brooklyn Botanical Garden
Brouwer Equipment Co.
Broyhill Co.
Charles Machine Works, Inc.
ChemLawn Corp.
Flymo, Inc.
Fox Valley Marking Systems, Inc.
Gandy Co.
Gulf States Paper Corp.
H. D. Hudson Manufacturing Co.
Homelite Division of Textron, Inc.
International Seeds, Inc.
Jacobsen Division of Textron, Inc.
Lesco Products
Lofts Seed, Inc.
Mallinckrodt, Inc.
Michigan Peat Co.
Milwaukee Metropolitan Sewerage District
Mott Corp.
New York State College of Agriculture and Life Science, Cornell University
New York State Turfgrass Association
Northrup King Co.
OMC Lincoln
Pennsylvania Turfgrass Council
Professional Grounds Management Society
Rain Bird Corp.
Richway Industries
Royal Coach/Buckner Corp.
O. M. Scott and Sons
Standard Golf Co.
Stanford Seed Co.
Sudbury Laboratory, Inc.
Tee 2 Green
Telsco Industries
The Toro Company
TUCO Agricultural Chemicals, Division of Upjohn Co.
Turfgrass Products Corp.
U.S. Department of Agriculture
Vicon Farm Machinery, Inc.
York Rakes

CHAPTER

1

The Turfgrass Industry

OBJECTIVES

After studying this chapter, the student should be able to

- Explain the role of turfgrasses
- Understand the diversity and importance of the turfgrass industry
- Describe the history of the turfgrass industry
- Discuss why turfgrasses have become the most widely used ornamental plants in the United States
- Describe career opportunities in the turfgrass industry
- List the skills needed by a turfgrass manager

INTRODUCTION

Turfgrasses are grasses that act as a vegetative ground cover. The turf areas they produce are usually mowed regularly (Figure 1-1). Turfgrasses serve a functional purpose by preventing soil erosion, but have an aesthetic purpose as well. A properly managed turf area is very attractive. Turfgrasses are also able to withstand hard use and some provide an ideal surface for sports fields and other recreational facilities.

Turfgrass is the major vegetative ground cover in the American landscape. In fact, it is the most widely used ornamental crop in this country. The turfgrass industry is a multibillion dollar a year business.

A few representative statistics illustrate how big the turfgrass industry has become. In the United States there are more than 50,000,000 lawns and 14,000 golf courses. In 1990 it was estimated that there were approximately 30 million acres (12 million hectares) of turfgrass in this country. This acreage is larger in size than the total combined area of New Hampshire, Vermont, Massachusetts, Connecticut, Rhode Island, and New Jersey. The National Golf Foundation estimates that more than 12% of the population in the United

FIGURE 1-1
Turfgrasses are generally mowed on a regular basis and often have a finer texture (narrower leaves) than other types of grasses.

States (over the age of 12) plays golf. There were over 25,000,000 golfers in the country in the early 1990's. In the more populous states, such as California, New York, and Florida, it is estimated that several billion dollars is spent annually to maintain turfgrass. TruGreen ChemLawn Corporation, the nation's largest lawn care company, maintains over one million laws. In New York State, on Long Island alone, there are approximately 4,000 horticultural businesses involved in lawn maintenance.

A BRIEF HISTORY OF THE TURFGRASS INDUSTRY

The gardeners of ancient civilizations appreciated the aesthetic qualities of grass. Lawns were included in gardens thousands of years ago in the Middle East and Asia. Turf was used in Europe for lawn bowling and cricket fields as early as the thirteenth century. The game of golf became popular in the 1400s in the British Isles.

During the Middle Ages many European villages established lawn areas called greens or commons which served as a meeting place and recreational area for the townspeople. Lawns also became common on the estates of wealthy landowners. The grass was cut with a scythe or "mowed" by grazing livestock such as sheep. The first mowing machine was invented in England in 1830 (Figure 1-2).

The birth of the modern turfgrass industry occurred in the United States after World War II. The rapid growth of the economy and the population resulted in a housing boom that had a revolutionary effect on the turfgrass industry. As millions of houses were built, millions of lawns were planted (Figure 1-3).

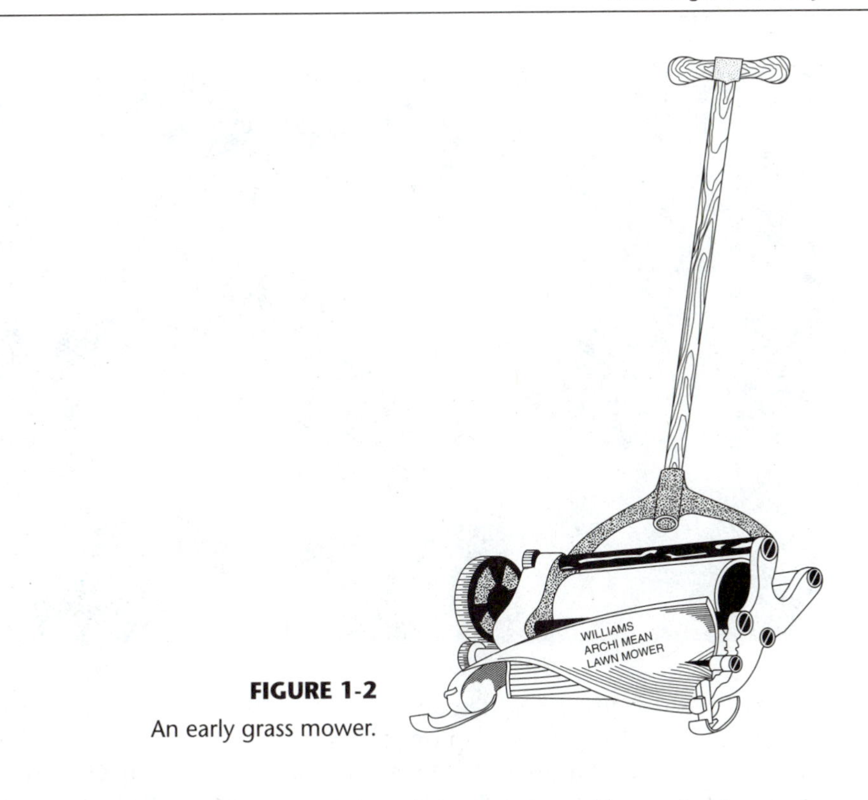

FIGURE 1-2
An early grass mower.

The economy continued to expand, and the number of hours in the working week shortened. Because people had more money and leisure time, recreational activities such as golf became increasingly popular.

These changes in the American lifestyle provided many challenges to the turfgrass industry. Millions of consumers wanted and could afford quality lawns

FIGURE 1-3
Lawns are an important part of every American community.

FIGURE 1-4
This soil would not erode if it had a cover of turfgrass. *(Courtesy of James Bates, photographer)*

and recreational turf. These demands stimulated the development of many new turf products in the 1950s. Improved grass cultivars, turf fertilizers, specialized equipment for turfgrass maintenance, and chemicals to control lawn weeds, diseases, and insects were introduced to the market. Today the turfgrass industry is one of the major agricultural industries in the United States.

WHY TURFGRASSES ARE SO POPULAR

There are many reasons why turfgrass is the leading ornamental plant material in the American landscape. It is a superior ground cover because it prevents soil erosion. Grass plants have extensive root systems which knit the soil together and hold it in place. The dense topgrowth produced by turfgrass also protects the soil. When soil is covered with turf, it cannot be washed or blown away. Soil erosion is a very serious problem in the United States because many crops do not have this ability to stabilize and conserve soil (Figure 1-4).

The value of grass as a ground cover can be appreciated by anyone who has moved into a new house before the lawn was installed. During dry periods dust storms occur, and after a heavy rain the yard is a muddy swamp. Healthy lawns absorb rainfall four to six times more effectively than do farm fields. Recent research has also shown that its dense root system and topgrowth enable turfgrass to prevent pesticides from moving down in the soil and contaminating groundwater.

FIGURE 1-5

Poor landscaping can significantly reduce the value of a property.

Turfgrass contributes beauty to the landscape. It has an attractive green color and a uniform appearance. Turfgrass can be cut to a short length for a neat appearance. Most ornamental plants would be severely injured or killed if they were mowed, but grass is unique in its ability to tolerate regular mowing and remain healthy. An attractive lawn and landscape increase property values by 5 to 15 percent (Figure 1-5).

A wide range of environmental conditions exists across the country and even on a single lawn. Fortunately, there are many different types of turfgrass available. This large selection allows the turf manager to select grasses that can grow in almost any location.

Turfgrasses are perennial plants and provide a permanent ground cover. A properly managed turf area will perform satisfactorily for many years. Grass is easy to grow and often tolerates improper maintenance practices. Many homeowners' lawns survive severe mismanagement.

Grass is an outstanding ground cover for athletic fields and recreational facilities (Figure 1-6). It produces a tough, durable surface that wears well. When injured, turfgrass can recover rapidly. It will spread and fill in damaged areas. This wearability also enables home lawns to serve as playgrounds. Turf also provides optimum footing for athletes and a relatively soft cushion when they fall on the surface of the field.

Turfgrass offers other advantages as well. The plants release significant amounts of oxygen into the air. A turf area 50 feet by 50 feet (15 meters by 15 meters) produces enough oxygen to meet the needs of a family of four. Turf acts like a rug and muffles sound. Grass, by a process known as transpiration,

FIGURE 1-6
Properly maintained turf produces an excellent surface for sports fields. *(Courtesy of Dakota Peat & Blenders)*

has a cooling effect on the environment. On a hot day, when the temperature of a sidewalk or street is well over 100°F (38°C), the temperature at the surface of a lawn will remain around 75°F (24°C). Roughly 50 percent of the sun's heat striking turf may be eliminated by transpirational cooling. Turfgrass also reduces glare from the sun and helps to remove air pollutants and dust particles from the atmosphere.

CAREER OPPORTUNITIES

The rapid growth of the turf industry in recent years has resulted in many turfgrass-related career opportunities. The job market for people trained in turfgrass management has been exceptionally strong. All predictions indicate that this healthy trend will continue in the future.

Golf courses are an important part of the industry (Figure 1-7). The golf course superintendent manages the golf course grounds. Larger courses may also employ assistant superintendents and have foremen supervising work crews. Skilled mechanics are needed to maintain and repair the sophisticated turf equipment.

Many other leisure activities are performed on turfgrass. Recreational facilities used for baseball, cricket, croquet, field hockey, football, frisbee, lawn bowling, lawn tennis, lacrosse, polo, rugby, soccer, and softball require supervisors trained in turfgrass maintenance. State and national parks provide a large number of turf management opportunities (Figure 1-8).

The lawn care industry has hired thousands of people in recent years. The

FIGURE 1-7

There are more than 20 million golfers and 13,000 golf courses in the United States. *(Courtesy of Tee 2 Green Corporation)*

FIGURE 1-8

Turfgrass management is an important part of a park manager's job.

FIGURE 1-9
Both small and large lawn maintenance businesses have been very successful in urban and suburban areas. *(Used with permission of The Toro Company. "Toro" is a registered trademark of the Toro Company, Minneapolis, Minnesota.)*

businesses this industry comprises have a great need for turf specialists. Lawn maintenance has been the fastest-growing sector of the turfgrass industry. There are also opportunities for people who want to be self-employed and operate their own lawn care businesses (Figure 1-9).

Grounds supervisors and employees are hired to maintain turfgrass at schools, colleges, universities, industrial complexes, hospitals, apartments, condominiums, governmental buildings, and other institutions.

Most landscaping and nursery businesses install and maintain laws. A landscape designer must be familiar with the principles of turfgrass management.

Sod growers and seed producers need turf specialists. A large part of a cemetery superintendent's job is turf management (Figure 1-10). Millions of acres of grass are grown along highways and roads. In Texas alone there are estimated to be 600,000 acres (243,000 hectares) of roadside turf. States employ turf experts to supervise these areas (Figure 1-11).

The companies that manufacture turfgrass equipment, irrigation supplies, fertilizers, seed, and pesticides hire numerous personnel. Their employees may conduct research, write technical reports, evaluate products, or be involved with advertising. Thousands of sales representatives are needed to sell their products (Figure 1-12).

Garden centers and other retail outlets that sell these turf supplies directly to the public also require personnel trained in turfgrass maintenance. Many homeowners rely on these employees for lawn care information.

The federal and state governments hire turf specialists to work in their

FIGURE 1-10

A cemetery is composed largely of turfgrass. *(Courtesy of 3M Agricultural Products)*

FIGURE 1-11

Large expanses of turf are found alongside highways.

FIGURE 1-12
A salesperson demonstrating a piece of turf equipment. *(Photo by Brian Yacur)*

cooperative extension programs. The extension specialist acts as a consultant and gives advice to people who have turfgrass problems (Figure 1-13).

Many high schools offer classes in ornamental horticulture, and teachers involved in these programs should be trained in turfgrass. Universities and colleges employ faculty members who teach turfgrass science courses. These institutions also need scientists and technicians to perform turfgrass research.

The turfgrass industry is huge and diverse. There are many career opportunities for people trained in turfgrass management.

THE TURFGRASS MANAGER

Of course, the turfgrass manager must know how to establish and maintain turf. But the job requires other important skills as well. Personnel management is a major responsibility. The manager spends a great deal of time directing the activities of employees under his or her supervision.

The turf specialist also has to prepare budgets, keep records, and purchase supplies. Business skills are necessary for the performance of these types of duties. Equipment maintenance and repair is an important part of the turf manager's job. Mechanical ability is very helpful (Figure 1-14).

Tall Fescues

Tall fescues are recommended for lawns, athletic fields, and low maintenance areas. The newer turf type cultivars are much more attractive than their predecessors, although compared to Kentucky bluegrass they still have a much coarser leaf texture and a more rapid vertical growth rate. Tall fescues have excellent drought avoidance and perform well under lower nitrogen fertility regimes. They are susceptible to winter kill, so we recommend that they be used only in southeastern New York and within the immediate vicinity of Lakes Erie and Ontario.

Tall fescue should be planted as a monostand. One season of growth is required before it can be used on recreational areas. Check with your seed dealer for availability.

	DESCRIPTION					ADAPTATION		
Cultivar	Color	Texture	Density	Growth Habit	Growth Rate	Shade	Close Mowing	Brown Patch*
Amigo	D	MF	M	ML	S	Mo	G	VG
Apache	MD	M	H	ML	-	VG	-	Mo
Arid	MD	M	MH	ML	-	Mo	-	Mo
Arriba	D	M	M	-	-	-	-	Mo-G
Austin	D	M	MH	M	-	-	Mo	VG
Avanti	D	MF	M	-	-	-	-	Mo
Aztec	D	M	M	-	-	-	-	Mo-P
Bonanza	D	M	MH	MH	-	Mo-P	Mo	Mo
Chieftain	M	M	M	M	-	Mo	Mo	G
Cimmaron	M	M	M	MH	-	P	P	Mo
Cochise	VD	MF	MH	ML	S	Mo-P	Mo-G	E
Crossfire	D	MF	MH	-	-	-	-	Mo
Eldorado	D	M	M	M	S	P	Mo	Mo
Guardian	VD	VF	H	ML	S	E	Mo	VG-G
Hubbard 87	VD	F	H	ML	-	E	VG	VG
Jaguar	ML	M	M	VEr	-	Mo-P	P	G
Jaguar II	M	MF	MH	-	-	-	-	Mo
Marathon	M	M	M	M	-	Mo-P	P	-
Maverick II	MD	M	H	-	-	-	-	Mo-G
Monarch	D	MF	M	L	-	Mo	-	Mo
Mustang	ML	M	M	-	-	Mo-P	Mo-P	G
Olympic	ML	M	M	Er	S	G	P	Mo
Olympic II	M	M	H	MH	S	G	Mo-P	VG
Phoenix	M	MC	H	Er	S	Mo-G	Mo-G	VG
Rebel	ML	M	M	VEr	-	Mo-G	P	E
Rebel II	ML	F	MH	Er	-	Mo-P	Mo	E
Safari	D	M	MH	M	-	E	VG	Mo
Shenandoah	D	F	H	M	S	G	VG	Mo
Silverado	D	F	MH	VL	S	Mo-G	-	Mo-G
Tribute	MD	M	M	ML	-	Mo	-	Mo
Wrangler	M	MF	MH	M	S	P	Mo-G	Mo

*Where brown patch resistance is desired select cultivars with a rating of G, VG, or E.

FIGURE 1-13

Extension specialists provide the turf manager with helpful information. *(Courtesy of New York State Turfgrass Association, Inc.)*

How does the manager attain these skills? Working in the industry is the best preparation for the job. There is no substitute for hands-on training. A tremendous amount of practical knowledge can be learned by working for and observing an experienced turf manager. During this apprenticeship, the turf worker gains many of the skills he or she will need to be a successful turf manager.

Formal education is also important. Many high schools have excellent horticulture programs, and turfgrass management is discussed in these classes. More than one hundred colleges and universities in the United States offer turfgrass courses and programs. The following list is a sample of the types of courses available to turfgrass students: introductory turfgrass science, advanced turfgrass science, golf course management, pathology (diseases), entomology (insects), soil science, soil fertility, soil and water conservation, soil chemistry, soil physics, soil microbiology, woody plant identification, landscaping, nursery management, weed control, turfgrass business operations, arboriculture (tree management), plant physiology, irrigation and drainage, recreational land management, botany, plant nutrition, agricultural engineering, landscape construction, outdoor power equipment, and small engine repair (Figure 1-15).

Business courses that teach accounting, budget preparation, record keeping, communication skills, personnel management, computer information

FIGURE 1-14
Today machinery is used to perform almost all turfgrass maintenance practices.

FIGURE 1-15

A student learning how to identify turfgrass diseases in a plant pathology class. *(Photo by Brian Yacur)*

systems management, keyboarding, and small-business management are very valuable to the turf specialist. Classes in writing, speech, mathematics, and psychology are also recommended.

Turfgrass managers continue their education after they have graduated from school by attending conferences and seminars (Figure 1-16). Each state has a turfgrass organization which sponsors turf conferences and field days. Companies that manufacture and supply turfgrass products offer training sessions throughout the country. The manager should read trade magazines such as *Landscape Management, Grounds Maintenance, Turf, Lawn and Landscape Maintenance, SportsTURF, Park Maintenance, Golf Course Management, Golf Course News,* the United States Golf Association's *Green Section Record,* and *Northern Turf Management, Western Turf Management,* or *Southern Turf Management.*

FIGURE 1-16
The Golf Course Superintendents Association of America holds a national conference every year. *(Used with permission of The Toro Company. "Toro" is a registered trademark of the Toro Company, Minneapolis, Minnesota.)*

SELF-EVALUATION

1. Why has turfgrass become the leading vegetative ground cover?
   a. Prevents soil erosion
   b. Aesthetic qualities
   c. Ability to tolerate hard use
   d. All of these
2. List career opportunities available in the turfgrass industry.
3. Explain how a person obtains the skills he or she needs to be a turf manager.
4. Approximately haw many people in the United States play golf?
5. Discuss the history of the turfgrass industry in the United States.

CHAPTER

2

Introduction to Turfgrasses

OBJECTIVES

After studying this chapter, the student should be able to

- Discuss the classification of grasses
- Explain how grass plants grow
- Describe the functions of important grass plant structures
- Explain how grass plants spread
- Describe how turf quality is measured
- Identify different levels of maintenance or cultural intensity

CLASSIFICATION

Plants are classified by placing those with similar characteristics in the same group. An example of the classification system is shown in Figure 2-1. It is important to understand some of the classification terminology. Angiosperms, the flowering plants, are separated into two major classes, the monocotyledons and the dicotyledons. These two groups are usually referred to as monocots and dicots, respectively. The monocots are primarily grasslike plants, and the dicots are broadleaf plants (Figure 2-2).

The true grasses are in the family *Poaceae,* which was formerly called *Gramineae.* The scientific name is the genus and species of the plant. For example, the scientific name for Kentucky bluegrass is *Poa pratensis. Poa* is the genus of all the bluegrasses; *pratensis* is the species name for Kentucky bluegrass. Other bluegrass species are *Poa annua* (annual bluegrass), *Poa trivialis* (rough bluegrass), *Poa compressa* (Canada bluegrass), and *Poa arachnifera* (Texas bluegrass).

A final subdivision is necessary. Each species contains a number of *cultivars.* Midnight, Touchdown, Vantage, and Adelphi are examples of Kentucky

Kingdom:	Plantae
Division:	Embryophyta
Subdivision:	Phanaerogama
Branch:	Angiospermae
Class:	Monocotyledoneae
Subclass:	Glumiflorae
Order:	Poales
Family:	Poaceae
Subfamily:	Pooideae
Tribe:	Poeae
Genus:	Poa
Species:	Pratensis
Cultivar:	Adelphi

FIGURE 2-1
Classification of Adelphi Kentucky bluegrass.

bluegrass cultivars. These cultivars are considered to be of the same species because they are very similar in structure and physiology. However, they differ in traits such as shade tolerance, leaf width, color, close mowing tolerance, cold hardiness, heat and drought tolerance, fertility requirements, insect and disease resistance, rate of growth, recuperative potential, ability to tolerate traffic, and establishment rate (Figure 2-3). Variety is another term for cultivar.

Some cultivars are produced by hybridization. This involves crossbreeding different varieties in an attempt to develop an offspring that exhibits the best characteristics of the parents. Many cultivars are the result of natural selection.

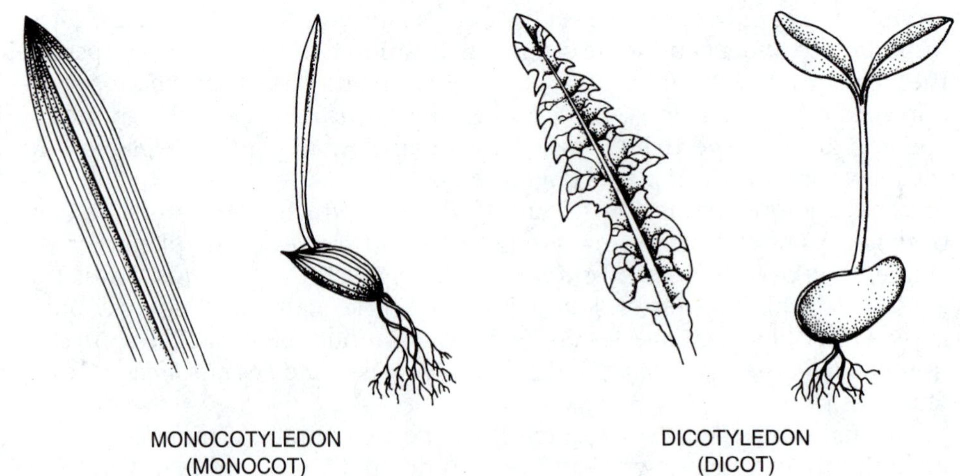

FIGURE 2-2
The grasses are monocotyledons. Broadleaf plants are dicotyledons. Monocots have a single seed leaf (cotyledon) present when the seedling emerges from the soil. Dicot seedlings have two cotyledons. Monocots have parallel leaf veins; dicot leaf veins branch from the midrib into a network of finer veins.

FIGURE 2-3
The leaf width of these two tall fescue cultivars differs greatly. *(Courtesy of Art Wick, Lesco Products)*

The original plants were discovered growing in a lawn, golf course fairway, cemetery, or some other type of grassy area and were collected for further evaluation. They were selected as possible new varieties or as parents of new varieties because they exhibited some outstanding characteristics such as improved disease resistance or excellent color retention at cold temperatures.

In the future genetic engineering techniques will be used to develop new cultivars. For example, genes that contribute to drought resistance in a desert plant could be inserted into another species to improve its performance during dry periods. Nevertheless, though there is great potential for improving grasses by gene insertion, it is still unclear how successful these techniques will be.

PLANT GROWTH

All living organisms are made up of very small units called *cells*. It is estimated that a human contains a million million cells. *Tissues* consist of large numbers of cells which perform the same function. The systems and structures of a grass plant are composed of these various tissues.

Meristematic Tissue

Meristematic tissue is responsible for plant growth. Cells in meristematic regions are able to divide or enlarge. Cell division and cell enlargement result in an increase in the size of the plant. As cells at the tip of a root actively divide, for example, the root grows deeper into the soil.

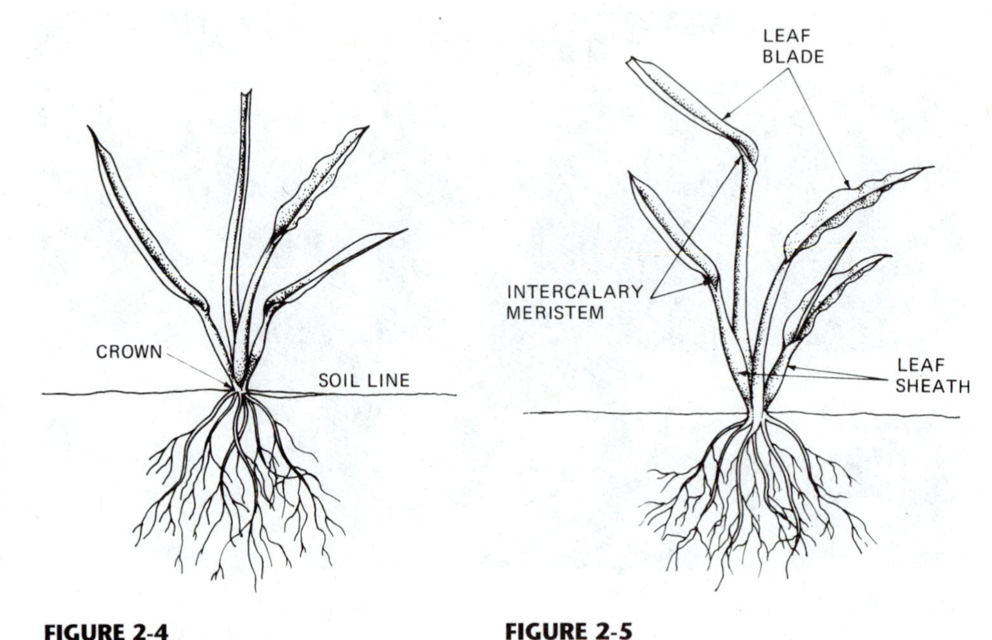

FIGURE 2-4

The crown is the vital meristematic region located near the soil surface.

FIGURE 2-5

The intercalary meristem is located at the base of the leaf blade.

The most important meristematic region in a grass plant is called the *crown.* All new growth is initiated at the crown. Its meristematic tissue produces the leaves, roots, and stems. The crown is located at the base of the plant near the soil surface where the leaves, roots, and stems join (Figure 2-4). This growing point is surrounded by the lower portions of several leaf sheaths.

Leaf blade elongation occurs from *intercalary meristem* tissue, which is located at the base of the blade (Figure 2-5).

Turfgrass can remain healthy despite frequent mowing because of continued leaf development after defoliation. Grass grows back after being cut because the meristematic tissue is located beneath the path of the mower blade. The tissue that is cut off is the oldest portion of the leaf.

The growing point of dicot plants such as trees and shrubs is located at the top of the stem (Figure 2-6). Cell division and new tissue occur at the tip. Mowing removes the growing point of dicots and may result in serious injury.

It is vital that the crown be protected. The plant can survive the loss of the majority of its leaves and roots if this growing point remains alive. New growth will be initiated to replace the dead tissue. If the crown is injured the plant may not recover. A healthy, viable crown is white and *turgid* (the cells are filled with water). If the crown is brown and dried out, then it and the plant are dead (Figure 2-7).

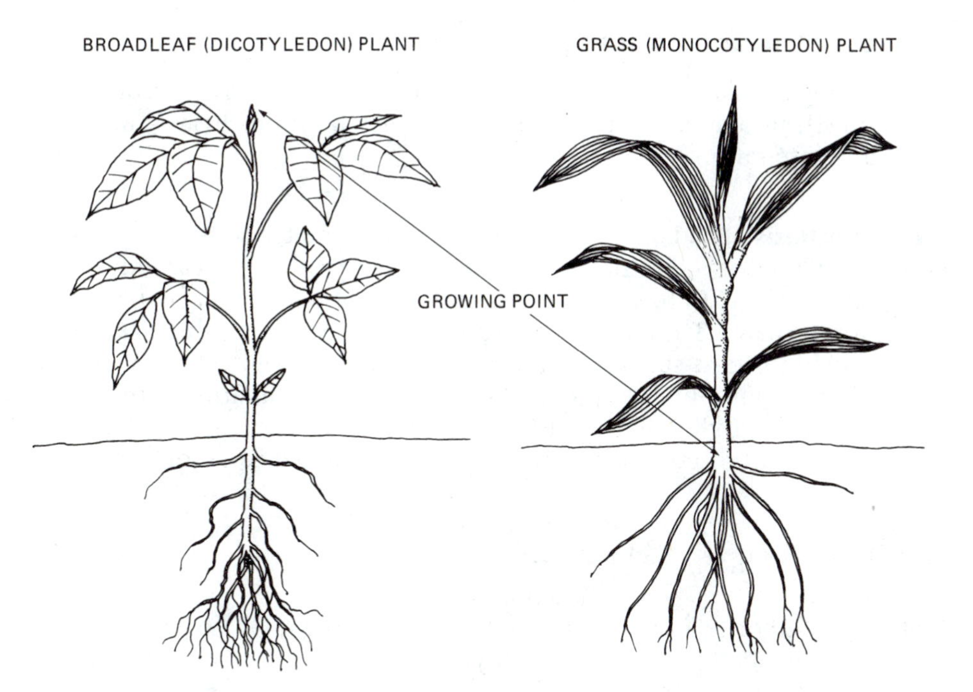

FIGURE 2-6
Growing point location for dicots and monocots.

FIGURE 2-7
The crowns of the plants in the foreground died because of winter injury.

Photosynthesis and Respiration

Many biological reactions occur in a plant. The food and energy required for plant growth are a direct result of two of these metabolic processes—photosynthesis and respiration.

Photosynthesis. Food is manufactured by a complex series of processes known as *photosynthesis*. Energy provided by sunlight causes carbon dioxide (CO_2) and water (H_20) to combine and form carbohydrates (sugars), the food source for the plant. The energy from sunlight is converted to chemical energy and is stored in these sugars. This reaction occurs in green cells that contain the pigment chlorophyll. Photosynthesis takes place primarily in the leaves, although chlorophyll is also found in stems.

The chemical reaction is described by this simplified equation:

$$12H_2O + 6CO_2 + \text{Light} \xrightarrow[\text{Reaction in green cells containing chlorophyll}]{\text{Yields}} C_6H_{12}O_6 + 6O_2 + 6H_2O$$

Water	Carbon dioxide		Reaction in green cells containing chlorophyll	Sugar	Oxygen	Water

The water used in the reaction is obtained from the soil by the roots. The carbon dioxide enters the leaves from the air. Normally the concentration of CO_2 in the air is approximately 0.03%, or 300 parts per million. The plant releases oxygen and water vapor into the air during the reaction. The other end product of photosynthesis, simple sugars, is either used immediately or stored for future use.

Respiration. The food is used by the plant to produce energy and to build cells and tissue. Energy is required for plant growth and development. This energy becomes available when carbohydrates manufactured during photosynthesis are broken down. The release of energy from sugars is the result of a process known as *respiration*. Respiration is basically the reverse of the photosynthesis reaction:

$$C_6H_{12}O_6 + 6O_2 \longrightarrow 6CO_2 + 6H_2O + \text{Energy}$$

Sugar	Oxygen	Carbon dioxide	Water	

Respiration takes place in all living cells. Food is consumed continually and converted to energy. Although respiration is necessary for a plant to survive, it is a wasteful reaction. Much of the energy released from the food is lost as heat during the reaction.

of photosynthesis is lower in the shade than it is in the full sun. Grass should be cut to a taller height in shady locations to compensate for the reduced amount of light. More green tissue is left on the plant when it is mowed higher. This extra leaf area increases the plant's ability to intercept sunlight and to produce food.

The turf manager who understands the basics of plant growth is better able to develop a proper management program.

PLANT STRUCTURES

The turf specialist should be familiar with the functions of the major structures of a turfgrass plant (Figure 2-11).

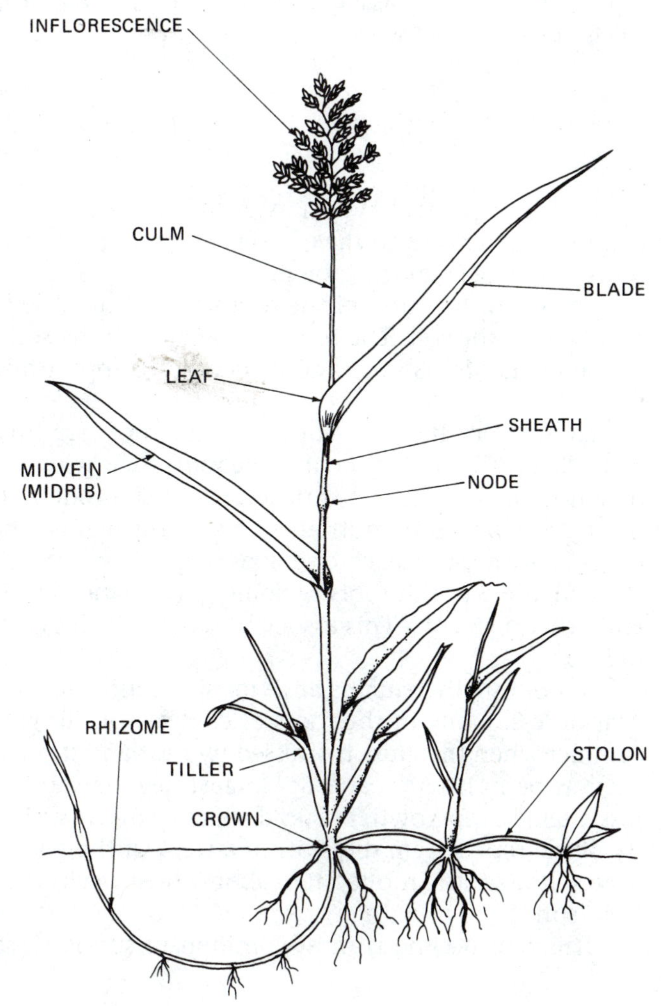

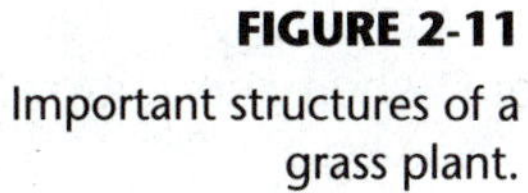

FIGURE 2-11

Important structures of a grass plant.

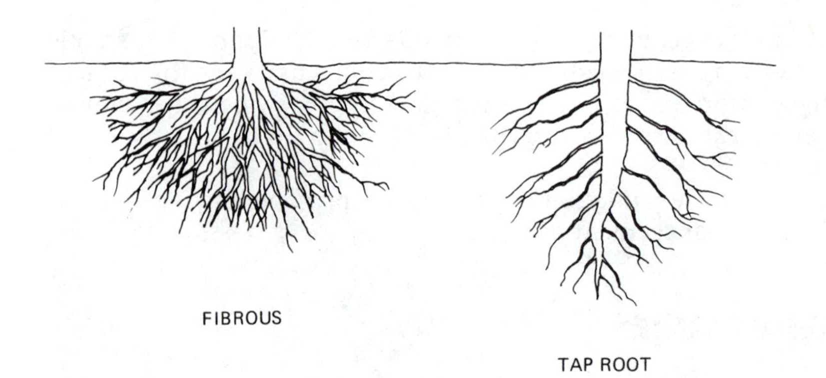

FIGURE 2-12 A grass plant has a fibrous root system composed of a mass of fine roots. All of the roots are basically the same size. The plant on the right has a large main root which produces smaller lateral roots. This is called a tap root system.

Roots

Healthy grass plants have an extensive fibrous root system (Figure 2-12). It is common for grasses to have several tons of roots per acre (0.4 hectare). The roots originate from the crown.

The major function of the root system is to absorb water and mineral nutrients from the soil. The roots grow through the soil searching for water and nutrients. Roots also provide mechanical support by anchoring the plant in the soil.

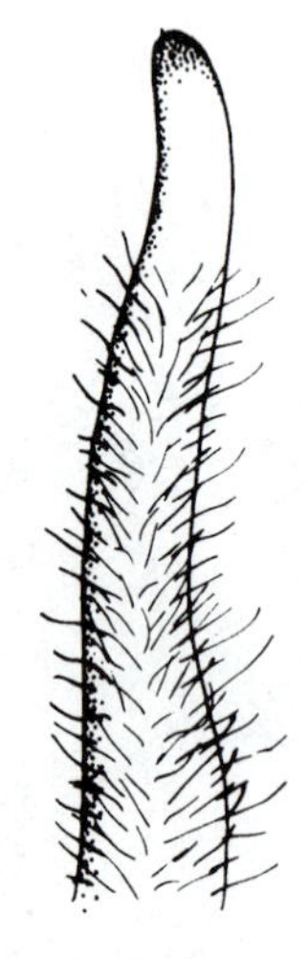

FIGURE 2-13 Root hairs are responsible for much of the water and nutrient absorption by a plant.

Located on the outer surface of the roots are tiny projections called *root hairs* (Figure 2-13). They enable the roots to be in contact with a larger amount of soil by increasing the surface area of the root system. Root hairs take up most of the water and nutrients used by the plants. The hairs must be replaced continually because each one lives only a few days.

Fungi live in the roots of some grass plants and help absorb phosphorus and other nutrients. This association between roots and fungi is called *mycorrhizae*.

A root usually remains alive for six months to two years. Most roots live a year or less. Many of the roots of cool season turfgrass plants die during the summer when the grass is stressed by heat and drought. Warm season species such as bermudagrass and St. Augustinegrass can experience extensive root loss after the topgrowth breaks dormancy in the spring. Some years the majority of the root system dies within a week of the appearance of the new, green leaves. It takes the root system about three weeks to recover completely from this "spring die-back."

This root decline is most common in spring seasons when shoot growth occurs early or follows a cold winter. There is temporarily not enough energy available to support both the topgrowth and the root system. When shoot "greenup" is slower and occurs later in the spring, the root loss is not significant.

The optimum soil temperature for root growth is around 55°F (13°C) for cool season turfgrasses and 80°F (27°C) for warm season turfgrasses. Best root growth for cool season grasses occurs in the spring and fall. Warm season grasses exhibit good root development in the late spring, summer, and early fall (Figure 2-14). Optimum temperatures for root development are lower than the optimum temperatures for topgrowth. Roots may continue to grow for as long as a month after leaves become dormant due to cold temperatures.

The bulk of the turfgrass root system is in the top 6 to 12 inches (15 to 30 centimeters) of soil. The roots of the warm season grasses are usually thicker and longer than those of cool season grasses. They may have some roots that grow down 5 feet (1.5 meters) or deeper into the soil. The roots of cool season grass species seldom grow deeper than 2 feet (0.6 meter). In the summer, when soil temperatures are high, their roots may be very shallow.

Root depth varies among species and is affected by several other factors. Soil conditions have a major effect on depth. Rooting is also influenced by maintenance practices such as mowing height, frequency of irrigation, aerification, and fertilization. These factors will be discussed in detail in later chapters.

Grass plants with extensive, vigorous root systems are considered stronger and better able to recuperate from stress. Deep roots make a plant more drought tolerant. Soil dries out from the surface down. Deep-rooted plants are able to obtain adequate amounts of water even when the upper inches of soil are dry. Shallow-rooted plants, however, are often better able to tolerate wet soil conditions.

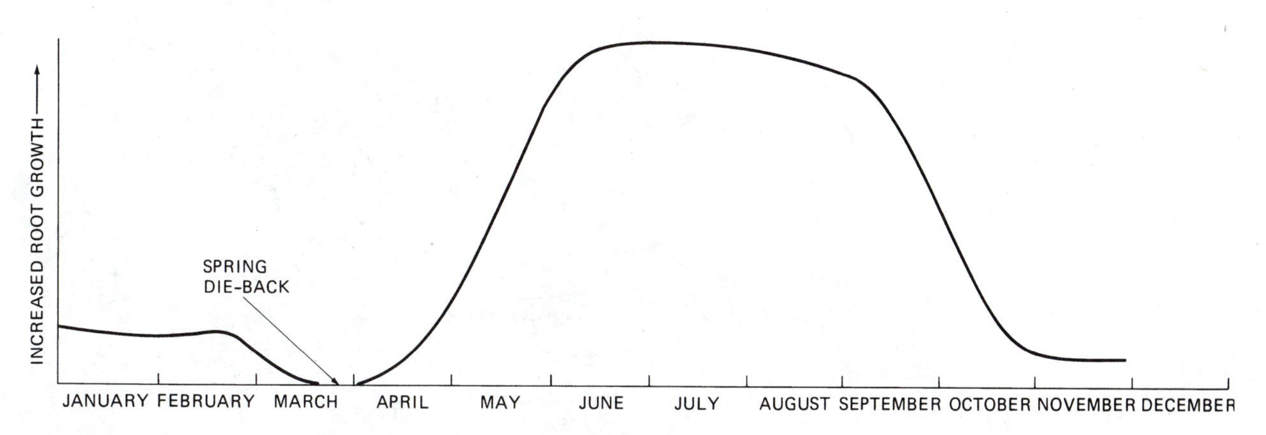

FIGURE 2-14 Growth cycle of the roots of warm season species.

Vascular System

The vascular system consists of tubelike tissue that connects the underground and aboveground plant structures. This system serves as a pipeline through which water, nutrients, and food move from one part of a plant to another. The movement of materials within a plant is called *translocation.* There are two types of vascular tissue—xylem and phloem.

Xylem. Water and minerals absorbed by the roots are carried up in the *xylem* to the stem and leaves. Water is pulled up fro the roots to replace water lost from the leaves. Much of the water in the leaves evaporates into the air. This water loss, called *transpiration,* causes the upward movement of water and dissolved nutrients in the xylem.

Phloem. Food manufactured in the leaves travels down in the *phloem* to stems and roots. These materials can also move up in the phloem to areas where growth is occurring.

Leaves

Many of the cells inside a leaf contain chlorophyll and can carry on photosynthesis. Leaves are the major site of food production.

The outer layer of leaf cells is called the *epidermis.* On the outer surface of the epidermal cells is a very thin waterproof covering called the *cuticle.* The waxy cuticle helps to keep water in the leaf.

Stomata (plural form of stoma) are openings in the epidermis (Figure 2-15).

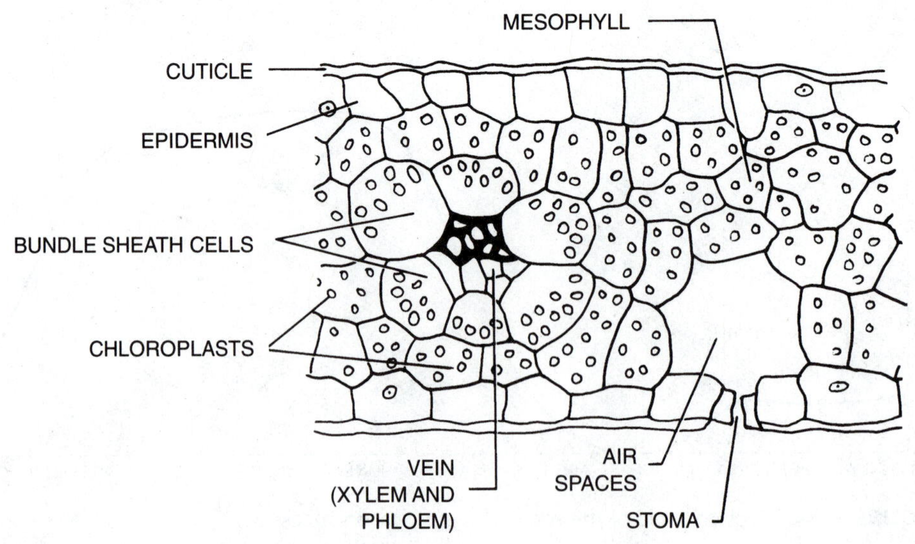

FIGURE 2-15
Cross-section of a leaf.

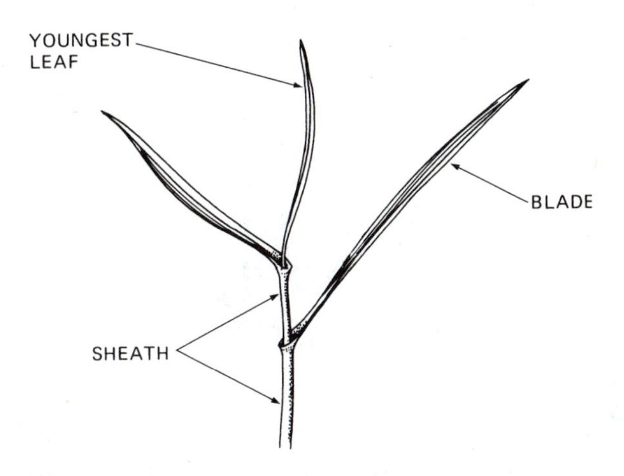

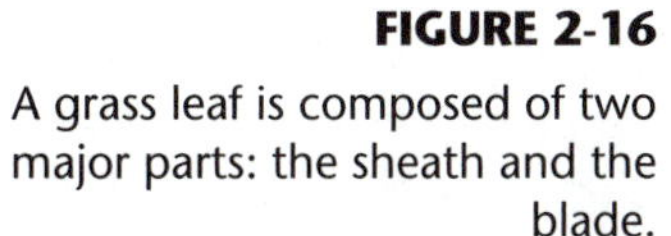
FIGURE 2-16
A grass leaf is composed of two major parts: the sheath and the blade.

Gases move in and out of the leaf through the stomata. The carbon dioxide necessary for photosynthesis and the oxygen used in respiration enter the leaf when the stomata are open.

During periods of drought, the stomata close to prevent water vapor from escaping from the leaf. When they close, photosynthesis stops because carbon dioxide no longer enters the leaf. Grass does not grow much when it is subjected to moisture stress. The stomata close at night because photosynthesis cannot take place without light.

The leaf veins are bundles of vascular tissue. The phloem and xylem in the veins connect with vascular tissue in the roots. The large vein running down the center of the leaf blade is called the *midrib*. A species such as perennial ryegrass has thick, tough vascular bundles and can be difficult to mow properly if the lawn mower blade is not sharp.

A grass leaf is divided into two major parts—the sheath and the blade (Figure 2-16). The *blade* is the upper, relatively flat part of the leaf. The *sheath* is the lower portion of the leaf. The bottom of each sheath is attached to the crown, where leaf growth is initiated. The sheaths are folded or rolled around each other and support the leaf blades, holding them above the ground so that they can intercept sunlight.

When an older leaf matures and dies, it is replaced by a new leaf. The new leaf develops within the sheath of the next oldest leaf and emerges at the top of the plant (Figure 2-17).

A leaf grows from meristematic tissue located at the base of the leaf. When grass is mowed, the mower blade removes the oldest tissue from each leaf that is cut. Grass needs to be mowed regularly because the leaves continue to elongate from the basal meristem. Mowing practices have a significant effect on leaf initiation and growth. These practices and effects will be discussed in Chapter 11.

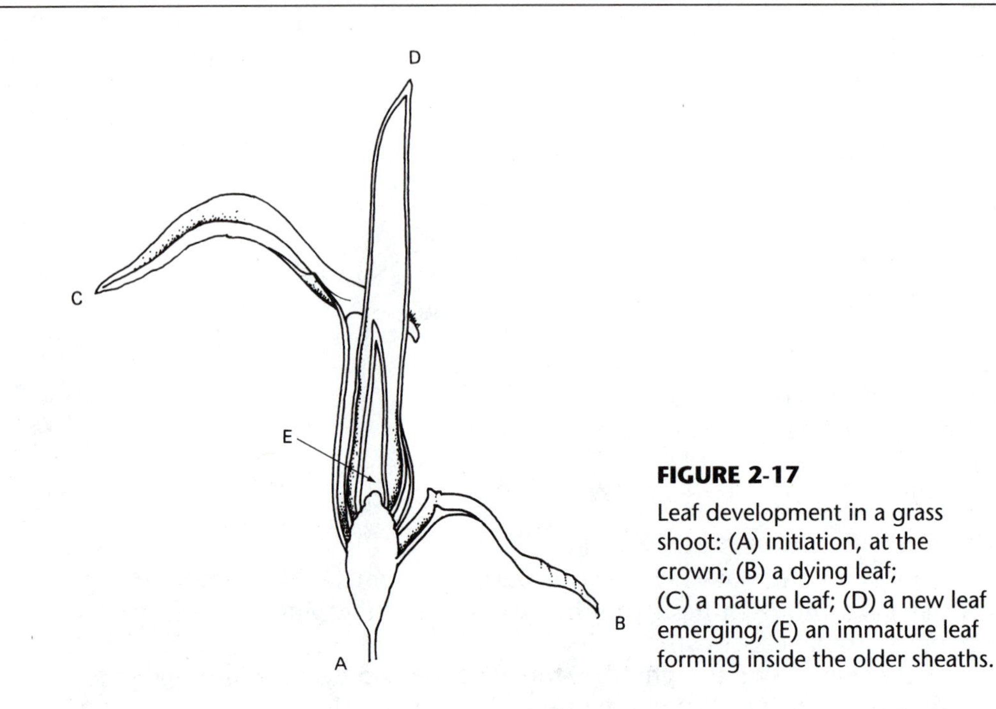

FIGURE 2-17
Leaf development in a grass shoot: (A) initiation, at the crown; (B) a dying leaf; (C) a mature leaf; (D) a new leaf emerging; (E) an immature leaf forming inside the older sheaths.

The rate of leaf growth and new leaf formation is also influenced by climatic conditions such as temperature, soil moisture levels, and light intensity. Nitrogen fertilizer stimulates the production of leaf tissue.

Leaves often turn brown or yellow when climatic conditions are unsatisfactory for plant growth. This change in color is due to a loss of chlorophyll. When favorable weather returns, new chlorophyll is produced and the leaves begin to photosynthesize again. Nutrient deficiencies also cause *chlorosis*—the lack of chlorophyll.

The cells in the leaves are turgid—filled with water. Turgor pressure makes the tissue stiff and gives it the strength to support the leaves. If the leaves lack water, they droop and bend over. Plant tissue that collapses or loses its stiffness because of a lack of water is said to be wilted.

Stems

There are three major types of stems associated with turfgrass: the crown, the flowering culm, and lateral or creeping stems. The *crown,* the principal meristematic region, is an unelongated stem. Vegetative growth occurs without a significant lengthening of stem tissue. Normally the topgrowth consists primarily of leaves. When a grass plant enters the reproductive stage, it produces an elongated stem which is called the *flowering culm* (Figure 2-18). The flowers and seeds appear at the top of the culm.

FIGURE 2-18
The flowering culm is an elongated stem formed during the reproductive stage.

Some turfgrass species have lateral or creeping stems called stolons and rhizomes. These stems elongate horizontally from the crown of the parent plant. *Stolons* grow along the surface of the ground, while *rhizomes* grow beneath the surface. Shoots and roots form at nodes on the horizontal stems (Figure 2-19). Rhizomes and stolons enable a plant to spread. These lateral stems are called *extravaginal growth* because they penetrate through the enclosing leaf sheaths and grow outside of them.

Kentucky bluegrass, creeping red fescue, and bahiagrass are examples of species that produce rhizomes. They are said to be *rhizomatous*. *Stoloniferous* grasses include St. Augustinegrass, creeping bentgrass, buffalograss, and rough bluegrass. Bermudagrass and zoysiagrass produce both rhizomes and stolons.

Creeping stems serve as a major storage area for reserve carbohydrates. Rhi-

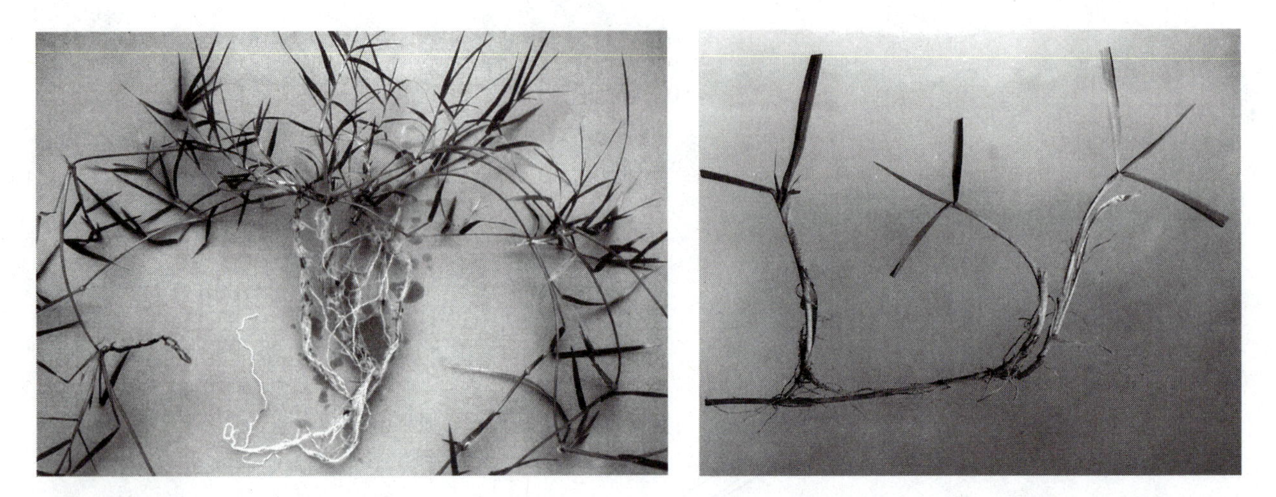

FIGURE 2-19 Stolons are aboveground runners (left), and rhizomes are underground stems (right). *(Courtesy of O.M. Scott and Sons Company)*

zomes and stolons allow a grass plant to spread and establish a dense turf. This spreading habit makes the grass better able to recuperate from stress and injury. It recovers by filling in thin turf or bare areas with new plants. The length and aggressiveness of creeping stems vary between species and cultivars.

Species such as perennial ryegrass and tall fescue do not produce either stolons or rhizomes. They spread very slowly and tend to grow in clumps or bunches. Their slow horizontal growth makes it difficult for them to replace damaged plants. It is important to use the correct seeding rate when establishing these bunch-type grasses. If they are seeded at too low a rate, the turf will be thin and may never attain a desirable density.

Tillers

Turfgrasses are perennials and continue to grow from year to year. Individual shoots, however, usually live only a year or less. They are replaced by new leaves and stems. These new shoots are called *tillers*.

Unlike rhizomes and stolons, tillers grow vertically. They develop at the crown and emerge from within the basal leaf sheath of the old stem. This is called *intravaginal growth*. The tiller then grows upward next to the parent shoot (Figure 2-20). The continual appearance of new tillers ensures the survival of grass plants from year to year.

Tillering results in an increase in the number of shoots in a turf area. As each plant produces tillers it spreads and gradually forms a very dense clump. Each clump enlarges in diameter until eventually the soil is completely covered by shoots. This is a lengthy process compared to the more rapid spread of rhizomes and stolons. If turf is primarily composed of species such as tall fes-

FIGURE 2-20
Tillers are produced by parent plants and also by daughter plants developed on rhizomes or stolons. Tillers increase shoot density.

cue and perennial ryegrass that spread only by tillering, it thickens up at a slow rate unless seeded at adequate rates initially.

The rate of tillering is usually greatest in the spring and fall. Frequent mowing to relatively short heights encourages tiller formation.

The Inflorescence

The *inflorescence* is the flowering part of a grass plant and is where seeds are formed. Regular mowing usually prevents seed production because the flowers develop at the top of the culm. At a normal cutting height the inflorescence is removed before seed heads can form. Some grasses, such as annual bluegrass and bermudagrass, can produce seed even at very short mowing heights.

Grass should not be allowed to go to seed. Shoot growth is inhibited by seed production because the plant uses its food and energy resources to develop the inflorescence at the expense of vegetative growth. The life span of a

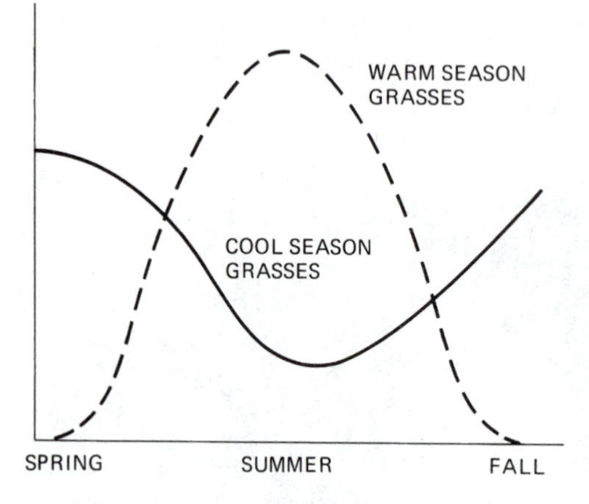

FIGURE 2-21
Warm season grasses attain their highest quality in summer. Cool season grasses look best in spring and fall.

tiller is reduced if it flowers and forms seeds. Seed heads also present an aesthetic problem. They look unattractive and spoil the uniform appearance of a turf area.

Seeds are produced in the spring or summer, depending on the species. The exposure to certain environmental factors stimulates reproductive growth. Day length and temperature have a major effect on the flowering of most species.

Commercial seed production will be discussed in the later chapters.

MEASURING TURF QUALITY

Why is one turf area considered to be of high quality while another is judged to be of poor quality? Four characteristics are generally used to measure turf quality—color, texture, density, and uniformity.

Color, texture, density, and uniformity vary among grass species and varieties. Climatic conditions affect these quality indicators (Figure 2-21). Grass is less attractive and less vigorous during periods of unfavorable temperatures or dry weather. Proper management is also necessary to ensure satisfactory turf. Fertilization and mowing practices have a major influence on quality.

Color is a measure of the light reflected by turfgrass. Generally, the greener the grass is, the more attractive it is considered (Figure 2-22). A dark green color is preferable to a light yellow-green. Poor color can be caused by a nitrogen deficiency, drought or temperature stress, diseases, insects, or other types of injury. It is normal for some species and varieties to be light green. A lack of dark green color does not necessarily mean that the grass is unhealthy.

Texture is primarily a measure of the width of the leaf blades (Figure 2-23). Finer-textured grasses that have narrow leaves are considered more attractive

FIGURE 2-22
A dark green color is considered an attractive characteristic.

FIGURE 2-23
Leaf textures vary from fine to coarse. The most attractive grasses have leaves 3 millimeters or less in width.

FIGURE 2-24
Some species and cultivars do not look good when grown together because of differences in leaf texture. *(Photo by Brian Yacur)*

than coarse-textured grasses with wide leaves. There are great differences in texture among species and varieties (Figure 2-24). Close mowing and increased density result in narrower leaves.

Perhaps the most important indicator of turf quality is density. *Density* is a measure of the number of shoots in an area. A dense, carpetlike turf is highly desirable (Figure 2-25). A spotty, thin turf is less satisfactory (Figure 2-26).

Density is also a measure of the ability of the grass to adapt to various conditions. A variety that is not shade tolerant will have less density if it is grown under trees. The turf on a football field will be sparse if the grasses do not wear well. High density cannot be achieved if a species is unable to resist diseases or other stresses. A thick sod cover is often a characteristic of grasses that produce vigorous rhizomes and stolons. Incorrect maintenance practices are a common cause of low density.

Uniformity is a combination of the other three quality indicators. An attractive turf has a uniform, consistent appearance. The color, texture, and density are similar throughout the turf area. All of the grass looks the same. Weeds, bare spots, diseases, or different textures and colors can spoil uniformity.

Turfgrass researchers use these visual observations to evaluate cultivars. Varietal trials are conducted by seed companies and universities. Varieties are grown in separate plots and are evaluated several times a year (Figure 2-27). A typical rating scale usually ranges from 1 (poorest quality) to 9 (highest quality). An example of varietal trial results appears in Table 2-1.

FIGURE 2-25

A dense turf is very attractive. *(Photo by Brian Yacur)*

FIGURE 2-26

Low shoot density greatly decreases turf quality. *(Photo by Brian Yacur)*

FIGURE 2-27
Research plots at a university.

TABLE 2-1 Quality Ratings for 12 Kentucky Bluegrass Cultivars Grown under Low-Maintenance Conditions*

Cultivar	April 1	May 4	June 8	July 15	September 2	October 17
Adelphi	4.0*	5.6	5.4	4.5	5.9	5.7
Baron	3.8	5.1	4.8	4.4	5.4	5.5
Bensun (A-34)	3.6	5.3	6.0	5.9	6.1	5.4
Glade	4.2	5.1	5.4	5.5	6.3	6.3
Newport	3.7	3.6	3.8	4.0	4.2	4.9
Parade	4.5	6.8	5.9	4.4	5.7	5.8
Ram 1	4.3	6.9	5.9	6.0	6.2	6.4
Sydsport	3.8	6.2	6.0	5.9	6.4	6.1
Touchdown	4.1	6.3	5.9	5.7	6.2	6.5

*Quality ratings can vary significantly from one region or location to another. A rating of 1 = no live turf, 9 = ideal turf, >5 = acceptable quality.

LEVELS OF MAINTENANCE

Different types of turf area require different levels of maintenance or cultural intensity. The level of maintenance that a turf area receives depends upon the level of turf quality desired and the amount of money available for maintenance.

A golf green, to be satisfactory, must exhibit outstanding density, uniformity, and color. The green must also be cut at an extremely short height to permit putting by golfers. To achieve such excellent quality, the grass is mowed, irrigated, fertilized, and treated with pesticides frequently. Regular performance of several other maintenance practices is necessary as well. A putting green is said to be a high-maintenance or high cultural intensity turf area because it requires much care and is very expensive to maintain.

Grass growing beside a highway usually receives little care. Its function is utilitarian, not aesthetic. The grass is there to prevent soil erosion. Someone driving past it at 55 miles per hour is totally unaware of turf density and uniformity. This type of utility turf is normally mowed only two or three times a year and may never be fertilized or irrigated. It is said to be a low- or minimum-maintenance area. Increasing the maintenance level to attain better turf quality would be a waste of money.

Most turfgrass maintenance programs fall somewhere between these two extremes. Many lawns receive medium (or moderate) levels of maintenance. Medium-high cultural intensity may be necessary to keep the grass on an athletic field in good shape.

Whether acceptable turf quality is achieved depends a great deal on the manager's skill. Sometimes even a skillful turf manager cannot attain the desired quality because of severe budget limitations or adverse environmental conditions.

SELF-EVALUATION

1. The name of the grass family is __________.
2. Another name for cultivar is __________.
3. Tissue is made up of small units called __________.
4. Cell division occurs in __________ regions.
5. Shoot and root growth are initiated at the __________.
6. Food is manufactured by a process known as __________.
7. Food is converted to energy by a process known as __________.
8. The function of the root system is to absorb __________ and __________.
9. Water travels up in the __________; food moves down in the __________.
10. Openings in the leaf epidermis are called __________.
11. Grass is green because of the presence of the pigment __________.

12. Horizontally growing underground stems are called __________.
13. Aboveground runners are called __________.
14. To live and grow, a plant needs sunlight, water, nutrients, mechanical support, certain temperatures, and two gases found in the air, __________ and __________.
15. Bermudagrass and St. Augustinegrass may lose much of their root system in the __________.
16. A leaf is composed of two major parts, the __________ and the __________.
17. The growing point of a __________ is located at the top of the stem.
18. Plant tissue that droops because of lack of turgor pressure is said to be __________.
19. The number of shoots per unit area is referred to as shoot __________.
20. The appearance of seed heads on grass plants disrupts turf __________.
21. The intensity of culture is also called the __________.
22. Discuss some factors that affect root growth.

CHAPTER

3

Warm Season Grasses

OBJECTIVES

After studying this chapter, the student should be able to

- Explain why certain turfgrasses are adapted to the warm season zone in the United States
- Describe the important characteristics of the warm season turfgrasses
- Explain where and for what purpose each warm season grass is used
- Discuss how a turf manager selects the best species and cultivar for a particular site

INTRODUCTION

The United States can be roughly divided into two major grass adaptation zones based on temperature (Figure 3-1). Regions 1A and 1B experience relatively cold minimum winter temperatures and are most suitable for the cool season turfgrasses. Consequently, they are called the cool season zones. Regions 2A, 2B, and 3 experience warmer temperatures and compose the warm season zone, where the warm season or subtropical turfgrass species are adapted.

It is important to remember that the boundaries on the map in Figure 3-1 are only approximate. In fact, it is impossible to establish a clear-cut line of division between the cool season and warm season zones. Tremendous variation occurs in California.

Climatic conditions in areas near the boundary separating the two zones are not ideal for either cool season or warm season turfgrasses. Winters may be too cold for warm season species and summers too hot for cool season species. These areas immediately north and south of the warm season–cool season zone

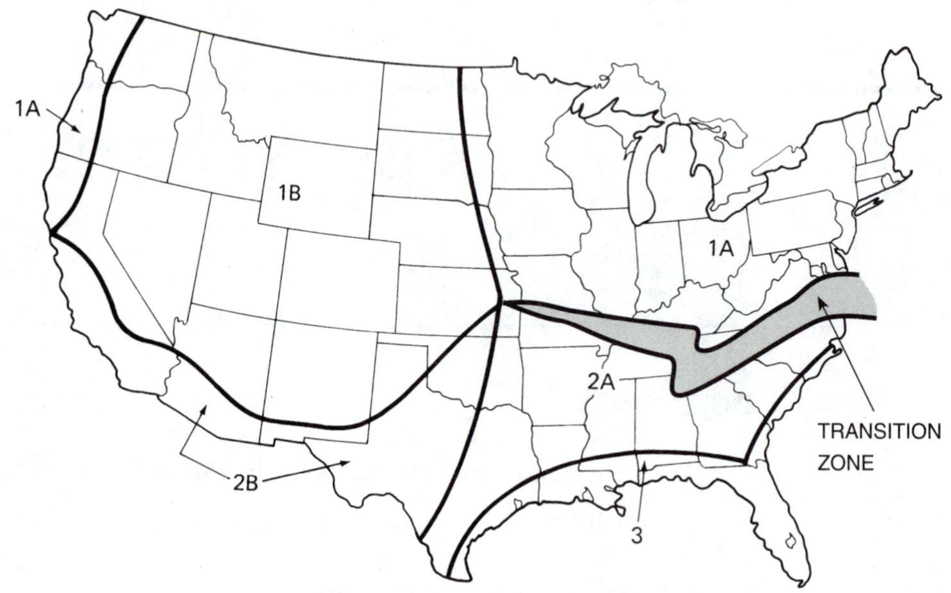

FIGURE 3-1
Turfgrass adaptation regions.

boundary are referred to as the transition zone. The transition zone in the eastern United States is a large area and is specifically identified in Figure 3-1.

Kentucky bluegrass, fine fescue, perennial ryegrass, bentgrass, and tall fescue are the primary turfgrass species grown in the cool season zone. Tall fescue is best adapted to the southern areas of the cool season zone and to the transition zone. The more cold hardy cultivars of two warm season species, bermudagrass and zoysiagrass, may be used in the transition zone and in some southern parts of regions 1A and 1B. Because of the lack of rainfall (semiarid conditions) in region 1B, irrigation is required to ensure the proper growth of these turfgrasses. If irrigation is not available, native dryland grass species such as crested wheatgrass, buffalograss, and blue gramagrass are used.

In regions 2A and 2B, bermudagrass and zoysiagrass are widely adapted. St. Augustinegrass, centipedegrass, and carpetgrass are grown in the southern part of these two regions. Cool season species such as Kentucky bluegrass, tall fescue, fine fescue, perennial ryegrass, and bentgrass, are used in some northern areas of the warm season zone. In region 2B, semiarid conditions exist, so these turfgrasses must be irrigated. If the turf area is not irrigated, then native dryland species such as buffalograss and blue grama should be planted. Region 3 experiences the warmest winter temperatures that occur in the warm season zone. St. Augustinegrass, bermudagrass, zoysiagrass, carpetgrass, centipedegrass, and bahiagrass can be grown in region 3.

Elevation has a significant effect on turfgrass adaptability. Temperatures tend to decrease as elevation increases. In Arizona, for example, bermudagrass is recommended for lawns in areas below an elevation of 4,000 feet (1,200 meters), but Kentucky bluegrass is planted on lawns above 4,000 feet.

The turfgrass species and cultivars discussed in this chapter are adapted to the southern United States. As a general rule, these subtropical grasses grow best when temperatures are 80°F (27°C) or higher. Their tolerance of cold weather is poor, and they may be killed by low winter temperatures or late spring frosts. Most of these grasses become dormant when the average daily temperatures drop to 50°F (10°C).

Low-temperature hardiness varies among the subtropical species and cultivars. Certain cultivars of bermudagrass and zoysiagrass can be used as far north as the southern areas of the cool season zone. St. Augustinegrass, carpetgrass, centipedegrass, and bahiagrass, however, are usually grown only in the warmest regions of the South.

When adequate seed supplies are available, turfgrass managers prefer to establish turf by seeding because this method is relatively easy and inexpensive compared to vegetative propagation.

However, several of the warm season turfgrasses produce only small quantities of seed or seed that is difficult to germinate. This makes it very expensive and impractical to establish turf by seeding with these species, so alternative methods of propagation are used. *Sod*—thin strips of turfgrass and soil—may be installed. *Plugs*—small pieces of sod—can be planted (Figure 3-2). *Sprigs*—pieces of stolons and rhizomes—are also used. New plants grow from the nodes on the sprigs. These types of vegetative propagation will be discussed in more detail in Chapter 9.

FIGURE 3-2
A warm season grass starting to spread from plugs.

FIGURE 3-3
Bermudagrass *(Cynodon dactylon)*.

WARM SEASON TURFGRASSES

Bermudagrass *(Cynodon dactylon)*

Characteristics. Bermudagrass is the most widely used turfgrass in the southern United States (Figure 3-3). It is especially popular in the middle and northern regions of the warm season zone. This grass is also grown extensively in the deep South. It usually stops growing when temperatures are below 60°F (16°C) and turns brown at 45°–50°F (7°–10°C). Low-temperature tolerance varies significantly among cultivars.

Other names for bermudagrass are couchgrass, quickgrass, wiregrass, and devilgrass. It was introduced into this country from eastern Africa in 1751. Bermudagrass hybrids are the highest-quality warm season grasses and require the highest level of maintenance. Low-growing with a fine texture and good color and density, these cultivars are considered the most attractive southern grasses (Figure 3-4). Bermudagrass is used on lawns, parks, cemeteries, athletic fields, airport runways, and golf course tees, fairways, and putting greens.

The grass spreads by aggressive stolons and, to a lesser extent, by rhizomes. It is not unusual for a vigorous stolon to grow 5 or 6 feet (1.5 to 1.8 meters) in

FIGURE 3-4
Bermudagrass can produce excellent turf. *(Courtesy of ChemLawn Corp.)*

a year (Figure 3-5). Bermudagrass has an excellent recuperative potential because of its ability to spread quickly. The improved cultivars (hybrids) are often heavy thatch producers. *Thatch* is a layer of partially decomposed or undecomposed plant debris located above the soil surface. The layer is composed of

FIGURE 3-5
Bermudagrass stolons. *(Courtesy of William Knoop)*

plant parts such as stems, rhizomes, stolons, and roots. Common bermudagrass is not a thatch producer.

Bermudagrass tolerates a wide range of soil conditions. Best growth occurs on fertile soils; it is less satisfactory on poorly drained soils. The grass has a rapid establishment rate. Pest problems include a number of diseases, nematodes, and a few insects. The most serious weakness exhibited by bermudagrass is its inability to grow in the shade. Bermudagrass is the least shade tolerant of the subtropical grasses.

Frequent mowing is necessary because this species has the most rapid vertical growth rate of the warm season grasses. The preferred cutting height for the improved hybrids is in the 0.5- to 1-inch (1.3- to 2.5-centimeter) range. A reel-type mower is superior for close mowing. Bermudagrass exhibits excellent drought tolerance, but has a high moisture requirement. During dry periods irrigation is necessary to maintain high turf quality.

Common bermudagrass has long stolon internodes and forms a less dense turf than the improved cultivars. It is also coarser textured, lighter green in color, and more disease susceptible, and it requires less maintenance than the improved bermudagrasses. Excessive seed head formation is characteristic of the common type. Establishment is by seeding. The best height of cut is 1.5 inches (3.8 centimeters). The majority of bermudagrass lawns are composed of common bermudagrass.

Cultivars. The improved cultivars produce the best-quality turf of any warm season grass. They are hybrids between *Cynodon dactylon* and *Cynodon transvaalensis* (African bermudagrass), or they are selections resulting from crosses within *C. dactylon*. The cultivars require the highest degree of maintenance, but the turf manager is rewarded with a fine-textured, dark green, dense, vigorous turf. Many of these varieties are propagated by sprigs, plugs, or sod because they do not produce viable seeds. Cultivars of bermudagrass include the following.

Ormond is a selection from a golf course fairway in Florida. It is aggressive, dark blue-green in color, and very drought tolerant, and produces a medium-dense turf. Ormond is used on lawns, athletic fields, and fairways in the warmer regions of the South.

Santa Ana is grown primarily in southern California. It produces very dense turf, has a dark blue-green color, and tolerates smog and saline (high levels of salt) soils. Santa Ana wears well and is used on lawns, golf courses, and sports fields.

Tiflawn is the earliest of the series of improved cultivars released by the Coastal Plain Experiment Station in Tifton, Georgia. Tiflawn was selected for its disease resistance, its drought and wear tolerance, and its ability to spread rapidly.

Tifway is an improvement over Tiflawn. It has a fine texture, and a dark green color, and forms few seed heads. It is widely used on lawns, fairways, and tees. Tifway II is a further improvement.

Tifgreen and Tifdwarf are widely used on golf greens because they are very low growing and can be cut to a height of 0.15 inch (3.8 millimeters). They are fine textured and have a dark green color. Both cultivars require a high intensity of culture. Tifgreen II is similar to Tifgreen, but has some improved characteristics.

Sunturf is very attractive and exhibits excellent salt, drought, and wear tolerance. It is primarily used on sports fields and lawns in the plains states. U-3 was the first major improved bermudagrass cultivar to be released. It has good cold tolerance and is still used in the upper South. Texturf 10 is popular for lawns, fairways, and athletic fields.

Midway and Midiron are cultivars with improved cold hardiness. They can survive the lower temperatures occurring in the northernmost areas of the warm season zone and in the transition zone. Midway is a popular choice for lawns and fairways. Midiron forms a vigorous, dense, wear-resistant turf and is recommended for athletic fields, tees, and fairways. Midlawn and Midfield are newer releases for the transition zone.

There are some improved cultivars that, like common bermudagrass, can be established from seed. However, these varieties usually have a finer texture and produce a denser turf than common. Examples are NuMex Sahara, Cheyenne, Jackpot, Sonesta, Sundevil, and Primavera.

St. Augustinegrass *(Stenotaphrum secundatum)*

Characteristics. St. Augustinegrass was introduced from the West Indies and is the least cold tolerant of the warm season grasses (Figure 3-6). It is used in the warmest areas of the subtropical zone. This species is adaptable to many soil conditions, but does best on moist, well-drained sandy soils. Irrigation is necessary during periods of dry weather because its drought tolerance is only fair.

FIGURE 3-6
St. Augustinegrass *(Stenotaphrum secundatum).*

FIGURE 3-7
St. Augustinegrass produces vigorous stolons. This is a variegated variety. *(Courtesy of William Knoop)*

St. Augustinegrass is a major lawn grass in the deep South. This species is used on approximately half of the lawns in Florida. It is coarse, but has good color and establishment rate. Propagation is primarily vegetative by sprigs or sod. Few seed heads are formed.

St. Augustinegrass has strong, thick stolons and produces a turf of medium density (Figure 3-7). Recuperation is good because of the aggressive stolons, but wear tolerance is only fair. A serious thatch problem is associated with this species. Salt tolerance is very good; shade tolerance is excellent. It performs better in the shade than any of the other subtropical species.

The maintenance requirement is medium, though the grass has a vigorous growth rate. Moderate fertilization is necessary. St. Augustinegrass is cut to a height of 1.5 to 3 inches (3.8 to 7.6 centimeters) with either a reel or rotary mower. St. Augustinegrass decline (SAD), a disease caused by a virus, is a severe problem. Chinch bugs also cause extensive injury, but can be controlled by pesticide applications.

Cultivars. The improved cultivars produce greater density and have a finer texture (Figure 3-8). Bitter Blue is a widely used variety that exhibits the best shade tolerance. Floratine is the finest textured of the cultivars and can be mowed close because it is low growing. Floratam shows resistance to the SAD virus and chinch bugs, but has a coarse texture, produces a lower-density turf,

FIGURE 3-8
An improved cultivar of St. Augustinegrass growing on a southern California lawn. *(Courtesy of Ted Bruetsch, photographer)*

and is not as low growing or shade tolerant as the other improved cultivars. Floratam II is also available.

Seville and Raleigh have SAD resistance. Raleigh is more winter hardy than Floratam and is very popular. Another variety, Floralawn, exhibits chinch bug tolerance.

Bahiagrass *(Paspalum notatum)*

Characteristics. Bahiagrass is native to South America and was introduced to the United States from Brazil. It is adapted to the warmer regions of the South. The grass has coarse, tough leaf blades. It spreads by short rhizomes and stolons (Figure 3-9). Recuperative ability is poor to fair. Bahiagrass wears well, but produces an open, low-density turf.

This species grows on a wide range of soils and tolerates poor soils. It is able to grow on dry, infertile, sandy soils. Bahiagrass is deep-rooted and very drought resistant. Shade tolerance is fair, and salt tolerance is poor. The grass develops numerous long seed heads. Propagation is mainly by seed, but sprigs and plugs are also used.

Bahiagrass is ideally suited to low-maintenance areas. It requires little care and is selected for use along roadsides and other sites where minimum maintenance is desirable. Fertilizer needs are low, and thatch accumulation is minimal.

FIGURE 3-9
Bahiagrass *(Paspalum notatum).*

The preferred mowing height is in the 1.5- to 3-inch (3.8- to 7.6-centimeter) range. A rotary mower is necessary to remove the many coarse, unattractive seed heads. This species is becoming more popular on lawns in spite of its poor turf quality because it is relatively resistant to pests. Nematodes are only a minor problem.

Cultivars. The variety Pensacola has narrower, erect leaves and improved density, but it is an abundant seed producer. It is grown along many of the highways in Florida. Argentine is typical of bahiagrass in that it is coarse textured and forms numerous seed heads. However, it has improved density and a more attractive, dark green color. Wilmington is an aesthetic advance because it has a finer texture and develops fewer seed heads than the other bahiagrasses. This variety must be established vegetatively because of limited commercial seed production.

Centipedegrass *(Eremochloa ophiuroides)*

Characteristics. Centipedegrass was introduced to this country from southern China in the early 1900s. It is used primarily in the warmest regions of the subtropical zone, but can be grown farther north than either St. Augustinegrass or carpetgrass. This species can tolerate a wide range of soil conditions and does well on acidic, infertile soils. Drought tolerance is poor because of a shallow root system. Irrigation may be required during dry periods.

Centipedegrass has a medium texture (Figure 3-10) and forms a turf of medium shoot density. It spreads slowly by stolons and has poor recuperative ability. Shade tolerance is fair, and salt tolerance is poor. It is the least wear resistant of the southern grasses. Establishment is mainly by sprigs, plugs, or sod, although seeding also occurs. Seed head formation is not an aesthetic problem.

Centipedegrass is a minimum maintenance species. It requires little fertilizer. *Chlorosis* (yellowing) due to a lack of iron can be a problem when centipedegrass is grown on alkaline soils or soils high in calcium. Mowing is infrequent because of its slow growth rate. Cutting height is 1 to 2 inches (2.5 to 5.1 centimeters) with either a rotary or reel mower. Centipede is used on low-maintenance lawns and other sites where quality is not important. Poor wear tolerance limits its use to areas that are not exposed to much foot

FIGURE 3-10
Centipedegrass *(Eremochloa ophiuroides).*

traffic. It has a relatively low susceptibility to diseases, but nematodes are a serious problem.

Cultivars. Oklawn is a finer-textured cultivar with improved temperature, drought, and shade tolerance. Centennial is a newer release. Georgia common is a seeded variety.

Zoysiagrass

There are three species of zoysiagrass used as turfgrasses. They are native to eastern Asia. Low-temperature hardiness varies among the species. Zoysias are found growing throughout the warm season and transition zones. Both stolons and rhizomes are produced. Propagation is by sprigs, sod, or plugs. Establishment is very slow; it may take two years before a dense turf is formed.

Establishment by seed has not been common because of poor seed germination. However, researchers are in the process of developing techniques that can significantly increase germination.

Zoysiagrass wears very well and has good shade, salt, and drought tolerance. Its ability to recuperate is poor because of the slow spreading rate. Adaptable to a wide range of soils, it grows best on well-drained soils. Nematodes can be a serious problem. Frequent irrigation may be necessary if the roots are attacked by these microscopic worms.

Zoysiagrass generally requires a medium level of maintenance. It is used for many purposes, and can produce good-quality turf on lawns and golf course fairways. Mowing height is 0.5 to 1 inch (1.3 to 2.5 centimeters) because of its low growing habit. The stem and leaf blades are stiff and tough, making the grass difficult to cut. A well-sharpened reel mower works best. Moderate fertility is recommended. Thatch buildup occurs.

Japanese Lawngrass *(Zoysia japonica)*

Characteristics. Also called Korean lawngrass, Japanese lawngrass is faster growing, coarser, and forms less dense turf than the other two species. It has the best low-temperature tolerance of the warm season grasses. A good choice for the transition zone and higher altitudes of the subtropical zone, it is also grown as far north as New Jersey and the coastal areas of New York and New England. This species can survive subzero temperatures.

However, the grass has a short growing season in the northern zone. It turns a straw color after the first heavy frost or when the average daily temperature drops to 50°F (10°C). Cool season species usually green up a month or more earlier in the spring than zoysia. Japanese lawngrass remains green all year in many areas of the South.

Cultivars. Meyer and Midwest are improved cultivars with good color and low-temperature hardiness. Meyer, the most widely used cultivar, has medium

FIGURE 3-11
Meyer zoysiagrass *(Zoysia japonica). (Courtesy of William Knoop)*

texture and density and a more vigorous growth rate (Figure 3-11). Wear and drought tolerance are excellent. Midwest has coarser-textured leaves and a lower shoot density, and spreads faster than Meyer. El Toro also has good tolerance as well as a low water-use rate.

Emerald is a hybrid resulting from a cross between *Z. japonica* and *Z. tenuifolia.* It produces a high-quality turf because of its fine texture, low growth habit, attractive color, and high density. Its low-temperature hardiness is poorer than that of Japanese lawngrass.

In the past zoysiagrass was not established by seed because of poor seed germination. However, researchers have developed chemical treatments that significantly increase germination rates. This has resulted in the appearance of a number of seeded cultivars in recent years. Examples are Compatability, Park Place, SR 9000, SR 9100, Sunrise, Zen 100, Zen 200CS, Zen 300CS, Zen 400CT, and Zenith.

Mascarenegrass *(Zoysia tenuifolia)*

Mascarenegrass is the least cold tolerant, most dense, slowest growing, and finest textured of the three species. It is adaptable to the warmer regions of the subtropical zone, but is not widely used.

FIGURE 3-12
Manilagrass *(Zoysia matrella).*

FIGURE 3-13
Carpetgrass *(Axonopus affinis)* develops tall seed heads.

Manilagrass *(Zoysia matrella)*

Manilagrass is a fine-textured, dense grass that has been a popular choice for lawns in the warm season zone (Figure 3-12). It is less cold tolerant and slower growing than Japanese lawngrass. Cashmere is an improved cultivar.

Common Carpetgrass *(Axonopus affinis)*

Carpetgrass is native to the West Indies and Central America and is grown in the warmest areas of the South. It and St. Augustinegrass are the two least low-temperature hardy turfgrasses. Best adaptation is to acidic, infertile, moist soils.

Carpetgrass is similar in appearance to St. Augustinegrass, but has a lighter green color. The leaf texture is coarse. The grass is stoloniferous and forms a turf of medium density. The recuperation rate is relatively slow. Wear and salt tolerance are poor. Partial shade is tolerated. Establishment is primarily by seed.

Carpetgrass is considered an unattractive species, partially because of the many tall, coarse seed heads developed throughout the summer (Figure 3-13). A rotary mower is required to remove them. Cutting height is 1 to 2 inches (2.5 to 5.1 centimeters). It is shallow rooted for a warm season grass, and drought tolerance is poor. Irrigation is necessary on well-drained soils.

Carpetgrass is most suitable for minimum maintenance areas and wet, poorly drained sites. Easily established by seeding, it is used along roadsides and on slopes for erosion control. This species is also grown on low-maintenance lawns that receive little traffic. Disease and insect problems are not usually severe. There are no improved cultivars.

Axonopus compressa is called tropical carpetgrass. It is similar in appearance to *A. affinis*, but is less winter hardy. Use is greatly restricted because of its extremely poor low-temperature tolerance.

Dichondra *(Dichondra* species)

Dichondra is a broadleaf plant that produces an attractive, low, dense ground cover (Figure 3-14). It is primarily grown on lawns in southern California, especially in the Los Angeles area. Low-temperature tolerance is poor. Dichondra is a native of the southeastern United States, but is not used there because of disease problems.

The plant has soft, pale green leaves and grows close to the ground (Figure 3-15). It is tolerant of partial shade. In the sun dichondra seldom grows taller than 3 inches (7.6 centimeters). Maximum height in the shade is 6 inches (15.2 centimeters).

A small-leaved, dense stand occurs when dichondra is mowed to a height of 0.5 to 1 inch (1.3 to 2.5 centimeters). A mowing height of 1.5 to 2 inches (3.8 to 5.1 centimeters) results in lower density, larger leaves, and increased drought tolerance. It has poor wear tolerance and should be planted only on sites where traffic is limited.

Dichondra does not grow well on wet, compacted soils. Regular watering is necessary, however. Moderate fertility is also required. Establishment is by seeding or plugs. Dichondra is a dicot and is injured by broadleaf weed killers such as 2,4-D. Pest problems include nematodes, cutworms, two-spotted mites, flea beetles, vegetable weevils, slugs, snails, and several diseases caused by fungi.

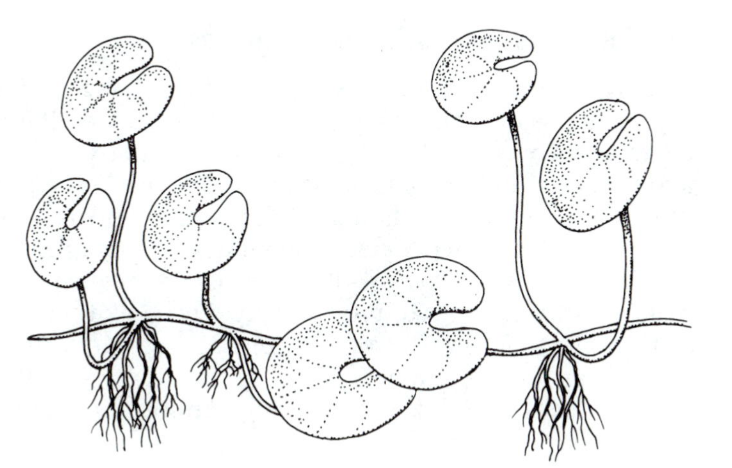

FIGURE 3-14
Dichondra *(Dichondra repens).*

FIGURE 3-15 Dichondra produces an attractive lawn. *(Courtesy of Ted Bruetsch, photographer)*

Kikuyugrass *(Pennisetum clandestinum)*

Kikuyugrass was introduced to California from Africa for use as a turfgrass, but it has a number of undesirable characteristics. Spreading by extremely aggressive, thick stolons, it tends to overwhelm more desirable species and completely take over turf areas. A dense, tough sod is produced. Kikuyugrass has poor color, is very thatchy, and coarse textured, and is difficult to mow. For these reasons, it is generally regarded as an unwanted weed. However, some turf managers like kikuyugrass because it is drought tolerant, wear resistant, and very recuperative. It winter kills in the upper South.

Buffalograss *(Buchloë dactyloides)*

Characteristics. Buffalograss is an important rangeland grass adapted to the dry, semiarid Great Plains region of the United States. It is native to the prairies of states such as Oklahoma, Texas, Arizona, Kansas, Colorado, the Dakotas, and Montana. Buffalograss is used as a turfgrass in these low-rainfall areas because of its excellent drought tolerance. It is the most drought-tolerant turfgrass in the United States and requires no irrigation to survive.

Buffalograss is *dioecious*—it has male and female plants (Figure 3-16). The male plant, if unmowed, produces a pollen stalk that looks like a seed head. Female plants form a lower-growing, denser stand and are preferred when buffalograss is used as a turfgrass.

This species exhibits excellent high-temperature hardiness and is also tol-

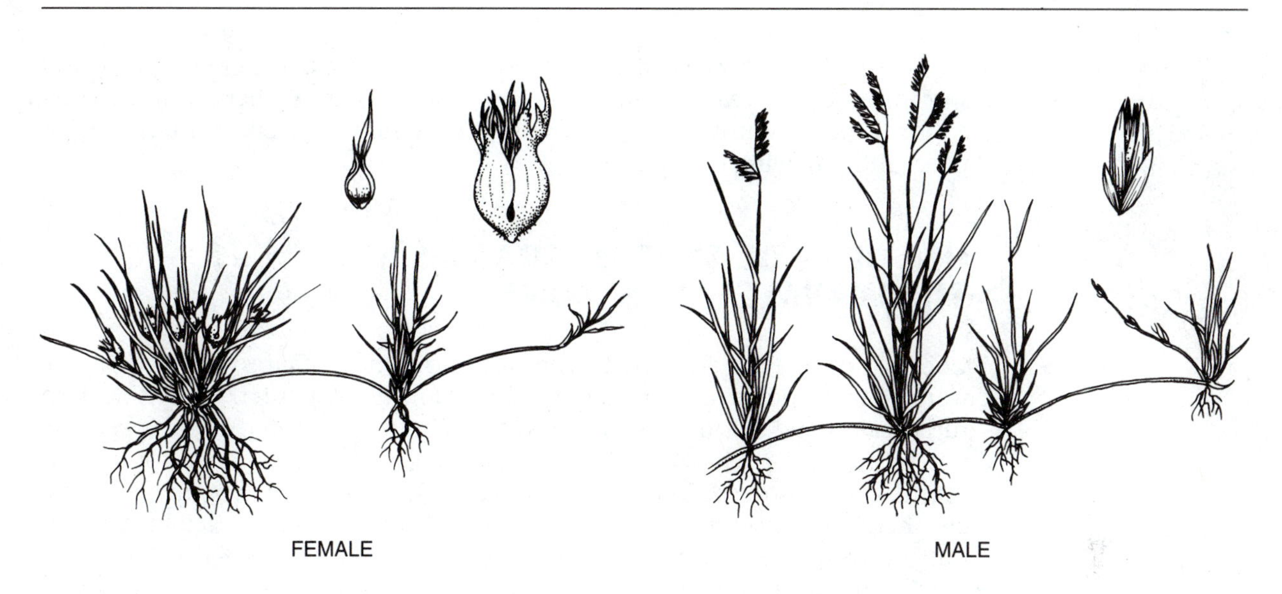

FIGURE 3-16 Buffalograss *(Buchloë dactyloides).*

erant to low temperatures. A very slow vertical growth rate, gray-green color, fine texture, and stolons are characteristic of buffalograss. If left unmowed it seldom grows higher than 4 or 5 inches (10.2 to 12.7 centimeters). It tolerates alkaline soil conditions (soil pH above 7.0) and grows best on heavier, finer-textured soils. Shade tolerance is very poor.

Buffalograss is used on unirrigated lawns and fairways and is virtually pest-free if not overirrigated or overfertilized. It is planted along roadsides and other minimum-maintenance sites for erosion control. Establishment is by seed, plugs, or sod. Cutting height is best at 2 inches (5 centimeters) or above. Little or no fertilizer is necessary, but light applications improve turf quality.

Cultivars. Buffalograss is viewed as an "environmentally friendly" grass because it requires minimal water, fertilizer, and pesticide use. Consequently, it has generated a great deal of interest among plant breeders, and a number of improved cultivars have been released in recent years. Prairie, Buffalawn, "609," NE315, and Bison are examples.

Blue Grama *(Bouteloua gracilis)*

Blue grama is commonly found growing in the same semiarid regions as buffalograss. However, the latter species is more widely used as a turfgrass. Blue grama is primarily grown as a soil conservation grass on dry, low maintenance sites. It will grow on sandier soils than will buffalograss.

Blue grama is low growing and fine textured, has grayish-green leaves, and spreads by short, weak rhizomes. Drought tolerance is excellent, but wear tolerance is poor. Maintenance requirements are minimal. Like buffalograss, it needs little mowing or fertilizer.

USE OF COOL SEASON GRASSES IN THE WARM SEASON ZONE

Cool season grasses adapted to the northern United States are used in the warm season zone to provide a green color during the winter. Many southern turf areas would normally be brown throughout the winter months because the subtropical grasses become dormant. To maintain a green color year-round, cool season species such as annual ryegass, perennial ryegrass, fine fescue, Kentucky bluegrass, rough bluegrass, and bentgrass are seeded into lawns, putting greens, tees, fairways, and other quality turf (Figure 3-17). This overseeding occurs in the fall at the onset of cooler temperatures.

Northern grasses are able to retain their green color all winter long in most southern areas. The temperatures are not cold enough to cause them to become dormant. In the spring the permanent warm season grasses begin to grow again and soon predominate. The cool season species are unable to survive the warmer spring weather and disappear.

Instead of overseeding, turf can be sprayed with a green colorant to solve the winter brown problem. The colorant contains a pigment that gives the

FIGURE 3-17

A golf tee in the South which has been overseeded with a cool season species. *(Courtesy of Art Wick, Lesco Products)*

dormant grass an acceptable green color. Proper application is necessary or the turfgrass can be injured. One application of a good colorant can keep the grass green until spring. As the grass begins to grow again, the leaf tissue coated with the colorant is removed when the turf is mowed. It is important to remember that grass sprayed with a colorant, although green, is still dormant and has very poor wear resistance.

TURFGRASS SELECTION

When selecting a turfgrass, several factors are considered. The selection is obviously limited to species and cultivars that are adapted to regional climatic conditions. Low- and high-temperature hardiness is the most important consideration. All of the subtropical species exhibit excellent heat tolerance. The southern turf manager must choose a warm season grass which will tolerate the coldest temperature occurring in his or her area. The amount of rainfall is also a factor, unless irrigation is provided. Buffalograss or blue grama is planted in the dry regions of the southwestern United States if irrigation is unavailable.

Other criteria such as the ability of the grass to adapt to local soil conditions are important as well. The site should be carefully evaluated for possible stress factors such as shade or large amounts of foot traffic. Resistance to insect, disease, and nematode problems should be considered. The quality expected and level of maintenance to be provided have a major effect on the selection (Figure 3-18). The turf manager must consider all of the strengths and weaknesses of a species or cultivar before making a choice (Table 3-1). It is wise to seek the advice of Cooperative Extension Service turfgrass specialists.

FIGURE 3-18

This variety is a poor choice for a quality turf area because of its unattractive appearance and poor mowing quality.

TABLE 3-1 General Characteristics (Some Cultivars May Differ) of Warm Season Turfgrasses

	Bermudagrass ***(Cynodon dactylon)***	**St. Augustinegrass** ***(Stenotaphrum secundatum)***	**Zoysiagrass** ***(Various species)***
Spreading habit	Stolons and rhizomes	Stolons	Stolons and rhizomes
Leaf texture	Fine to medium	Coarse	Fine to medium
Shoot density	High	Medium	Medium to high
Soil type required	Wide range; fertile best	Wide range	Wide range
Establishment rate	Very fast	Fast	Very slow
Recuperative ability	Excellent	Good	Poor—will recover but slowly
Wear resistance	Very good to excellent	Fair	Excellent
Cold tolerance	Poor to fair	Least tolerant	Fair to good
Heat tolerance	Excellent	Excellent	Excellent
Drought tolerance	Excellent for common; good for hybrids	Fair	Excellent
Shade tolerance	Very poor	Excellent	Good
Salt tolerance	Good	Good	Good
Submersion tolerance	Excellent	Fair	Poor
Maintenance level	Medium to high	Medium	Medium
Fertility requirement	High for hybrid; medium for common	Medium	Medium
Mowing height (inches)	0.5–1 0.2 on greens 1–2 for common	1.5–3	0.5–2
Seed head production	Few to many	Few	Few
Thatching tendency	High for hybrids; low for common	High	Medium to high
Disease potential	High for hybrids; low for common	High	Medium
Nematode problems	Severe	–	Severe

Bahiagrass *(Paspalum notatum)*	**Carpetgrass** *(Axonopus affinis)*	**Centipedegrass** *(Eremochloa ophiuroides)*	**Buffalograss** *(Buchloë dactyloides)*
Short stolons and rhizomes	Stolons	Short stolons	Stolons
Coarse	Coarse	Medium	Fine
Low	Medium	Medium	Medium
Wide range; tolerates poor soils	Wet, acidic best	Wide range; grows well on sandy, acidic, infertile soils	Wide range; best on fine-textured, alkaline soils
Moderate	Moderate	Moderate	Slow
Poor to fair	Poor to fair	Poor	Poor
Good	Poor to fair	Poor	Good
Poor	Very poor	Very poor	Good
Excellent	Excellent	Excellent	Excellent
Excellent	Poor	Poor	Most tolerant
Fair	Fair	Fair to good	Poor
Poor	Poor	Poor	Very good
Good	Fair	Poor	Poor
Low	Low	Low	Low
Low	Low	Low	Low
1.5–3	1–2	1–2	2.0
Many, tall and conspicuous	Many, tall and conspicuous	Inconspicuous	At base of female plant
Very low	Low	Medium	Low
Low	Low	Low	Low
–	–	Severe	–

A few examples will help to illustrate this selection process. If the site is shady, the manager must select a shade-tolerant grass. In the deep South, St. Augustinegrass is an excellent choice. It is the most shade-tolerant subtropical species. However, St. Augustinegrass is not adapted to the lower winter temperatures common in the cooler regions of the warm season zone. In these areas, both bermudagrass and zoysiagrass are successfully grown, but Japanese lawngrass *(Zoysia japonica)* is the favored shade grass because of the extremely poor shade tolerance exhibited by bermudagrass. Tall fescue is a popular choice for the shade in the upper South.

When the manager wants to establish a high-quality turf, improved bermudagrass cultivars are usually the first choice if the site is not shaded. They are very attractive and respond well to higher levels of maintenance. The varieties differ in cold hardiness, so a cultivar that is suited to the local climate should be selected.

Bahiagrass is an unattractive species because of its very coarse texture, low shoot density, and numerous unsightly seed heads. It would be unacceptable where better-quality turf is expected. However, bahia is an excellent choice for minimum-maintenance areas where appearance is relatively unimportant.

Only the highest turf quality is satisfactory on golf greens. To maintain a proper putting surface, the grass is cut daily to a height of 0.25 inch (6.4 millimeters) or less. Tifdwarf or Tifgreen, bermudagrass cultivars that grow low to the ground, are selected because they can tolerate this close mowing and retain their beautiful appearance. The varieties also wear and recuperate well.

A homeowner who planted Tifgreen or Tifdwarf in the hope of having a lawn with the quality of a golf green would likely be very disappointed by its performance. The average homeowner would not be able to provide the intensive culture that these two cultivars receive when growing on a putting green. Without this high level of maintenance, they will perform poorly. In fact, the homeowner's lawn may look less attractive than a neighbor's low-maintenance, low-cost, bahiagrass lawn.

The turf manager must give careful consideration to all relevant factors before selecting a turfgrass for a specific site. A poor choice will result in the need for increased maintenance, endless headaches for the manager, and unsatisfactory turf quality. The turf manager should be aware of new cultivars and their characteristics. New varieties are developed and released because they exhibit improved qualities. They are often better than the cultivars that the manager has used in the past.

A subtropical turf area is usually composed of one turfgrass species only. Mixing of species is uncommon because of differences in aggressiveness, color, texture, density, spreading habit, maintenance requirements, and tolerance to environmental stresses. This incompatibility results in a severe lack of uniformity when species are grown together.

SELF-EVALUATION

1. Subtropical turfgrasses grow best when temperatures are ___________ or higher.
2. Floratam is a cultivar of ___________.
3. The common name of *Paspalum notatum* is ___________..
4. The two turfgrasses grown in low-rainfall, semiarid areas when irrigation is not possible are ___________ and ___________.
5. The broadleaf plant used as a ground cover on lawns in southern California is ___________.
6. The turfgrass with the slowest establishment rate is ___________.
7. Oklawn is a cultivar of ___________.
8. The most shade-tolerant warm season grass is ___________.
9. Improved ___________ cultivars are used on high-quality turf areas throughout the South.
10. Pieces of stolons or rhizomes are called ___________.
11. The genus of mascarenegrass is ___________.
12. The most cold-tolerant of the warm season turfgrass species is ___________.
13. St. Augustinegrass and ___________ are the least cold-tolerant turfgrasses.
14. The two cultivars commonly used on putting greens are ___________ and ___________.
15. Pensacola is a cultivar of ___________.
16. Which subtropical species is least shade tolerant?
17. Which subtropical species has the poorest wear tolerance?
18. What is SAD?
19. For what reason are cool season grasses grown in the warm season zone?

CHAPTER

4

Cool Season Grasses

OBJECTIVES

After studying this chapter, the student should be able to

- Explain why certain turfgrasses are adapted to the cool season zone
- Describe the important cool season turfgrasses
- Discuss the characteristics and uses of these species
- Explain how a turf manager selects the proper species and varieties for a particular site

INTRODUCTION

The cool season or temperate zone consists of roughly the northern two-thirds of the United States (Figure 4-1). The zones and regions of grass adaptation are explained in detail in the introduction to Chapter 3. The turfgrasses adapted to the cool season zone grow best at temperatures in the 60°–75°F (16°–24°C) range.

The cool season grasses all exhibit above-average low-temperature hardiness, although it varies between species and cultivars. The bluegrasses and bentgrasses can tolerate the harsh Canadian winters, while tall fescue and perennial ryegrass are better adapted to those regions of the cool season zone where cold temperatures are not extreme.

Heat tolerance is generally poor to fair. Some of these cool season grasses, such as tall fescue, creeping bentgrass, and Kentucky bluegrass, can be grown in the transition zone and at higher elevations in the warm season zone. Many are overseeded into southern turf to provide green color in the winter when the permanent warm season grasses are dormant. They can be used in the dry, semiarid, and arid regions of the cool season zone if irrigation is supplied.

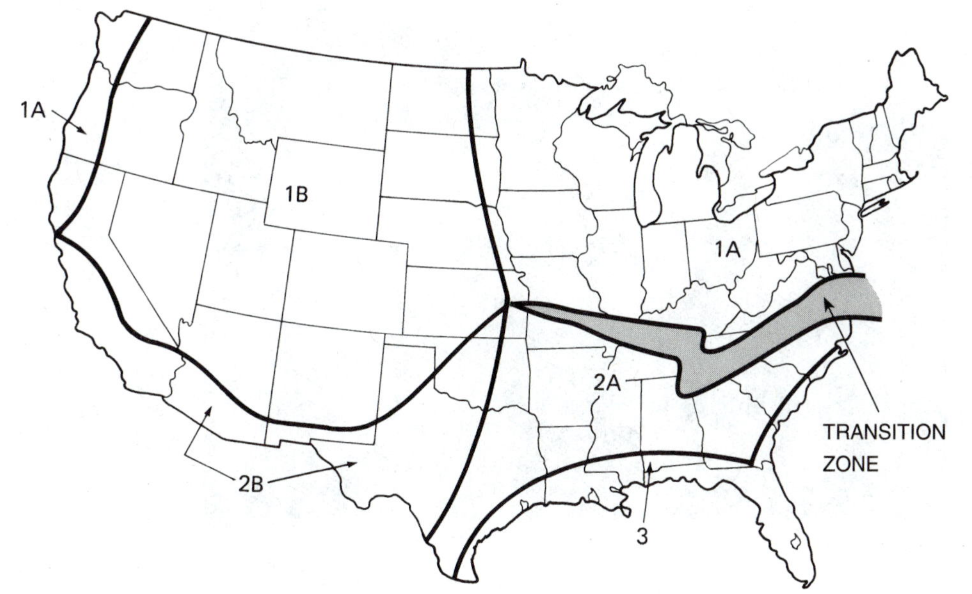

FIGURE 4-1
Turfgrass adaptation regions.

When irrigation is unavailable in these areas, extremely drought-tolerant species such as crested wheatgrass and smooth bromegrass should be planted.

The cool season turfgrass species are native to northern Europe and Asia. Except for a few creeping bentgrass cultivars, they are all established from seed. Most grow best when the soil pH is slightly acid—6.0 to 7.0.

COOL SEASON TURFGRASSES

The following are the most important turfgrasses species used in the cool season zone. Both the common type and improved cultivars will be discussed.

Kentucky Bluegrass *(Poa pratensis)*

Characteristics. Kentucky bluegrass was introduced to this country from Europe in the 1600s. It is now the most important and widely used northern turfgrass. This multipurpose species is planted throughout the cool season zone, is common in the transition zone, and is also used at higher elevations in the subtropical zone. Lawn and fairway use is especially extensive (Figure 4-2). It is difficult to make generalizations about Kentucky bluegrass because there are many cultivars available and characteristics can vary significantly.

Kentucky bluegrass spreads by strong rhizomes and produces a turf of medium to high density (Figure 4-3). It is used on athletic fields because the vigor-

FIGURE 4-2
Kentucky bluegrass is widely used on cool-season-zone lawns. *(Courtesy of O. M. Scott and Sons Company)*

FIGURE 4-3
Kentucky bluegrass *(Poa pratensis).*

FIGURE 4-4
Rhizomes enable Kentucky bluegrass to recuperate from stress.

ous rhizomes allow good recovery (Figure 4-4). Leaf texture is medium to medium-fine, depending upon the variety. The grass is best adapted to well-drained, fertile soils with a pH in the 6 to 7 range. Germination and establishment rates are slow for many cultivars, but some, such as Baron and Mystic, establish fairly rapidly.

Drought tolerance is good. The species becomes dormant during extended periods of high temperatures and drought in the summer, but recovers when cool, moist weather returns. Shade tolerance is relatively poor, although some cultivars, such as Enmundi, Eclipse, Bristol, Glade, A-34 (Bensun), Touchdown, America, and Chateau, have exhibited acceptable quality in moderate shade. Kentucky bluegrass exhibits increased tolerance to shaded conditions in warmer regions. The grass is fairly disease resistant, although susceptibility to each turf disease varies greatly between cultivars. Leaf spot, summer patch, necrotic ring spot, and stripe smut are the most common problems. Thatching tendency is medium. The maintenance level is medium–low to medium-high.

Cultivars. The common, nonimproved varieties such as Delta, Kenblue, Newport, Park, and South Dakota Certified require lower levels of maintenance. They grow more erect and are generally less disease resistant than the improved cultivars. Improved, higher-quality cultivars include Adelphi, Alpine, Aspen, Banff, Barsweet, Blacksburg, Bonnieblue, Bristol, Challenger, Chateau, Cobalt, Columbia, Destiny, Eclipse, Enmundi, Glade, Indigo, Julia, Liberty,

FIGURE 4-5
Improved cultivars of Kentucky bluegrass can produce outstanding lawns. *(Used with permission of The Toro Company. "Toro" is a registered trademark of the Toro Company, Minneapolis, Minnesota.)*

Limousine, Midnight, Miracle, Monopoly, Mystic, Noblesse, NuStar, Opal, Princeton, Ram I, Shamrock, SR 2000, Sydsport, Touchdown, Unique, Victa, and Washington. Preferred mowing height is 1.5 to 3.0 inches (3.8 to 7.6 centimeters). A few new varieties can be cut to 1 inch (2.5 centimeters) or less.

Extensive research and breeding efforts have resulted in the release of over 200 cultivars. All of these varieties have different strengths and weaknesses. Kentucky bluegrass is adaptable to a wide range of environmental conditions and maintenance programs because of the wide selection of cultivars. Some are very attractive and suitable for the finest lawns, while other common types are used for minimum-maintenance sites such as roadsides (Figure 4-5).

Rough Bluegrass *(Poa trivialis)*

Characteristics. This species is called rough, or roughstalk, bluegrass because the sheath at the base of the stem is bumpy and feels rough to the touch. The grass spreads by short, aboveground stolons (Figure 4-6). Its recuperative potential is only fair. The leaves are a shiny, light yellow-green color and cause turf to have a patchy, nonuniform appearance when rough bluegrass is mixed with other grasses. Cold tolerance is excellent. Density is high.

A number of problems are associated with rough bluegrass. It is the least heat tolerant and wear resistant of the cool season grasses. Drought tolerance is very poor because of its shallow root system. These weaknesses greatly limit its use.

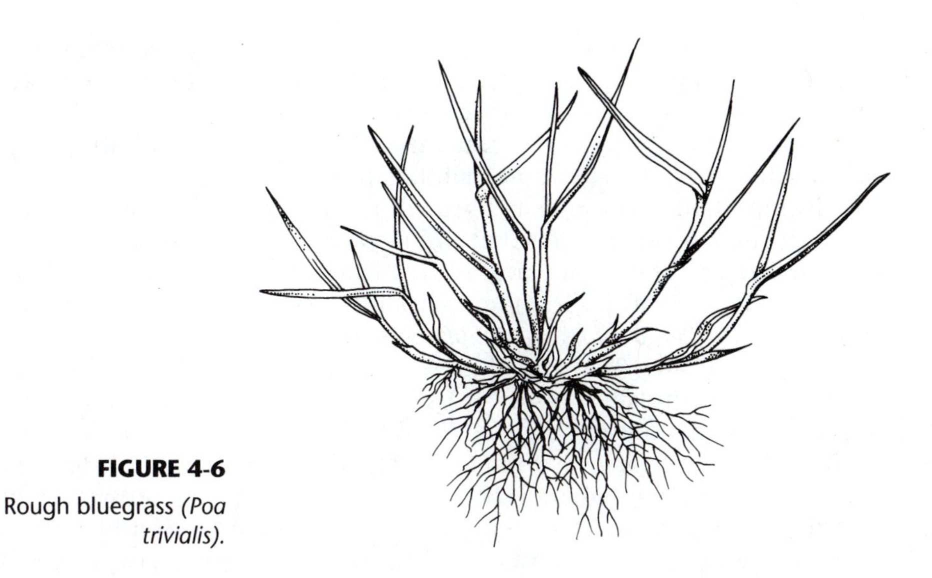

FIGURE 4-6
Rough bluegrass *(Poa trivialis).*

Rough bluegrass has two strengths—it exhibits good shade tolerance and can grow on wet, poorly drained soils. The other shade-tolerant cool season species, the fine fescues, do not grow well on moist soils. Rough bluegrass is used when a shaded area is also wet.

The establishment rate is slow. Disease susceptibility and maintenance level are medium. The grass is cut to a height of 0.5 to 2 inches (1.3 to 5.1 centimeters). The thatching tendency is medium-low. It is sensitive to lawn herbicides such as 2,4-D and may be injured.

Cultivar. Sabre is an improved cultivar that has a somewhat darker green color than the common type. It is recommended for moist shade in the North and for winter overseeding of golf course greens in the South.

Canada Bluegrass *(Poa compressa)*

Canada bluegrass produces an inferior-quality turf and is used for soil conservation purposes on minimum maintenance sites. It is an extremely stemmy grass—the leaves appear high on the stem and are sparse. This species is weakly rhizomatous and forms an open, blue-green sod. Cold tolerance is excellent. It is grown in the northern areas of the cool season zone and is well adapted to droughty, acidic, infertile soils. Reubens and Canon are improved cultivars.

Annual Bluegrass *(Poa annua)*

Characteristics. Annual bluegrass is considered a weed in most turfs. It is a widespread problem in highly maintained turf areas throughout the humid regions of the cool and warm season zones. Annual bluegrass, referred to by many turf managers as "Poa," is included in this chapter on cool season turfgrasses for two reasons. First, because it is such a common, persistent weed and is very difficult to control, turf managers often have to live with it. Understanding the characteristics and maintenance requirements of annual bluegrass is important because Poa is sometimes treated as if it were a desirable turfgrass. The second reason is that turf researchers are attempting to develop an improved type of annual bluegrass that will be more acceptable for turf use.

There are two subspecies of annual bluegrass. *Poa annua* ssp. *annua* is a winter annual, *Poa annua* ssp. *reptans* persists as a perennial in wetter, more highly managed sites. The annual subspecies produces seeds in May and June, which germinate in late summer. It has shallow roots and more erect growth than the perennial subspecies.

Poa annua ssp. *reptans* is found on sites that are frequently irrigated and mowed to a very short height. It has stronger roots and more prostrate growth than the annual type. The perennial subspecies produces seed heads several times during the growing season, and the seeds can germinate at any time when the soil is warm enough. The perennial type is more common in the cool season zone, while the annual type dominates in the warm season zone.

An irrigated, moist, well-maintained turf may be composed of 90 percent *Poa annua* although the species is not intentionally planted. This is the result of its remarkable seed-producing ability. Seed production is so prolific that as many as 60 seedlings have been observed growing per square inch. Seeds are developed even when the grass is mowed lower than 0.25 inch (6.4 millimeters). The seed is remarkably hardy.

Annual bluegrass is basically a bunch-type, noncreeping grass, but close mowing results in the development of short, weak stolons. The leaves are light green, 2 to 3 millimeters wide, and soft. Clumps of *Poa annua* are easily identified by the presence of seed heads (Figure 4-7). These seed heads are unattractive and disrupt the uniformity of the turf.

Poa annua requires a high level of maintenance. Frequent irrigation is necessary because of its shallow roots. The root system is especially weak when the plant is producing seed. On hot days the grass is cooled by watering it for a few minutes in the early afternoon. This practice is called *syringing*. Fungicide treatments are necessary to control diseases. *Poa annua* becomes highly susceptible to disease when it is weakened by heat and drought stress. Fertility requirements are high. The preferred cutting height is 1 inch (2.5 centimeters) or less.

This grass species invades better-quality turf that is well maintained. It is common on golf greens and fairways and high-quality lawns. The greatest problem with annual bluegrass is its inability to tolerate heat and drought stress. The grass dies out or goes dormant during the hotter, drier parts of the summer, resulting in dead or weak brown patches throughout the turf. Daily irrigation is often necessary to prevent this from happening. This frequent watering is very expensive and impossible where water use is restricted.

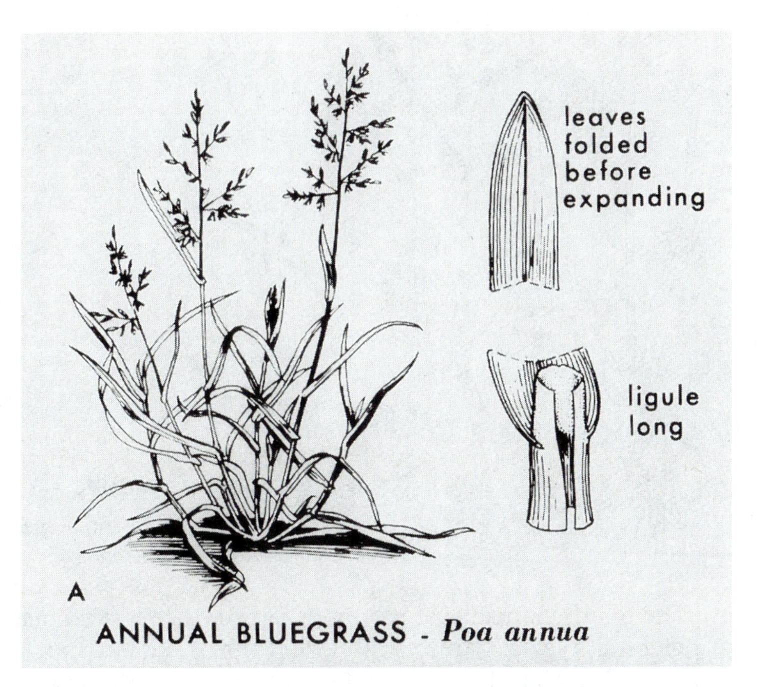

FIGURE 4-7
Annual bluegrass *(Poa annua). (Courtesy of New York State Turfgrass Association, Inc.)*

Poa annua is also susceptible to cold weather injury. It often dies or is severely weakened when the turf is covered with a layer of ice in the winter.

Unfortunately, *Poa annua* does not disappear permanently, when it suffers from heat and drought stress. The numerous seeds in the soil germinate or the dormant grass recovers when the turf is irrigated again or a significant rainfall occurs. Poa grows best in the cooler, more humid spring and fall.

Control strategies will be discussed in the chapter on weeds.

Perennial Ryegrass *(Lolium perenne)*

Characteristics. Perennial ryegrass is a long-lived perennial in the cool season regions which experience mild winters and cool, moist summers. This species does not tolerate prolonged drought or very high or low temperatures. In areas where temperature extremes occur, perennial ryegrass may act as a short-lived perennial, sometimes surviving for only a few years. Improved cultivars generally exhibit better resistance to cold, heat, and drought.

Perennial ryegrass is a noncreeping, bunch-type grass (Figure 4-8). It forms a turf of medium density. Leaf texture is fine to medium. Best adaptation is to moist, moderately fertile soils with a pH of 6.0 to 7.0, although it tolerates a wider range of soil conditions. Shade tolerance and recuperative potential are relatively poor. Wear resistance is good. There are a large number of varieties, and characteristics vary.

FIGURE 4-8
Perennial ryegrass *(Lolium perenne).*

The maintenance level required, fertility needs, and disease susceptibility of perennial ryegrass are medium. Red thread, dollar spot, brown patch, leaf spot, and *Pythium* blight are the most common disease problems. Thatching tendency is low. The grass is generally cut to a height of 1.5 to 3.0 inches (3.8 to 7.6 centimeters), though some cultivars tolerate very close mowing and are used on fairways. A major problem with common perennial ryegrass has been its poor mowing quality. The thick, tough vascular bundles in the leaf blades are difficult to cut, and the leaf tips may have a torn, ragged appearance after mowing. The leaves of the improved cultivars are less likely to fray and shred when cut.

Perennial ryegrass is used to broaden the genetic base of a seed mixture. For example, it is not susceptible to summer patch and necrotic ring spot, two root diseases that can devastate some Kentucky bluegrass cultivars. Another reason for its addition is that it has a rapid establishment rate and protects the seedbed from erosion and other problems until the slower germinating species such as Kentucky bluegrass appear. The species comes up quickly and prevents soil and the remaining seed from washing away. This characteristic is especially helpful when mulch is not placed on a seedbed or if a slope is seeded.

Seed companies often add a small percentage of perennial ryegrass to a seed mixture for this purpose. Its seeds can germinate in a few days, and the seedlings have a fast vertical growth rate. In comparison, Kentucky bluegrass seeds may take as long as three weeks to germinate. Large percentages of perennial ryegrass in a mixture should be avoided because it is so competitive. If too much perennial ryegrass seed is used in a mixture, the aggressive seedlings will take over the entire seedbed. As a result, more desirable species which have a slower establishment rate, such as Kentucky bluegrass, will not have a chance to fill in.

Perennial ryegrass is selected for use on athletic fields because its wear resistance is good. It is also used as a repair grass by overseeding it into sports turf that has been damaged. The overseeding machine drops seed into grooves that it slices into the turf. The seed germinates quickly, and the perennial ryegrass fills in the bare spots in a short time.

Perennial ryegrass is a very popular choice for winter overseeding in the subtropical zone. It is also planted to provide a temporary ground cover in the cool season zone. For example, if a golf course green is unplayable in the spring because of winter injury, a temporary green can be established using perennial ryegrass.

Cultivars. Improved, turf-type cultivars include Advent, Affinity, All Star, Assure, Blazer, Blazer, Brightstar, Charger, Citation II, Commander, Dandy, Dasher II, Delaware Dwarf, Derby, Dimension, Envy, Fiesta II, Gettysburg, Legacy, Manhattan II, Morningstar, Nighthawk, Omega II, Palmer II, Pennant, Prelude II, Premier, Prizm, Ranger, Repell, Repell II, Riviera, Saturn, SR 4000, SR 4100, and Yorktown II. They have been selected for desirable characteristics such as a greater tolerance to environmental stresses, a more satisfactory mowing quality, and better disease resistance (Figure 4-9). These varieties also have a finer leaf texture and darker green color than common perennial ryegrass (Figure 4-10). The common type, because of its unacceptable traits, should be restricted to minimum-maintenance sites. Many of the turf-type cultivars are very attractive and can be used as a component of high-quality turf. They mix well with Kentucky bluegrass.

FIGURE 4-9
The turf-type perennial ryegrass in the center has improved mowing quality comparable to the Kentucky bluegrass cultivar on the left. The ryegrass cultivar on the right exhibits the poor mowing quality typical of the nonturf-type varieties. *(Courtesy of Northrup King Co.)*

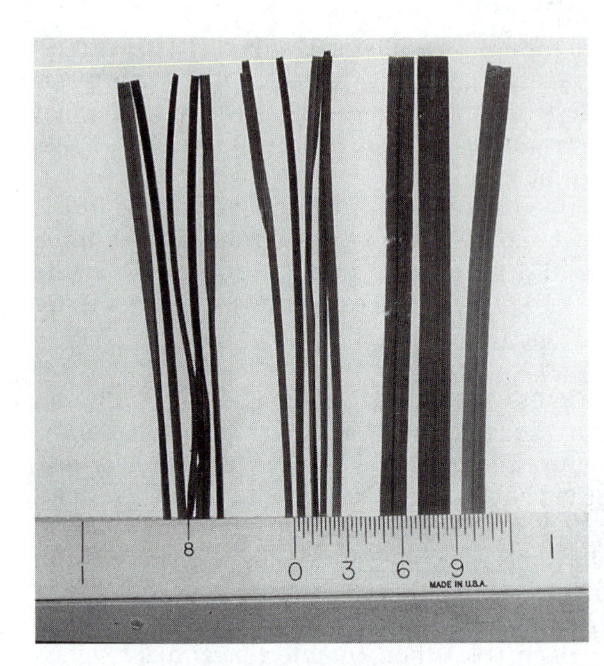

FIGURE 4-10
The nonturf-type perennial ryegrass on the right has a coarser texture than the typical turf-type ryegrass cultivar in the center or the Kentucky bluegrass variety on the left. *(Courtesy of Northrup King Co.)*

Some of the cultivars, such as All Star, Citation II, Commander, Dandy, Dasher II, Pennant, Repell, Repell II, Saturn, SR 4000, and SR 4100, exhibit resistance to surface-feeding insects because of the presence of endophytic fungi. An endophyte is a fungus that lives inside the leaves and stems of a grass plant. It gives off toxins that inhibit or repel insects.

Varieties high in endophytes are not usually attacked by above-ground feeders such as chinch bugs, sod webworms, billbug adults, aphids and cutworms. Roots, however, do not contain high enough levels of endophytes to be protected against white grubs and other insect pests that feed below ground.

In some cases the presence of endophytic fungi seems to make grass more tolerant of environmental stresses such as drought. They may also cause an increase in plant vigor.

Annual Ryegrass *(Lolium multiflorum)*

Annual, or Italian, ryegrass has coarse, shiny, light green leaves. It is a bunch-type grass without rhizomes or stolons. It is an annual or short-lived perennial because of its inability to tolerate temperature extremes. For example, in the cool season zone it does not generally survive the winter. Annual ryegrass is a very poor choice for seed mixtures. Weeds often fill in the bare spots when it disappears. Seed mixtures containing annual rye are usually inexpensive because its seed is cheap. Such mixtures should be avoided unless turf quality is unimportant.

Annual ryegrass exhibits a more rapid germination and vertical shoot growth rate than perennial ryegrass. It is extremely aggressive and competitive. Its major use is for winter overseeding in the South. Annual ryegrass is sometimes sold as a "first aid" or repair grass in the cool season zone because of its ability to provide a quick, temporary cover when bare spots appear in a turf.

Intermediate Ryegrass *(Lolium hybridum)*

Intermediate ryegrass is a new species developed by crossing perennial ryegrass and annual ryegrass. The resulting annual/perennial hybrids exhibit some desirable characteristics of each species. Turfgrass breeders are developing cultivars that combine the low seed cost and rapid germination of annual ryegrass with some of the turf-type qualities of improved perennial ryegrasses. Intermediate ryegrasses are primarily used for overseeding dormant warm season turf.

Creeping Bentgrass *(Agrostis palustris)*

Characteristics. Creeping bentgrass requires the highest level of maintenance of the cool season grasses and is used primarily on golf course putting greens, tees, bowling greens, and grass tennis courts. It is also used on fairways at golf courses with higher maintenance budgets. Creeping bentgrass produces a fine-textured, soft, extremely dense, carpetlike sod. Its quality surpasses that of any other northern turfgrass (Figure 4-11).

FIGURE 4-11

Creeping bentgrass exhibits excellent quality when properly maintained. *(Courtesy of Art Wick, Lesco Products)*

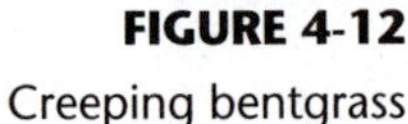

FIGURE 4-12
Creeping bentgrass *(Agrostis palustris).*

Creeping bentgrass is the ideal grass for putting greens because of its beauty and ability to recuperate and tolerate close mowing. Strong, vigorous stolons enable it to recover quickly when injured (Figure 4-12). This excellent recuperative potential more than compensates for its poor wear resistance. Creeping bentgrass can be cut as low as 0.125 inch (3.2 millimeters) because of its prostrate, low-growth habit.

This species is very cold tolerant. However, for a cool season grass it is slow to green up in the spring and loses its color early in the fall. Heat tolerance is fair, salinity tolerance is good. It does not grow well on compacted soils. Soil mixes for greens usually contain high percentages of sand to help reduce the rate of compaction. Optimum soil pH is in the 5.5 to 6.5 range.

Creeping bentgrass is not recommended for lawns. If mowed at typical lawn heights (1.5 to 3 inches) its visual quality is poor. When cut very short, creeping bentgrass is highly attractive but becomes shallow rooted. Constant care is necessary because of its shallow root system, rapid shoot growth rate, and aggressive spreading habit. Frequent, perhaps daily irrigation is required during periods of dry weather. Syringing—a brief midday watering—is advisable when temperatures are high. Fertility demands are high. Greens may be fertilized once every three weeks. Creeping bentgrass is very disease susceptible at close cutting heights. Fungicide treatments are necessary to protect this species from injury.

Greens are mechanically aerated to alleviate compaction. The frequency of this core cultivation depends on the amount of foot traffic that the green receives. Creeping bentgrass has a high thatching tendency. Vertical mowing and topdressing are two practices recommended to reduce thatch buildup. The grass is sensitive to herbicides.

Creeping bentgrass is usually cut to a height of 0.125 to 0.5 inch (3.2 to 13 millimeters). Greens are mowed very close (3 to 6 millimeters) to ensure good putting quality. A reel-type mower is essential because of the low cutting height. Daily or slightly less frequent mowing is required. The grass is very competitive and often spreads off a green and dominates irrigated or wet locations.

Cultivars. Cultivars propagated vegetatively include Toronto (C-15), Pennpar, Cohansey (C-7), Washington (C-50), and Evansville. These varieties are poor seed producers and are established by planting stolons. Penncross, Penneagle, Seaside, Prominent, Emerald, Pennlinks, Putter, Providence, Procup, Crenshaw, Cobra, Regent, Southshore, and Lopez are propagated by seed. The less aggressive varieties are used on fairways. Penncross is widely used on golf greens and has been the most popular cultivar. Providence and Pennlinks are two newer cultivars that are very popular choices for greens. The seeded cultivars have largely replaced the vegetatively propagated cultivars because establishment by seed is easier and cheaper.

Colonial Bentgrass *(Agrostis tenuis)*

Characteristics. Colonial bentgrass is a very attractive grass that produces a high-density, fine-textured turf. It spreads by short, weak stolons and rhizomes and is much less aggressive than creeping bentgrass. Drought and heat tolerance are poor. Its use is largely restricted to coastal regions along the Atlantic Ocean in the Northeast and the Pacific Ocean in the Northwest.

The maintenance level is fairly high but less than that required by creeping bentgrass. Disease susceptibility, fertility needs, and thatching tendency are medium-high. Fungicide treatments may be necessary, and frequent irrigation is required during dry periods. The grass is low growing and the preferred cutting height is in the 0.4- to 1-inch (1- to 2.5-centimeter) range. Cold tolerance is good, shade tolerance is fair, wear resistance is poor, and its recuperative potential is fair to poor. Colonial bentgrass is best adapted to sandy soils with a pH of 5.5 to 6.5. It is sensitive to herbicides.

This species is occasionally used on golf fairways, tees, and better-quality lawns because of its attractiveness and ability to tolerate close mowing. Colonial bentgrass does not achieve the quality of creeping bentgrass, nor can it be mowed as low, but it is less time-consuming and expensive to maintain. The availability of improved Kentucky bluegrass and perennial ryegrass cultivars that can be cut to 1 inch (2.5 centimeters) or less has resulted in a diminished role for colonial bentgrass.

Colonial bentgrass seed sometimes contains small quantities of creeping bentgrass seed. This accidental contamination is a serious problem because the aggressive creeping bentgrass will spread and eventually dominate the site. Colonial bentgrass will outcompete Kentucky bluegrass and dominate when the turf is intensely fertilized, closely mowed, and frequently irrigated.

Cultivars. Two recommended cultivars are Exeter and Highland. Exeter is very attractive and has excellent low-temperature hardiness. Highland has improved drought tolerance.

Velvet Bentgrass *(Agrostis canina)*

Velvet bentgrass is considered the most beautiful turfgrass. The extremely fine-textured, almost needlelike leaves are soft and velvety. The grass produces a turf of very high density. Spreading by stolons, it is less aggressive than creeping bentgrass, but has a more vigorous creeping habit than colonial bentgrass. Uniformity is excellent.

Velvet bentgrass is used on golf greens and other high-quality, high-maintenance turf sites. It is adapted to the mild, humid coastal regions of New England and other northern areas where the ocean has a moderating influence on temperature. Best growth occurs on acid (pH 5.0 to 6.0), sandy, well-drained soils.

Wear resistance and recuperative potential are relatively poor. Velvet is more drought tolerant than creeping and colonial bentgrass, but still requires frequent irrigation. Disease susceptibility and thatching tendency are high. Shade tolerance is very good. The preferred mowing height is 0.2 to 0.4 inch (5 to 10 millimeters). Kingstown is an improved cultivar developed at the University of Rhode Island.

Redtop *(Agrostis alba)*

Redtop is an unattractive, stemmy grass with coarse-textured, grayish-green leaves. Its leaves and stems are more erect than those of the other *Agrostis* species. The grass spreads by strong, short rhizomes and produces an open, low-density turf. It is named redtop because of the red color of the seed heads.

Redtop germinates relatively quickly and for this reason was once widely used in seed mixtures. The seed is not usually included in mixtures today because clumps of the grass can persist for many years and will detract from the quality of the turf. Perennial ryegrass is a much more desirable nursegrass. Redtop tolerates a wide range of soil conditions, and does well on wet, poorly drained sites. It is still added to seed mixtures that are to be planted on minimum-maintenance areas such as roadsides if the soil is wet or poorly drained. This grass should not be cut shorter than 2 inches (5.1 centimeters).

Tall Fescue *(Festuca arundinacea)*

Characteristics. Tall fescue, although considered a cool season grass, is susceptible to low-temperature injury. In the colder regions of the cool season zone, it can be a short-lived perennial, thinning out after a few or several years as a result of low-temperature stress. This species is generally a poor choice for high elevations and the most northern areas in the cool season zone, though some of the newer cultivars have improved cold tolerance. Tall fescue has the best heat tolerance of the cool season grasses and is a long-lived perennial when grown in the transition zone and the southern regions of the cool season zone.

Tall fescue may develop a few short, weak rhizomes, but it spreads primarily by tillering and has a bunch-type growth habit. This species is not particularly compatible with other grasses because it grows in clumps and is often coarser textured than other species (Figure 4-13). When used in a seed mixture, the mixture should contain at least 80% tall fescue seed by weight. Tall fescue is regarded as a weed when coarse clumps of it are found in a high-quality, dense, finer-textured turf.

FIGURE 4-13
A coarse-textured cultivar of tall fescue *(Festuca arundinacea)*.

This species has some excellent low-maintenance characteristics. Drought tolerance is very good because of its extensive and deep root system. Disease problems are not common. It tolerates a wide range of soil conditions and will grow on infertile sand, compacted clay, and alkaline and saline soils. Although tall fescue can tolerate low fertility, it responds favorably to moderate levels of fertilization.

Tall fescue does not tolerate close mowing. A cutting height of 2 inches (5 centimeters) or higher is usually best, though new, lower-growing cultivars can be cut shorter. Adaptation to shade is fair to good. Thatching tendency is low. Establishment is by seed, and the establishment rate is good.

Tall fescue wears very well because of its tough leaves and sturdy root system. It has been a popular choice for athletic fields and other sites that receive heavy traffic and hard use. Despite its superior wear tolerance, tall fescue is not the ideal sports field grass. It is slow to recover when injured. The lack of vigorous rhizomes or stolons severely limits its recuperative potential. The grass does not achieve good wearability until the second year after establishment and may be badly damaged if played on during the first season. Improved perennial ryegrass cultivars and wear-tolerant Kentucky bluegrass cultivars are often used in place of tall fescue on recreational fields today, especially in the colder regions of the cool season zone.

Cultivars. There are two types of tall fescue. The older, coarse-leafed varieties such as Kentucky 31 and Alta are unattractive compared to the other cool season grasses. They have very coarse, erect, stiff leaves and produce a turf that has a low shoot density. The newer, turf-type cultivars have leaves that are denser and finer textured than the older type (Figure 4-14).

The coarse-leafed varieties are planted on minimum-maintenance sites to promote soil stabilization. They are widely used along roadsides and highways. Kentucky 31 is the most popular coarse-textured variety.

Amigo, Austin, Avanti, Bonanza, Cimmaron, Cochise, Crossfire, Eldorado, Guardian, Jaguar II, Marathon, Monarch, Mustang, Olympic, Olympic II, Rebel, Rebel II, Shenandoah, Silverado, Tribute, and Wrangler are examples of the finer-textured tall fescues. These turf-type cultivars produce a much more attractive, denser turf than the course type and can be mowed to a lower height. Some varieties such as Rebel II and Shenandoah contain endophytes.

The turf-type cultivars are popular lawn grasses in the cooler parts of the warm season zone, the transition zone, and in the warmer areas of the cool season zone.

Fine Fescues

The fine fescues are various *Festuca* species and subspecies that have very fine-textured leaf blades (Figure 4-15). The leaves are usually narrower than 2 millimeters and in some cases are only 0.5 millimeter wide. The leaves tend to roll up and often have a bristlelike, wiry appearance. The color is good.

FIGURE 4-14

The turf-type tall fescue cultivar on the left has much finer leaf texture than the coarse-textured variety on the right. *(Courtesy of Lofts Seed, Inc.)*

The fine fescues have a slow, nonaggressive spreading habit, although the ability to spread varies among species and subspecies. Eventually, however, as they fill in the shoot density becomes quite high. Wear resistance is fair, recuperative potential is fair to poor. The fine fescues tolerate low light intensities and are the best of the cool season grasses for shaded sites. They are widely used in seed mixtures because of their superior shade tolerance (Figure 4-16).

The fine fescues are also excellent low-maintenance grasses for lawns. Unlike many grasses that require little care, they are relatively attractive. They are drought tolerant because of their very dense root system and ability to roll their leaves during dry periods. As the blades curl, less leaf surface area is exposed to the air, resulting in less water loss due to transpiration. Fertility requirements are low. High rates of nitrogen encourage disease problems, and this increased susceptibility is a serious threat to the grass. Frequent irrigation also thins it out.

The fine fescues exhibit a slow vertical growth rate and do not require frequent mowing. Preferred cutting height is in the 1.5- to 3.0-inch (3.8- to 7.6-centimeter) range. They will grow on a wide range of soils, but do not tolerate wet, poorly drained soil conditions. Establishment rate is moderately fast compared to that for Kentucky bluegrass. The fine fescues are winter hardy and can be grown throughout the cool season zone. Disease problems do occur and can be severe.

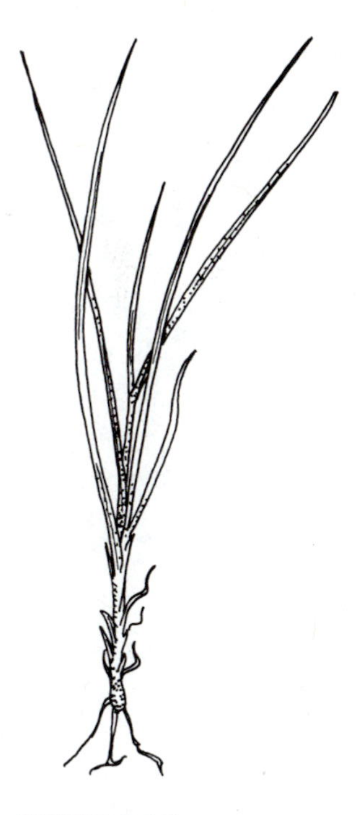

FIGURE 4-15

Fine fescue *(Festuca rubra)*.

The fine fescues are usually not grown alone in monostands. Instead, they are often mixed with Kentucky bluegrass and occasionally with perennial ryegrass or colonial bentgrass. Turfs composed of Kentucky bluegrass and fine

FIGURE 4-16
Fine fescue growing in partial shade.

fescue are common. However, with the availability of new Kentucky bluegrass cultivars that can tolerate shade, droughty soil conditions, and low levels of maintenance, the fine fescues may be used less.

The following species are grouped together as fine fescues. The three subspecies of *Festuca rubra* are commonly called red fescues and are closely related. Some of the newer fine fescue cultivars contain endophytes.

Slender Creeping Fescue *(Festuca rubra* ssp. *trichophylla).* This grass is often referred to as creeping red fescue. It is fine leafed, produces a dense turf, and is weakly rhizomatous. It spreads slightly by small, short rhizomes. Dawson, Polar, Barcrown, Marker, Smirna, and Napoli are improved cultivars.

Strong Creeping Fescue *(Festuca rubra* ssp. *rubra).* Strong creeping fescue is also referred to as creeping red fescue, and some turf specialists call it spreading fescue. It is coarser than slender creeping fescue, forms a less dense sod, and has longer, thicker rhizomes. It exhibits a more vigorous spreading habit than the other fine fescues, although it is not as aggressive as other cool season species such as creeping bentgrass and Kentucky bluegrass. Examples of improved cultivars are Fortress, Pennlawn, Boreal, Ruby, Durlawn, and Flyer.

Chewings Fescue *(Festuca rubra* ssp. *commutata).*Chewings is very similar to slender creeping red fescue. However, it lacks rhizomes and forms a somewhat denser turf. It is low growing and fine textured and is considered the most attractive of the fine fescues by many turf managers. The creeping red fescues exhibit better high- and low-temperature tolerance. Agram, Atlanta, Camaro, Checker,

Dignity, Enjoy, Highlight, Jamestown, Jamestown II, Longfellow, Southport, SR 5000, Tiffany, Trophy, Victory, and Waldorf are improved cultivars.

Hard Fescue *(Festuca longifolia).* Hard fescue is a low-growing, noncreeping, bunch-type grass. Drought tolerance and disease resistance are better than for the red fescues. It grows very slowly vertically and is well adapted to shade and poor soils. Hard fescue is generally used for soil erosion prevention on utility sites, but Aurora, Biljart, Discovery, Reliant, Scaldis, Spartan, SR 3000, SR 3100, Waldina, and Warwick are improved cultivars which perform satisfactorily on lower-maintenance lawns.

Sheep Fescue *(Festuca ovina).* Sheep fescue is used for soil stabilization and erosion control. It requires the lowest level of maintenance of the fine fescues and produces a poor-quality, clumpy turf. Leaf blades are stiff and blue-green in color. Sheep fescue exhibits excellent drought tolerance and is best adapted to sandy or gravelly soils. It grows very slowly and is a good choice for difficult-to-mow areas. Fertilization is not necessary. It is a noncreeping, bunch-type grass. Bighorn and MX 86 are cultivars.

Other Turf Grasses

Smooth Bromegrass *(Bromus inermis).* Smooth bromegrass is coarse textured and has an extensive root system. The grass spreads by vigorous rhizomes and produces an open, low-quality turf. Wear resistance is poor, but heat, cold, and drought tolerance are good. It is used primarily as a soil stabilization grass on minimum-maintenance sites in the dry, semiarid region.

Timothy *(Phleum* species*).* Timothy is an important forage and pasture crop with limited usefulness as a turfgrass. It is occasionally used for soil erosion prevention along roadsides. Timothy is grown on athletic fields in northern Europe, and researchers in the United States and Europe are attempting to develop improved turf-type cultivars.

Crested Wheatgrass *(Agropyron cristatum).* Crested wheatgrass, also called fairway wheatgrass, is a coarse-textured, noncreeping, bunch-type grass. It is used on unirrigated lawns and fairways in the drier, colder regions of the cool season zone. The grass is best adapted to the semiarid, northern areas of the Great Plains. Quality is low, but it is able to survive lengthy drought conditions without irrigation. Drought tolerance is excellent because of an extensive and deep root system.

A mowing height of 1.5 to 3.0 inches (3.8 to 7.6 centimeters) is recommended. Low to moderate fertility is required. Irrigation should be minimal because heavy watering will stress crested wheatgrass.

TURFGRASS SELECTION

The turfgrass manager must select turfgrasses that are adapted to the climatic conditions in his or her area. The grasses should exhibit resistance to the area's common diseases. The turf quality expected and level of maintenance have a great influence on the turf manager's decision. Special problems associated with the site to be established, such as shade or heavy amounts of foot traffic, must be considered. The selection process requires that the turf manager evaluate the strengths and weaknesses of the cool season species and varieties to determine which are most suitable (Table 4-1). Recommendations from turf specialists working with the Cooperative Extension Service are very helpful.

If, for example, a turf manager plans to establish a minimum-maintenance turf along a highway, tall fescue is a possible selection. In the northern regions of the cool season zone, however, it would be a poor choice because of its intolerance of very cold temperatures. In the southern regions of the zone tall fescue would be a good choice because the winters are less severe and it is adaptable to the hotter summers.

Disease resistance is very important. The turf manager should select varieties that are able to tolerate the diseases that are a serious threat in his or her area. For example, Fylking is an attractive Kentucky bluegrass cultivar appropriate for lawns in colder regions of the temperate zone. However, in warmer regions where summer patch disease is common, Fylking is much less satisfactory because of its susceptibility to this disease.

There are some cultivars that are resistant to most of the turfgrass diseases in the cool season zone. But these varieties may lose their ability to resist if the microorganism that causes a disease changes as a result of sexual reproduction or mutation. These changes can enable the microorganism to overcome the resistance of the grass. While turf managers may prefer the uniform appearance created by planting a single variety, a monostand has an increased risk potential. If the cultivar succumbs to an environmental stress, the entire turf is lost. A one-variety monoculture requires careful maintenance to ensure its survival.

It is generally recommended that the turf manager use a *blend* if only one species is planted. A blend is a combination of seed from cultivars of the same species. If a disease or some other environmental stress thins out one cultivar, the entire turf will not be destroyed. The surviving cultivars will spread and eventually fill in the bare spots.

A *mixture* is a combination of seed from different species. Using a mixture guarantees the permanence of a turf because it is highly unlikely that all the species will succumb to the same stress. Kentucky bluegrass is usually the major component of cool season zone seed mixtures. It is widely used because of its attractive appearance, strong rhizomatous spreading habit, good tolerance of most environmental and soil conditions, and the large selection of improved cultivars available. Kentucky bluegrass is compatible with fine fescues and perennial ryegrass and is often grown with these two species. Kentucky bluegrass cultivars should compose the largest percentage because they gener-

TABLE 4-1 General Characteristics (Some Cultivars May Differ) of Cool Season Turfgrasses

	Colonial Bentgrass *(Agrostis tenuis)*	**Creeping Bentgrass** *(Agrostis palustris)*	**Kentucky Bluegrass** *(Poa pratensis)*	**Perennial Ryegrass** *(Lolium perenne)*	**Fine Fescue (various *Festuca* species)**	**Tall Fescue** *(Festuca arundinacea)*
Spreading habit	Short stolons and rhizomes	Strong stolons	Strong rhizomes	Bunch type	Bunch type or short rhizomes	Bunch type
Leaf texture	Fine	Fine	Medium-fine to medium	Fine to medium	Very fine	Coarse to medium
Shoot density	High	Highest	Medium to high	Medium	High	Low to medium
Soil type required	Well-drained, sandy best	Sandy, moist, fertile best	Wide range; well-drained, moist best	Wide range; moist, fertile best	Infertile, well-drained	Wide range
Establishment rate	Medium	Medium	Slow to medium	Fast	Medium-fast	Medium-fast
Recuperative ability	Poor to fair	Best	Good	Poor to fair	Poor to fair	Poor to fair
Wear resistance	Poor	Poor	Fair	Fair to good	Fair	Good
Cold tolerance	Good	Excellent	Excellent	Poor to fair	Good	Fair
Heat tolerance	Fair	Fair to good	Fair	Poor to fair	Fair	Good
Drought tolerance	Poor	Poor	Good	Fair	Very good	Very good
Shade tolerance	Fair to good	Fair to good	Poor	Poor	Excellent	Fair to good
Salt tolerance	Poor	Best	Poor	Fair	Poor	Good
Submersion tolerance	Fair	Excellent	Fair	Fair	Poor	Good
Maintenance level	Medium to high	High	Medium to medium-high	Medium	Low to medium	Low to medium
Fertility requirement	Medium-high	Highest	Medium to medium-high	Medium	Low	Medium
Mowing height, in inches	0.5–1	0.125–0.5	1.5–3.0	1–3.0	1.5–3.0	1.5–3
Thatching tendency	Medium-high	High	Medium	Low	Medium	Low
Disease potential	Medium-high	High	Medium	Medium	Medium	Low

ally have more desirable characteristics. Kentucky bluegrass is also the only northern lawn grass that spreads and recuperates well. Fine fescue is added to a seed mixture because it tolerates shade, drought, and low maintenance. Perennial ryegrass increases the adaptability of the mix and holds the soil in place until the other two species germinate.

When Kentucky bluegrass and fine fescue are grown together, one species eventually predominates. Kentucky bluegrass will be more competitive if the area is sunny, moderate or high levels of fertilizer are supplied, and irrigation occurs. Fine fescue has the advantage when the area is shaded and not watered, little or no fertilizer is applied, and the soil is droughty.

Fine fescues are not normally used on better-quality turf sites. They have some serious disease problems and a fairly poor recuperative ability. Moreover, they produce a clumpy, less attractive turf than Kentucky bluegrass. Blends of straight Kentucky bluegrass can be used on sunny, high-maintenance sites. Fine fescue and perennial ryegrass are not usually planted without Kentucky bluegrass, though golf courses may seed turf-type perennial ryegrasses alone on fairways and roughs.

The turf manager must consider the level of maintenance to be practiced and the turf quality expected when selecting turfgrasses. Creeping bentgrass is perfect for a golf course putting green, but would be a nuisance on a home lawn because of its high maintenance requirements. A mixture of common Kentucky bluegrass and fine fescue might be satisfactory on an average lawn, but would be inadequate for one where excellent quality was desired.

Any special problems that the grass will encounter will affect the turf manager's decision. Fine fescue or shade-tolerant Kentucky bluegrass varieties are favored if the site is shaded. If the site is both shaded and wet, rough bluegrass may be chosen. An athletic field will be seeded with a mixture of perennial ryegrass and wear-resistant Kentucky bluegrass cultivars. Tall fescue may be used in the warmer regions of the cool season zone.

Kentucky bluegrass and perennial ryegrass cultivars that tolerate close mowing are used on fairways cut at 0.75 inch (19 millimeters) or higher. Creeping bentgrass is normally used on fairways cut to a height less than 0.75 inch.

The turf manager should be aware of new varieties and their characteristics. New cultivars are developed and released because they have improved qualities. They are often better than the varieties the manager has used in the past. A cultivar that was outstanding ten years ago may not be the best choice today. It may have lost its disease resistance or may have been surpassed by a newer cultivar with superior traits. Merion Kentucky bluegrass, for example, was an excellent cultivar for many years, but is now susceptible to several diseases.

SELF-EVALUATION

1. The common name of *Poa trivialis* is ___________.
2. Which species is often identified by the presence of seed heads?
3. Which species is commonly used for golf course putting greens?
4. The cool season species that is included in seed mixtures because of its rapid establishment rate is ___________.
5. Fine fescues exhibit good drought and ___________ tolerance.
6. The northern turfgrass with the coarsest leaves is ___________.
7. Which grass is planted in wet, shaded locations?
8. Kentucky bluegrass spreads by ___________.
9. Creeping bentgrass spreads by ___________.
10. Olympic, Cochise, and Rebel are examples of turf-type ___________.
11. Which cool season grasses are used primarily for winter overseeding in the South?
12. Unsatisfactory mowing quality is most likely associated with which species?
13. Adelphi, A-34, Glade, Limousine, and Touchdown are cultivars of ___________.
14. Fine fescue and ___________ are often planted together.
15. Crested wheatgrass is grown in semiarid regions on sites where ___________ is unavailable.
16. Develop a seed mixture for a high-quality area in the full sun.
17. Develop a seed mixture for a dry, shady area.

CHAPTER

5

Introduction to Soils

OBJECTIVES

After studying this chapter, the student should be able to

- Describe the components of soil
- Define the important soil characteristics such as texture and structure
- Discuss how soil texture and structure affect plant growth
- Describe and illustrate a soil profile

INTRODUCTION

The turf manager must have an understanding of soil science because soil conditions have a major impact on plant growth. Grass plants obtain nutrients and water from the soil. The soil provides mechanical support by anchoring the roots of the plants. How well the plants grow depends a great deal on how well the soil is managed. Poor soils can be improved with proper management, and good soils can be ruined by incorrect practices.

Soil is a mixture of four components—air, water, organic matter, and minerals (Figure 5-1). Air and water are found in the pore spaces between the solid particles. Larger pores contain air; smaller pores usually contain water. The organic matter portion is mainly composed of plant debris and microorganisms such as bacteria and fungi. Organic material is most abundant near the soil surface.

In simplest terms, the mineral particles may be thought of as rock dust. Over a period of hundreds or thousands of years, rock is disintegrated by a process known as weathering. Warming and cooling, freezing and thawing, wetting and drying, wind and water erosion, glacial ice movement, and root penetration are all examples of mechanical forces that cause the disintegration of rock. Chemical processes such as the reaction of minerals with water and

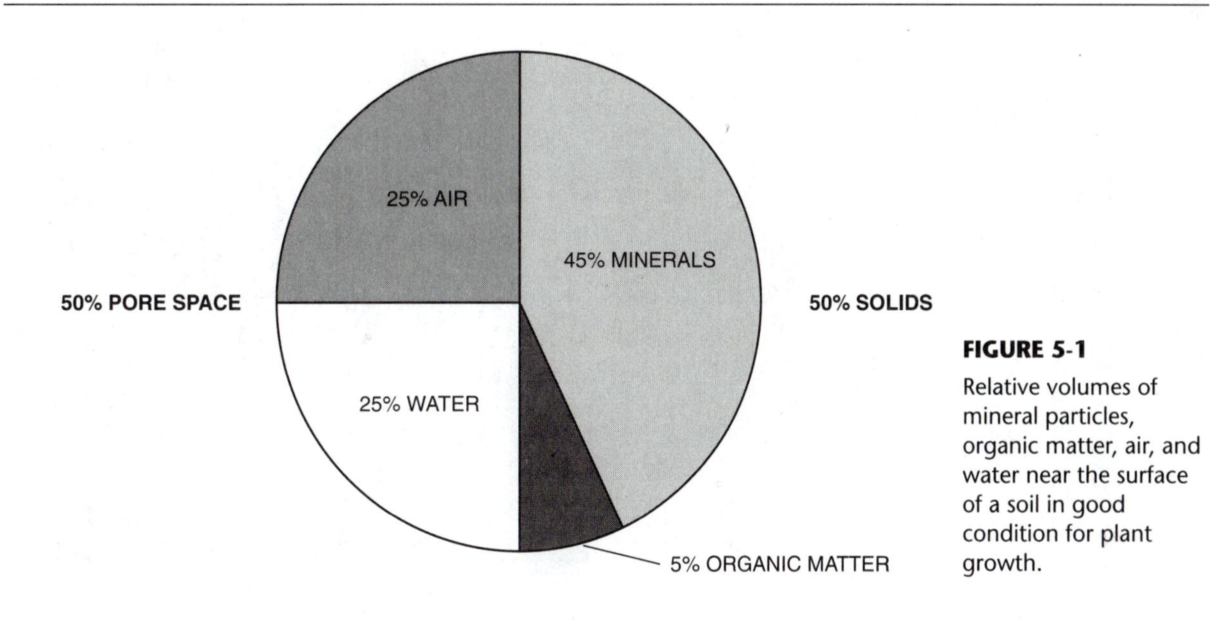

FIGURE 5-1

Relative volumes of mineral particles, organic matter, air, and water near the surface of a soil in good condition for plant growth.

acids contribute to the decomposition. As time passes the parent rock is broken down into smaller rocks and eventually into individual minerals.

The type of soil formed is influenced by the mineral composition of the original parent rock, the length of time that the rock has been subjected to weathering processes, the topography of the area, the native vegetation, and the climate. Some soils were transported from one place to another by glaciers, rivers, or wind. In the United States thousands of different soil types have been classified and described by soil scientists. Turf managers can obtain local soil survey maps from the Soil Conservation Service office in their county (Figure 5-2).

SOIL PROPERTIES: TEXTURE

Soil particles vary greatly in size and are placed into groups known as *separates* (Table 5-1). The five types of sand separates are generally referred to simply as sand. It should be remembered, however, that the term *sand* describes a wide range of particle sizes. Sand particles are larger than silt, and clay particles are the smallest of all the separates. The diameter of a clay particle is less than 1/10,000 of an inch.

The amount of each separate present in a soil depends upon the extent of the weathering and the type of parent rock. Coarse-grained rocks tend to produce a sandy soil, and fine-grained rocks usually form a clay soil.

Soil *texture* refers to the amount or proportion of sand, silt, and clay in a soil. The percentage by weight of each of these three separates in a soil deter-

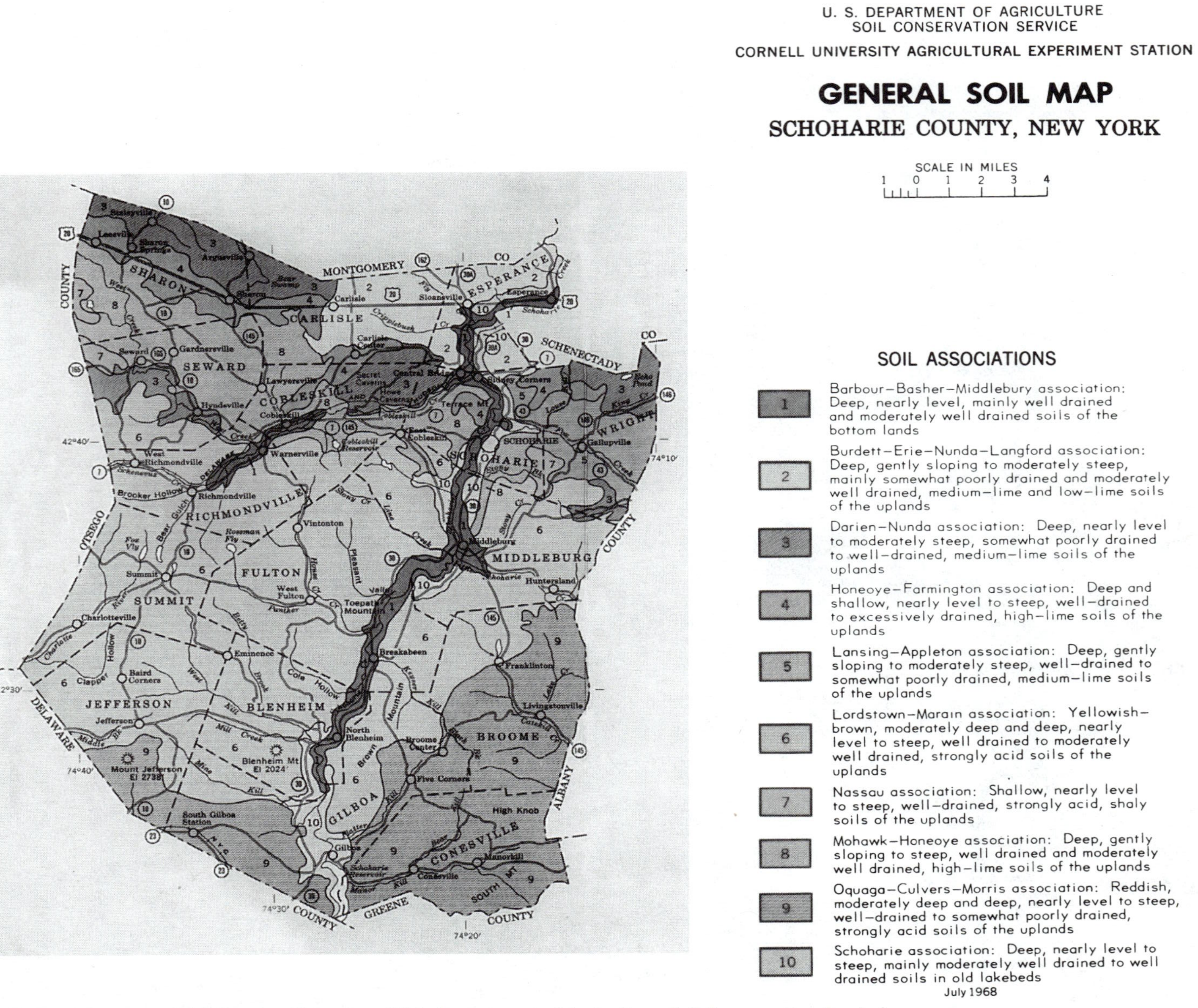

FIGURE 5-2 A general soil map. *(Courtesy of U.S. Department of Agriculture, Soil Conservation Service)*

TABLE 5-1 Classification of Soil Particles According to Size Using the U. S. Department of Agriculture System

Separate Name	Diameter (millimeters)	Particle Size Analogy
Very coarse sand	2.00–1.00	Ball 8 ft in diameter
Coarse sand	1.00–0.50	Ball 4 ft in diameter
Medium sand	0.50–0.25	Medicine ball
Fine sand	0.25–0.10	Basketball
Very fine sand	0.10–0.05	Softball
Silt	0.05–0.002	Golf ball
Clay	Less than 0.002	Buckshot or popcorn kernel

mines its textural class (Figure 5-3). For example, a soil containing 70 percent sand, 15 percent silt, and 15 percent clay is considered to be a sandy loam.

Textural class is of great significance to the turf manager. Many of the physical and chemical properties of a soil which affect plant growth are strongly influenced by its texture.

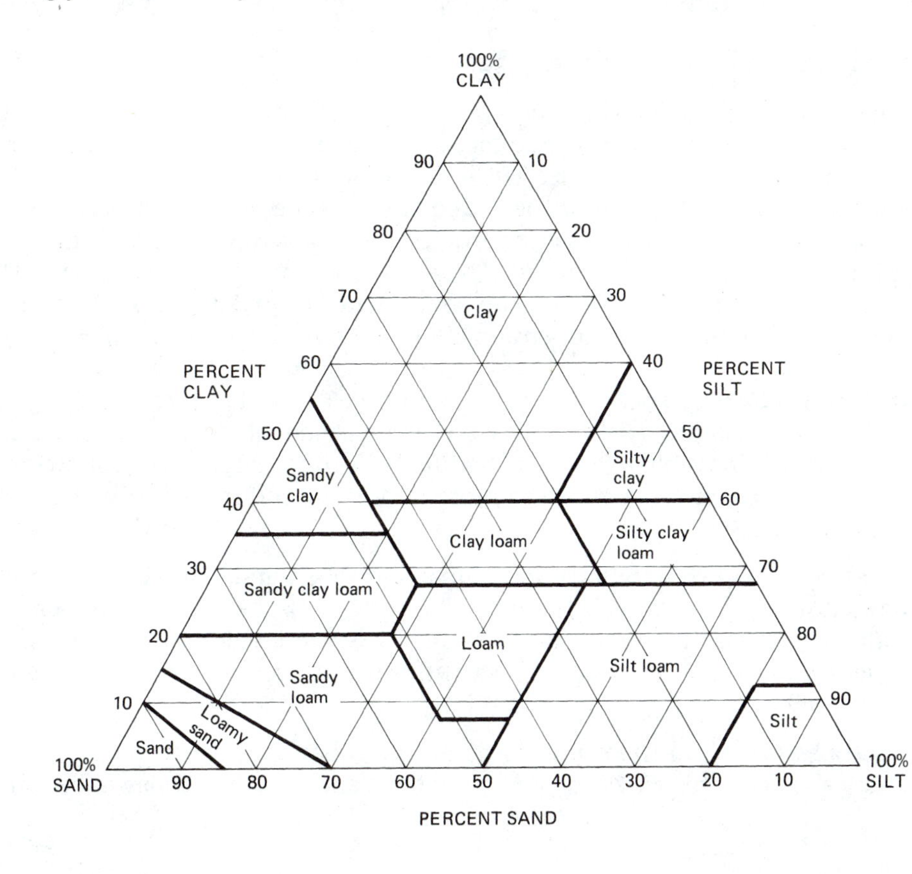

FIGURE 5-3
Soil textural triangle.

STEP	PROCEDURE
1	Fill a quart jar two-thirds full with water.
2	Add one teaspoon of dish*washer* detergent. The dishwasher detergent acts as a dispersing agent, breaking soil aggregates into individual particles.
3	Add soil until the water level is almost to the top of the jar. Screw the jar lid on tightly and shake the jar vigorously for five minutes

OBSERVATIONS:

The sand will begin to sink to the bottom of the jar immediately after the shaking stops. It will be completely settled out in less than one minute.

All of the silt particles should settle out within a few hours and will form a second layer on top of the sand layer.

The small clay particles will remain in suspension for a long time.

FIGURE 5-4
The sand and silt separates in the soil can be observed by using the simple method of separation described.

Methods of Determining Texture

Laboratory Analysis. The most precise determination of soil texture is by laboratory analysis. A widely used method of textural analysis involves mixing the soil with water in a cylinder. The separates settle out of suspension at different rates of speed because of their different sizes. Because larger particles fall through water faster than smaller particles, the sand sinks to the bottom of the cylinder first, followed by the silt. The small clay particles settle out last. The percentages of sand, silt, and clay can be calculated by measuring the percentage of the soil that remains in suspension after each separate settles out (Figure 5-4).

Determination by Feel. Sending a soil sample to a soil testing laboratory for particle-size analysis is the most accurate method of determining texture. However, it is possible to estimate texture by feeling the soil. The following is a list of the characteristics of the more important and most easily identified textural classes.

Sand. Sand particles are large in comparison to silt and clay; therefore, sand has a distinctive gritty feel. The individual grains can be felt when the sand is rubbed between the thumb and forefinger. Wet sand will hold together when squeezed in the hand. However, once the pressure is released it crumbles easily if touched.

Sandy Loam. A sandy loam consists of more than 50 percent sand particles and feels very gritty. However, it also contains enough silt and clay to be some-

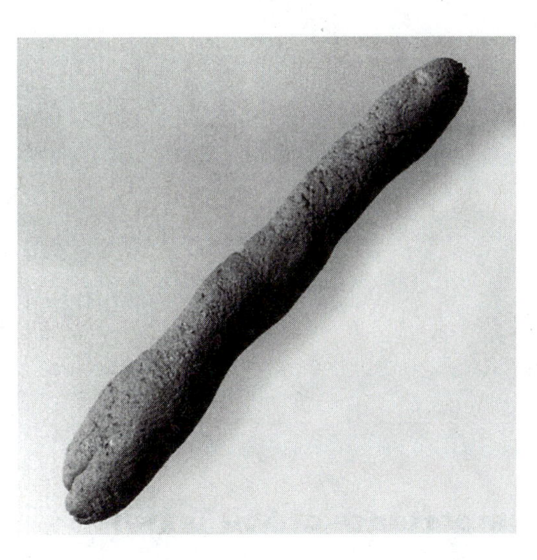

FIGURE 5-5
A clay ribbon.

what cohesive. This means that if the soil is squeezed when wet, it will stick together and can be handled without breaking apart.

Loam. A loam soil is similar in cohesiveness to a sandy loam, but it feels different. Approximately 25 to 50 percent of a loam soil is sand, so it has a gritty feel. There are also enough of the smaller silt and clay separates in the soil to give it a smooth feel. A loam feels both gritty and smooth.

Silt Loam. A silt loam soil is at least 50 percent silt. Silt has a soft, smooth feel. A dry silt loam soil feels like flour or talcum powder.

Clay Loam. Approximately 30 to 40 percent of this soil is clay. The finer particles give it a smooth feel, but there is also enough sand present to make it feel gritty as well. A clay loam soil can be distinguished from the previously mentioned textures by its greater cohesiveness. The clay becomes very sticky when wet, and the soil can be shaped into a thin "ribbon" when rolled between the thumb and forefinger. Only soil textures with a high clay content such as clay loam or clay form good ribbons (Figure 5-5).

Clay. A clay soil is composed of at least 40 percent clay and is very fine textured. A wet clay soil can easily be rolled into a long ribbon.

The ability to determine texture by feel is a valuable skill because it enables the turf manager to quickly estimate the soil texture while on the site. However, experience is required for accurate determinations. Turf managers can test their accuracy by practicing the feel method on soils of known texture. With practice they will be able to recognize, for example, that if a soil feels both gritty and smooth but does not form a good ribbon, it is a loam.

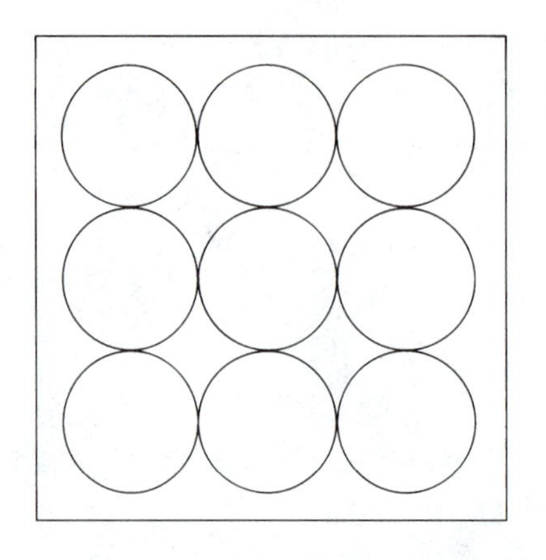

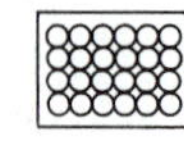

FIGURE 5-6
The larger the particles, the larger the spaces between them.

Importance of Soil Texture

Different soil textures require different management programs. Although some textural classes are considered more desirable than others, most soils can support satisfactory turfgrass if they are properly managed.

Coarser soils with a high sand content behave differently from finer soils containing a significant amount of clay. Because sand particles are very large in comparison to the silt and clay separates, sandy soils have a greater number of large pore spaces.

To understand this concept, imagine a room filled with basketballs and a room filled with marbles (Figure 5-6). The basketballs represent sand particles and the marbles represent clay particles. The empty spaces between the basketballs are much larger than the spaces between the marbles. Large spaces between sand particles are called *macropores*. The very small spaces between clay particles are called *micropores*.

Sandy Soils

Macropores are abundant in sandy soils. Sandy soils are well drained because of these large-sized pore spaces. Water moves relatively quickly out of the macropores and down through the soil. It is replaced by air. Macropores are also referred to as *aeration pores* because they allow good air movement in the soil. Aeration is important because roots need oxygen for respiration and energy production. The open, loose character of the soil allows deep root penetration. Sandy soils are referred to as "light" soils because they are easily worked with a rototiller or plow.

There are problems associated with sandy soils. Sometimes they drain too well and dry out too quickly. During dry periods they are often incapable of supplying plants with enough water to ensure adequate growth (Figure 5-7). Frequent irrigation may be necessary.

Sandy soils should also be fertilized more frequently than finer-textured

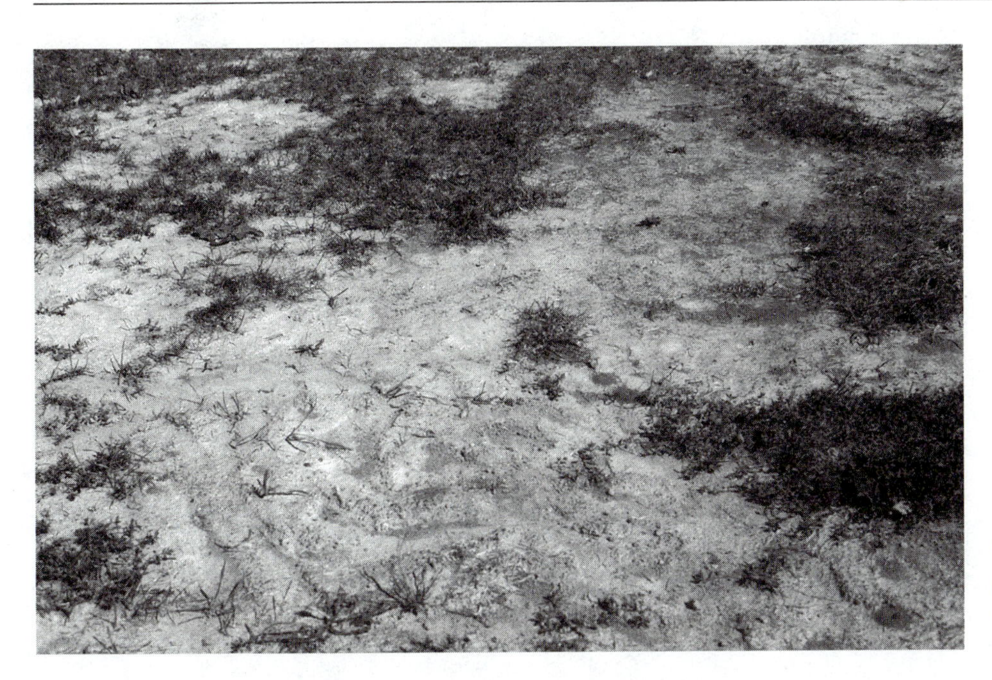

FIGURE 5-7
Poor growth on a sandy soil due to lack of water.

soils. Sand is less able to hold and store nutrients. Water washes the nutrients down through the soil and out of reach of plant roots. This removal of nutrients is called *leaching*. Both rainfall and sprinkler irrigation cause leaching.

All sandy soils should not be expected to behave exactly the same. Their characteristics will vary, partly because sand particle size can vary significantly.

Clayey Soils

Soils containing significant amounts of clay also have advantages and disadvantages.

The smaller-sized micropores are ideal for holding water. They continue to retain moisture long after water drains out of the macropores. The water-holding capacity of a clayey soil is also greater because clay particles have a larger surface area than the other separates. The clay particles can be covered with more water than silt or sand particles.

Figure 5-8 explains why smaller objects such as clay particles have a greater external surface area per gram than larger objects such as silt and sand particles. A gram of clay particles (454 grams = 1 pound) has at least 1,000 times as much surface area as a gram of sand (Table 5-2).

The loss of plant nutrients from clays by leaching is very small compared to the losses that occur in a sandy soil. The clay soil holds more water, so fewer nutrients are leached below the root zone. Nutrients are held on sites on the surface of soil particles. The greater surface area in a clay soil allows the particles to be covered with a large amount of nutrients.

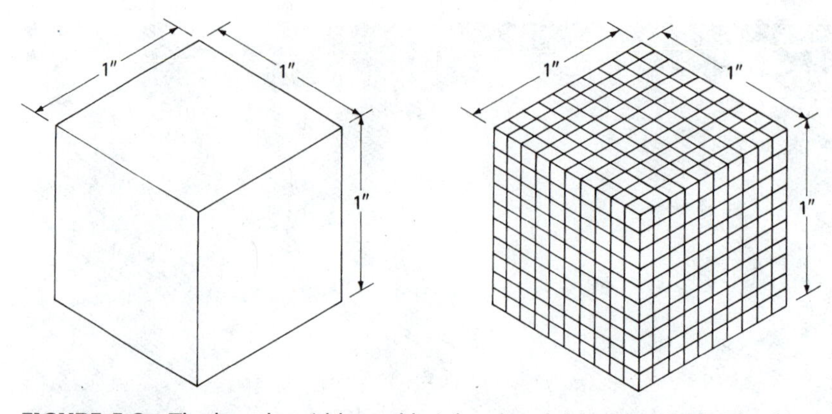

FIGURE 5-8 The length, width, and height of each block is 1 inch. The block on the left has 6 sides. The surface area of each side is 1 in^2 (1 in x 1 in). The total surface area is 6 in^2 (6 sides x 1 in^2). The block on the right is divided into 1,000 smaller blocks, each with a length, width, and height of 0.1 inch. The total number of sides is now 6,000 (1,000 blocks x 6 sides per block). The surface area of each side is 0.01 in^2 (0.1 in x 0.1 in). The total surface area is now 60 in^2 (6,000 sides x 0.01 in^2).

Some clay particles are platelike in structure and have internal surface area between the plates. This further increases their capacity to absorb and store nutrients.

Too much clay in a soil can cause serious problems. The soil may drain poorly and remain too wet. Wet clay is very sticky and forms clods if rototilled or plowed. At the other extreme, clayey soils become hard and bricklike when they dry out.

TABLE 5-2 Comparison of Surface Area per Gram Exhibited by Sand, Silt, and Clay Separates

Soil Separate	Approximate Number of Particles per Gram	Approximate Surface Area per Gram (cm^2/g)
Very coarse sand	100	10
Coarse sand	700	25
Medium sand	6,000	50
Fine sand	45,000	100
Very fine sand	750,000	230
Silt	5,700,000	450
Clay	90,000,000,000	8,000,000

Because water penetrates finer-textured soils slowly, puddles may occur after a heavy rain. Inadequate air movement and root penetration are also a problem with clayey soils.

The best textures do not have extreme proportions of any one separate. A loam soil is very satisfactory for plant growth. This texture contains 25 to 50 percent sand and 10 to 25 percent clay. A loam has enough sand to be well drained and well aerated. It also has enough clay to hold desirable levels of water and nutrients. However, less ideal textures can be very satisfactory when they are managed properly. Excellent turfgrass can be grown on a sandy soil, for example, if irrigation is provided. A good structure helps to compensate for too much sand or clay.

SOIL PROPERTIES: STRUCTURE

Soil structure refers to the arrangement of the mineral particles. Separates will stick together and form *aggregates*—clusters of soil particles. This can be very beneficial.

A clayey soil without a good structure consists mainly of micropores. If clay particles bind together, however, the individual aggregates may be as large or even larger than a sand particle. This results in an increase in larger-sized pores and improved drainage and aeration.

A granular structure is the best structure for growing turfgrass. The granules (rounded aggregates) do not fit tightly together, and the soil is open and loose.

Structure develops if particles adhere to each other and form aggregates. Organic matter is essential. Decaying plant material and microorganisms release compounds that cement clay particles together. A granular structure cannot exist without this organic "glue."

Organic matter also has very beneficial effects on a sandy soil. It acts like a sponge and holds water. Humus is extensively decomposed organic material. It has tremendous surface area and increases the ability of a sandy soil to store nutrients.

Root penetration opens up a soil and contributes to structure formation. Freezing and thawing, wetting and drying, and the soil-mixing activities of earthworms and other soil organisms also promote aggregation.

Negative electrical charges are common on the surface of clay particles. Calcium is positively charged. Because unlike charges attract each other, the calcium tends to bind clay particles into aggregates. Both lime and gypsum contain calcium, and their addition to the soil may improve structure.

Turfgrass helps to develop a good soil structure because it has a dense root system and produces a large amount of organic matter.

A good structure can solve many of the problems caused by an unsatisfactory texture. Texture is a permanent characteristic—particle size does not readily change. Soil structure, however, is very changeable.

FIGURE 5-9
The soil on the left has a granular structure and supports healthy plants. The soil on the right is compacted. The loss of the larger pore spaces results in poor drainage, inadequate aeration, a shallow root system, and weak plants.

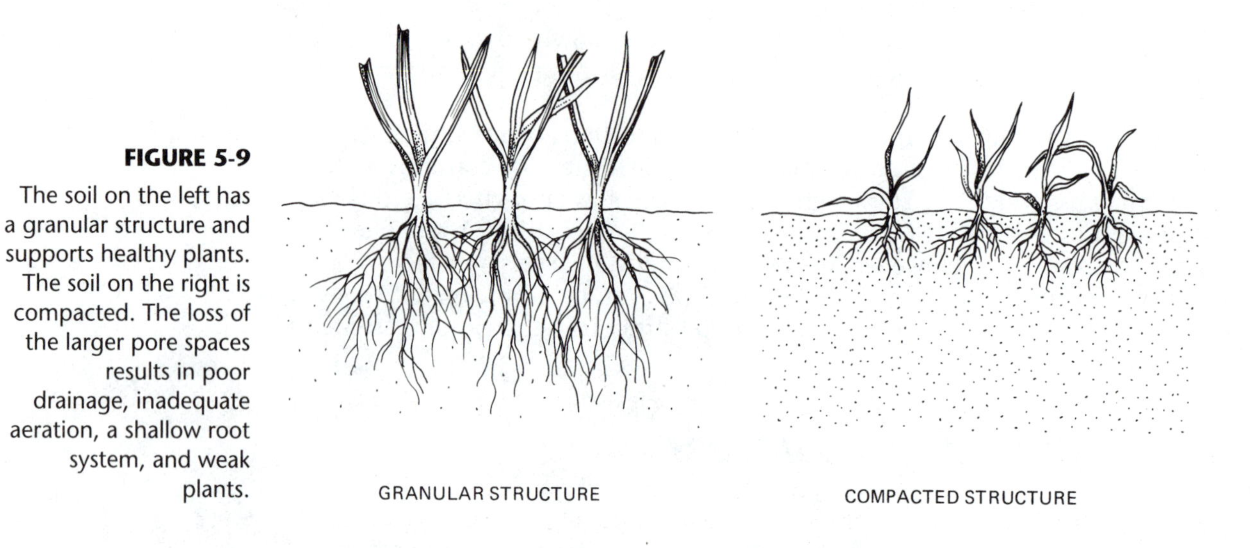

A good structure can be destroyed by improper management. Excessive traffic on a clayey soil forces the particles closer together, causing *compaction*. The granular structure collapses, eliminating the larger-sized pores (Figure 5-9).

A granular structure that has taken years to develop may be ruined in minutes if the soil is rototilled when it is too wet. One method of determining whether the soil is dry enough to work is by crushing a handful of soil into a ball. If the ball sticks together when it is dropped to the ground, the soil is too wet.

SOIL DEPTH

The best way to determine soil depth is to dig a hole and examine the soil profile. As the hole is dug, different layers are observed. The soil profile is composed of these layers, which are called *horizons* (Figure 5-10). Many soils beneath turf areas have been disturbed or modified, so all of the horizons may not be present or distinctive.

O Horizon

The layer above the soil surface is the *O horizon* and consists of undecomposed or partially decomposed organic matter. The organic material is primarily living and dead plant tissue. On turf areas the layer is composed of roots, rhizomes, stolons, and stems, as well as leaf clippings that fall to the ground when the grass is mowed. The O horizon produced by turfgrass plants is called thatch. If soil is mixed with the thatch, the layer is called mat. This mixture is usually the result of topdressing, a maintenance practice discussed in Chapter 18.

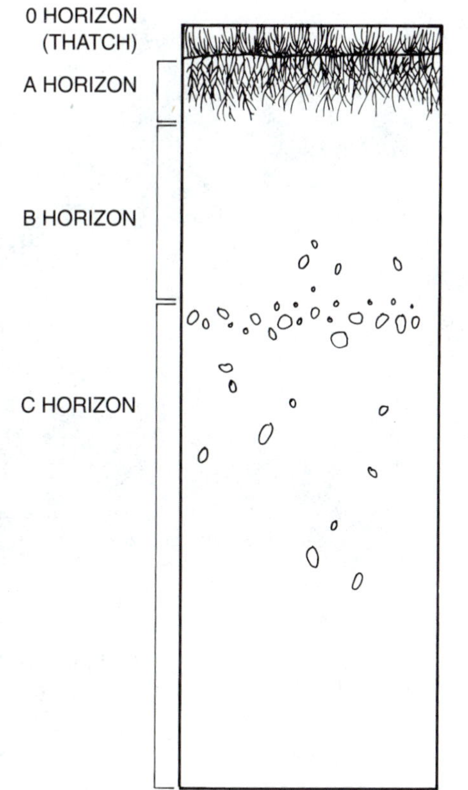

FIGURE 5-10
A soil profile.

A Horizon

The *A horizon* is the topsoil. It is a combination of mineral particles and decomposed plant material. The plant residues in this layer are decomposed by soil microorganisms. The humus gives the A horizon a dark, rich color and helps to develop a good soil structure. Well-structured topsoil is an excellent medium for plant growth.

B Horizon

The *B horizon* is less desirable. It is called the subsoil and has a much lower organic matter content than the topsoil. Many soils have higher percentages of clay in their B horizons than in their A horizons. The very small clay particles can be washed out of the topsoil and will accumulate in the B horizon. The subsoil is often compacted and poorly structured. Water, air, and roots have difficulty penetrating it.

C Horizon

The *C horizon* is a zone of partially weathered parent material. Solid bedrock is often found beneath the C horizon.

FIGURE 5-11
This grass is growing above a septic tank which is very near the surface. The plants are suffering from moisture stress because the roots cannot obtain enough water from the shallow layer of soil.

Turfgrass soils may not contain all of these horizons. For example, when houses are built the topsoil is sometimes buried or removed from the construction site. To minimize compaction, soils on golf course greens or athletic fields are often modified by the addition of a sand layer. A young soil may contain A and C horizons, but not yet have formed a B horizon.

A deep layer of well-structured topsoil is very desirable. This helps to ensure adequate aeration and drainage. The deeper rooting encouraged by a thick layer of topsoil results in good mechanical support. It also enables plants to obtain water when the soil near the surface dries out. The extensive root system that develops can retrieve nutrients that would have leached below the reach of shallow-rooted grass.

Problems can occur if the topsoil is shallow or if solid bedrock is only several inches beneath the soil surface. During droughty periods the upper few inches of soil dry out readily and roots cannot find water (Figure 5-11). During wet periods water drains through the shallow topsoil, but is stopped by impenetrable subsoil or bedrock. If the rain continues, the topsoil may be flooded.

Turfgrass with a shallow root system is poorly anchored in the soil and does not wear well. The plants are more easily damaged or torn out of the soil when they are walked on or driven over with maintenance equipment.

ORGANIC SOILS

A few soils are classified as organic soils. They contain at least 20 to 30 percent organic matter, and often 80 percent or more.

When plants die the debris is usually decomposed by soil microorganisms. These microorganisms are normally aerobic—they need oxygen to live. Organic soils develop in wet environments such as marshes, swamps, and bogs. The water prevents air from entering the soil, and the plants that grow in these areas are not rapidly decomposed when they die. Eventually the soil contains a large amount of organic matter.

Only a small percentage of the soils occurring in the United States are organic soils. The majority are mineral soils. Peat is one type of partially decomposed plant material found in organic soils. Peat can be added to mineral soils to improve their structure.

SELF-EVALUATION

1. Soil consists of mineral particles, organic matter, water, and ___________.
2. Sandy soils feel ___________, and clay loam and clay soils can be rolled into a ___________.
3. The larger pore spaces found between sand particles are called ___________.
4. The organic layer produced by turfgrass that is located above the soil surface is called ___________.
5. The A horizon is also referred to as ___________.
6. The addition of ___________' improves the structure of many soils.
7. When mineral particles are forced close together, resulting in the elimination of large pore spaces, the soil is said to be ___________.
8. Which of the following problems are associated with a sandy soil?
   a. Poor drainage
   b. Poor aeration
   c. Inadequate moisture retention
   d. Leaching of nutrients
9. Which of the following problems are associated with a soil containing significant amounts of clay?
   a. Compaction
   b. Inadequate nutrient and moisture retention
   c. Poor aeration
   d. Poor drainage
10. Why is a granular structure desirable?
11. Which is most likely to change—soil texture or structure?
12. List the three soil separates that determine soil texture.
13. What is the smallest soil separate?
14. What is the textural class of a soil containing 30 percent sand, 20 percent silt, and 50 percent clay?
15. Why does a shallow soil make it difficult to grow high-quality turf?
16. Why is a loam texture desirable?

CHAPTER

6

Soil Modification

OBJECTIVES

After studying this chapter, the student should be able to

- Explain how undesirable soil conditions can be improved by modifying the soil
- Describe the most effective methods of soil modification
- Identify organic matter sources which can be used to improve soil structure
- Discuss the types of drainage systems that are designed to remove excess water from the soil

INTRODUCTION

Turfgrass plants are perennial and provide a permanent ground cover if properly managed. The turf manager's best opportunity to improve a soil is before the grass is planted.

The soil should be carefully examined prior to turf establishment. Shallow or poorly textured soils may cause serious problems in the future (Figure 6-1). Grass will survive in spite of such conditions, but its quality may never be satisfactory. The difficulty of maintenance is increased if soil conditions are undesirable.

Some areas do not require nice-looking, high-quality turfgrass. Grass is grown on these low-maintenance areas to prevent soil erosion and to give the site a green color. When attractive, vigorous turf is not necessary, soil improvement is seldom required.

If better-quality turfgrass is expected or required, the manager may consider improving a poor soil. There are several ways to modify undesirable soil characteristics. The turf manager must analyze the cost of these approaches and determine which method is most appropriate. The decision is based on two considerations: how serious the problem is, and how much money is available to correct it.

FIGURE 6-1
Continual replanting and repair work may be necessary when grass is grown on a soil which has undesirable characteristics.

COMPLETE MODIFICATION

If the existing soil is unfavorable, a soil with a desirable texture can be placed over it. Bringing in a new soil for the root zone is called *complete modification*. It is most effective when a layer of soil 8 to 12 inches (20 to 30 centimeters) deep is added. This alternative is considered if the original soil is too shallow or too fine textured.

The soil that is used for complete modification should not contain too much clay. Loams and sandy loams are often the preferred textural classes. In some cases a media containing 80 to 100 percent sand is used.

High percentages of sand reduce the likelihood of compaction problems. A sand media is recommended for golf course greens, athletic fields, and other areas where heavy traffic is common. Medium- to coarse-textured sand (0.25 to 1.2 millimeters in diameter) is best.

Covering an area with a desirable soil is an ideal solution to soil problems, but it is very expensive. Extensive labor is required to bring in, spread, and grade the new layer. It may be necessary to remove some of the old soil from the site to prevent the soil level from being raised too high. The media itself is expensive. Usually the turf manager can afford to use complete modification for small areas only.

Topsoil is sold by the cubic yard or ton. A cubic yard of soil is 3 feet high, 3 feet long, and 3 feet wide. A cubic yard of soil contains 27 cubic feet (3 ft × 3 ft

$\times$ 3 ft). It weighs 1 to 1.5 tons, depending upon the texture, how wet it is, and other variables.

The cost per cubic yard depends upon the availability of topsoil locally, its quality, and the number of yards purchased. The price varies greatly from one location to another.

The following problem illustrates how the cost of topsoil is calculated. The area to be covered is 100 feet long and 81 feet wide:

$$100 \text{ ft} \times 81 \text{ ft} = 8{,}100 \text{ ft}^2$$

Eight inches of topsoil is needed, or 0.67 foot.

$$8{,}100 \text{ ft}^2 \times 0.67 \text{ ft} = 5{,}400 \text{ ft}^3$$

$$5{,}400/27 = 200 \text{ yd}^3$$

If the topsoil is \$10.00 per yd^3, the cost is:

$$\$10 \times 200 \text{ yd}^3 = \$2{,}000$$

If the supplier sells soil by the ton, multiply the weight of a cubic yard by the number of cubic yards needed. A cubic yard of sand, for example, weighs 1.5 tons. If 200 yd^3 were needed, this would be equivalent to 300 tons (200 yd^3 $\times$ 1.5 tons).

Because the supplier knows how many cubic yards a fully loaded truck holds, the amount of topsoil delivered can be measured. A load of soil on the truck is very loose and porous. When it is unloaded and spread it has a tendency to settle. Because the volume decreases, the depth of the new layer may be less than desired. Some turf managers increase their topsoil order by 20 percent to compensate for this shrinkage.

Problems can occur at the interface between the new layer and the original soil if the two textures differ substantially. For example, the downward movement of water may be stopped at this point. To eliminate interface problems the turf manager may mix some of the topsoil into the upper few inches of subsoil so that there is a more gradual transition between the two layers. This is usually not necessary if the top layer is sufficiently deep or if a subsurface drainage system is installed.

PARTIAL MODIFICATION

If money is limited or the soil problems are not too severe, *partial modification* is recommended. Partial modification involves incorporating materials into the existing soil to improve its structure or texture. Improving structure is a popular modification technique because it is effective and relatively inexpensive.

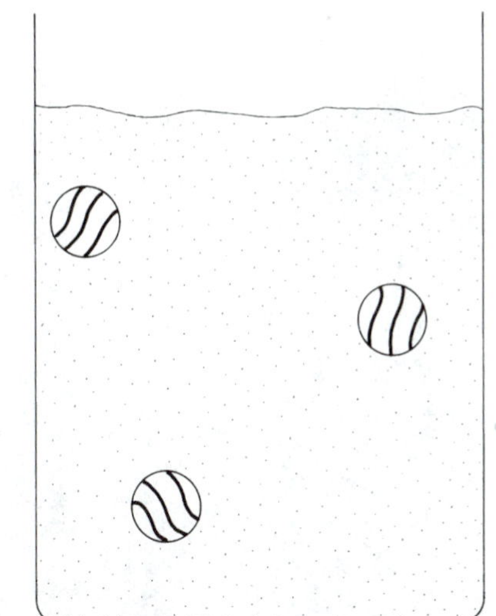

FIGURE 6-2
It is a common belief that the addition of sand to a clay soil will result in the formation of large pore spaces. Working small amounts of sand into the clay is comparable to sticking a few marbles into a jar of flour. Porosity is not improved.

Inorganic Materials

Sand is sometimes added to clayey soils in an attempt to improve aeration and drainage. Unfortunately, sand is usually an ineffective amendment (Figure 6-2). It can even make clay soils worse by producing a cementlike mixture. Often a tremendous amount of sand must be mixed into the root zone before positive effects are obtained.

The incorporation of a loam soil high in organic matter can have a favorable effect on both clay and sandy soils. To be beneficial, 2 to 4 inches (5 to 10 centimeters) of the loam must be uniformly mixed into the upper 6 to 8 inches (15 to 20 centimeters) of soil (Figure 6-3). The greatest benefit is achieved when texture is modified to a depth of 10 to 14 inches (25 to 36 centimeters).

Other inorganic amendments are also used. These are generally coarse materials incorporated into a soil to increase the number of large pores. The result is better drainage and aeration.

Calcined clays are clay minerals that have been calcined (heated at very high temperatures). This process results in an expanded, hard, very porous particle that helps to improve water movement into and through the soil. Calcined clays also absorb much water; however, the water is held so tightly within the particles that only a small amount of it is available to plants.

Diatomaceous earth is also calcined. It is composed of diatoms, a type of algae that has hard cell walls. When heated at high temperatures the material becomes ceramiclike and porous. When added to sand or soil it behaves very much like calcined clay.

Zeolite is a mineral formed from volcanic rock. One type of zeolite, called clinoptilolite, is hard and very porous and can be ground into sand-sized par-

FIGURE 6-3
A rototiller is often used to mix the loam topsoil into the root zone.

ticles. It has many internal pores and a high cation exchange capacity. When even a small amount of zeolite is added to a sand, it causes a significant reduction in leaching losses by adsorbing nutrients.

Organic Materials

In Chapter 5, the effects of organic matter on soil structure were discussed. Organic matter helps to develop a granular structure in clayey soils, resulting in larger pore spaces and improved drainage and aeration. Organic matter is also very valuable in sandy soils. Extremely sandy soils remain structureless even after the addition of organic matter. However, the organic matter enables sands to hold and store nutrients and water. Organic matter will produce a granular structure in sandy loam soils.

Most organic materials contain small amounts of nutrients. When the organic matter is decomposed the nutrients are released into the soil and can be used by the grass.

When complete modification is unnecessary or too expensive, many turf managers improve their soil by incorporating organic matter. This is often a better solution than attempting to improve soil texture by partial modification.

FIGURE 6-4
It is difficult to incorporate the material deep into the root zone by hand. *(Courtesy of Michigan Peat Co.)*

The function of the organic matter is to improve the soil temporarily until the grass is well established. The organic matter eventually decomposes, but it will be replaced by plant tissue produced by the grass. The grass begins to contribute substantial amounts of organic matter to the soil several months after it is planted. The root system will become extensive and help to maintain a granular structure.

The organic material should be spread evenly over the soil. To be effective, it must then be mixed thoroughly into the upper 4 to 6 inches (10 to 15 centimeters) of soil (Figure 6-4). Incorporating 2 inches (5 centimeters) or more of the organic material results in the greatest benefit.

Sources of Organic Matter. Peat is an excellent soil amendment (Figure 6-5). It is widely used because of its availability. Peat can be purchased at garden centers and even hardware stores everywhere in the United States. It is easy to obtain, but it is more expensive than many other sources of organic matter. Sphagnum peat moss is the most common type of peat available. It is composed of sphagnum moss plants and decays quickly. Reed-sedge peat is usually preferred as a soil amendment because it has a slower decomposition rate and its beneficial effects last longer.

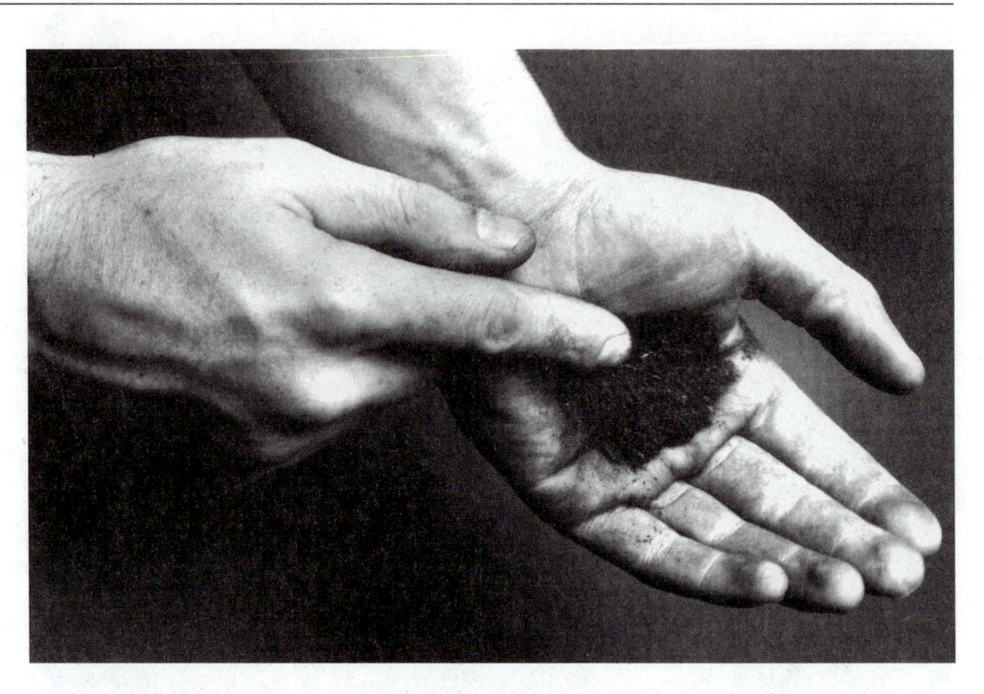

FIGURE 6-5

There are several types of peat moss. This is reed-sedge peat. *(Courtesy of Michigan Peat Co.)*

Peat is commonly sold by the cubic foot in bales. For larger jobs it can be purchased by the cubic yard (Figure 6-6). From 40 to 160 ft^3 (1.1 to 4.5 m^3) of peat are added per 1,000 ft^2 (93 m^2). This amount is equivalent to spreading a layer approximately 0.5 to 2 inches (1.3 to 5 centimeters) deep. The larger amounts are much more effective. The peat should be moistened before it is worked into the soil.

FIGURE 6-6

Large bags of peat. *(Courtesy of Dakota Peat & Blenders)*

FIGURE 6-7
A large compost pile composed of leaves, grass clippings, and soil. It is hidden behind the hedge and building.

The turf manager may be able to obtain manure from a local farm. Well-rotted manure is a valuable soil amendment. Fresh manure should not be added to the soil immediately before seeding because it may injure the grass. One drawback to manure is that it often contains weed seeds.

Compost is an inexpensive and easily obtained source of organic matter. Organic waste materials such as leaves, grass clippings, straw, sawdust, and even garbage can be composted. The materials are placed in layers in a pile and allowed to rot (Figure 6-7). The material decays more rapidly if it is shredded and placed in layers alternated with a layer of soil.

The microorganisms that are responsible for the decomposition require air, water, and nutrients. The pile should be turned and mixed once or twice a month to improve aeration. If it is not turned frequently the pile should be no higher than 6 feet (1.8 meters) to ensure that the bottom of it receives air. The compost pile should be fertilized and kept moist.

The compost is ready for use when it is well rotted and can be easily worked into the soil. This may take from several weeks to a year.

Some companies and municipalities are composting large amounts of organic waste rather than dumping it in landfills. This compost is often given away free of charge.

Decomposition will occur even if the materials are simply dumped into a pile and turned over once or twice a year. However, the decomposition rate will be very slow.

Digested sewage sludge, sawdust, and seaweed are used as soil additives. Because incorporating sawdust causes a temporary nitrogen deficiency in the soil, nitrogen fertilizer should be added along with the sawdust. Seaweed contains plant hormones that stimulate root growth.

Also available is a product that consists of single-celled plants that will improve soil structure. The organisms grow in the soil and produce a residue that bonds individual particles together into aggregates.

IMPROVING DRAINAGE

Soils that have a high clay content often drain too slowly. Inadequate drainage can occur even in coarse-textured soils if they are subjected to heavy traffic and become compacted.

Poorly drained soils can cause serious problems. Wet soils tend to be compacted more easily, and it is difficult to maintain a desirable structure. Soils that are very wet may be 5–15°F (3–8°C) colder than dry soils. Plant growth is inhibited when the temperature of the root zone is too cool.

The decomposition of organic matter occurs at a slower rate in waterlogged soils, since there are fewer beneficial organisms because of the lack of air.

Many nutrients are in chemical forms that cannot be absorbed by plants. It is difficult for grass plants to absorb even available forms of nutrients because the roots cannot respire and produce the energy necessary for nutrient uptake. Roots growing in poorly aerated soils also have a reduced ability to take in water.

After irrigation or a rain, puddles of water may remain on the surface of a soil that does not drain rapidly. If the puddles occur on a golf course green or athletic field, they will interfere with play. Diseases are also more common where the soil remains very wet and the grass is damp.

Improving texture and structure may provide only a partial solution to drainage problems. Serious problems can be corrected by installing a drainage system.

Subsurface Drainage Systems

Installing a subsurface drainage system results in the rapid removal of excess water from the root zone. This type of system is expensive to install, but it is a very effective method of improving soil drainage.

In the past clay or concrete tiles were used in subsurface drainage systems. The drainage line consisted of individual tiles laid end to end. Excess water entered the system through the joints between the tiles (Figure 6-8).

Today plastic tubing, which is also called tile, is widely used, because it is light and easy to work with. Long pieces of flexible plastic tubing have perforations (holes) through which water enters (Figure 6-9).

The newest type of tile is made of geocomposite drainage materials. The tile has a plastic, waffle-like core surrounded by filter fabric (Figure 6-10). Also called sheet drains, these tiles are installed narrow side up, and water enters through the fabric. They can be buried with minimal trenching because the

FIGURE 6-8

Ceramic drainage tiles in a trench. *(Courtesy of Charles Machine Works, Inc.)*

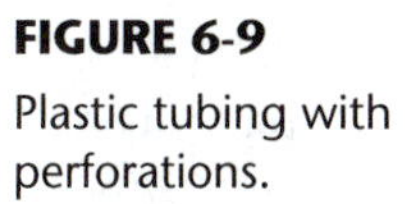

FIGURE 6-9

Plastic tubing with perforations.

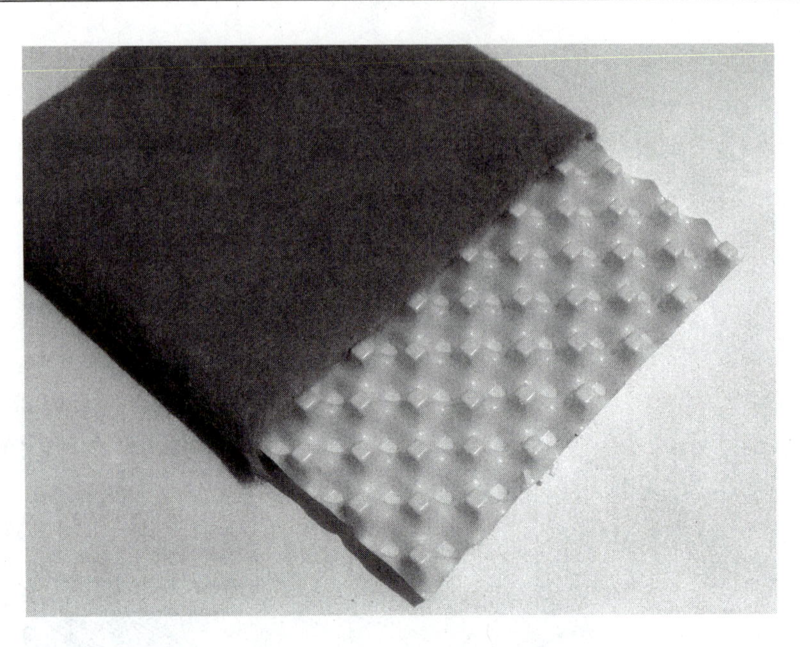

FIGURE 6-10
Geocomposite drainage material, also called waffle-core drainage. *(Photo by Brian Yacur)*

geocomposite material is very thin. In some soils there is a possibility that the fabric may become plugged with silt or other fine soil particles.

The tile lines that collect the water are called *laterals*. The water flows through the laterals into a main drain and then drains out of the system through an opening at the end of the main tile line. The laterals and main lines are often arranged in a herringbone pattern or a continuous lateral design (Figure 6-11).

The lines must be laid on a slope to allow the water to flow through the system to the outlet. A grade or fall of 0.5 percent or more is preferred, which

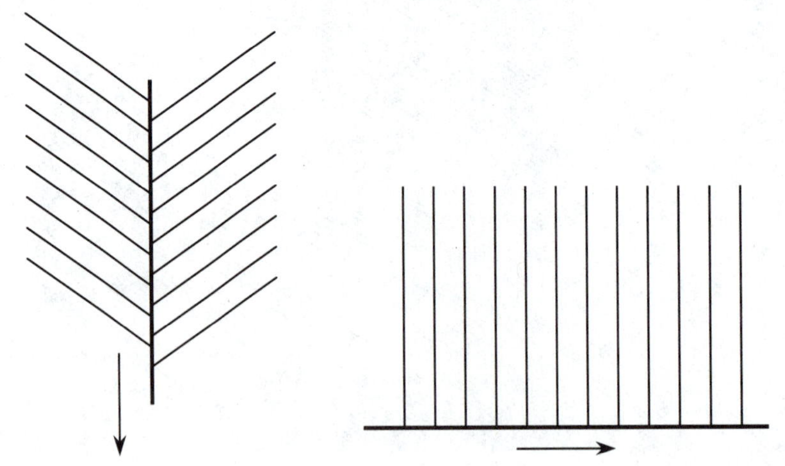

FIGURE 6-11
Two popular drainage system designs.

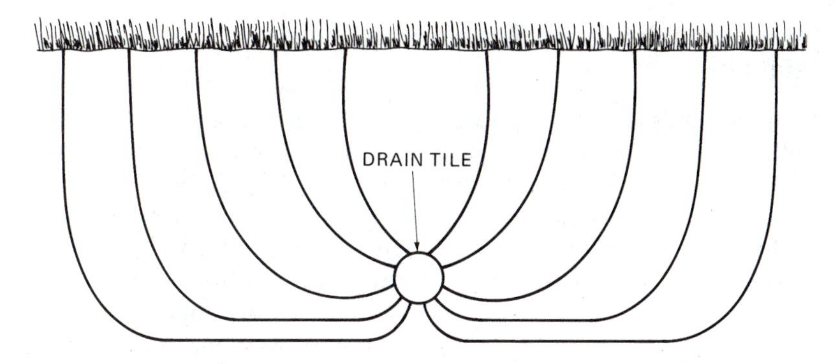

FIGURE 6-12

Water moves through the soil in a horizontal pattern to a drainage tile.

is equivalent to a drop of at least 6 inches (15 centimeters) per 100 feet (30 meters) of tile.

Some soil will also enter the lines. If the slope is insufficient, the water will not move rapidly enough to carry the sediment out of the system. Eventually the lines may become clogged with soil.

If clogging is expected to be a problem, the laterals should enter the main line at 45° angle. If the laterals enter the main at a 90° angle, the water flow is slowed and sediment may not be removed.

The correct distance between the lateral lines is determined by the horizontal permeability of the soil (Figure 6-12). The poorer the horizontal permeability, the closer together the laterals must be placed. In a tight clay subsoil, it may be necessary to place the lines as close as 20 feet (6.1 meters) apart. A 100-foot (30-meter) spacing for the lines may be adequate in a sandier subsoil. When draining sports turf, lines are usually placed 15- to 20 feet (4.6- to 6.1 meters) apart. This is to ensure that the soil surface will not remain overly wet and interfere with play. The closer spacings are used even when the athletic field or putting green has a sandy texture.

The depth at which the lines are installed also depends upon how well water moves horizontally through the soil. Normally tile or plastic pipe is placed at least 2 feet (0.6 meter) deep in the soil. Lines that are too shallow may be damaged if heavy equipment is frequently driven over the turf.

The laterals and main drain lines are often covered with several inches of gravel or coarse sand before the trenches are filled with soil. This increases the ability of water to move to the drainage lines.

The outlet pipe must be kept cleaned out. If it becomes clogged with sediment, the entire system loses much of its effectiveness. Outlets often empty into lakes, streams, storm sewers, ditches, or large holes filled with gravel. Serious environmental problems can occur if pesticides leach into the system and are discharged into lakes or streams.

When turf managers plan a subsurface drainage system, they should consult with the Soil Conservation Service or other drainage specialists. These experts will study the location and soil conditions and recommend an appropriate design.

Other Methods of Removing Excess Water

Athletic fields are often sloped to encourage water runoff. The middle of the field is higher and is called a crown. Water flows from the crown in the center of the field down a 1- to 2 percent slope to the sidelines.

Lawn areas around buildings should also have a 1 percent slope from the building to the edges of the lawn. This allows surface water to move away from the building. If the soil is not properly graded around a building, the basement may be flooded during heavy rains or when snow melts in the spring.

Dry wells are useful for draining low spots where water tends to accumulate. These wells are holes 4 to 5 feet (1.2 to 1.5 meters) deep and 2 to 3 feet (0.6 to 0.9 meter) wide that are filled with large stones. Grass will grow on top of the dry well if the stones are covered with a layer of soil. The soil should be coarse textured so that surface water can drain down into the dry well.

Trenches filled with gravel or crushed stone can be used to remove or collect excess water. They are called French drains or slit trenches and are primarily dug in wet, poorly drained areas (Figure 6-13). The width is commonly 4 to

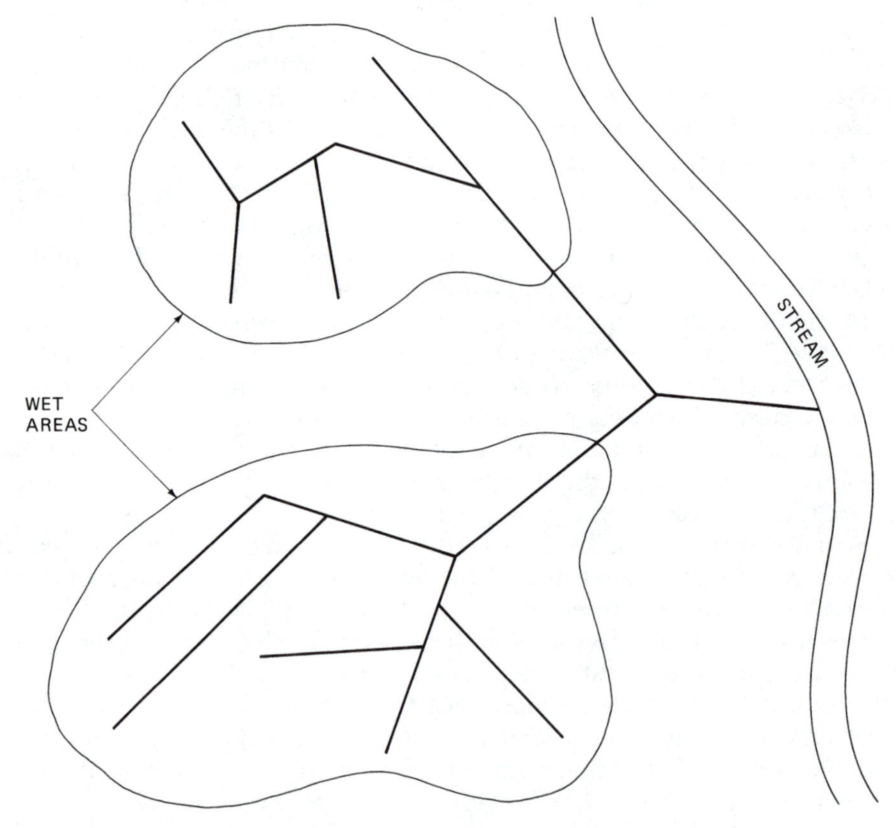

FIGURE 6-13
The series of trenches should be designed to follow the pattern of the wet areas.

FIGURE 6-14
A trencher machine digging a drainage trench. *(Courtesy of Charles Machine Works, Inc.)*

8 inches (10 to 20 centimeters), and the depth is usually 2 to 4 feet (0.6 to 1.2 meters) (Figure 6-14). On sites where excess water is a serious problem, drainage tile or tubing may be placed at the bottom of the trench. A gradual slope is necessary for water to move through the trenches.

The surface of the gravel or stone layer should be kept open. If it is covered with soil or sod, water will not be able to move down into the trench as quickly.

Wider trenches can be dug at the base of slopes to intercept surface water before it runs onto a turf area. This type of drainage is called an *interceptor drain* (Figure 6-15).

Another drainage technique is called *sand injection.* A specialized machine makes slits in the soil 0.75 to 1 inch (1.9 to 2.5 centimeters) wide and 9 inches (23 centimeters) deep; it then fills the slits with sand. Typically these vertical bands of sand are placed 20 inches (51 centimeters) apart. They serve as a channel through which water can drain from the area. Small-diameter plastic tubing may be installed at the bottom of the slits. Sand injection is used primarily on golf greens and sports fields constructed with fine-textured soils.

FIGURE 6-15
An interceptor trench filled with crushed stone.

SELF-EVALUATION

1. The price of topsoil is $18.00 per yd^3. How much will it cost to cover a 1,080-ft^2 area with 12 inches of topsoil?
2. Which of the following statements are true? The addition of organic matter
   a. Does not improve a sandy soil.
   b. Causes clay particles to form aggregates.
   c. Increases the capacity of a sandy soil to store nutrients.
   d. Improves a clay soil by holding water.
   e. Increases aeration in a clayey soil.
3. List the problems associated with a waterlogged soil.
4. Nitrogen should also be added to the soil when which of these organic materials is used?
   a. Compost
   b. Well-rotted manure
   c. Sawdust
   d. Peat
5. A common subsurface drainage design is the ___________ pattern.
6. Water enters a drainage system through tile lines that are called ___________.
7. Athletic fields should have a ___________ percent slope from the center to the sidelines.
8. Incorporating peat improves the ___________ of a soil.
9. Mixing small amounts of ___________ into a clay soil can result in reduced aeration and drainage.
10. A deep layer of almost pure sand may be used to modify a site completely when ___________ is a potential problem.
11. Animal manure is an excellent soil conditioner, but it does contain ___________.
12. Trenches filled with ___________ or ___________ can be used to remove excess water from an area.
13. Topsoil, sand, and gravel are sold by the cubic yard or ___________.
14. Discuss the differences between partial and complete modification.

CHAPTER

7

Soil Chemistry

OBJECTIVES

After studying this chapter, the student should be able to

- Discuss the basic principles of soil fertility
- Explain the effect of soil pH on plant growth
- Describe the materials used to raise or lower soil pH
- Discuss soil salinity

INTRODUCTION

Soil chemistry is a very complex subject. It is not necessary for the turf manager to understand every chemical reaction that occurs in the soil. However, the knowledge of some basic soil chemistry is very helpful. Soil fertility and soil reaction (pH) have a major effect on plant growth.

SOIL FERTILITY

Nutrients in the Soil

An actively growing turfgrass plant is 75 to 85% water. Water is absorbed from the soil by plant roots. The remaining 15 to 25% of the plant's weight is dry matter. This dry portion is composed of 16 essential elements (Figure 7-1). They are known as essential elements because a plant cannot successfully complete its life cycle if any one of these nutrients is lacking.

A major portion of the plant dry matter content consists of three elements—carbon, hydrogen, and oxygen. Plants obtain carbon and oxygen from the air. Carbon dioxide (CO_2), a gas, enters the leaves through the stomata. Water (H_20) taken in by the roots supplies hydrogen and oxygen. These three elements are so common in a plant's environment that abundant quantities are always available for plant use.

The other 13 essential elements are acquired by plants from three sources.

ESSENTIAL ELEMENTS USED IN RELATIVELY LARGE AMOUNTS BY PLANTS:	ESSENTIAL ELEMENTS USED IN RELATIVELY SMALL AMOUNTS BY PLANTS:
Carbon	Iron
Hydrogen	Manganese
Oxygen	Boron
Nitrogen	Molybdenum
Phosphorus	Copper
Potassium	Zinc
Calcium	Chlorine
Magnesium	
Sulfur	

FIGURE 7-1
The 16 essential elements.

As minerals in the soil weather and break down, nutrients are released. Fertilizer applications supply some of the essential elements. When plant tissue decomposes, nutrients are returned to the soil.

The major portion of nutrients found in the soil are not available for plant use. They are part of the structure of complex, insoluble compounds. Plant roots primarily absorb only nutrients that are in simple, soluble forms dissolved in water. The *soil solution* consists of soil water containing dissolved nutrients.

Insoluble forms of nutrients that are "tied up" or tightly bound in minerals and organic matter are not permanently unavailable. Essential elements found in minerals are slowly released by weathering processes. As the minerals are broken down, nutrients become available for plant use. Microorganisms decompose organic matter and convert complex molecules into simpler, soluble forms. The nutrients released can be absorbed by plant roots.

These simple, available forms are called *ions*. Some are positively charged and are known as *cations*. Others are negatively charged and are called *anions* (Figure 7-2). For example, calcium is taken in by plants when it is in the cation form Ca^{++}. Nitrogen is primarily used in the anion form NO_3^-.

The reason why clay and humus particles have a large surface area was explained in Chapter 5. These surfaces are primarily negatively charged. Negatively charged nutrients—anions—are not held on these surfaces to any great extent because like charges repel each other. Cations, however, are absorbed to these surfaces because unlike charges attract each other.

Nutrient anions dissolved in the soil solution are either taken in by plant roots or leached—washed down through the soil by rain or irrigation water. They are not stored in the soil in available forms.

This is one of the reasons grass often experiences nitrogen deficiencies if nitrogen fertilizers are not applied regularly. Most of the nitrogen in the soil is

ELEMENT	CHEMICAL SYMBOL	IONS USED BY PLANTS
Nitrogen	N	NO_3^-, NH_4^+
Phosphorus	P	HPO_4^{--}, $H_2PO_4^-$
Potassium	K	K^+
Calcium	Ca	Ca^{++}
Magnesium	Mg	Mg^{++}
Sulfur	S	SO_4^{--}
Iron	Fe	Fe^{++}, Fe^{+++}
Manganese	Mn	Mn^{++}
Boron	B	$H_2BO_3^-$ and others
Copper	Cu	Cu^{++}
Zinc	Zn	Zn^{++}
Molybdenum	Mo	MoO_4^{--}
Chlorine	Cl	Cl^-

FIGURE 7-2
Chemical symbols and available forms of the essential elements.

locked up in insoluble compounds in organic matter. When the available form, nitrates (NO_3^-), is released it enters the soil solution. Much of the nitrate that is not absorbed by roots is leached away because the anion is not held by the soil.

Cations, however, are stored in the soil. Only a small portion of the nutrients in available cation forms are found in the soil solution. Many are loosely bound to the surfaces of clay and humus particles (Figure 7-3). Despite the fact that they are held on these negatively charged exchange sites, they can still easily become dissolved in the soil solution. These stored cations are protected from leaching until they enter the soil solution. The ability of a soil to absorb exchangeable cations is referred to as *cation exchange capacity* (CEC).

Nutrient Absorption by Roots

Turfgrass plants have extensive fibrous root systems that are ideal for locating nutrients. The roots primarily absorb nutrients from the soil solution. However, roots in direct contact with soil particles can also take in cations held on the surface exchange sites.

To understand nutrient uptake, it is necessary to understand how water moves into roots. Water enters roots by a process known as *osmosis*. Water tends to move from regions of high water concentration to regions of low water concentration.

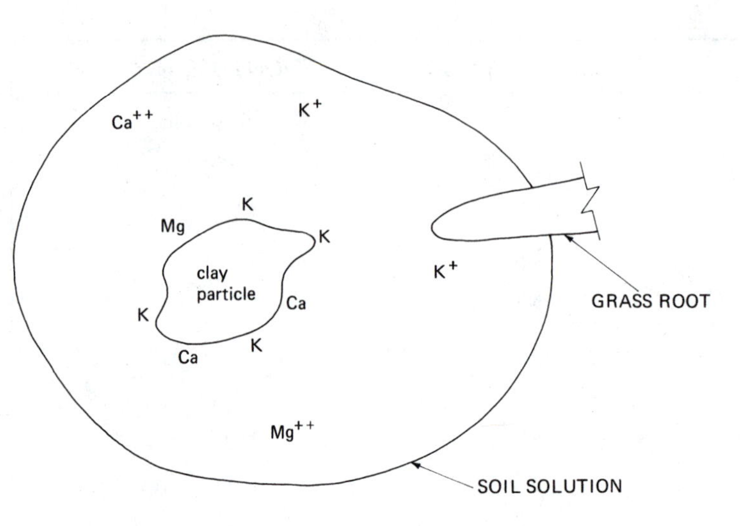

FIGURE 7-3
Nutrients are held on exchange sites and are dissolved in the soil solution. Roots absorb nutrients from the soil solution.

Plants "attract" water by accumulating dissolved substances (solutes) in root cells. The greater the solute concentration, the lower the water concentration. The dissolved materials in root cells cause water to flow into the roots.

This concept also explains fertilizer "burn" and the injury caused by high levels of salts in the soil. Water continues to enter the roots as long as the concentration of water in the soil solution exceeds the concentration of water in root cells. If too much fertilizer or salt is applied, the water outside the plant may have a greater concentration of dissolved substances than the water inside the plant. The direction of water movement reverses, and water leaves the root cells. This lack of water may kill the plant.

If the turf manager applies too much of some fertilizers by mistake, a fertilizer "burn" may be the result. He or she should immediately water the area. A heavy irrigation will wash the fertilizer off the leaves and will help to leach the fertilizer salts below the root zone (Figure 7-4).

Under normal soil conditions the concentration of dissolved substances is higher in root cells than in the soil solution. Because nutrients are more abundant in root cells, there is a natural tendency for them to move out into the soil solution. Membranes surrounding the cells prevent nutrients from leaving the roots.

This raises an obvious question—if the nutrients are unable to pass through the membrane, how were they able to get into the cell? The cells produce chemical molecules that bind with the nutrient ions on the outside of the membrane. These carrier molecules then transport the nutrients through the membrane. After being absorbed the nutrients enter the xylem and are distributed throughout the plant.

Several factors influence the rate of nutrient uptake. Sufficient quantities of nutrients in available forms must be present in the soil. A well-developed

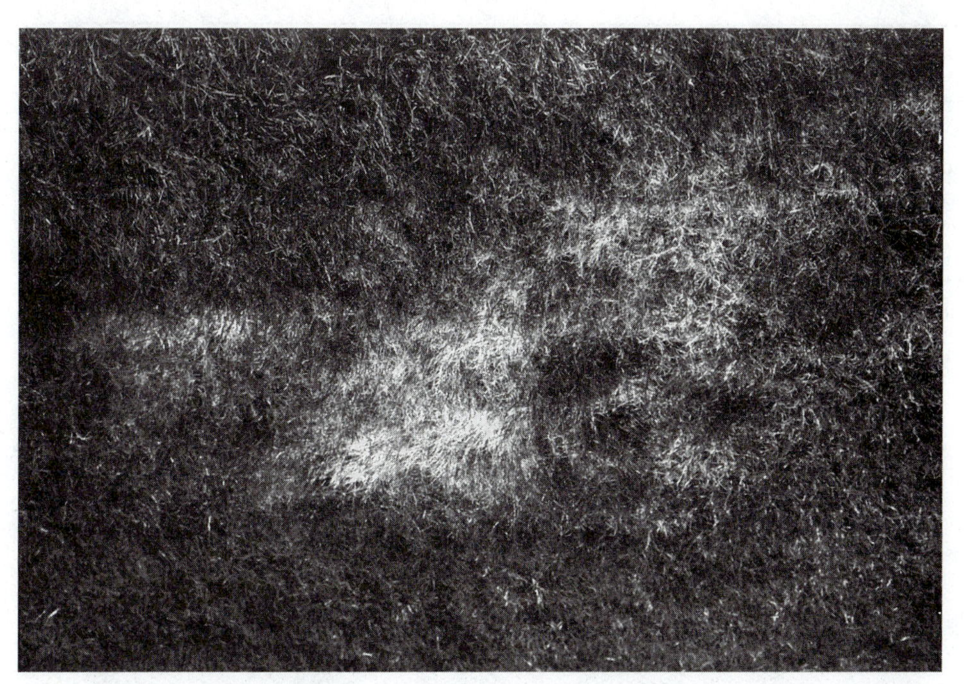

FIGURE 7-4
The result of a fertilizer burn. *(Courtesy of William Knoop)*

root system enables the plant to obtain essential elements. Adequate soil moisture is necessary because ions move through the soil solution to the roots.

Soil temperature affects nutrient absorption. Plants use energy to carry ions in through cell membranes. This energy is produced by root respiration. Very low or high soil temperatures inhibit respiration.

Roots need oxygen (O_2) for respiration to occur. In waterlogged or compacted soils low O_2 levels can limit nutrient uptake.

SOIL REACTION

Soil reaction is the degree of acidity or alkalinity of a soil. The pH scale is used to measure this acidity or alkalinity. A pH of 7.0 is considered to be neutral—neither acidic nor alkaline. Values lower than 7.0 indicate acid soils, and values higher than 7.0 indicate alkaline (basic) soils (Figure 7-5).

Soil pH is actually a measure of the concentration of hydrogen ions (H^+) in the soil solution. An increase in the number of hydrogen ions in the soil solution decreases the pH—the soil becomes more acidic. The H^+ concentration increases or decreases ten times for each unit change in pH. A pH of 5.0, for example, is ten times as acidic as a pH of 6.0.

Soil pH varies greatly throughout the United States—it may be as low as 3.5 and as high as 9.5 Most turfgrass soils are in the 4.5 to 8.0 range.

pH	5.5	6.0	6.5	7.0	7.5	8.0	8.5	pH
← Strongly Acid	Medium Acid	Slightly Acid	Very Slightly Acid	Very Slightly Alkaline	Slightly Alkaline	Medium Alkaline		→ Strongly Alkaline

FIGURE 7-5 The pH scale.

The major factor influencing soil pH is precipitation. Calcium and magnesium are minerals that neutralize acidity and raise pH. The greater the rainfall, the larger the amount of calcium and magnesium that is leached from soils.

Arid, dry regions of the western United States usually have soil pH in the alkaline range. Less rainfall results in less leaching of neutralizing minerals. The effect of precipitation on soil pH is illustrated in Figure 7-6. Regions experiencing minimal rainfall tend to have alkaline soil conditions. In the more humid regions acid soils occur.

Other factors influence soil pH. Soils formed from limestone contain large amounts of calcium, and the pH does not drop even if rainfall is abundant. In arid regions sodium (Na) contributes to soil alkalinity. Soils with high levels of organic matter may have low pH values because acids are released from organic materials. Plant roots give off acid-forming substances. Nitrogen fertilizers may have an acidifying effect.

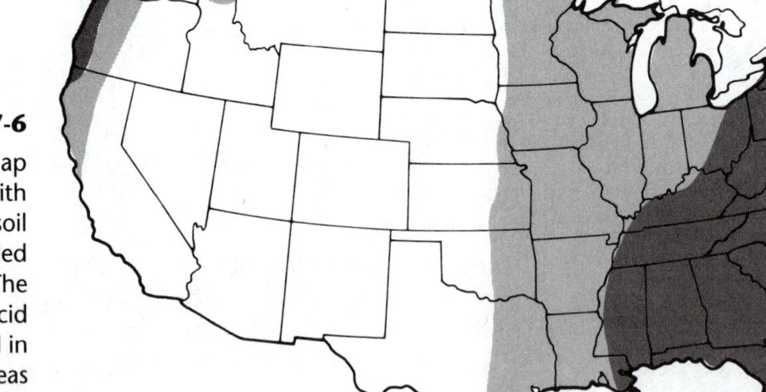

FIGURE 7-6 A generalized map showing regions with acid and alkaline soil pH. The nonshaded area is alkaline. The shaded areas have acid soils. Soils located in the darker shaded areas are generally more acidic than those in the lightly shaded areas.

TURFGRASS SPECIES	pH RANGE	TURFGRASS SPECIES	pH RANGE
Bahiagrass	6.5–7.5	Centipedegrass	4.5–5.5
Bentgrass		Fescue	
Colonial	5.5–6.5	Fine	5.5–6.8
Creeping	5.5–6.5	Tall	5.5–7.0
Bermudagrass		Grama, blue	6.5–8.5
Common	5.7–7.0	Ryegrass	
Improved	5.7–7.0	Annual	6.0–7.0
Bluegrass		Perennial	6.0–7.0
Annual	5.5–6.5	St Augustinegrass	6.5–7.5
Kentucky	6.0–7.0	Wheatgrass, crested	6.0–8.0
Rough	6.0–7.0	Zoysiagrass	
Buffalograss	6.0–7.5	Japanese lawngrass	5.5–7.5
Carpetgrass	5.0–6.0	Manilagrass	5.5–7.5

FIGURE 7-7 Optimum soil pH ranges for the major turfgrass species. Species may tolerate wider ranges than those listed.

Most turfgrasses grow best in soils having a pH of 6.0 and 7.0. Other species such as centipedegrass and carpetgrass, prefer more acidic conditions. Some species, such as crested wheatgrass and blue gramagrass, can tolerate greater alkalinity. Optimum pH ranges for the primary turfgrass species are listed in Figure 7-7.

The major reason 6.0 to 7.0 is the preferred pH range is that all of the essential elements are in available chemical forms in this range. Significant amounts of each nutrient are available for plant use when the soil pH is around 6.5 (Figure 7-8).

Nitrogen, for example, is normally found in the soil as part of the structure of organic matter. Essential elements such as nitrogen are released from the organic material and become available for plant use when the material is decomposed by microorganisms. Extreme acidity or alkalinity inhibits microorganisms and results in decreased nitrogen availability.

Plants have difficulty obtaining phosphorus at high or low soil pH values. Below a pH of 6.5, it combines with iron, aluminum, or manganese and forms insoluble compounds. As the pH increases above 7.0, phosphorus reacts with calcium or magnesium and is, again, "fixed" in unavailable compounds.

Micronutrients such as iron and manganese become more available at pH values lower than 6.5; however, this is not usually a problem because plants need only small amounts of these elements. Manganese and aluminum may become so soluble in strongly acidic soils that they have a toxic effect on plants.

When soil is maintained at the ideal pH, plants can obtain nutrients they need and growth is increased. If the soil pH is too acidic or alkaline, plants may

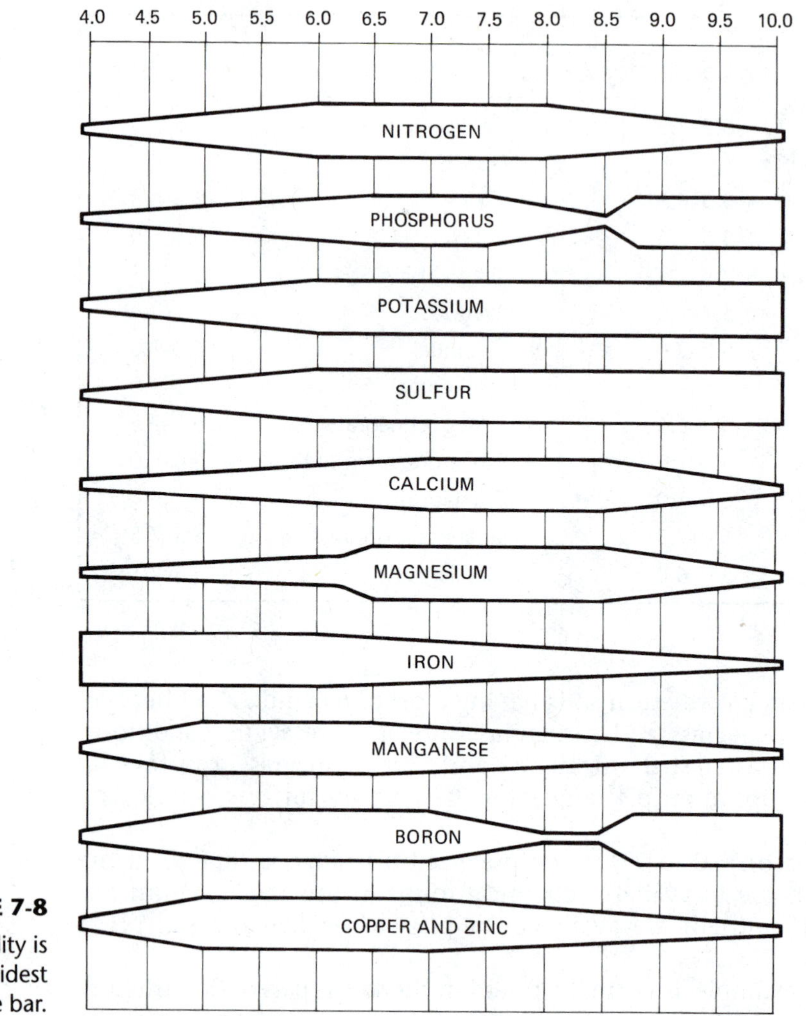

FIGURE 7-8
Maximum availability is indicated by the widest part of the bar.

experience nutrient deficiencies. In such cases, the soil may contain abundant quantities of these essential elements, but they are locked up in complex, insoluble compounds and are unavailable to plants.

Neutralizing Soil Acidity

Lime materials containing calcium or magnesium are applied when the soil pH is too low. Calcium and magnesium neutralize acidity and raise the soil pH. Adding lime to a soil is also beneficial because calcium and magnesium are plant nutrients and they help to improve soil structure.

The most common liming material is calcium carbonate ($CaCO_3$). It con-

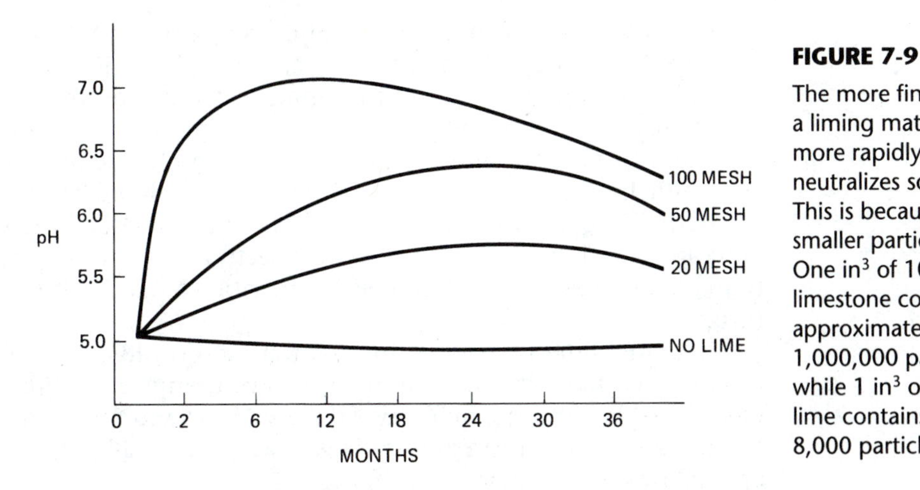

FIGURE 7-9

The more finely ground a liming material is, the more rapidly it neutralizes soil acidity. This is because of the smaller particle size. One in^3 of 100-mesh limestone contains approximately 1,000,000 particles, while 1 in^3 of 20-mesh lime contains about 8,000 particles.

tains 40 percent calcium and is called ground or agricultural limestone. Calcium carbonate is the most widely used type of lime because it is very inexpensive.

This material is produced by grinding limestone rock. The rate at which calcium carbonate corrects acidity depends upon how finely the material is ground. The finer the lime, the quicker it works. Ground limestone that passes through a 100-mesh sieve (100 holes per linear inch) raises pH much more rapidly than material that can only pass through a 20-mesh sieve (20 holes per linear inch) (Figure 7-9).

MINIMUM GUARANTEED ANALYSIS	
Calcium Oxide (CaO)	30.0%
Elemental Calcium (Ca)	21.4%
Magnesium Oxide (MgO)	20.0%
Elemental Magnesium (Mg)	12.0%
NEUTRALIZING VALUE	
Total Calcium Carbonate ($CaCo_3$) Equivalent	104.0%
Calcium Carbonate ($CaCo_3$) Equivalent from Magnesium Sources	50.0%
Calcium Oxide (CaO) Equivalent	57.0%
FINENESS	
% Through No. 20 U.S. Standard Sieve	98.0%
% Through No. 60 U.S. Standard Sieve	60%
% Through No. 100 U.S. Standard Sieve	40.0%

FIGURE 7-10

The fineness of the liming material is stated on the bag.

Lime used for turf should be ground finely enough to allow all of the particles to pass through a 20-mesh sieve, and at least 50 percent should pass through a 100-mesh sieve. This information is stated on the bag (Figure 7-10).

In areas where the soil is low in magnesium, dolomitic limestone should be used. It is ground limestone that contains both calcium and magnesium. Dolomitic lime is also very slow working and should be finely ground.

Burned or quicklime raises the soil pH more quickly. It is very water soluble and can "burn" the grass if applied incorrectly. Burned lime is more expensive than ground limestone and can be disagreeable to handle. It is seldom used on turfgrass.

Hydrated lime is faster acting than ground or dolomitic lime but slower than burned lime. It is also more expensive than ground lime and can have objectionable handling qualities. Hydrated lime and ammonia-type nitrogen fertilizers react and form ammonia gas, which can kill grass. They should be applied two weeks apart.

Rates of hydrated lime greater than 25 pounds (11.3 kilograms) per 1,000 ft^2 (93 m^2) can injure turfgrass. Even at lower rates it should never be applied when the grass is wet. If the particles stick to the foliage, they can burn (dehydrate) the tissue. Hydrated lime is not normally recommended for turfgrass use.

The amounts of ground limestone required to raise soil pH are listed in Table 7-1. Dolomitic limestone is applied at approximately the same rates.

Table 7-1 indicates that clay soils need more lime than sandy soils need. This is because finer-textured soils have more particle surface area and exchange sites. They can hold more hydrogen ions than can coarser, sandier soils. Finer-textured soils or soils with significant amounts of organic matter

TABLE 7-1 Approximate Amounts of Calcium Carbonate Required per 1,000 Ft^2 to Raise the Soil pH to 6.5*

	Pounds of Limestone/1,000 ft^2 (93 m^2)		
If pH	Sandy Soil	Loamy Soil	Clayey Soil
6.0	20	35	50
5.5	45	75	100
5.0	65	110	150
4.5	80	150	200
4.0	100	175	230

*To convert rates to pounds per acre, multiply by 43.5. It is best to limit each application to 50 pounds per 1,000 ft^2 when lime is applied to the surface of an established turf. Larger amounts can result in excessive alkalinity near the surface of the soil before the lime eventually moves downward. Applications should be spaced at least a few months apart.

have reserve acidity and are said to be buffered because they resist pH change. Eventually the stored H^+ ions will become dissolved in the soil solution and contribute to soil acidity. More calcium or magnesium must be added to a clayey soil because it contains more H^+ ions.

Sometimes soil testing laboratories report "buffer pH" as well as soil pH. The soil pH number indicates the pH of the soil solution, but does not tell anything about the soil's buffering capacity. To determine the lime requirement it is necessary to know the soil texture and refer to a chart such as the one in Table 7-1.

The buffer pH method determines the lime requirement by adding a buffering solution (pH 7.5) to the sample. The change in soil pH that occurs after the buffering solution is added indicates the buffering capacity and determines how much lime is needed.

Ground or dolomitic lime will not burn grass. However, excessive application rates can raise the soil pH too much. If the soil becomes too alkaline, nutrients will remain in unavailable forms. Overliming is most likely to cause problems in sandy soils. Lime raises the pH of a sandy soil faster than it changes the pH of a loam or clay soil.

Lime is applied with a fertilizer spreader. It is normally sold as a finely ground powder. The small particles must be spread evenly. Lime moves straight down in the soil—it does not move laterally (horizontally). Soil beneath "missed" spots will not be neutralized.

The ground and dolomitic limestones move down through the soil very slowly. This is not a problem if the correct amount of lime is spread before the grass is planted. The lime can be rototilled or plowed into the soil. It should be mixed uniformly 6 to 8 inches (15 to 20 centimeters) deep. The pH of the root zone can often be raised to 6.0 or higher in a few months.

When lime is applied to an established lawn, it cannot be worked into the soil. Lime may move down through the soil only 0.5 to 1 inch (1.3 to 2.5 centimeters) a year. Surface applications can take two or more years to increase the pH of the root zone to the proper level. The turf manager must not allow the pH to drop too low before adding lime.

In humid areas soil pH tends to become acidic because of the leaching of calcium and magnesium. Plants also remove these two neutralizing elements from the soil. The nutrients are not returned to the soil if the clippings are removed when the grass is cut. Many nitrogen fertilizers have an acidifying effect.

The turf manager must test the soil to determine whether lime is needed. The soil should be tested before the grass is planted, and then tested again every two or three years. If test results show that the pH has dropped to 6.0 (or to a minimum optimum pH value), lime should be added. For the majority of turfgrasses lime should be applied at 6.0 Soil testing will be discussed in detail in Chapter 8.

Ground or dolomitic lime and fertilizer should not be applied to turfgrass at the same time. If they are put down together, nitrogen may be lost. It is converted into a gas and escapes into the air. Phosphorus is "fixed" by the

FIGURE 7-11
Pelletized and nonpelletized lime.

calcium and magnesium. It is bound up in insoluble, unavailable compounds. When lime is spread, a week should be allowed to pass before fertilizer is applied, and vice versa. However, lime and fertilizer can be worked into a seedbed together if incorporation is not delayed.

Lime is very inexpensive and can be applied at any time during the year. In acid soils, applications are often necessary every two or three years. Turfgrass quality is greatly diminished if the soil pH becomes too acidic.

Very finely ground 200-mesh lime particles can raise soil pH one unit in several weeks. However, it is difficult to spread the dustlike particles evenly when they are in a dry, powdered form. The particles can be suspended in water and applied with a pesticide sprayer. When lime is mixed with water and sprayed on turf it is called *slurry* or *fluid lime*. Very finely ground lime can also be purchased in a pelletized form (Figure 7-11). The pellets are easier to spread but more expensive than the powdered forms. After application the pellets dissolve into powder when they come into contact with moisture.

Correcting Soil Alkalinity

Moderately alkaline pH in the range of 7.5 to 8.5 may develop in soils with an excess of sodium, calcium, or magnesium. This usually occurs in arid or semiarid regions where rainfall is infrequent and leaching is minimal. Less leaching is also associated with poorly drained soils. Irrigation water may contain sodium, calcium, and magnesium and contribute to alkalinity.

At alkaline pH values, micronutrients such as iron, manganese, copper, zinc, and boron become unavailable to plants. Alkalinity is corrected by applying

TABLE 7-2 Approximate Amounts of Elemental Sulfur Required per 1,000 Ft^2 to Lower the Soil pH to 6.5*

Lowering pH	Pounds of Elemental Sulfur/1,000 ft^2 (93 m^2)	
If pH	Sandy Soil	Clayey Soil
8.5	35–45	45–60
8.0	25–35	35–50
7.5	10–15	20–25

*To convert rates to pounds per acre, multiply by 43.5. The rate of a single application should not exceed 5 pounds per 1,000 ft^2 when sulfur is applied to the surface of an established turf.

sulfur (Table 7-2). Elemental sulfur is commonly used. Aluminum sulfate is also effective, but is toxic to grass if improperly applied. Fertilizers containing acidifying chemicals such as ammonium sulfate and iron sulfate help to lower pH.

Turf managers have to be careful when applying sulfur to established turfgrass. Rates should not exceed 5 pounds per 1,000 ft^2 (2.3 kilograms per 93 m^2) per application. Sulfur can injure turfgrass even at low rates if the plants are stressed. For example, it should not be applied to cool season grasses during periods of higher temperatures.

Some soils have a high pH because they are formed from alkaline parent material. Because of their buffering capacity they may be very resistant to pH changes. If the soil pH cannot be lowered to the desired level, it is a good idea to apply larger amounts of the nutrients that are less available at alkaline pH levels.

SALTED SOILS

In semiarid or arid regions high levels of salts in the soil can be a serious problem. These salts are generally chlorides and sulfates of sodium, calcium, and magnesium. In high rainfall areas excess salts are leached from the root zone, but extensive leaching does not occur in regions where total annual precipitation is less than 20 inches (51 centimeters). Consequently, the salt content of the soil can build up to injurious levels in these drier areas. Salted soils are common in the southwestern United States.

High salt concentrations, as explained earlier in this chapter, reduce the ability of plants to absorb water and nutrients. Salts may also enter the plant and cause toxicity. The symptoms of salt injury shown by turfgrass include wilting, a blue-green appearance to the leaves, a stunting of growth, and tip burn (tip of the leaf is yellow). Grasses that lack salt tolerance may die.

Salted soils are placed in one of three categories based on their soluble salt content and exchangeable sodium percentage. *Saline* soils contain levels of soluble salts high enough to reduce plant growth. Their pH is usually below 8.5 because only a small amount of exchangeable sodium is present. *Sodic* soils contain sufficient amounts of sodium to interfere with plant growth. Their pH is generally higher than 8.5 because of the high levels of sodium. *Saline-sodic* soils contain large enough quantities of both soluble salts and exchangeable sodium to cause plant injury.

The soluble salts in saline soils are readily leachable, so frequent, deep irrigation will flush them beneath the root zone if the soil is permeable and well drained. Satisfactory turf can be grown on saline soils if excess salts can be leached by irrigation. Good drainage can be encouraged by mechanical aeration and the installation of drain lines. If salts cannot be removed from the root zone because the irrigation water has a high salt content, not enough irrigation water is available, or drainage is poor, the turf manager must use salt-tolerant grasses. Bermudagrass, zoysiagrass, St. Augustinegrass, creeping bentgrass, tall fescue, and alkaligrass *(Puccinellia distans)* exhibit good to excellent salt tolerance.

High levels of sodium are toxic to turf and create an unsatisfactory physical condition in the soil. Sodium causes the dispersion of clay particles, which means that aggregates or clumps of particles are not present. For this reason sodic soils are tight, structureless, and impermeable. Poor aeration and low water infiltration are common characteristics.

These problems may be corrected by adding gypsum (calcium sulfate) or sulfur to a sodic soil. Either amendment will cause the replacement of sodium on the exchange sites and promote the formation of clay particles into aggregates. After sodium has been replaced on exchange sites and has become dissolved in the soil solution, irrigation will help to leach it from the root zone. Gypsum is more widely used as a treatment for sodic soils than sulfur because it works faster. Sulfur, however, is very effective and will reduce the extreme alkalinity as well. Any practice such as core cultivation or drain line installation that encourages good drainage is beneficial.

All types of salted soils often experience nutrient unavailability, especially of iron and phosphorus, because of their high pH. Acidifying the soil with sulfur increases nutrient availability and improves drainage characteristics by dissolving some calcium carbonates which accumulate on the outside of sand particles and plug the soil.

SELF-EVALUATION

1. Plants are composed of 75 to 85 percent ___________.
2. Which is not an essential element for plants?
   a. Aluminum   c. Iron
   b. Nitrogen   d. Boron
3. K is the chemical symbol for
   a. Phosphorus
   b. Kryptonite
   c. Potassium
   d. Manganese
4. Which of the following is a cation?
   a. K^+ c. Cl^-
   b. N_2 d. NO_3^-
5. Why does the available form of nitrogen readily leach from the soil?
6. What is the soil solution?
7. Fertilizer burn is similar to drought injury. Why?
8. The preferred soil pH range for most turfgrasses is
   a. 4.0–5.0   c. 6.0–7.0
   b. 5.0–6.0   d. 7.0–8.0
9. A pH of 6.5 is
   a. Alkaline
   b. Slightly acid
   c. Neutral
   d. Extremely acid
10. Which plant nutrient also raises soil pH?
    a. Nitrogen
    b. Sulfur
    b. Calcium
    d. Potassium
11. Which lime material contains magnesium?
    a. Hydrated lime
    b. Ground lime
    c. Dolomitic lime
12. Which lime material changes pH most quickly and can burn grass?
    a. Hydrated lime
    b. Ground lime
    c. Dolomitic lime
13. How many pounds of ground lime should be spread per 1,000 ft^2 to raise the pH of a clayey soil from 5.5 to 6.5?
14. To increase acidity and lower the pH the turf manager should apply
    a. Iron
    b. Phosphorus
    c. Manganese
    d. Sulfur
15. How can saline soil conditions be improved?
16. What factors result in a soil becoming more acidic?

CHAPTER

8

Soil Testing

OBJECTIVES

After studying this chapter, the student should be able to

- State why soil testing is important
- Collect a representative soil sample
- Describe how often soil tests should be performed
- Explain how soil test kits are used
- Describe the types of analysis performed by soil testing laboratories
- Discuss tissue testing

INTRODUCTION

Soil testing plays an important role in a successful turf management program. The tests measure the soil pH and the amount of nutrients available to the grass plants. Turf managers use this information to determine the lime or sulfur and fertility requirements of the soil.

COLLECTING A SOIL SAMPLE

The soil sample to be analyzed can be collected whenever the ground is not frozen. It is best to wait at least two weeks after fertilizing before sampling. Many turf managers prefer to take their samples at the same time each year so that seasonal differences do not affect the test results.

The soil sample must be characteristic of the entire area from which it is collected. The few ounces of soil that are analyzed may represent thousands or even millions of pounds of soil. If the sample does not accurately represent the true soil conditions, the test results will be misleading. Consequently, the fertilizer and lime or sulfur recommendations based on these results will be incorrect. To collect a "typical" soil sample, the proper sampling technique must be used.

The sample to be tested should be composed of soil taken from at least a dozen different locations. The individual samples are mixed together to form a "composite" sample. If only a few samples are gathered to form the composite, one atypical sample may have a significant effect on the test results. By collecting and mixing 12 or more individual samples, there is a much better chance that the composite sample will be representative of the entire area. Fewer than 12 individual samples is acceptable if the area is very small.

The samples should be collected randomly from all over the area that is to be tested. They should not be taken from just one small section. However, it is wise to avoid locations with environmental or soil conditions that are different from those commonly found in the area.

For example, an individual sample collected from a low spot may contain

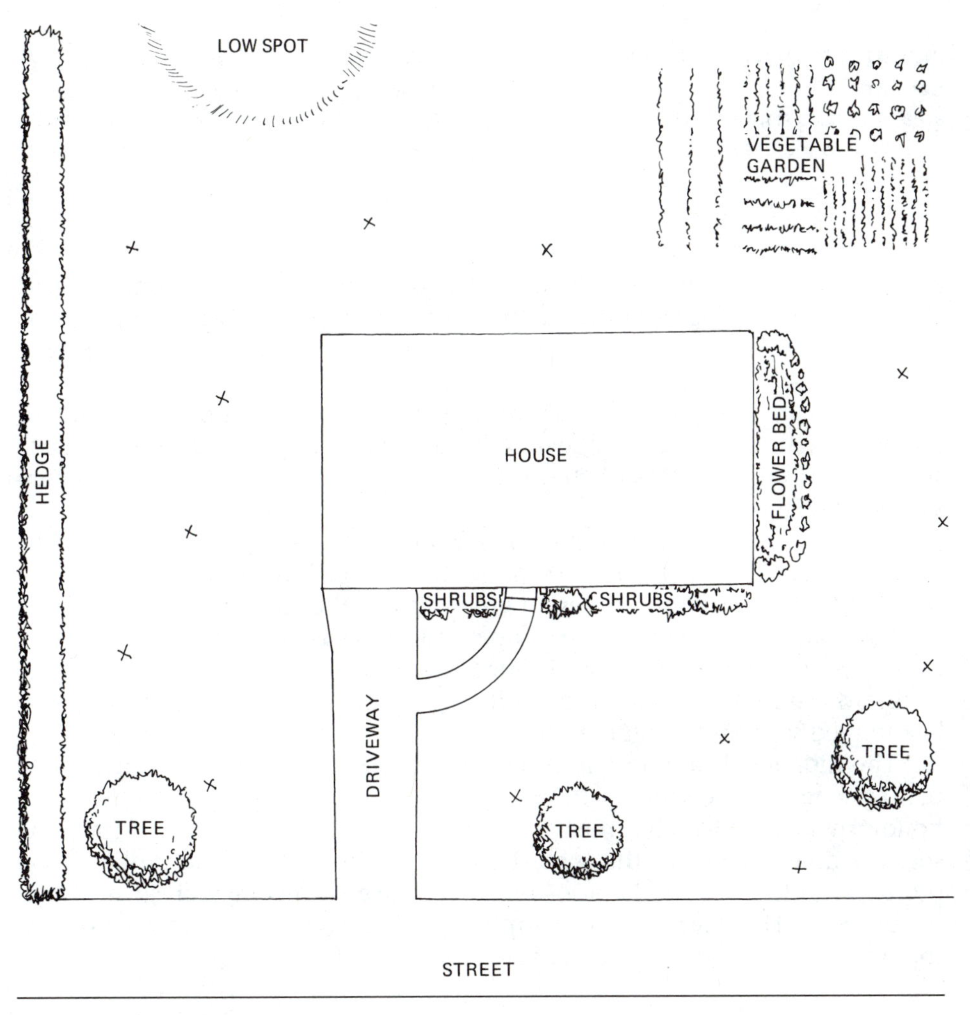

FIGURE 8-1

This is an example of proper sampling technique. Individual samples were collected at each spot marked with an X. The samples were then mixed together to make a composite sample.

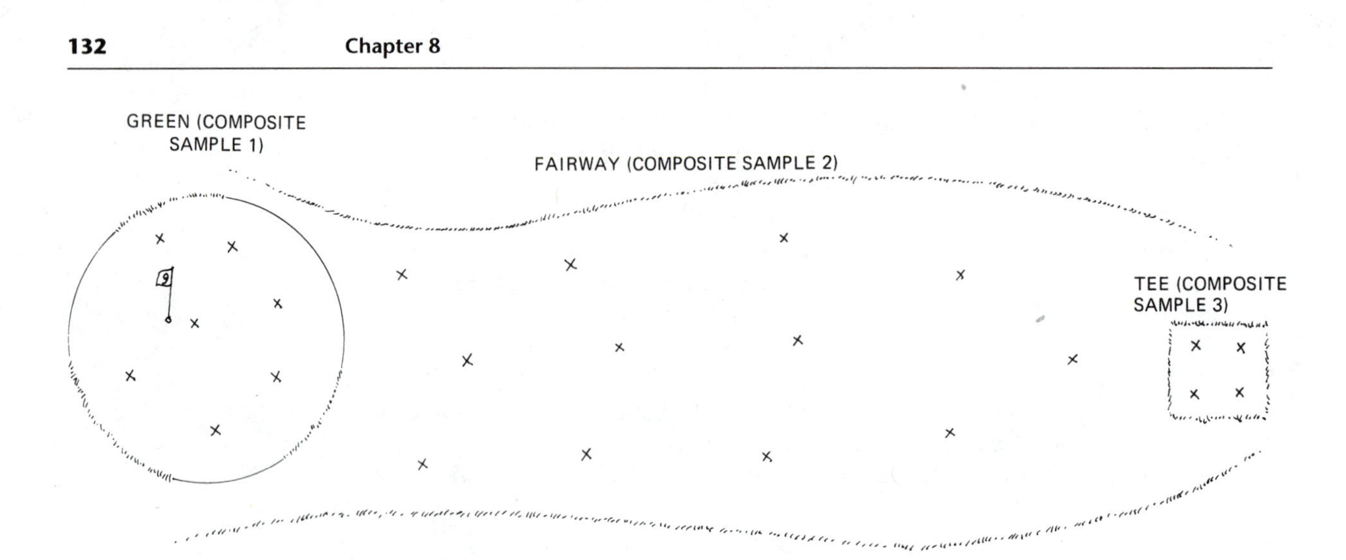

FIGURE 8-2 Three composite samples should be collected from this golf course hole because the green, fairway, and tee are fertilized, mowed, and irrigated differently. Each area also has a different soil texture. Fewer individual samples are necessary when a small area such as a tee is sampled.

an abnormally large amount of nutrients. Water running down from adjacent slopes may wash fertilizer into the low spot. Fertility levels near a tree may be atypically low because both the tree and the grass are removing nutrients from the soil.

Any small, unusual, or unique sites, such as a high spot, slope, low spot, or a flower or shrub bed should be avoided as a sampling area. It is also best to stay away from the edges of the turf area and driveways or sidewalks when sampling (Figure 8-1).

However, if a dissimilar area is large enough to be significant, it should be sampled separately. A composite sample should be collected from each sizable or important area that has different environmental or soil conditions.

For example, one composite sample cannot be used to represent an entire lawn accurately if the front yard is filled with trees and heavily shaded but the backyard is not. Areas that receive different types of maintenance should also be sampled separately (Figure 8-2).

The individual samples are often called core samples because many turf managers use a core sampler to collect them (Figure 8-3). The sampler has a hollow metal tube which is pushed into the soil. When it is removed from the soil a sample remains in the tube. This tool is also called a soil probe. A soil profile sampler is a similar tool which removes a rectangular slice of soil (Figure 8-4). The width of the sample makes it easy to observe root depth, topdressing layers, and other characteristics.

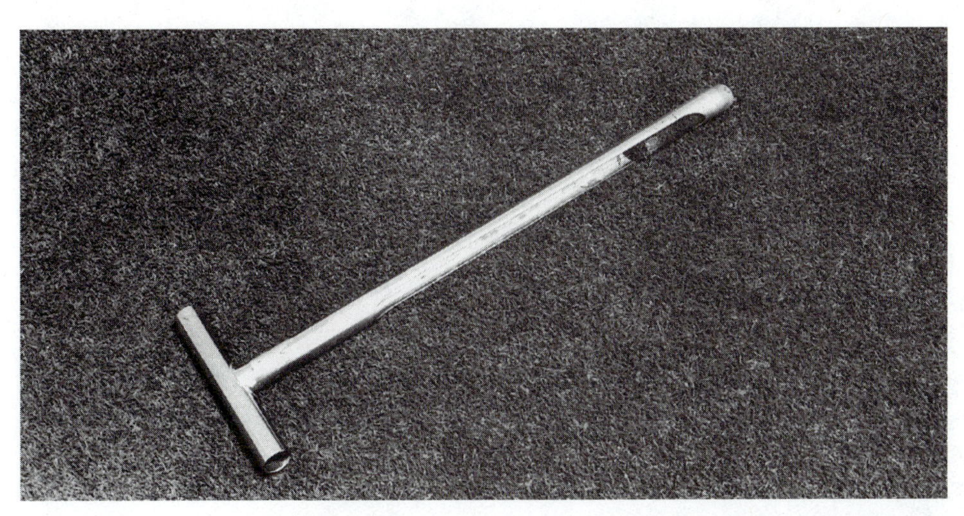

FIGURE 8-3
A core sampler or soil probe.

FIGURE 8-4
A soil profile sampler. *(Courtesy of Turfgrass Products Corp.)*

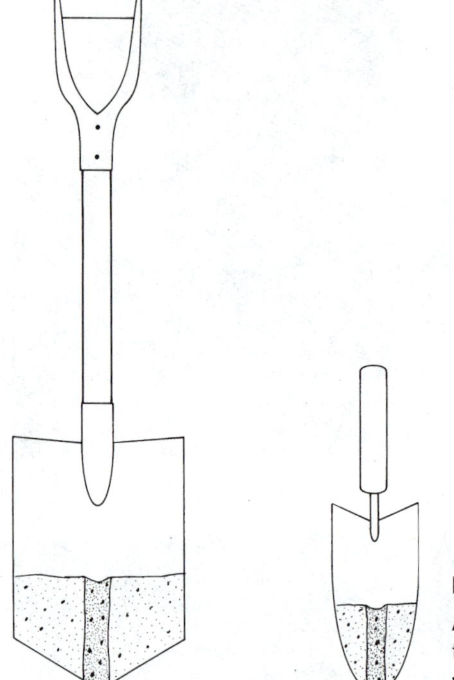

FIGURE 8-5

A sample 1 inch wide should be removed from the middle of the slice of soil, which is dug with a shovel or trowel.

Soil can be sampled with a shovel or trowel. A slice of soil 1 inch (2.5 centimeters) wide should be collected if either of these tools is used (Figure 8-5).

Established turf areas are sampled to a depth of 3 inches (7.6 centimeters). This is the recommended sampling depth because most of the turfgrass root system is usually in the top 3 inches of soil. Areas under construction are sampled to a depth of 6 inches (15.2 centimeters). The soil should always be tested before the grass is planted. This gives the turf manager the opportunity to work lime and fertilizer deep into the soil by mechanical means. The sample should be taken after soil amendments (topsoil, sand, peat moss, and so on) are incorporated.

The sample should not contain any grass blades or green, vegetative tissue. Normally, the thatch is removed from the sample as well. However, soil testing laboratories request that the thatch layer be included if the area is to be tested for chemical injury or pesticide residues. The thatch should also be included with the soil or sampled separately when the layer is so excessively thick that the majority of the root system is growing in it.

The core samples are placed in a clean paper or plastic bag or a cardboard container. Metal buckets should not be used. The core samples are mixed together thoroughly. If drying is necessary, the sample is spread on clean newspaper and allowed to air dry. It should not be baked in an oven. Heating the soil causes certain chemical reactions to occur which affect the soil test results.

One cup (1/2 pint) of soil is needed to perform a complete soil test. Each composite sample is identified with a label that describes the location from which it was collected.

Newly established areas should be tested annually for a few years. Once the grass is well established, the soil is tested every two to three years. However, the soil should be tested immediately if the turf quality becomes unsatisfactory. When a problem area is to be tested, a composite sample is also taken from an adjacent, healthy area. The soil conditions at both locations can be compared, and this may help to identify the problem in the area of concern.

SOIL TESTING

Turf managers can perform the tests personally or they can send the samples to a professional soil testing laboratory for analysis.

If the turf manager tests a large number of samples each year, he or she can save money by purchasing a soil test kit. The results can be obtained immediately if the turf manager tests the samples rather than sending them to a laboratory. However, the soil testing lab will provide more accurate results.

Using Soil Test Kits

Many different types of soil test kits are available. A basic test kit designed for homeowners measures phosphorus and potassium levels and soil pH (Figure 8-6). Tests for most of the plant nutrients and pH can be performed with the

FIGURE 8-6
Inexpensive test kits are available at garden centers and hardware stores. *(Courtesy of Sudbury Laboratory, Inc.)*

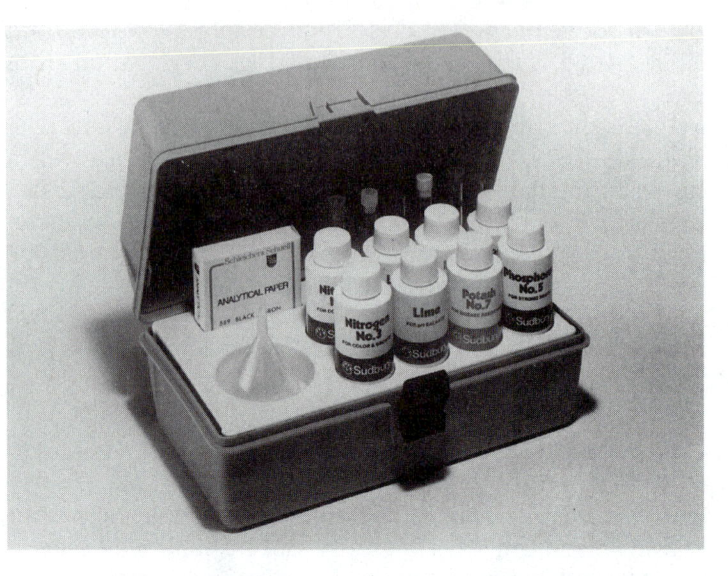

FIGURE 8-7
Professional turf managers generally prefer to use more sophisticated test kits such as the one shown here. *(Courtesy of Sudbury Laboratory, Inc.)*

more expensive kits (Figure 8-7). Local soil conditions determine the specific tests that are most appropriate. Turf managers should select a kit which allows them to test for all of the nutrients that are commonly deficient in their soils. Soils should always be tested for phosphorus, potassium, and pH.

Nitrogen test results are usually unreliable, although many kits measure nitrate nitrogen levels. Nitrogen in nitrate form is absorbed by plant roots. The majority of nitrogen in the soil is tied up in organic matter and is unavailable to plants. It is released when soil microorganisms decompose the organic material. The rate of release depends on microbial activity, which is affected greatly by weather conditions—soil temperature and moisture. The amount of nitrate nitrogen in the soil varies with the weather and is unpredictable.

There are two other reasons why soil nitrate levels change frequently. Nitrates in the soil solution are often taken up by plants as soon as they are formed. Levels also change after a rain or irrigation. Nitrates (NO_3^-) are readily leached from the soil because they are anions and are not held on the cation exchange sites.

Other nutrient levels in the soil are more stable, and test results can accurately measure the future availability of these nutrients. How much nitrogen fertilizer should be applied is determined by observing the color, density, and growth rate of the grass. The amount of nitrogen needed depends on the grass species and variety, the frequency of irrigation, the soil texture, and the level of turf quality desired by the manager. Nitrogen fertilization is discussed in more detail in Chapter 10.

Soil tests do not measure the total amount of nutrients in the soil. Only a small fraction of the total amount is available to plants. The turf manager needs to know the amount of phosphorus or calcium or magnesium that can be taken up and used by the plant. Soil testing kits and laboratories determine how many pounds of each nutrient are available per 1,000 ft^2 (93 m^2) or acre (43,560 ft^2 or 0.4 hectare).

All of the soil test kits work on the same principle. An extracting solution containing an acid is added to the soil sample. This leaching agent is mixed with the soil by shaking the sample. The acid extracts the available nutrients.

The sample is then filtered to remove the nutrient solution from the soil. The liquid is filtered into a glass tube, and a reagent is added to the solution. This reagent produces a color change. The color in the tube is then compared to a color chart. The pounds of available nutrients per 1,000 ft^2 (93 m^2) or acre (0.4 hectare) are shown on the chart next to the color block that matches the solution in the glass tube.

This information is then used to calculate the number of pounds of the nutrient that should be applied to the turfgrass. Included with the test kit is a sheet or booklet which explains how to estimate fertilizer needs based on the soil test results.

Color indicator reagents are also used to determine the soil pH. The reagent changes color when it is mixed with the soil and is compared with color blocks on a pH chart (Figure 8-8).

FIGURE 8-8

A typical soil pH test kit. *(Courtesy of Sudbury Laboratory, Inc.)*

Soil test kits can be satisfactory if the turf manager carefully follows the directions and avoids contaminating the reagents. Fresh reagents should be purchased every year. The most accurate results, however, are obtained by sending soil samples to a soil testing laboratory.

Soil Testing Laboratories

Most turf managers use professional soil testing laboratories. Each state has a soil testing lab located at the state agricultural college. The address and information on mailing procedures can be obtained from any county agricultural extension office. There are also a number of private soil testing laboratories in the United States. Some fertilizer and seed companies and garden centers will test a customer's soil free of charge.

Soil testing labs analyze samples using sophisticated equipment which gives precise readings of soil nutrient values. Some of the instruments are so sensitive that concentrations of an element as small as 1 part per billion can be measured. Electronic digital pH meters are accurate to within 0.1 of a pH unit (Figure 8-9). The instruments are operated by skilled lab technicians. A computer is often used to analyze the test results and recommend fertilizer rates.

If turf managers choose to do their own testing but are inexperienced, they should send the first samples they test to a professional lab as well. They can

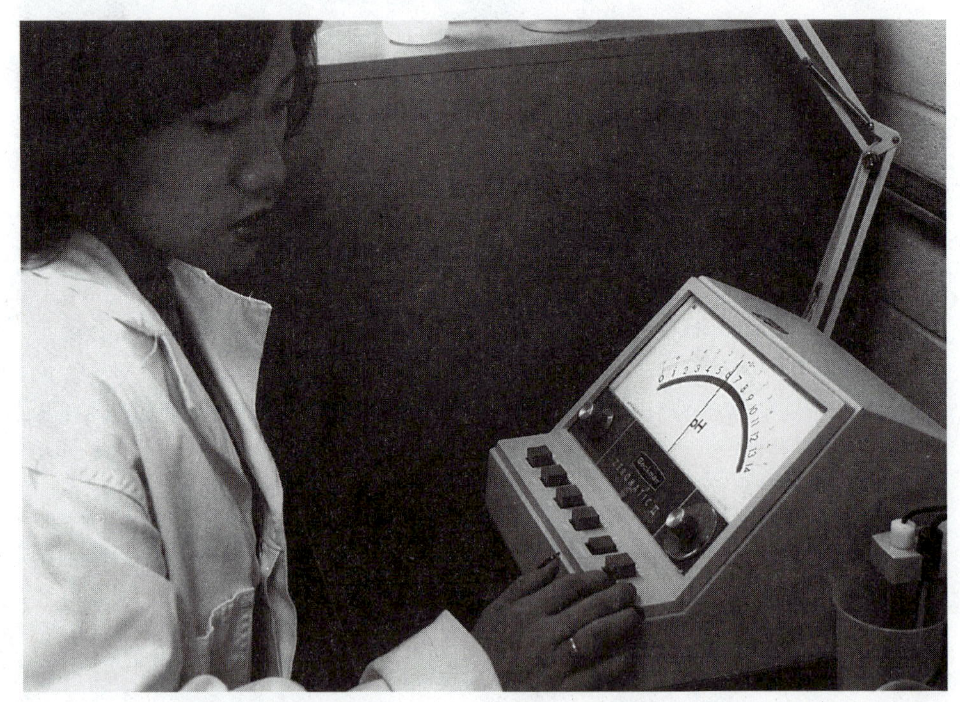

FIGURE 8-9
An electronic pH meter is highly accurate. *(Courtesy of James Bates, Photographer)*

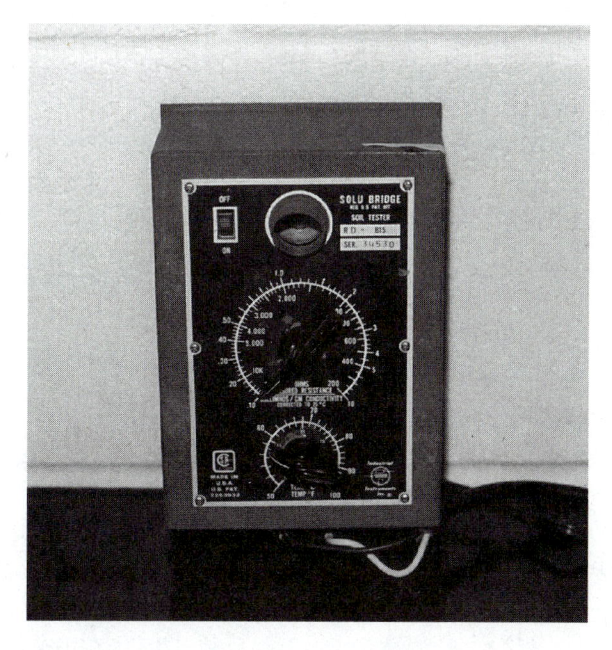

FIGURE 8-10

A solubridge measures the flow of electricity through the soil.

then compare their test kit results with the fertilizer recommendations made by the lab. If both recommendations are similar, the turf manager will know that he or she has performed the tests correctly.

Even when the turf manager is competent at soil testing, he or she should periodically have a professional lab do a complete soil analysis. For each sample provided, the lab can measure and determine the percentage of organic matter, texture, compaction (bulk density), chemical toxicities, cation exchange capacity (ability of the soil to store nutrients), and soil salinity.

The soil should be tested for salinity in regions of the United States where a high salt content is common. In other areas, salt injury may be a problem because of an overapplication of fertilizer or salt spread to melt ice. Instruments known as solubridges or electrical conductivity meters measure the electrical conductivity of the soil (Figure 8-10). The higher the salt level, the higher the electrical conductivity.

TISSUE TESTING

There are also instruments which will measure the amount of each element in the turfgrass plant by analyzing grass clippings. This is called tissue, or foliar, testing. Foliar test results can be used to determine the nutritional requirements of the plant, the ability of the roots to absorb nutrients, and the availability of nutrients in the soil. Tissue testing is becoming popular because it is the most accurate method of evaluating nutrient availability.

SELF-EVALUATION

1. How many square meters are there in 1,000 ft^2?
   a. 26
   b. 78
   c. 93
   d. 129
2. Established turf areas should be sampled to a depth of
   a. 1 inch
   b. 3 inches
   c. 6 inches
   d. 12 inches
3. How many square feet are contained in an acre?
   a. 28,610
   b. 43,560
   c. 57,106
   d. 111,840
4. Well-established turf areas should be soil tested every __________ years.
   a. 2 to 3
   b. 4 to 5
   c. 6 to 7
   d. 8 to 10
5. A solubridge is used to measure soil __________.
6. Determining the nutrient content of grass clippings is called __________.
7. A composite sample should be composed of at least __________ individual samples, unless the area is very small.
8. The soil sample should not contain __________.
9. The amount of available nutrients is determined by comparing the __________ of the nutrient solution with __________ blocks on a chart.
10. Why are nitrogen soil test results unreliable?
11. Discuss how to properly collect soil samples from a golf course.

CHAPTER

9

Establishment

OBJECTIVES

After studying this chapter, the student should be able to

- Explain how a turf manager decides the species and cultivars which are to be planted on a site
- Discuss how the site should be prepared before planting
- List and define the information on a seed label
- Identify the most appropriate times of the year for turfgrass establishment
- Discuss the four methods of turfgrass establishment—seeding, sodding, sprigging, and plugging
- Describe how the new planting should be cared for until the turfgrass is well established

INTRODUCTION

Proper establishment is the first and most critical step in a successful turfgrass management program. If the site is prepared and planted correctly, the result will be a permanent stand of grass that exhibits satisfactory quality. Improper establishment practices often result in an inferior stand of grass initially and can cause serious problems in the future. For example, acceptable quality may never be achieved if the species and cultivars that are planted cannot adapt to the environmental conditions existing on the site. Failure to improve a poor soil is another example of an establishment error that has undesirable long-term effects on turf quality.

Turfgrass establishment consists of species and cultivar selection, site preparation, planting, and postplanting care. Performing each of these practices correctly is essential. Mistakes made at the time of establishment may continually frustrate the turfgrass manager's future attempts to produce a satisfactory turf.

FIGURE 9-1
Special seed mixtures are formulated for shady sites. *(Courtesy of Stanford Seed Co.)*

SPECIES AND CULTIVAR SELECTION

The process of selecting the appropriate turfgrasses for a specific site is discussed in detail in Chapters 3 and 4. It is most important that the grass selected be adaptable to the growing conditions at the location to be established. It is essential that the grass have the ability to tolerate any stress factors associated with the site. For example, if trees are present, shade tolerance is a necessity (Figure 9-1). Turfgrasses grown on an athletic field should exhibit good wear tolerance and recuperative potential. Resistance to diseases common in the area is a major criterion.

The degree of turf quality desired and the level of maintenance that will be provided also strongly influence the decision. Density, texture, and color are the primary aesthetic considerations. The size of the maintenance budget, the amount of time and labor available, the type of equipment to be used, and the expertise of the turf manager determine the level of maintenance that can be performed. High-quality hybrid bermudagrasses and creeping bentgrasses can produce a turf of unsurpassed beauty, but the maintenance requirement is substantial. Centipedegrass, common bermudagrass, bahiagrass, carpetgrass, and nonturf-type tall fescue need minimal care, but produce a turf of significantly lower quality.

The suitability of a grass is determined by considering many factors, including soil conditions, climate, diseases that are likely to occur, intended use of the site, the degree of shading, the appearance of the grass, and the amount of maintenance it will require. The turf manager should make a careful evaluation of the strengths and weaknesses of a cultivar before deciding to use it. When the manager is uncertain about the appropriateness of a grass, he or she should consult the Cooperative Extension Service.

Cool season turfgrass species are often planted together (Figure 9-2).

Classic Sunny Mixture™

#009689

FEATURES

ADAPTS TO VARIABLE CONDITIONS
VERY GOOD DISEASE RESISTANCE
DROUGHT AND WEAR TOLERANT
DARK GREEN COLOR
CONTAINS COMPATIBLE VARIETIES
QUICK ESTABLISHMENT

INGREDIENTS

50% Julia and Dawn or Shamrock Kentucky Bluegrasses
30% Assure and Commander Perennial Ryegrasses
20% Shademaster Creeping Red Fescue

DESCRIPTION

Classic Sunny Mix™ is designed for professional use where a premium quality general purpose mixture is desired. Adaptability to a broad array of environmental conditions make it ideal for establishing a long-lasting high quality turf. Classic Sunny Mix™ will thrive in full sun or light shade. This mixture contains varieties with proven premium performance.

GEOGRAPHICAL ADAPTATION

Classic Sunny Mix™ is well adapted within the cool humid and cool arid zones.

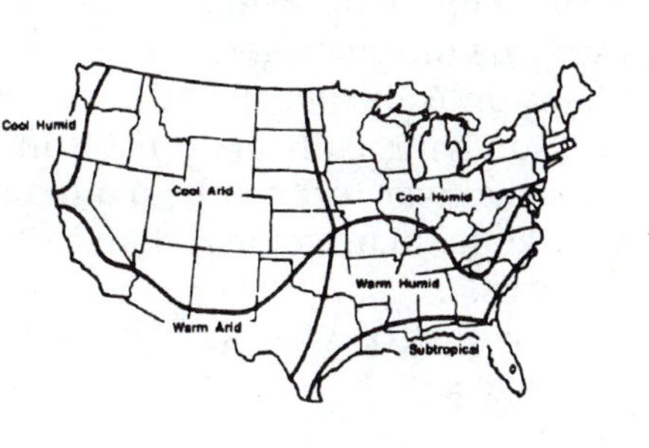

FIGURE 9-2
This product contains a mixture of three species. *(Courtesy of Lesco Company)*

A typical cool season seed mixture for lawns consists of Kentucky bluegrass, fine fescue, and perennial ryegrass. A mixture is a combination of two or more species and allows greater adaptability. Kentucky bluegrass excels in sunny locations, spreads by rhizomes, and usually requires at least medium maintenance. The fine fescues tolerate shaded and droughty conditions and perform satisfactorily at low maintenance levels. A turf composed of both of these species has a broader genetic base and will tolerate a wider range of conditions. Two or more cultivars of each species are often present in the seed mixture.

Perennial ryegrass is added to the seed mixture to increase the mixture's adaptability. This species is also used because of its rapid germination and seedling growth rate. It emerges first and protects the seedbed from erosion until the slower-germinating Kentucky bluegrass and fine fescue cultivars emerge. A seed mixture usually should not contain more than 20 percent perennial ryegrass by weight. The species can be so aggressive that large percentages of it in a mixture result in a stand almost totally composed of perennial ryegrass. Smaller percentages ensure that it will not dominate the site by outcompeting slower-to-establish seedlings for sunlight, nutrients, and water.

In sunny areas it is possible to use a blend of Kentucky bluegrass cultivars. A blend is a combination of different cultivars of the same species. A single-species planting results in greater uniformity and a more attractive turf. Kentucky bluegrass blends are popular because numerous outstanding cultivars are available. Through proper blending a very versatile, adaptable turf can be established. The large selection of Kentucky bluegrass cultivars allows the turf manager to develop a seed blend that will be able to tolerate wide range of growing conditions. Though there are shade- and drought-tolerant cultivars, it is best to use a mixture with fine fescues as well when the site is shaded or dry and not irrigated.

The ideal blend consists of cultivars that are similar in texture and color, resistant to the diseases that are likely to cause problems, and able to tolerate environmental conditions existing on the area to be established. The blend should be composed of at least three cultivars (Figure 9-3).

Planting only one cultivar of a cool season species is usually avoided except on golf course greens and a few other special type of turf site. It is true that the greatest uniformity is achieved by using a single cultivar, but there are some serious drawbacks to a monostand. If the cultivar loses its resistance to a disease or is not adaptable to the growing conditions on the planting site, the entire stand of turf will be weakened and may die. When a blend or mixture is used, the turf manager is better protected from such devastation because it is highly unlikely that all of the cultivars present will succumb to the same stress. The surviving cultivars can fill in and replace any injured or dead grass. A single-cultivar turf can be maintained successfully, but it requires more care and skillful management than a site composed of different turfgrass species or cultivars.

Mixtures of warm season turfgrass species are unusual. Normally only one

Quality Blue Blend™

#001715 | **Poa pratensis**

FEATURES

BROAD GEOGRAPHIC ADAPTATION
EXCEPTIONAL DARK GREEN COLOR
SPRING AND FALL VIGOR AND COLOR
GOOD SOD TENSILE STRENGTH
VERY GOOD DISEASE RESISTANCE
EXCELLENT HEAT, DROUGHT, AND WEAR TOLERANCE

INGREDIENTS

Contains 25.0% of four of the following Kentucky Bluegrass varieties: Julia, Dawn, Merit, Midnight, Glade, Shamrock, and Wildwood.

DESCRIPTION

Quality Blue Blend contains four compatible elite Kentucky Bluegrass varieties combining a wide variety of strengths in tolerance or resistance to diseases, drought, heat, cold, and wear. These grasses exhibit fine dark green leaf texture. Quality Blue Blend produces a dense dark green sod of premium quality. This blend has been produced with the highest quality certified seed to ensure the grower maximum results. Seed analysis reports are available on request.

GEOGRAPHICAL ADAPTATION

Quality Blue Blend is adapted throughout the cool humid and cool arid zones.

FIGURE 9-3 This product contains four Kentucky bluegrass varieties. *(Courtesy of Lesco Company)*

species is planted. Warm season species are often incompatible, and a mixture results in a patchy, nonuniform turf. Most of the warm season species are stoloniferous, and grasses that spread by stolons tend to segregate when in a mixture rather than form a homogenous turf. Also, several of the species are established vegetatively, not by seed, so it would be difficult to mix them when planting.

SITE PREPARATION

Proper site preparation is essential for successful turfgrass establishment. If soil conditions need improvement, possible corrective measures include bringing in topsoil, adding organic matter to the existing soil, and installing a subsurface drain tile system. (Soil modification is discussed in detail in Chapter 6.) The turf manager may also wish to install an underground irrigation system. On many building sites the topsoil is removed before construction. It is redistributed if it was saved. The soil that will comprise the root zone should be tested to determine how much fertilizer, lime, or sulfur should be incorporated.

Weeds growing on the planting site should be eradicated. This is especially true if perennial grasses are present. Unlike broadleaf weeds, perennial grasses cannot be readily controlled with a herbicide without damaging the turf. These grassy weeds are a serious problem in a turf area because herbicides that control them will also kill the desirable turfgrasses. Glyphosate is commonly used to eradicate all of the vegetation on the establishment site. The turf manager should wait seven days after the application before disturbing the soil.

In some cases, the soil is chemically fumigated. Fumigation results in a partial sterilization of the soil. Weeds, weed seeds, insects, disease-causing microorganisms, and nematodes are killed. Fumigation is very effective, but is also expensive, time-consuming, and can be dangerous because some of the chemicals used are quite toxic. If a gaseous fumigant is used, the area must be covered with polyethylene for 24 to 48 hours to prevent the chemical from escaping from the soil.

Methyl bromide has been the most widely used fumigant. However, today there are many restrictions on its use. For example, it must be applied by someone who is licensed and certified in turf fumigation.

It is important to clean up the area. All rocks, trash, and debris should be removed from the planting site. Rocks left in the soil can cause moisture stress problems. Plants situated above a rock will have a shallow root system if there is only a thin layer of soil between the rock and the soil surface. Wood and tree stumps that remain in the soil may contribute to the development of fairy rings and mushrooms (Figure 9-4). Even stones should be removed from the surface soil whenever possible because they can interfere with future maintenance operations such as coring or vertical mowing.

A proper grade must be established. The soil should gently slope away from buildings to prevent water from accumulating around the foundation or in the basement. A 1- to 2-percent slope is normally recommended. Steep slopes should be avoided wherever possible. They are subject to serious erosion problems, hazardous to mow, and droughty because water runs off. In northern areas grades around sidewalks and driveways should be lower than the pavement or slope away from it to avoid "ice rinks" in the winter.

Before planting, the soil is loosened mechanically to ensure good water,

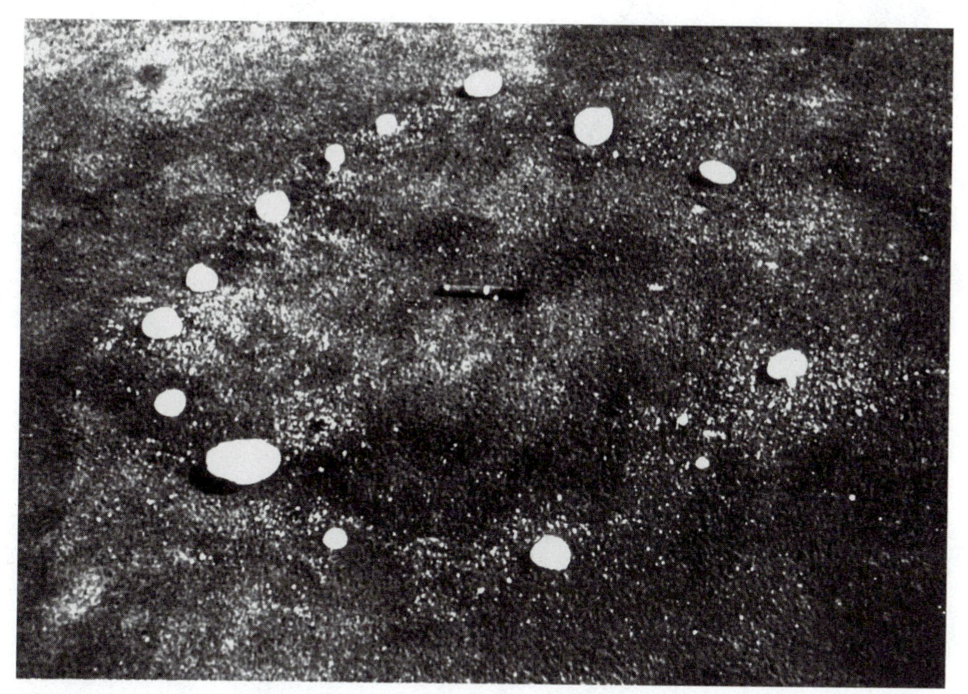

FIGURE 9-4
A fairy ring.

air, and root penetration. The soil should be worked to a depth of at least 6 to 8 inches (15.2 to 20.3 centimeters). This can be accomplished with a plow or rototiller. Shallower tillage can be performed by disking or harrowing the soil (Figure 9-5). The soil should not be worked when too wet or too dry because its structure can be ruined.

Fertilizer, lime, and organic material for improving structure are added next. They are normally incorporated into the upper 4 inches (10.2 centimeters) of soil. The amount of fertilizer and lime applied depends upon the results of the soil test. It is very important to work sufficient amounts of phosphorus and lime into the prospective root zone if the soil test shows that these materials are deficient, because both move so slowly through the soil. Application to the soil surface after the turf is established is considerably less effective because of the slow downward movement of phosphorus and lime. Adequate levels of available phosphorus are essential for seedling development.

It is common to use a fertilizer that contains greater amounts of phosphorus and potassium than nitrogen. Mixing large amounts of nitrogen into the soil can be wasteful because soluble forms of the nutrient are very prone to leaching. The frequent irrigation required for successful turfgrass establishment can result in the nitrogen being quickly washed beneath the reach of the

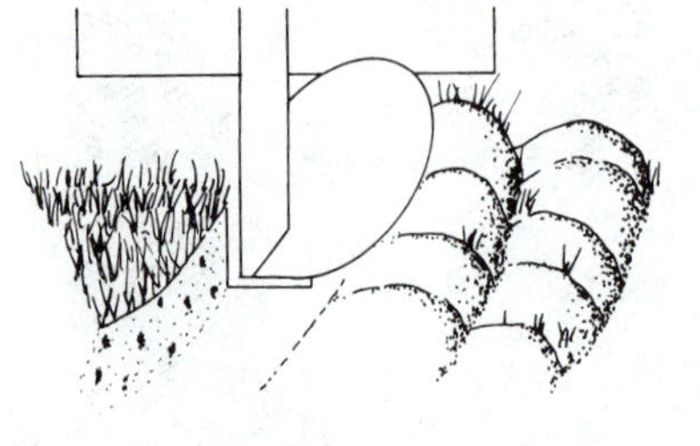

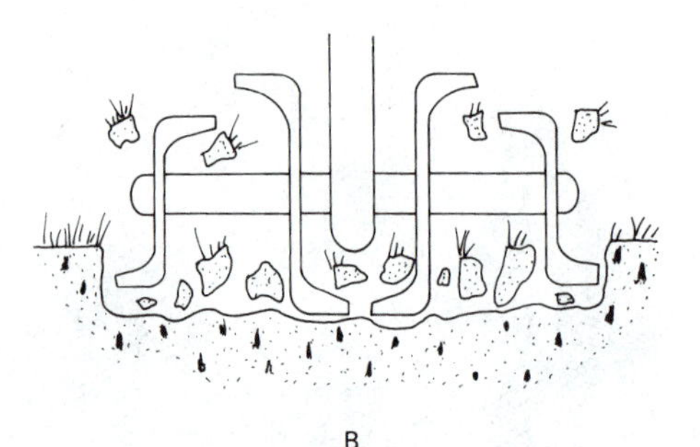

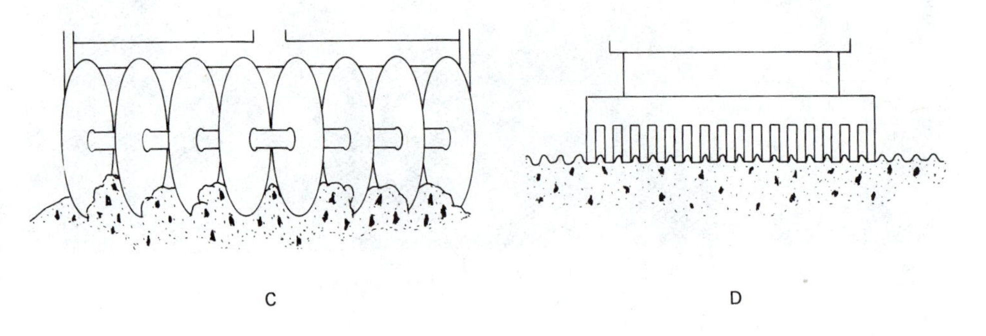

FIGURE 9-5
Tillage can be performed with a plow (A), a rototiller (B), a disk cultivator (C), or a harrow (D).

small roots of the new plants. Additional nitrogen is often applied a few weeks after planting.

The final steps in soil preparation involve leveling and firming the soil. After the soil has been loosened by tillage, the surface must be smoothed. If the soil has many high and low spots, a number of problems can result. Seed will tend to wash into depressions, leaving the higher areas bare. An uneven surface will be difficult to mow properly once the grass is established. A rough surface on athletic fields and golf greens can disrupt sports activities.

Before this final grading can occur, the soil must be firm. When the soil is loosened mechanically, it tends to expend. As equipment is driven over it and people walk on the planting bed, sunken wheel marks and footprints appear, destroying the uniformity of the surface. It is not unusual for a fine-textured soil to settle 1 to 2 inches (2.5 to 5.1 centimeters) after it has been plowed or rototilled. Settling can be speeded up by irrigating the site. The quickest method of firming the soil is to roll it. Rolling should not be performed when the soil is overly wet.

On larger areas a tiller rake attached to a tractor is used to smooth out surface irregularities (Figure 9-6). Steel chain mats or plank drags made from boards will also do the job. A section from an extension ladder can be pulled

FIGURE 9-6
A tiller rake.

across the seedbed. Smaller areas are often leveled by hand-raking. Final grading also breaks up any crust on the soil surface which could hinder the emergence of seedlings.

A common error made during site preparation is overworking the soil. People often think that the soil should be pulverized into a fine dust. This excessive tillage destroys the structure of the soil and is very detrimental. Large clumps or clods of soil should be broken up; however, soil particles in a correctly prepared seedbed should range in size from as small as a pea to an inch in diameter.

PLANTING

There are four methods of turfgrass establishment—seeding, sodding, sprigging, and plugging. The latter three methods are types of vegetative propagation. Almost all cool season grasses and some warm season grasses are propagated by seed, which is the cheapest method of establishment. Sodding is expensive, but provides an "instant" turf. Sprigging and plugging are the common ways of propagating some grasses that produce little or no seed, or poor-quality seed (Figure 9-7).

SPECIES	PROPAGATION METHODS
Bahiagrass	Seed
Bermudagrass	
Common	Seed
Improved	Vegetative and seed
Blue gramagrass	Seed
Buffalograss	Vegetative and seed
Carpetgrass	Vegetative and seed
Centipedegrass	Vegetative and seed
St. Augustinegrass	Vegetative
Zoysiagrass	
Japanese lawngrass	Vegetative and seed
Manilagrass	Vegetative
Bentgrass	
Colonial	Seed
Creeping	Seed (vegetative for a few cultivars)
Bluegrass	
Kentucky	Seed (sod is popular)
Rough	Seed
Fine fescues	Seed
Ryegrass	
Annual	Seed
Perennial	Seed
Tall fescue	Seed
Wheatgrass	Seed

FIGURE 9-7
Primary methods of propagation for the major turfgrass species.

ESTABLISHMENT BY SEED

Seeding is the preferred method of establishment because it is the least expensive. Seed can also be stored for long periods, unlike the vegetative materials that must be planted relatively quickly after they are harvested. Seed will remain viable for several years if it is kept in a dry location and not exposed to excessive heat.

Seed Labels

The turf manager must be able to understand the information on a seed label in order to evaluate the contents of a bag or package of seed. Each state has specific laws that require the labeling of turfgrass seeds. Unfortunately, these requirements vary somewhat from state to state, so there is a lack of uniformity in seed-labeling regulations. State laws often pertain more to agricultural crops other than turfgrass, and some important information may not be revealed.

Purity is the percentages by weight of pure turfgrass seed, crop seed, weed seeds, and inert matter in the container. The sum of these percentages should equal 100 percent. The percentage of each turfgrass species and cultivar present in the package is listed (refer to the labels shown in Figures 9-2 and 9-3). The percentage by weight of seeds of plant species that are grown as agricultural crops other than the turfgrasses already listed are included under the general heading of "crop seed." These seeds are not specified by name. This can be a problem because a number of perennial grasses that can cause serious weed problems in turf fall into this category. Examples are bentgrass, tall fescue, redtop, timothy, orchardgrass, and bromegrass.

Inert matter includes contents such as chaff (empty seed coverings), soil, stems, nonviable seeds, and stones. The percentage of weed seeds is listed on the label, but how many weeds are present or what they are is not identified. The exception is noxious weeds—species that are considered particularly undesirable. The number of noxious weed seeds per pound or ounce must be stated on the label. However, this information is not always valuable because of the lack of uniformity among state labeling laws. *Poa annua* (annual bluegrass) can be a very bad turfgrass weed, but many states do not classify it as a noxious weed.

Despite the inconsistencies in seed-labeling laws, the information concerning purity is essential to the turf manager. The turfgrass species and cultivars in the container are determined by examining the label. Generally, the lower the percentage of crop seed, weed seeds, and inert matter, the higher the quality of the mixture or blend (Figure 9-8). As the turfgrass industry continues to grow and becomes an increasingly significant economic factor, the laws regulating turfgrass seed labeling will undoubtedly improve. Some states now require a listing of any seeds which would produce plants that are undesirable in turf. They are classified on the label as "undesirable" species.

The other important information stated on the label is the *germination* rate. The percentage of pure seed that is alive and will germinate is determined by the seed company and verified by state seed laboratories. All purity and germination figures listed on the label are checked by seed labs in each state in which the seed is sold.

The turf manager should note the viability testing date stamped on the label. Most states require that the germination rate be retested every nine months in case changes have occurred. A poor germination rate may be because of extended storage of the seed or improper harvesting and processing.

The germination rate is tested in the laboratory under controlled, ideal environmental conditions. This viability is often significantly lower when the seed is planted outside under field conditions.

The longer the turf manager stores the seed, the lower the germination

[1] FINE-TEXTURED GRASSES	[2] GERMINATION
29.74% Glade Kentucky Bluegrass	90%
33.58% Kentucky Bluegrass	85%
26.93% Shadow Chewings Fescue	85%
COARSE KINDS	
[3] 8.25% Perennial Ryegrass	88%
OTHER INGREDIENTS	
[4] 0.00% Crop	
[5] 1.5% Inert Matter	
[6] 0.00% Weeds	
[7] **NO NOXIOUS WEEDS**	
[8] Tested 9/94	

NOTES:

[1] Finer-textured grasses are preferred in most situations.
[2] The higher the percentage the better.
[3] This is a common, coarser type.
[4] Should be 0.00% or close to it.
[5] The lower the better.
[6] Should be 0.00% or close to it.
[7] Should not contain any noxious weeds.
[8] Test date should be recent—normally no more than nine months old.

FIGURE 9-8
A sample seed mixture analysis.

rate. The germination rate can drop off rapidly if the seed is kept in a hot or damp location.

Determining the percentage of pure live seed (PLS) in the bag or package is also helpful to the turf manager when evaluating and comparing seed. If a box of seed contains 95 percent Kentucky bluegrass seed and this seed has a germination rate of 80 percent, then 76 percent (0.95×0.80) of the contents by weight is pure live seed.

Certified seed is seed that is guaranteed to be true to type genetically. A blue certification tag attached to a container of seed ensures that the seeds will produce plants which have characteristics identical to those associated with the cultivar. To guarantee genetic trueness, the seed is inspected by state personnel in the production field and after it has been harvested. Certified seed is often more expensive than noncertified seed, but is generally worth the extra cost.

Best Time to Seed

The best time to seed is before the longest period of optimum temperature and moisture conditions that occurs during the growing season. Seeding at this time allows the seedlings to become well established before unfavorable environmental conditions occur. If the new plants have the opportunity to develop an adequate root system they have a better chance of surviving the stresses associated with unfavorable growing conditions. The degree of weed competition must also be considered when selecting a seeding time (Figure 9-9).

The preferred time for seeding cool season turfgrasses is late summer. In late August and September soil and air temperatures are ideal for seed germination and seedling growth. Moisture conditions are also favorable. The greatest stress period for cool season grasses is in midsummer. Seeding in the late summer allows the plants to grow for the maximum length of time before they are confronted with the hot, dry conditions of midsummer.

After growing through the fall and during the following spring, the plants should be hardy enough to survive heat and drought stress. Weed competition is reduced in the late summer and autumn because seeds of many weed species germinate only earlier in the growing season.

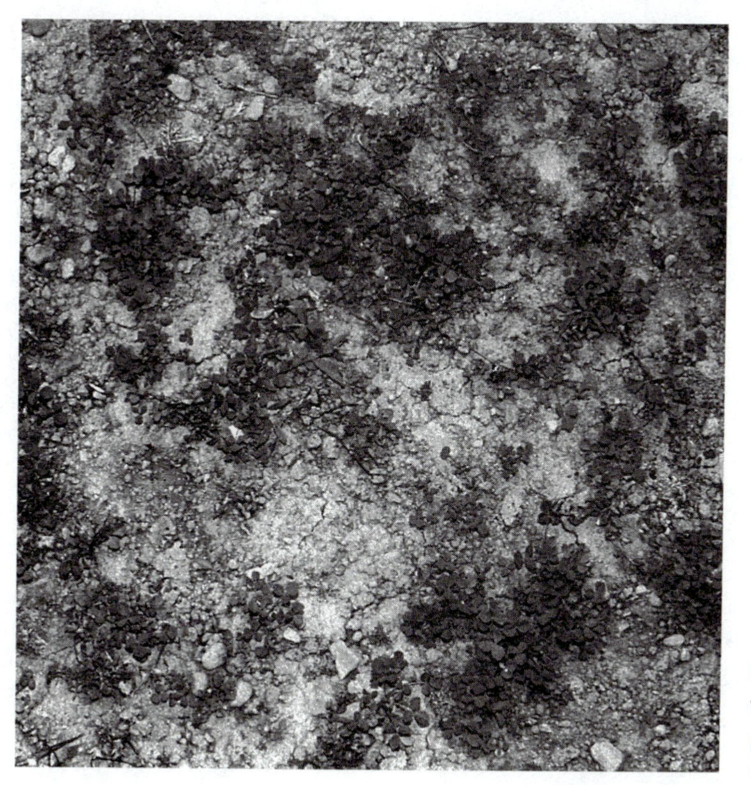

FIGURE 9-9
Weeds can be a serious problem in a seedbed. *(Photo by Brian Yacur)*

Seeding in the spring is less desirable because of the shorter duration of favorable environmental conditions before summer stress. Seed may take longer to germinate in the spring than in the late summer because of cooler soil temperatures. Another drawback to spring seeding is the extensive weed competition that occurs at this time.

The seed of cool season grasses is occasionally planted in late fall before the soil freezes. At the time of seeding the soil temperatures should be consistently lower than 50°F (10°C). This practice is known as *dormant seeding* because the low soil temperatures inhibit seed germination. The seed does not germinate until the soil warms up the following spring. Dormant seeding is most commonly performed on wet, poorly drained sites that would not dry out until late in the spring. Consequently, the soil could not be cultivated and planted early enough in the spring to establish sufficient growth before summer. Seeding in late fall solves this problem.

For a dormant seeding to be successful, it is essential that the seed does not germinate before spring. If the seeds are planted too early in the fall and germinate immediately, many of the small seedlings may be killed. They will have such a limited root system that they may be heaved out of the soil during the winter. The seedlings then desiccate because their roots are not in contact with the soil. A continuous snow cover is helpful in preventing seed loss due to erosion.

Dormant or early-spring seedings are often best in shaded areas because the seedlings have the opportunity to receive maximum sunlight before the leaves appear on deciduous trees. Seeding in early or midsummer is not generally recommended because of the high temperatures, unfavorable moisture conditions, and excessive weed competition. However, seeding in the summer can be satisfactory if the site is frequently irrigated and a vigorous weed control program is initiated.

Warm season grasses should be established by seeding in late spring or early summer. They grow best during the high temperatures occurring in the summer. A late-spring or early-summer seeding allows these warm season grasses to germinate rapidly and become well established before their period of greatest stress in the winter. Late-summer or fall seedings are not recommended because of the unfavorably cool soil temperatures existing at that time. The optimum temperature range for seed germination of warm season turfgrasses is 70°–95°F (21°–35°C).

Seeding Rate

The seeding rate depends upon various factors, including the species and cultivar, the percentage of germination, whether or not environmental conditions are favorable for germination, how well the seedbed is prepared, postplanting care, and the cost of the seed. The number of seeds per pound is important in determining the amount of seed that is sown (Table 9-1). For example, creeping bentgrass seeds are very small, and there may be 7,000,000 seeds per

TABLE 9-1 Approximate Number of Seeds per Pound and Seeding Rates for the Major Turfgrass Species

Species	Approximate Number of Seeds per Pound	Pounds of Seed per 1,000 ft^2 (93 m^2)
Bahiagrass	160,000–200,000	3–8
Bentgrass		
Colonial	6,000,000–8,500,000	0.5–1.5
Creeping	5,000,000–7,000,000	0.5–1.5
Bermudagrass (hulled)	1,200,000	1–2
Bluegrass		
Kentucky	1,000,000–2,200,000	1–2
Rough	2,000,000	1–2
Buffalograss	40,000–70,000	3–7*
Carpetgrass	1,200,000	1.5–5
Centipedegrass	500,000	0.25–2*
Fescue		
Fine	350,000–600,000	3–5
Tall	170,000–300,000	5–9
Grama, blue	900,000	1.5–2.5
Ryegrass		
Annual	200,000	5–9
Perennial	250,000	5–9
Wheatgrass, crested	320,000	3–6
Zoysiagrass	1,000,000–1,300,000	1–3

*The higher rates are best, but lower rates are commonly used because the seed is expensive.

pound. A seeding rate of 1 pound (0.45 kilogram) per 1,000 ft^2 (93 m^2) is usually sufficient to produce enough plants to provide adequate density. Tall fescue seeds are large in comparison to creeping bentgrass seeds, and there are only 200,000 seeds per pound (Figure 9-10). If tall fescue seed is applied at 1 pound per 1,000 ft^2, the number of plants that appear will be too small to provide good coverage. Consequently, tall fescue may be seeded as heavily as 9 pounds (4 kilograms) per 1,000 ft$_2$ (93 m^2).

Seed count per pound should be remembered when the turf manager is evaluating seed mixtures. Using a seed mixture that contains 50 percent Kentucky bluegrass seed and 50 percent fine fescue seed by weight will not result in an equal number of plants of both species. The mixture may actually contain as much as 80 percent Kentucky bluegrass seed and as little as 20 percent

FIGURE 9-10
Creeping bentgrass seed is considerably smaller than tall fescue seed.

fine fescue seed because there can be as much as four times as many Kentucky bluegrass seeds per pound (2,000,000 compared to 500,000). The result would be a turf that is predominantly Kentucky bluegrass.

The degree of postplanting care also has a major effect on seeding rate. The primary reason for seeding failures is a lack of water. If the turf manager will not be able to irrigate the seedbed as frequently as he or she believes is necessary, a higher seeding rate should be used.

Seeding at too light a rate results in too few plants and poor density. The stand has an unsatisfactory appearance, and weeds can flourish in the bare areas. However, grasses that have a strong spreading habit because of stolons or rhizomes can eventually achieve a proper density if maintained correctly. The weeds can be chemically controlled, and the turfgrass will fill in the bare spots in a relatively short time if they are not too large. This is not true for bunch-type grasses that spread only by tillering. They may take years to achieve proper density if the seed rate is not high enough.

It is generally better to use a bit too much seed rather than not enough, but excessive rates are a waste of money and can cause problems. Unnecessarily high rates resulting in excessive density lead to increased competition among seedlings for sunlight, moisture, and nutrients because of the greater number of plants. The seedlings are weakened by the competitive stress and become very susceptible to diseases. Common bermudagrass is more likely to be injured during its first winter if the seeding rate is too high.

FIGURE 9-11
Drop-type fertilizer spreaders are also excellent seeders.

Seeding

Seed is often applied with a fertilizer spreader. Gravity- or drop-type spreaders are preferable (Figure 9-11). The broadcast types, which are also called centrifugal or rotary spreaders, throw the seed over a greater width, but the distribution pattern is less satisfactory. Seeds are too light to be spread accurately, especially if there is wind. When a mixture is used, the species are not distributed uniformly with a rotary spreader because of the differences in seed size and density.

To ensure uniform coverage, the amount of seed to be used should be divided in half and sown in two directions. For example, one-half could be spread in an east-west direction and then the other half in a north-south direction. When very low rates are sown, as in the case of centipedegrass, soil or sand can be mixed with the seed.

The next step is to cover the seed with a shallow covering of soil. It can be raked in by hand or with a tiller rake or dragged in with a steel chain mat. The seed should be worked into the top 0.25 to 0.33 inch of soil. Another alternative is to leave the seed on top of the soil and cover it with a light topdressing of soil.

Seed beneath a thin covering of soil is less likely to dry out or wash away. Good seed-soil contact better enables the seed to absorb water. However, seed should not be incorporated too deeply in the soil. If it is too deep, the food

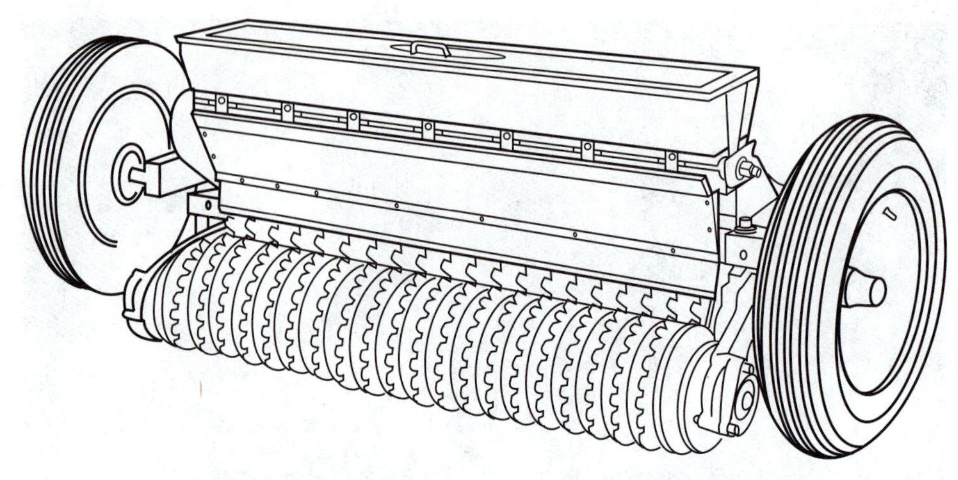

FIGURE 9-12
A cultipacker seeder.

reserves in the seed may be depleted before the seedlings are able to reach the soil surface and photosynthesize. After the seed is covered with a shallow layer of soil, rolling may be performed. Pressing the soil down with a light roller increases seed-soil contact.

A cultipacker seeder is a piece of equipment used for seeding large areas (Figure 9-12). The machine is attached to a tractor and has a 5- to 10-foot (1.5- to 3.0-meter) seeding width. The cultipacker levels the soil, places seed in the soil at the proper depth, and then firms the soil around the seed. Ten acres (4.0 hectares) can be seeded per day with this type of machine.

A hydroseeder is a machine that sprays a seed and water mixture. Fertilizer and mulch are often mixed in the tank with the seed and sprayed at the same time (Figure 9-13). Hydroseeders were originally used for seeding steep slopes and large sites such as the areas beside highways. Today they are used to seed lawns and many other types of turf areas.

Another name for this method of establishment is hydraulic planting. When water, seed, and mulch are applied together, some people refer to the process as hydromulching.

A hydroseeder is the fastest method of seeding. However, the machine is very expensive and requires a ready supply of large volumes of water. Another problem is that the seed is not incorporated in the soil and is subject to moisture stress and erosion. The mulch material helps to prevent the seed from washing away or desiccating. It is often shredded newspaper that is dyed green. Wood fiber products are also available.

If a stand of grass is needed in a shorter period of time than the normal seed germination rate allows, there are techniques to speed up the process. Seed can be soaked in water for a few days until roots or shoots appear outside the seed coats. Planting this pregerminated seed can reduce germination time by as much as 50 percent.

FIGURE 9-13
Establishing a lawn by hydroseeding. *(Courtesy of William Knoop)*

For successful seed pregermination, the water must be changed at least once a day. Warm water works best. The water should be aerated. Many turf managers use a fish tank aerator (aquarium pump). For large quantities of water it may be necessary to run a hose from an air compressor into the water. Some grass species require light to germinate.

Athletic field managers often pregerminate seed when they need to repair an area quickly. Most Kentucky bluegrass cultivars are slow to germinate, so they can be pregerminated to make them more competitive in the seedbed. This is very helpful when mixing Kentucky bluegrass with a fast-germinating, aggressive species such as perennial ryegrass.

Once seed is pregerminated it must be handled carefully and planted quickly. The seed is delicate because the roots and shoots are outside the protective seed coat.

If the pregerminated seed is going to be applied with a hydroseeder, it can be placed directly into the tank. Otherwise, it is spread out and allowed to dry enough that it can be mixed with sand, calcined clay, vermiculite, Milorganite (an organic fertilizer), or some other carrier. The mixture is then applied with a drop spreader.

Another technique for speeding up germination is called priming. With pregerminating, the seed is covered with water. With priming, only 1 gallon of water is added for every 10 pounds of seed. This is enough water to initiate germination but not enough to cause the roots and shoots to break through the seed coat. Chemicals such as salt and polyethylene glycol can also be added to water to limit the amount that enters the seed.

Primed seed is not as advanced as pregerminated seed, but it will germinate faster than untreated seed. Primed seed can be applied with a spreader without having to be dried or mixed with a carrier. It does not have to be used immediately.

Mulching

The turf manager should consider using mulch on the seedbed. It helps to conserve moisture and reduce seed loss. A thin layer of mulch shields the soil surface from the drying effects of the sun and wind and can decrease water loss by more than 50 percent. This is extremely beneficial because, for seeding to be successful, the seedbed should be continually moist until the grass is established.

Mulch also protects the seed and soil from washing. A heavy rain or irrigation can erode the seedbed and result in a poor stand of grass. This problem is minimized when the drops of water strike a layer of mulch. The mulch acts as a protective barrier, absorbing the impact of the water.

Mulching increases the likelihood of seed germination and enables the turf manager to establish a uniform, dense stand of grass. The extra expense is normally justified. Mulch is essential on sloped sites and when adequate postseeding irrigation cannot be guaranteed.

There are several types of mulching materials that can be used. Straw has been a popular mulching material for many years. Wheat straw is used primarily, but oat and barley straw are also good. The straw should be free of grain seed heads. If they are present, the bales can be fumigated to destroy the seeds.

The recommended application rate is 2 bales per 1,000 ft_2 (93 m^2). This is equivalent to approximately 80 to 100 pounds (36 to 45 kilograms) of straw. The rate per acre (0.4 hectare) is 2 tons (1,814 kilograms) or approximately 90 bales. Straw can be spread by hand on small areas. Mechanical mulch blowers are available for mulching large areas. These machines chop up the bales and uniformly distribute the straw. Because straw is light, it can be blown easily by the wind after it has been spread. An asphalt binder or tackifier can be used to help hold the straw in place. The straw can also be covered with netting or cheesecloth to prevent its movement. If the straw is wetted immediately, it is less likely to blow. Rolling also helps.

There is a tendency to bury the seedbed under too thick a mulch layer. Mounds of straw should be avoided (Figure 9-14). Straw is biodegradable and there is no need to remove it if less than 50 percent of the soil surface is covered. If heavier rates are used, one-half of the straw should be removed when the new grass is 1 inch (2.5 centimeters) high.

Grass hay is not usually as desirable a mulching material as straw. It is less

FIGURE 9-14
Too much straw mulch.

likely to be clean—free of weed seeds. Salt hay, however, makes an excellent mulch.

Peat moss has some undesirable characteristics, but it offers the advantage of being readily available to turf managers because it is sold at most garden centers. Peat mulch can be costly and must be wetted thoroughly when applied or it can absorb large amounts of water from the soil. Moistening the peat also prevents it from blowing around. A layer of peat 0.13 to 0.25 inch (3.2 to 6.4 millimeters) thick is best.

Wood products are used as mulches. These materials include wood chips, shredded bark, wood shavings, and wood cellulose fibers. Pine needles have been used successfully for mulching seedbeds.

Various types of net mulches are available. Examples are jute and tobacco netting, cheesecloth, coarse-mesh burlap, and spun polyester. These materials must be removed before the seedlings become too large. They are reusable. Some erosion control fabrics consist of paper strips woven through yarn and are completely biodegradable (Figure 9-15).

Clear polyethylene may be used to cover a seedbed when the soil temperatures are too low for rapid germination. The polyethylene covering traps heat beneath it in the same manner that a greenhouse retains heat. This raises the soil temperature and speeds seed germination. The covering must be removed if the temperature becomes too high.

FIGURE 9-15
Spreading an erosion control fabric. *(Courtesy of Gulf States Paper Corp.)*

Seed mats are strips of organic mulch which contain seed and fertilizer. They are simply laid out on the seedbed and watered. Seed mats have not been widely used because the establishment expense is greater than for traditional seeding methods.

Postplanting Care

Keeping the seedbed moist until the grass is well established is the key to successful establishment. The top 0.5 inch (1.3 centimeters) of soil must not be allowed to dry out. Germinating seed requires a constant moisture supply, and seedlings have such shallow root systems that they will desiccate if the surface soil becomes dry. Heavy irrigations should be avoided because erosion can occur. Light, gentle waterings are essential. During hot, dry periods irrigating as many as three times a day may be necessary. Frequent watering should continue for at least three weeks after planting.

A common mistake when seed mixtures contain significant amounts of perennial ryegrass is to stop watering too soon. Perennial ryegrass may be up in a week or less, while some Kentucky bluegrass cultivars take three weeks to germinate. The rapid germination and vertical growth of the ryegrass seedlings deceives people into thinking that all of the grasses in the mixture are established. They stop irrigating frequently, and the slower-establishing species such as Kentucky bluegrass perish.

Mowing is necessary when the seedlings reach a height 50 percent taller than the height to which the grass will be cut. The mower should be correctly adjusted and the blade sharp or the delicate seedlings can be seriously injured. Watering should be decreased before the first mowing so that the soil is firm or the mower wheels will sink into the surface of the seedbed. The use of heavy mowing equipment is avoided until the grass is well established.

A light application of nitrogen after the seedlings emerge is beneficial. Seedlings are very sensitive to herbicides and are susceptible to herbicide injury. The turf manager should delay herbicide applications as long as possible, preferably for at least a month after the seedlings appear in the seedbed or until the new grass has been mowed twice. On a new common bermudagrass lawn, for example, it is best to wait several months before using an herbicide If weed control becomes necessary early, it is important to select the herbicide that is least injurious to seedlings.

Sometimes pesticide labels recommend applying the herbicide on new seedings at lower than normal rates. Mowing removes many of the weeds present in a new seeding.

Traffic should be kept off the seedbed until the grass is established. A "Please Don't Walk on the New Grass" sign or a barrier of stakes and twine around the seedbed is advisable (Figure 9-16).

FIGURE 9-16

Twine and stakes help to keep people off a seedbed. (*Courtesy of James Bates, photographer*)

VEGETATIVE ESTABLISHMENT

The site preparation for vegetative establishment is identical to that required when seeding. Postplanting care is very similar as well. Planting sites established vegetatively are usually not mulched. The best times for seeding are also the ideal times for vegetative propagation.

Sodding

Sodding is a practice which involves transplanting large pieces of established turf (Figure 9-17). The entire site is covered with three pieces of mature, high-quality turf, which can be composed of many different species and cultivars. Sodding is expensive, but offers the advantage of providing an "instant" turf. Installing sod usually costs five to seven times as much as seeding, but less postplanting care is necessary. A seeding may take up to a year to produce an equally dense turf. On slopes where seed and soil erosion would be a serious problem, sodding is an ideal solution. A covering of sod eliminates weed competition on the planting site.

Three months to two years are required to produce good sod. The exact length of time depends upon the species and cultivar used, the soil, the cli-

FIGURE 9-17
Sod on a forklift.
(Courtesy of Brouwer Turf Equipment, Ltd.)

FIGURE 9-18
A small sod cutter which harvests 12-inch (0.3-meter) wide pieces.

mate, the sod production method, and the maintenance program. The turf must be dense and strong before it can be harvested. Sod strength is due to rhizomes, stolons, and roots knitting together with the soil. This tensile strength is very important because weak sod will pull apart and tear when handled, resulting in a sodding job of inferior quality. Species that have aggressive rhizomes and stolons, such as Kentucky bluegrass and bermudagrass, can be harvested in a shorter time than slower-spreading species. Sod production is also discussed in Chapter 21.

Sod is harvested with a mechanical sod cutter, which has a cutter knife that slices through the soil (Figure 9-18). The width of the pieces usually varies from 12 to 24 inches (0.3 to 0.6 meter), depending upon the type of machine used. The more sophisticated machines used by large sod growers can cut the pieces to any length desired and then roll or fold and stack the sod in one operation (Figure 9-19).

The surface soil should be moist when the sod is harvested. It is difficult to cut sod if the soil is dry. The cutting depth is quite important. Thin-cut sod is generally preferable—the thinner the sod, the quicker it roots. Thick sod has a greater amount of soil and roots and is more drought tolerant. Thin sod has such a limited root system that its roots must grow at a faster rate and quickly enter the underlying soil. Thick sod is also heavier and more tiring to install. It is best to cut the sod deeper than normal if the turf manager is not able to

FIGURE 9-19
A large sod harvester. *(Courtesy of Brouwer Turf Equipment, Ltd.)*

water it as much as he or she thinks is necessary. Usually the soil attached to the sod should be 0.3 to 1 inch (0.8 to 2.5 centimeters) deep. However, the sod should not be cut so thin that it falls apart easily. Cutting sod thinner also benefits the sod grower. The less soil that is removed at each harvest, the greater the number of years that the field can be used to produce sod.

The turf manager should inspect sod carefully before purchasing it. The sod should be strong, uniform, dense, free of insects, diseases, and weeds, composed of appropriate species and cultivars, and fresh. Sod must be installed within 24 to 48 hours after it is harvested. When sod remains stacked on pallets for longer periods its temperature increases because of the heat given off during respiration. High temperatures cause desiccation and can injure the sod. If the turf manager is unable to install the sod immediately after purchase, it should be stored in a cool place and be kept moist.

Before laying sod, the planting site should be irrigated. The rooting rate is increased significantly if the underlying soil is moistened before installation. The sod is then normally placed on the site by hand. A faster system involves unrolling long pieces of sod several feet wide from a bar attached to a tractor. The strips are fitted tightly together in a staggered manner so none of the ends of adjacent pieces are in line. It is important to handle the sod gently enough to avoid stretching or tearing the pieces. They should be rolled or folded when

FIGURE 9-20
Tamping sod.

carried, rather than held by one end. If the sod is stretched, it will shrink as it dries, leaving gaps between the pieces.

On a steep slope, wooden pegs can be hammered through the sod to anchor the strips until the roots knit into the soil. All sod should be tamped or rolled after installation to improve the contact between the sod and the underlying soil (Figure 9-20).

The last step is irrigation. One good soaking per day during the first week is recommended. Sod usually roots into the underlying soil in one to two weeks. Irrigation frequency can be decreased once this occurs.

A recent innovation is washed sod. Water is used to wash the soil from the sod. This soil-less sod is lighter and quicker rooting than the conventional type. It also eliminates potential layering problems. When placed on top of a sand base, the soil layer on the sod may interfere with the downward movement of water and roots. This is because water and roots sometimes have trouble moving from a finer-textured layer into a coarser layer.

Washed sod is more expensive than conventional sod and requires more irrigation when first installed.

Sprigging

Sprigging is one of the methods of vegetative establishment used to establish turfgrasses that produce poor quality seed or not enough seed to make seeding

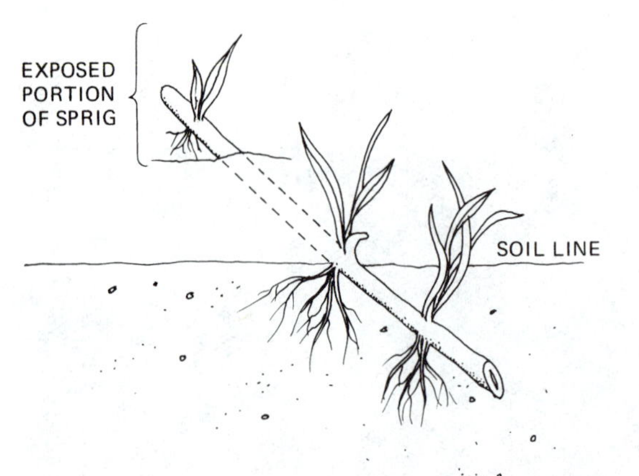

FIGURE 9-21
Part of the sprig should be exposed after planting. Leaves and roots grow from each node on the sprig.

practical. Examples are zoysiagrass, hybrid bermudagrasses, and some cultivars of creeping bentgrass. Stoloniferous warm season grasses are commonly propagated by sprigging. The turf manager should buy sprigs that are certified to be genetically pure or true to type.

One type of sprigging involves the planting of pieces of stolons or rhizomes in holes or furrows. These pieces are machine-cut into small lengths, usually 3 to 6 inches (7.6 to 15.2 centimeters) long. Each sprig should have at least two nodes. After planting, shoots and roots are produced at the nodes, and eventually the new plants spread and cover the site. The pieces of stolons and rhizomes are called sprigs.

Sprigs can be planted by machine or hand. The first step is to make furrows (trenches) in the soil 6 to 18 inches (0.15 to 0.45 meter) apart and 1 to 3 inches (2.5 to 7.6 centimeters) deep. The sprigs are soaked in water and then planted 4 to 6 inches (10.2 to 15.2 centimeters) apart in the furrows. For the most rapid establishment, sprigs are placed end to end in rows which are 6 inches (15.2 centimeters) apart. The soil is then pushed into the furrow and firmed around the sprig with a roller. The sprigs should not be covered completely when the furrows are filled in. Part of the sprig, approximately 25 percent of its length, should remain above the soil surface (Figure 9-21). This encourages topgrowth.

Using a machine to sprig enables the turf manager to complete all of the planting steps in one operation (Figure 9-22). One to four bushels of sprigs are required per 1,000 ft^2 (93 m^2), depending upon the spacing. Frequent irrigation is necessary after planting. Once the sprigs start to grow, a light topdressing of soil encourages the stolons to spread.

Another method of sprigging is to insert the sprigs into the soil without making furrows. A special tool that has a notched, blunt tip can be used to push sprigs into loose soil. One end of the sprig is pressed into the soil; the other remains above the surface exposed to light.

Sprigs are often broadcast over the entire site onto a moist soil surface (Figure 9-23). The stolons are then partially covered by rolling or disking them

FIGURE 9-22

A machine that plants sprigs in one operation. *(Courtesy of U.S. Department of Agriculture)*

FIGURE 9-23

Spreading sprigs by hand over the planting site. *(Courtesy of Art Wick, Lesco Products)*

FIGURE 9-24
A machine implanting sprigs. *(Courtesy of Art Wick, Lesco Products)*

into the soil, or by topdressing. This technique is called stolonizing by some turf managers. Machines can be used to spread the stolon pieces and press them into the soil (Figure 9-24). A layer of topdressing 0.25 inch (6.4 millimeters) deep is very effective. A portion of the stolon should not be covered.

Hydrosprigging is another method. The sprigs are broadcast and then covered with mulch applied by a hydromulcher.

Sprigs can be purchased by the bushel or obtained by buying sod and pulling it apart into separate sprigs. This can be accomplished by hand or with a shredding machine. Shredding one square yard of dense sod produces approximately one bushel of sprigs.

Normally 5 to 10 bushels of stolons per 1,000 ft^2 (93 m^2) are required. The lower rates are used on lawns, athletic fields, and fairways. The higher rates are used on golf course greens and tees. Broadcasting sprigs results in faster establishment than planting sprigs because of the higher stolon application rate (Figure 9-25). However, the loss of stolons due to desiccation is more likely with stolonizing. Mortality can be decreased by spreading sprigs over small areas and then topdressing immediately before broadcasting further stolons. Frequent irrigation is necessary to ensure successful establishment.

Plugging

Plugging is the planting of plugs of mature turf into the soil. These small pieces of sod may be round or square and are usually 2 to 4 inches (5.1 to

FIGURE 9-25

An example of a newly rooted sprig pulled from a newly established area. *(Courtesy of William Knoop)*

10.2 centimeters) wide and 0.75 to 4 inches (1.9 to 10.2 centimeters) deep (Figure 9-26). Plugs are produced by cutting sod into appropriate-sized pieces or by removing plugs from an established turf. One square yard of dense sod provides 324 2-inch (5.1-centimeter) diameter plugs or 84 4-inch (10.2-centimeter) plugs. Zoysiagrass, St. Augustinegrass, centipedegrass, and buffalograss are examples of species that may be established by plugging. Normally

FIGURE 9-26

A zoysiagrass plug.

FIGURE 9-27
Planting plugs by hand.

only stoloniferous grasses are propagated with plugs. It is the most common method of establishing zoysiagrass.

Plugging can be performed with a machine or by hand (Figure 9-27). The plugs are placed in holes on 6- to 18-inch (0.15- to 0.46-meter) centers, (Figure 9-28). Two-inch diameter plugs are usually planted on 6-inch centers, while larger plugs are placed farther apart. The wider the plug and the shorter the distance between plugs, the more rapidly the site is covered with a dense turf. The site should be rolled after planting. Large areas are often plugged mechani-

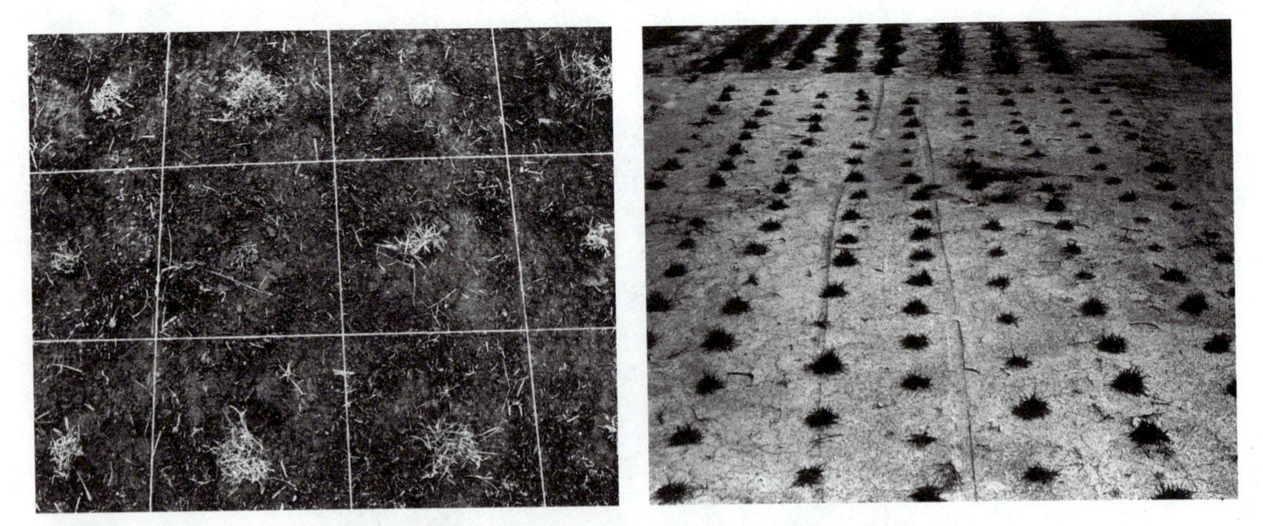

FIGURE 9-28 Establishment by plugging. *(Courtesy of U.S. Department of Agriculture)*

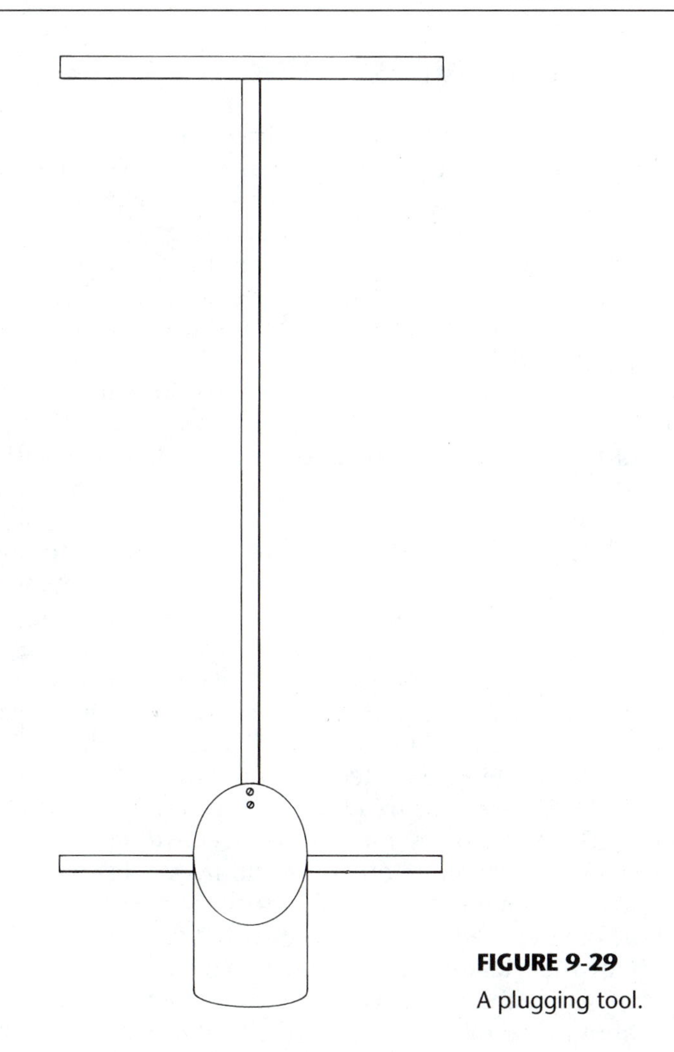

FIGURE 9-29
A plugging tool.

cally, but small areas can be done manually. A plugging tool similar to a cup cutter or large bulb planter is used (Figure 9-29).

Plugging is the slowest method of establishment. Zoysiagrass, the major species propagated by plugging, may take two years to fill in completely between the plugs unless they are planted close together. Zoysia usually spreads about 6 inches (15.2 centimeters) per growing season.

Plugs should be irrigated frequently until the roots have spread into adjacent soil. The site should be rolled again one or two weeks after planting in case the plugs have "floated" up out of the ground because of the irrigation.

WINTER OVERSEEDING

Winter overseeding is the practice of seeding cool season grasses "over" or into a warm season turf before the existing grass becomes dormant because of cold temperatures. The cool season grasses provide a green, vigorous, wear-tolerant turf during winter periods, when the permanent grass grows slowly or turns brown. In the spring, with the return of temperatures favorable to the warm season turf, the temporary grasses thin out and disappear. Overseeding is most common on golf course greens and tees, but also occurs on fairways, lawns, and athletic fields.

Many cool season species are used for winter overseeding. Annual rye has been widely used because of its low seed cost and rapid germination rate, but turf-type perennial ryegrasses produce a better-quality turf and are quite popular today. Perennial ryegrass, *Poa trivialis*, or creeping bentgrass are used on golf greens.

The best date for overseeding varies greatly throughout the South. It may occur as early as September 15 in northern areas of the warm season zone and in November or later in southern areas. Proper timing is important because the permanent grasses are still too competitive and disease problems too severe if overseeding is performed early. If overseeding occurs late, the cooler soil temperatures will result in slower seed germination. Overseeding is usually performed 20 to 60 days before the first expected killing frost. Many turf managers overseed when the soil temperature 4 inches (10 centimeters) deep is in the mid-70s (F) or when the midday air temperatures drop to the low 70s (F).

There are many methods of overseeding. Thatch control is important because seed germination and seedling establishment are difficult when a thick thatch layer is present. Also, nitrogen fertilization should be stopped four weeks before overseeding to slow down the growth of the warm season grass. Plant growth regulators can be applied to make the permanent grass less competitive. Generally the area is vertical-mowed (dethatched or verticut) in several directions before overseeding and then cut at a short height. After the seed is spread it is worked into the turf with a rake or drag mat. A heavy seed rate is necessary Topdressing—spreading a thin layer of soil over the area—increases seed germination. The area is irrigated frequently until the new grass is up. A less effective but common method is simply to spread the seed and water or rake it in.

Pest problems must be controlled. Diseases are a severe threat to the temporary grasses. Fungicide-treated seed is recommended, and fungicide applications may be necessary. A preemergence herbicide for annual bluegrass control may be applied two to three months before overseeding.

When temperatures warm up in the spring, the overseeded grasses begin to die out at the same time that the permanent warm season turf begins to green-up and grow. The decline of the cool season grasses can be encouraged by cutting back on fertilizer before the transition, and cutting the turf as close as possible without injuring the permanent grass. Irrigation may be decreased because warm season grasses often need less water than the cool season grasses.

SELF-EVALUATION

1. Seedlings are sensitive to ___________.
2. A ___________ is used to firm the seedbed.
3. Wheat ___________ is a widely used mulching material.
4. To ensure successful establishment, frequent ___________ is required.
5. The planting site should not be too moist when the first ___________ occurs.
6. Slopes that are seeded should be ___________ to prevent the erosion of soil and seed.
7. Before installing sod, the soil should be ___________ to encourage faster rooting.
8. The ___________ sod is, the quicker it roots into the underlying soil.
9. Sod should be installed within ___________ hours after it is harvested.
10. Sprigs are small pieces of ___________.
11. A zoysiagrass turf is vegetatively established by planting ___________.
12. Vegetative propagation is commonly necessary when a species is a poor ___________ producer.
13. One square yard of dense sod provides approximately ___________ bushel(s) of sprigs.
14. ___________ is often performed on a turf area before winter overseeding occurs.
15. What important information is found on a seed label?
16. Discuss the advantages and disadvantages of establishment by sodding.

CHAPTER

10

Fertilization

OBJECTIVES

After studying this chapter, the student should be able to

- Identify the nutrients required by turfgrass plants
- Discuss the different types of fertilizers
- List the factors that influence the selection of an appropriate fertility program
- Explain why nitrogen is the key nutrient in a turfgrass fertility program
- Distinguish between fast release and slow release nitrogen carriers
- Discuss the rate and frequency of fertilizer application for turfgrass
- Describe methods of fertilizer application

INTRODUCTION

As explained in Chapter 7, turfgrass plants require 16 essential elements to grow and complete their life cycle. These nutrients are divided into three groups based on the amounts of each needed by plants (Figure 10-1). Carbon, hydrogen, and oxygen are obtained from air or water and are readily available to turfgrass. The other 13 elements are removed from the soil by plant roots.

The three elements that receive the greatest attention from turfgrass managers are nitrogen, phosphorus, and potassium because they must be added to the soil at regular intervals. They are referred to as primary nutrients because turfgrass requires larger quantities of nitrogen, phosphorus, and potassium than it does of the other ten elements obtained from the soil. They are also called fertilizer nutrients because of their presence in most fertilizers. Of the three, nitrogen is most likely to be deficient in the soil.

Calcium, magnesium, and sulfur are called secondary elements. Calcium and magnesium are normally supplied to the soil by the application of liming materials. Sulfur is added to the soil by using fertilizers that contain the nutrient or acidifying materials for lowering soil pH such as elemental sulfur and

OBTAINED FROM AIR AND WATER

Carbon
Hydrogen
Oxygen

OBTAINED FROM THE SOIL

	Element	Chemical Symbol	Available Form
Primary (fertilizer) nutrients	Nitrogen	N	NO_3^-, NH_4^+
	Phosphorus	P	$H_2PO_4^-$, HPO_4^{--}
	Potassium	K	K^+
Secondary nutrients	Calcium	Ca	Ca^{++}
	Magnesium	Mg	Mg^{++}
	Sulfur	S	SO_4^{--}
Minor or micronutrients	Iron	Fe	Fe^{++}, Fe^{+++}
	Manganese	Mn	Mn^{++}
	Copper	Cu	Cu^{++}
	Boron	B	$H_2BO_3^-$ and others
	Zinc	Zn	Zn^{++}
	Chlorine	Cl	Cl^-
	Molybdenum	Mo	MoO_4^{--}

FIGURE 10-1
Essential elements required by turfgrasses.

aluminum sulfate. Some sulfur ends up in the soil because of air pollutants (sulfur dioxide) or the application of sulfur-containing pesticides.

Micronutrients or *trace elements* are materials required in very small amounts by plants. These minor nutrients are found in plants in quantities measured in parts per million. However, they are just as essential to turfgrass growth as the primary nutrients, no matter how minute the amount required by the plant (Figure 10-2). In many areas of the United States it is not necessary to apply micronutrients. With the exception of iron, micronutrient deficiencies are rare.

FERTILIZER

Fertilization is the practice by which nutrients are supplied for plant growth. Most fertilizers are applied in a dry form with a spreader, but occasionally nu-

ELEMENT	CONCENTRATION IN DRY TISSUE	DEFICIENCY SYMPTOMS
Nitrogen	2.5–6.0%	Older leaves yellow-green, reduced shoot growth
Potassium	1.0–4.0%	Interveinal yellowing especially on older leaves, leaf tips and margins scorched
Phosphorus	0.2–0.6%	Older leaves dark green first, then appear purple or reddish
Calcium	0.2–1.0%	Deficiency rare, new leaves reddish-brown and stunted
Magnesium	0.1–0.5%	Interveinal chlorosis, a striped appearance, cherry-red margins
Sulfur	0.2–0.6%	Yellowing of older leaves
Iron	50–500 ppm	Interveinal yellowing of new leaves
Manganese	Very small amounts	Rare, similar to iron deficiency
Copper	Very small amounts	Never a problem
Zinc	Very small amounts	Rare, growth stunted, thin and shriveled leaves, appears desiccated
Boron	Very small amounts	Rare, cholorotic, stunted growth
Molybdenum	Very small amounts	Rare, older leaves pale green
Chlorine	Very small amounts	Never a problem

FIGURE 10-2

Relative concentrations of the essential elements in turfgrass plants and nutrient deficiency symptoms.

FIGURE 10-3

A large pendulum-type spreader used for dry application of granular fertilizers. *(Courtesy of Vicon Farm Machinery, Inc.)*

To convert P_2O_5 to P multiply by 0.44
To convert P to P_2O_5 multiply by 2.29
To convert K_2O to K multiply by 0.83
To convert K to K_2O multiply by 1.2

FIGURE 10-4
Useful fertilizer conversions.

trients are mixed with water and sprayed on turf (Figure 10-3). This latter practice is commonly performed by many lawn care services. Most fertilizers contain nitrogen, phosphorus, and potassium and are called complete fertilizers. If, for some reason, it is not necessary to apply one of these primary nutrients, the turf manager may use a fertilizer that contains the other two. Individual carriers that supply nitrogen, phosphorus, or potassium alone are also available. In regions where secondary or micronutrients are deficient in the soil, fertilizers often contain these elements as well.

The fertilizer analysis states the percentage of nutrients in a fertilizer. Generally the analysis percentages are rounded down to whole numbers and called the grade. The *grade* is the minimum guaranteed analysis of a fertilizer. The three numbers on a bag of fertilizer represent its grade. The first number in the sequence indicates the percentage of elemental nitrogen (N), the second is available phosphoric acid (P_2O_5), and the third is soluble potash (K_2O).

A 20-5-10 fertilizer contains by weight 20 percent nitrogen, 5 percent P_2O_5, and 10 percent K_2O. A 50-pound bag of this fertilizer contains 10 pounds of nitrogen (0.20×50), 2.5 pounds of P_2O_5 (0.05×50), and 5 pounds of K_2O (0.10×50). The actual amounts of elemental phosphorus and potassium are 44 percent and 83 percent of the grade numbers. For example, 4.4 percent of the weight of a bag of 30-10-10 fertilizer 0.44×0.10) is elemental phosphorus (Figure 10-4).

Ratio is another important fertilizer term. A 30-10-10 fertilizer contains three parts of nitrogen for each one part of P_2O_5 and K_2O. It is said to have a 3:1:1 ratio. The 20-5-10 fertilizer mentioned earlier has a 4:1:2 ratio. Rather than recommending a specific grade, turfgrass specialists often suggest using a fertilizer that has a certain ratio. If a 2:1:1 ratio is recommended, for example, the turf manager may choose a 20-10-10, 10-5-5, 14-7-7, or 18-9-9 fertilizer.

The correct fertilizer ratio depends on the quantity of available nutrients in the soil. The quantities of phosphorus and potassium that are available for plant use can be accurately determined by testing the soil. As discussed in Chapter 8, soil tests results enable the turf manager to develop the most effective and efficient fertility program. The quantity of essential elements in the soil depends upon the amount and type of fertilizer applied previously, the level of nutrients that occur naturally in the soil, and the extent of nutrient losses from the soil.

Nutrient Losses

Losses are due to four causes. A large amount of nutrients is removed from the soil by the plants themselves. If grass clippings are allowed to remain on the turf, then most of these nutrients are returned to the soil and can be used by the turfgrass again. However, when the clippings are collected and removed from the site, the nutrients are no longer available to the plants. Some nutrients are converted to gaseous forms and diffuse into the atmosphere. Leaching by water from rain or irrigation results in nutrients being moved below the reach of turfgrass roots. A further, temporary loss occurs when nutrients become "fixed" in the soil. Because of various chemical reactions they are converted to insoluble, unavailable forms.

Effect of Maintenance on Nutrition Program

The level or degree of turfgrass maintenance has a major effect on the nutrition program. Low-maintenance, utility sites usually receive minimal fertilization because quality is relatively unimportant. One light fertilizer application per year may be sufficient. The turf manager desires a vigorous, dark green, dense turf on a higher-maintenance site. Better-quality turf can only be achieved by supplying greater amounts of fertilizer.

Unsatisfactory turfgrass is often the direct result of insufficient fertilization. The quality of many turf areas could be improved dramatically by increasing the fertilization rate and number of fertilizer applications. There has been a tendency to supply smaller amounts of nutrients to turfgrass in recent years, primarily because of the higher price of fertilizer. The greater expense is the result of increased energy costs for the manufacture of nitrogen carriers and the increasing scarcity of phosphorus. Reduced fertilization is beneficial when the manager applied too much nitrogen in the past and encouraged problems such as greater disease incidence. However, the turf manager must be careful not to go to the opposite extreme and starve his or her turfgrass. Despite the higher costs, fertilizer is still the best investment dollar for dollar in a successful turf management program.

NITROGEN

Nitrogen is the key nutrient in a turfgrass fertility program. With the exception of carbon, hydrogen, and oxygen, plants require more nitrogen than any other essential element. On a dry weight basis, a healthy turfgrass plant is composed of 3 to 5 percent nitrogen. This nutrient is more likely to be deficient than the other 15 essential elements. There are several reasons for this. Abundant levels of nitrogen are not normally found in soils. When nitrogen-containing fertilizer is applied to the soil, significant quantities of this nutrient may be lost because of leaching and volatilization. As mentioned in Chapter 7, the nitrate ion (NO_3^-) is the chemical form most commonly used by plants. It is very susceptible to leaching because it has a negative charge and is not stored in the soil on cation exchange sites. Volatilization is the loss of nitrogen to the atmosphere in a gaseous form.

Nitrogen losses are especially severe when the turfgrass is growing on a sandy soil and irrigation occurs frequently. Sandy soils are very prone to leaching, and heavy irrigation results in a larger amount of nitrogen being washed down beneath the root zone. Clipping removal also leads to nitrogen depletion.

Nitrogen has many important functions in a turfgrass plant. It is a component of chlorophyll, proteins, amino acids, enzymes, and numerous other plant substances. The effects of nitrogen fertilization are readily seen. Shortly after the fertilizer is applied the plants turn a darker green color and vertical shoot growth increases significantly.

Because nitrogen is the key nutrient in turfgrass nutrition, fertility programs are normally expressed in terms of how many pounds of nitrogen per 1,000 ft^2 (kilograms per 93 m^2) are applied. Unfortunately, soil tests are not particularly helpful in determining nitrogen fertilization rates. (The reasons for this have been discussed in detail in Chapter 8.) Consequently, unless tissue testing is performed, the turf manager must base his or her nitrogen fertility program on visual observations and general recommendations from local turfgrass specialists.

The need to apply nitrogen can be determined by two quality indicators—color and density. The nutrient is an important component of the chlorophyll molecule, and nitrogen-deficient plants turn a yellowish-green color. This condition is known as *chlorosis*. When grass becomes chlorotic and loses its desirable green color, the manager may want to apply fertilizer. However, before fertilizing, he or she should be certain that the turf has not turned color because of environmental stress or pest problems. It is also important to remember that turf does not have to be dark green to be healthy.

Density is the most important indicator. A thin, open turf populated with weeds is often caused by inadequate nitrogen fertility. A third indicator sometimes used to assess nitrogen levels is clipping yield. Low levels of available nitrogen result in a slower vertical growth rate and a reduced mowing frequency.

Nitrogen fertilization, based on these quality indicators, is greatly influenced by the degree of turf quality that is desired. A low-maintenance area may receive as little as 1 pound of nitrogen per 1,000 ft^2 a year, while the rate for a high-quality bermudagrass turf may be as much as 12 to 16 pounds of nitrogen per 1,000 ft^2 a year. Other factors that affect how much nitrogen is applied include the soil texture, the amount of irrigation, whether clippings are removed, the type of nitrogen source used, the length of the growing season, the species and cultivars that compose the turf, the degree of shading, and the size of the maintenance budget. A sports turf usually requires extra nitrogen because the grass must be aggressive. Foliar analysis (tissue testing) can be used to establish fertilizer needs. In most cases turf managers base their nitrogen rates on the appearance and performance of the turf.

Overfertilization can be just as detrimental as not applying enough nitrogen. Excessive rates cause physiological changes in a plant such as thinner cell walls, more tender, succulent tissue, and reduced food reserves. These changes result in decreased heat, drought, cold, and wear tolerance. Disease resistance is also diminished. The application of too much nitrogen significantly increases the need for mowing because of the rapid shoot growth it stimulates.

The nitrogen found in most fertilizers is produced by nitrogen in the atmosphere reacting with natural gas (methane). The result of this reaction is the formation of ammonia, which is then combined with other chemicals such as nitric acid, sulfuric acid, phosphoric acid, and carbon dioxide. The end products of these reactions are four common nitrogen carriers—ammonium nitrate (33-0-0), ammonium sulfate (20-0-0), ammonium phosphate (11-48-0, 20-50-0), and urea (45-0-0). The form of nitrogen in these carriers is 100 percent water soluble and is immediately available for plant use. These carriers are said to be quickly available or fast-release sources of nitrogen. They are widely used on many agricultural crops.

Water-soluble nitrogen can be absorbed by turfgrass roots as long as there is sufficient moisture in the soil. Such high solubility has both advantages and disadvantages. Turfgrass responds rapidly after their application but the response is short term. This is because much of the nitrogen is immediately taken up by plants, leached below the root zone, or lost to the atmosphere as a gas. These forms of nitrogen are less expensive than more complex, insoluble carriers. Fast-release nitrogen is not greatly affected by soil temperature. However, it has a high burn potential and can injure turfgrass if applied incorrectly.

Some of the disadvantages associated with water-soluble carriers post special problems for the turf manager. Turfgrass is one of the few crops that has fertilizer applied directly onto its foliage. This increases the likelihood of foliar burn. The short-term plant response characteristic of these carriers means that frequent fertilization may be required. Another problem is the inefficiency of the water-soluble forms. Nitrogen is wasted because the turfgrass plants are able to take up more nitrogen than they need and leaching and gaseous losses may be significant. This can be a serious problem if the nitrates leach down into the groundwater.

Slowly available nitrogen carriers were developed to solve the problems associated with the quickly available forms. They are also referred to as slow-release or controlled-release. A certain percentage of the nitrogen in these carriers is not immediately soluble in water and therefore not initially available for plant use. The result is a lower burn potential, a long-term plant response, and greater efficiency (Figure 10-5). A steadier release pattern reduces the likelihood of plants absorbing more nitrogen than they need and decreases leaching losses. Slowly available nitrogen carriers are more expensive than soluble forms because the manufacturing process is more costly.

Ureaformaldehyde (UF), sulfur-coated urea (SCU), and isobutylidene-diurea (IBDU) are examples of slowly available nitrogen carriers. All are pro-

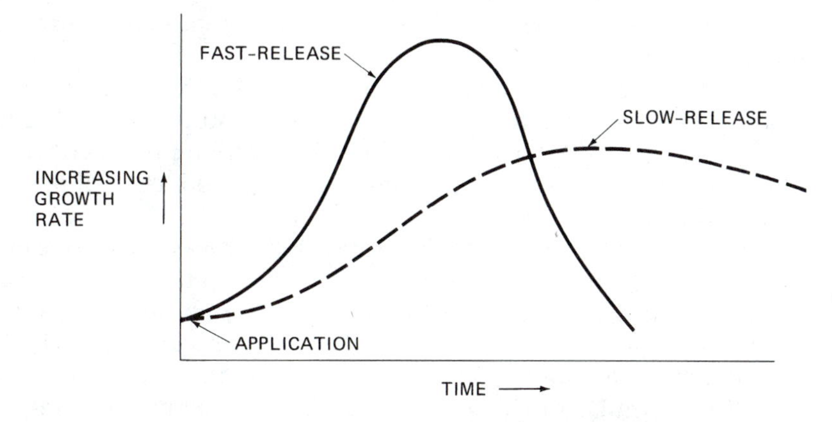

FIGURE 10-5

Typical release patterns of fast-release and slow-release fertilizers. Grass fertilized with the slow-release material exhibits a more uniform growth response.

duced by chemical processes which result in a certain percentage of the nitrogen in urea becoming temporarily unavailable for plant use.

Ureaformaldehyde is synthesized by combining urea and formaldehyde. When a urea:formaldehyde (U:F) ratio of 1.3:1.0 is used, the resulting product has approximately 67 percent slowly soluble nitrogen. The other 33 percent is called cold-water-soluble nitrogen (CWSN). It is composed of unreacted urea and low molecular weight, short-chain methylene ureas that immediately provide nitrogen to turfgrasses. The rest of the nitrogen does not become available until the larger molecules are broken down into smaller units by soil microorganisms, freeing the urea. The cold-water-insoluble nitrogen (CWIN) that is soluble in hot water is intermediate or moderately slow release, and the fraction that is insoluble in hot water (HWIN) is very slowly soluble.

One problem with UF is its dependence on higher temperatures for adequate nitrogen solubility. Microbial degradation of the slowly soluble forms is minimal at cool soil temperatures. Consequently, nitrogen availability is significantly reduced, and enough nitrogen may not be supplied by this carrier at certain times of the year. The activity index (AI) is a measure of relative solubility. It is the percentage of cold-water-insoluble nitrogen (CWIN) that is soluble in hot (212°F; 100°C) water. The higher the AI, the more rapidly the slowly available nitrogen becomes soluble. A UF material should have an AI of at least 40 percent to ensure that sufficient quantities of nitrogen will be supplied to turfgrass during the year following application.

The solubility of a UF fertilizer is controlled by the U:F ratio. A 1.9:1.0 ratio product has 67 percent CWSN and provides a much greater quantity of soluble nitrogen.

Sulfur-coated urea is simply urea granules that have been sprayed with sulfur and wax. The urea is water soluble, but the sulfur coat produces an insoluble barrier. The granule is also usually coated with a thin layer of wax to seal any cracks or defects in the sulfur coat. Eventually the coating is burst open because of the diffusion of water into the granule, and the nitrogen is released.

The coating thickness varies among particles. Those that have a thin or imperfect coat will release first. Approximately 30 percent of the nitrogen in most SCU products is quickly available. By blending particles with thin, medium, and thick coats, manufactures can make products that release nitrogen steadily for 10 to 15 weeks.

Sulfur-coated urea is popular because it is the least expensive of the slowly available nitrogen sources. However, problems can result if the fertilizer is damaged during handling. If the sulfur coats are cracked, the particles are no longer slow release. One method of preventing this and improving the overall performance of SCU is to add a polymer coat on the outside of the sulfur coat.

IBDU is similar to UF but is minimally affected by soil temperature. The nitrogen dissolves in water slowly, but does not depend upon microbial decomposition. Particle size has the major effect on solubility. Smaller particles dissolve more readily and release soluble nitrogen faster than larger particles. A blend of different-sized particles results in a fertilizer that releases nitrogen relatively uniformly for three or four months. However, high soil pH can impede nitrogen release.

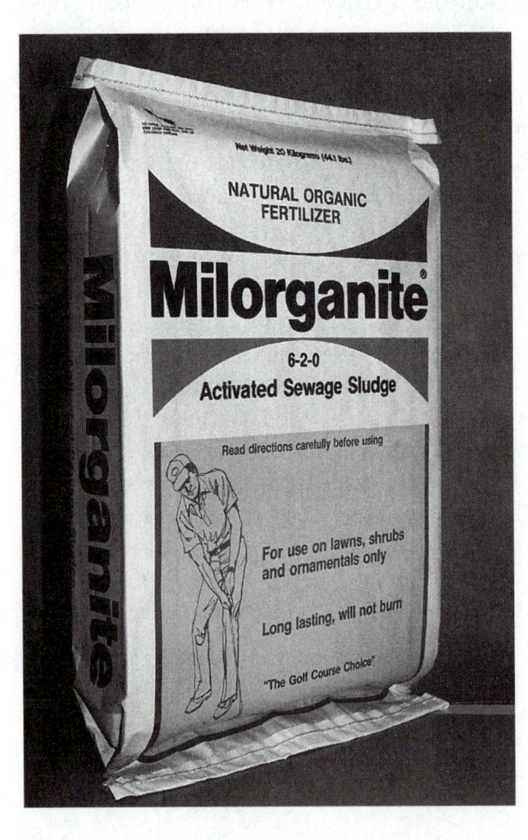

FIGURE 10-6

Milorganite is a natural organic fertilizer. *(Courtesy of Milwaukee Metropolitan Sewerage District)*

Another method of producing slowly available nitrogen is to surround urea with a resin, polymer, or plastic coating. These ultra-thin coats are tough and durable. Moisture is absorbed by the coating and dissolves the urea; then the solution diffuses out into the soil. Most of these products are relatively new. They have a very uniform release rate, which depends primarily on the thickness of the coating. Organic matter releases nitrogen gradually because the nutrient is tied up in complex molecules. The nitrogen becomes available after microorganisms break down the complex molecules into simpler forms.

Milorganite is a natural organic fertilizer produced from sewage sludge by the Milwaukee Sewage Commission (Figure 10-6). Other products are derived from dried blood, bone and seed meal, fish scraps, poultry manure, and compost. Besides being slowly available nitrogen sources, natural organic fertilizers also provide the other essential nutrients and have been shown to suppress turf diseases.

Fast-release and slow-release nitrogen carriers have advantages and disadvantages. Many companies that manufacture turfgrass fertilizers mix soluble and insoluble carriers to ensure both a rapid and a long-term response (Figure 10-7).

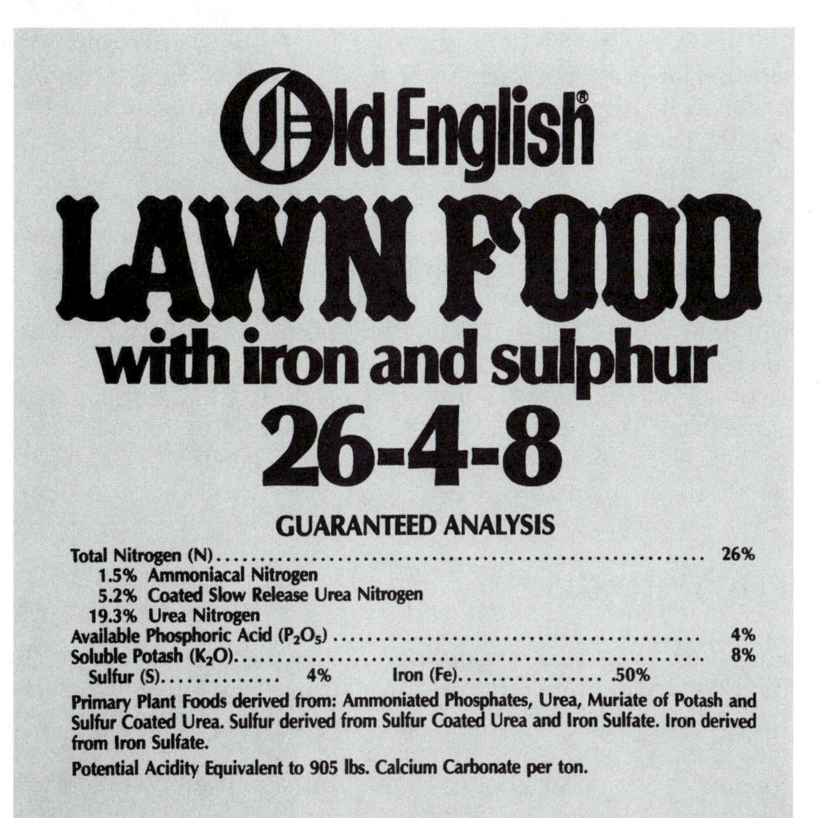

FIGURE 10-7

This label shows a combination of quickly available and slowly available nitrogen. *(Courtesy of Stanford Seed Co.)*

FIGURE 10-8

Salt index values for various chemicals applied to turfgrass. The higher the salt index value, the higher the burn potential.

CHEMICAL	SALT INDEX VALUE	CHEMICAL	SALT INDEX VALUE
Muriate of potash	114	Ureaformaldehyde	10
Ammonium nitrate	105	Gypsum	8
Sodium nitrate	100	Superphosphate	8
Urea	75	IBDU	5
Ammonium sulfate	69	Milorganite	4
Potassium sulfate	46	Dolomitic limestone	1
Magnesium sulfate	44		

POTASSIUM

Potassium has numerous functions in a plant. It counteracts many of the negative effects of nitrogen such as decreased plant tolerance to cold, heat, drought, and diseases. If turfgrass is supplied with adequate amounts of potassium, the ability of the plants to tolerate stress is increased.

Muriate of potash (potassium chloride) is the carrier commonly added to a fertilizer. It has a high burn potential (Figure 10-8). Potassium sulfate has a lower burn potential and releases potassium more gradually than muriate of potash. Though more expensive, potassium sulfate is the best choice for quality turf areas.

Potassium is susceptible to leaching. The problem, however, is more severe with nitrogen. Plants will absorb significantly greater amounts of potassium than they require. This is known as "luxury consumption." Large quantities of potassium may be removed if clippings are collected. This is more of a problem with muriate of potash than it is with potassium sulfate. Soil tests can be used to estimate potassium needs. Turf fertilizers with a 2:1 $N:K_2O$ ratio are often recommended unless soil test results indicate that less potassium is needed. Some turf managers use 1:1 ratios in an attempt to increase drought tolerance of their grass. Potassium is very important in regulating water absorption and retention in a plant.

PHOSPHORUS

Phosphorus has many important roles in a plant. A prominent characteristic is its ability to promote rooting. Adequate levels of phosphorus are crucial at the time of establishment. The nutrient should be incorporated into the soil before planting, and then, after planting, the new seedlings or sod should be topdressed with phosphorus fertilizer. All application should be based on soil test results.

The major problem with phosphorus is its "fixation" in the soil. At a soil pH below 6.0 and above 7.0 phosphorus becomes tied up in unavailable, insoluble forms. Maintaining the correct soil pH increases phosphorus availability. Superphosphate was for many years the primary phosphorus carrier. Today, ammonium phosphates are used in most fertilizers (Table 10-1).

TABLE 10-1 Characteristics of Nutrient Sources

	Approximate Nutrient Percentage			
Source	**N**	**P_2O_5**	**K_2O**	**Comments**
Ammonium nitrate	33	0	0	Fast-release nitrogen source
Ammonium sulfate	21	0	0	Fast release, strong acidifying effect, contains 24% sulfur
Urea	45	0	0	Fast release
IBDU (isobutylidene-diurea)	31	0	0	Slow release
Urea formaldehyde	38	0	0	Slow release
Methylene urea	38	0	0	Similar to UF but has more water-soluble nitrogen
Sulfur-coated urea	32	0	0	Slow release
Polymer-coated urea	39–44	0	0	Slow release
Milorganite	6	2	0	Activated sewage sludge, slow release
Ringer	6	1	3	Blood, bone, and seed meals, slow release
Sustane	5	2	4	Composted turkey waste, slow release
Monammonium phosphate	11	48	0	Phosphorus source in many fertilizers
Diammonium phosphate	20	50	0	Higher N than monoammonium phosphate
Superphosphate	0	20	0	Contains calcium and sulfur
Triple (treble) superphosphate	0	45	0	Contains calcium and sulfur
Potassium chloride (muriate of potash)	0	0	60	Most common potassium source, high burn potential
Potassium sulfate	0	0	50	Acidifying effect
Potassium nitrate	13	0	44	
Ferrous sulfate	0	0	0	Contains 20% iron and 18% sulfur, usually foliarly applied
Ferrous ammonium sulfate	7	0	0	Contains 14% iron and 15% sulfur, usually foliarly applied
Chelated iron	0	0	0	Contains 6–7% iron, longer residual response than the other sources

FERTILIZER PROGRAMS

Fertilizer application rates and the number of applications are determined by many factors, such as the desired level of quality, weather conditions, the length of the growing season, soil texture, the amount of irrigation provided, and whether clippings are removed. Environmental conditions affect the fertilizer program. Shaded turf, because it grows slower, is usually fertilized less than grass growing in the full sun. Turf use influences fertility requirements. An athletic field often needs more fertilizer than a lawn because sports turf has to recover from wear injury.

The species and cultivars that compose the turf have a major effect on fertilizer programs. Creeping bentgrass and improved bermudagrass are considered heavy feeders because they grow vigorously and need abundant amounts of nutrients to support this growth. Grasses that have a slow growth rate such as the fine fescues, centipedegrass, and carpetgrass need significantly less fertilization (Figure 10-9). Soil test results help the turf manager to develop the program.

Application dates also depend on many of the same variables listed earlier.

Species	Pounds of Nitrogen Needed per 1,000 ft^2 per Growing Month	Species	Pounds of Nitrogen Needed per 1,000 ft^2 per Growing Month
Bahiagrass	0.1–0.4	Centipedegrass	0.1–0.3
Bentgrass		Fescue	
Colonial	0.5–1.0	Fine	0.1–0.4
Creeping	0.5–1.3	Tall	0.4–1.0
Bermudagrass		Grama, blue	0.1–0.3
Common	0.5–1.0	Ryegrass	
Improved	0.7–1.4	Annual	0.4–1.0
Bluegrass		Perennial	0.4–1.0
Annual	0.5–1.0	St. Augustinegrass	0.5–1.0
Kentucky	0.4–1.0	Wheatgrass, crested	0.2–0.5
Rough	0.4–1.0	Zoysiagrass	
Buffalograss	0.1–0.4	Japanese lawngrass	0.5–0.8
Carpetgrass	0.1–0.4	Manilagrass	0.5–0.8

FIGURE 10-9 A comparison of the amounts of nitrogen needed by each turfgrass species to produce a satisfactory turf. A range is given because the exact amount depends on variables such as the cultivar being grown, weather conditions, the desired level of quality, soil texture, the irrigation program, and whether clippings are removed.

Whether conditions have the greatest influence on timing. Fertilization should occur at the beginning of or during periods when temperature and moisture conditions favor active turfgrass growth. The grass needs the nutrients when it is growing vigorously. Fertilization should be avoided at times when environmental and disease stress occurs. For example, a midsummer fertilizer application can be detrimental to cool season turf because it results in decreased heat, drought, and disease tolerance.

The most important time to fertilize cool season grasses is in late summer. A late fall fertilization in November is also beneficial. The fertilizer is usually applied when the rate of shoot growth has slowed significantly but the grass is still green. It promotes root growth and earlier spring green-up. A third application in the spring is recommended as well for better-quality turf. Often a lower nitrogen rate is used at this time.

The most important time to fertilize warm season grasses is in late spring. A second application in the summer is recommended. Early spring and late summer fertilization may also be necessary (Table 10-2).

TABLE 10-2 Examples of Fertilizer Application Schedules Recommended in Four Regions of the United States

	Jan.	Feb.	Mar.	Apr.	May	June	July	Aug.	Sept.	Oct.	Nov.	Dec.
Central Texas Lawns												
Common bermudagrass				X		X		X	X			
Hybrid bermudagrass				X	X	X	X	X	X			
Buffalograss					X				X			
Tall fescue			X						X			
Southern California (minimum fertilization schedule that will produce an acceptable lawn)												
Warm season grass				X	X				X	X		
Cool season grass			X		X					X	X	
Southern Florida Bermudagrass lawn												
High maintenance	X	X	X	X	X	X	X	X	X	X	X	X
Low maintenance		X		X		X	X		X			
Carpetgrass lawn												
Higher maintenance		X				X			X			
Low maintenance			X									
New York State												
Good-quality lawn					X				X		X	
Low-maintenance lawn									X			

The most common fertilizer application rate is 1 pound of nitrogen per 1,000 ft^2 (2.2 kilograms per 93 m^2). Turf requires fewer applications when slow-release nitrogen sources are used rather than fast-release products. Because of less nutrient storage capacity and greater leaching potential, sandy soils are fertilized at lower rates and more frequently than loam and clay soils.

The turf manager should use soil test results and Cooperative Extension Service recommendations as the basis for a fertility program. Nitrogen, phosphorus, and potassium receive the most attention, but other essential elements should not be overlooked. Iron deficiencies occur, especially on soils that are alkaline, sandy, or high in organic matter (Figure 10-10). Sometimes the application of iron causes grass to become a darker green color. Nitrogen also improves color. However, iron will not stimulate the rapid shoot growth that nitrogen will. Consequently, iron can be used to give turf a better color without increasing its maintenance requirements or causing the grass to be less tolerant of environmental stress. Relatively specific fertilizer programs for golf courses, lawns, athletic fields, and other turf areas are discussed in Chapters 20 and 21.

FIGURE 10-10

Iron is added to fertilizers in regions where deficiencies may occur. *(Courtesy of Stanford Seed Co.)*

APPLICATION METHODS

Granular Fertilizers

The majority of the fertilizer used on turfgrass is in a dry or granular form. It is distributed with a drop, rotary, or pendulum type spreader. The drop, or gravity-type, spreader has a series of openings at the bottom of the hopper through which granular fertilizer drops a few inches to the ground directly beneath (Figure 10-11). The rate of application can be changed by adjusting the size of the openings. Drop spreaders provide a very precise, uniform distribution pattern.

The width of homeowner-type drop spreaders is usually 2 feet (0.6 meter) or less, but wider models are available (Figure 10-12). Drop spreaders are normally preferred for the application of fine or very light particles such as ground limestone or granular pesticides that must stick to the foliage. Too much overlapping or misses between application swaths can result in streaking because of uneven nitrogen distribution.

Rotary spreaders are also called centrifugal, broadcast, or cyclone spreaders. Most have a plate, called an impeller, which is attached beneath the hopper and spins as the spreader wheels turn. When fertilizer drops through the adjustable openings at the bottom of the hopper, it falls onto the rotating impeller and is thrown away from the spreader in a semicircular pattern (Figure 10-13).

Application is faster with a rotary spreader because it broadcasts granular

FIGURE 10-11
A drop-type spreader. *(Courtesy of Gandy Co.)*

FIGURE 10-12

A large drop spreader. *(Courtesy of Gandy Co.)*

FIGURE 10-13

A rotary spreader. *(Courtesy of OMC Lincoln)*

materials over a wider area than the drop type. The spreading width normally ranges from 6 feet (1.8 meters) for small spreaders to 60 feet (18.3 meters) for very large ones. Streaking is less likely with rotaries because the swaths are overlapped and the edge of the distribution pattern is not as sharp as that produced by a drop spreader.

A rotary spreader does not provide as accurate and uniform an application as a drop spreader, but the distribution can be quite satisfactory if the proper overlap is used. Spreading mixed materials of different sizes is a problem because larger, heavier granules are thrown farther than smaller, lighter particles. Foliar-applied granular pesticides are not usually spread with a rotary because of unsatisfactory distribution. Ground limestone will often drift when applied with a rotary spreader. The speed at which the spreader is pushed or driven has a major impact on application rate.

Pendulum-type spreaders have a spout that moves from side to side (Figure 10-3). They are pulled by a tractor or turf vehicle, have a large hopper capacity, and can throw dry materials a great distance when the spout moves rapidly.

Spreaders should be thoroughly cleaned after use. They must be accurately calibrated (openings set at the proper size) to ensure that the correct amount of material is applied. Spreader care and calibration are discussed in Appendix C.

Liquid Fertilizers

The application of fertilizer in a liquid form has become popular with many lawn care companies and some golf course superintendents. The fertilizer is sprayed on the turf and is often mixed with pesticides. Liquid application is usually less expensive than granular applications, though the initial cost of the sprayer equipment is quite high compared to the cost of a spreader. Generally 3 to 5 gallons (11.4 to 19.0 liters) of the fertilizer-water mixture are applied per 1,000 ft^2 to ensure that the fertilizer is washed into the root zone. Urea is the most widely used fertilizer material because it is soluble in water. Unfortunately, urea has a high burn potential and releases most of its nitrogen in a few weeks. Products are now available that have a lower burn potential and somewhat longer-lasting effects.

The application of small amounts of nutrients in low spray volumes (less than 0.5 gallon per 1,000 ft^2) directly to the foliage is called *foliar feeding*. Much of the fertilizer is absorbed by the turfgrass leaves. Foliar feeding is primarily used to supply micronutrients such as iron.

Fertigation is the application of nutrients through the irrigation system. Minute amounts of fertilizer are regularly metered into the irrigation lines and distributed along with the irrigation water through the sprinkler heads. The irrigation system must be capable of distributing water very uniformly. The advantages of fertigation include a more efficient plant use of nutrients, a steadier growth rate, and a savings on labor costs. Fertigation occurs on some golf courses, but is not yet widely used.

FERTILIZER BURN

Fertilizer burn can be a serious problem if fertilizers are applied improperly. Soluble, fast-release nutrient sources in both liquid and granular forms should be watered in following application. Irrigation moves the fertilizer off the foliage into the soil and prevents burning of the leaves. It does not occur after foliar feeding because very low fertilizer rates are applied. The foliage should be dry when granular fertilizers are applied. Fast-release nitrogen generally should not be applied at rates higher than 0.5 pound of nitrogen per 1,000 ft^2.

If an overapplication of fertilizer occurs, heavy irrigation is recommended. If a spill of granular fertilizer occurs, as much fertilizer as possible should be picked up before irrigating.

SELF-EVALUATION

1. The two nutrients generally supplied by liming materials are __________ and __________.
2. Nitrogen losses on a sandy soil can be severe because of __________.
3. The ratio of a 30-15-15 fertilizer is __________.
4. The second number in a fertilizer grade represents the percentage of __________ in the bag.
5. A 30-pound bag of 25-12-12 fertilizer contains __________ pound(s) of nitrogen.
6. Water-soluble nitrogen is said to be quickly __________.
7. SCU is the abbreviation for __________.
8. __________ is more temperature dependent than SCU and IBDU.
9. An example of a fast-release nitrogen carrier is __________.
10. __________ increases the ability of a plant to tolerate stress.
11. The major problem with __________ is its conversion to unavailable forms in the soil.
12. If cool season turfgrasses were to be fertilized only once a year, the best time would be in __________.
13. Sandy soils are fertilized at __________ rates and more __________ than loam or clay soils.
14. __________ spreaders distribute fertilizer more uniformly than __________ spreaders.
15. The micronutrient that is most likely to be deficient is __________.
16. Establish a fertilizer program for a quality lawn in your area.
17. Discuss the advantages and disadvantages of water-soluble (immediately available) nitrogen sources such as urea.
18. Why are natural organic fertilizes becoming popular?

CHAPTER

11

Mowing

OBJECTIVES

After studying this chapter, the student should be able to

- Explain why correct mowing practices are important to the quality of the turf
- Explain the effects of mowing on turfgrass plants
- List the correct cutting height for turfgrass species and cultivars
- Identify the factors that influence the selection of the correct cutting height
- Describe the factors that determine how often turfgrass should be cut
- Describe the advantages and disadvantages of different types of mowers
- Compare the advantages and disadvantages of collecting grass clippings
- Distinguish between safe and unsafe mowing practices
- Discuss the use of plant growth regulators

INTRODUCTION

Turfgrass is mowed regularly for both aesthetic and functional reasons. Generally, the shorter grass is cut, the more attractive it appears. Regular mowing results in a more uniform turf surface and a finer leaf texture. Few people would prefer the appearance of a field to that of a closely mowed turf. Tall grass on golf courses and athletic fields interferes with sports activities. A lawn soon loses its usefulness if it is not cut.

As explained in Chapter 2, grass can be mowed regularly and survive because almost all of the meristematic tissue is located below the path of the mower blade. Leaf formation continues at the crown, which is found at the base of the plant near the soil surface. Cell division and elongation in mer-

FIGURE 11-1
The wider leaves on the left are from a bentgrass plant that has not been mowed. The leaves on the right are on bentgrass plants mowed regularly at putting green height.

istematic regions causes leaves to extend in length and enables the grass plant to "grow back."

Mowing has several effects on the growth habit of turfgrass plants. It results in increased shoot density because of greater tillering, but causes a decrease in root and rhizome growth. Less food is available to support the growth of roots and rhizomes because it is needed for the production of new shoot tissue. Food storage is also reduced. Mowing causes a decrease in leaf width but increased leaf succulence (Figure 11-1).

Mowing has an influence on pest problems as well. Weed populations usually decrease because of increased turf density and the continual removal of the terminal growing point from broadleaf weeds. These dicots have meristematic tissue at the top of the stem. When a broadleaf weed has its growing point removed repeatedly, the plant is in a state of constant stress and is eliminated eventually. Disease problems, however, often increase because of mowing. When shoot tissue is cut off, an open wound remains. Fungi that cause turfgrass diseases can gain entrance into the leaf through the wound before it heals.

Some effects of mowing, such as reductions in the root system and food reserves, are detrimental to the turf. Individual turfgrass plants would probably be healthier if they were allowed to grow to full height. However, overall turf quality suffers severely when the grass is not cut. If mowing is performed correctly, an attractive, functional, and healthy stand of turfgrass can be main-

FIGURE 11-2
The strip in the center has been scalped—cut too low. *(Photo by Brian Yacur)*

tained. Improper mowing practices have an especially devastating impact on turf quality because grass is cut frequently. When errors are continually repeated once or twice a week, serious injury will occur. The major reason for the decline of many turf areas is poor mowing practices such as cutting the grass too low or too infrequently, or using a dull blade (Figure 11-2).

MOWING HEIGHT

The mowing height is the height of the topgrowth immediately after the grass is cut. The selection of the most appropriate mowing height is influenced by several factors. Each species has a recommended mowing height range (Table 11-1). This range is determined by the growth characteristics of the species or cultivar.

As will be explained in greater detail later in this chapter, no more than one-third of the leaf growth should be removed at each mowing. Grasses that produce leaves and stems which grow horizontally have a lower growth habit than grasses that produce more erect, upright leaves and stems. The more horizontal the direction of growth, the lower the plant can be cut without remov-

TABLE 11-1 Mowing Heights Normally Recommended for Turfgrass Species*

Species	Mowing Height Range (in)	(cm)
Bahiagrass	2.0–4.0	5.0–10.2
Bermudagrass		
Common	1.0–1.5	2.5–3.8
Hybrids	0.25–1	0.6–2.5
Carpetgrass	1.0–2.0	2.5–5.0
Centipedegrass	1.0–2.0	2.5–5.0
St. Augustinegrass	2.0–3.0	5.1–7.6
Zoysiagrass	0.5–2.0	1.3–5.0
Creeping bentgrass	0.2–0.5	0.5–1.3
Colonial bentgrass	0.5–1.0	1.3–2.5
Fine fescue	1.5–3.0	3.8–7.6
Kentucky bluegrass	1.5–3.0	3.8–7.6**
Perennial ryegrass	1.5–3.0	3.8–7.6**
Tall fescue	1.5–3.0	3.8–7.6
Crested wheatgrass	1.5–2.5	3.8–6.4
Buffalo grass	2.0–3.0	5.0–7.6
Blue grama	2.0–2.5	5.0–6.4

*Certain cultivars may be exceptions.

**Some cultivars will tolerate closer mowing.

ing over one-third of the leaf tissue or injuring the crown. The height from the ground of the lower leaves also affects the mowing height (Figure 11-3).

If the grass is mowed too low, the crown can be damaged and too much topgrowth is removed. The ability of the plants to photosynthesize is severely limited when an excessive amount of leaf tissue is removed. The result is a reduction in the size of the root system, a partial depletion of stored foods, and a weakening of the turfgrass. Turf managers refer to this type of extreme defoliation as *scalping*. After a severe scalping, only brown stubble and bare soil may remain. Many people would like their turf to have the closely mowed appearance of a golf course putting green. Unfortunately, most turfgrass species will not tolerate such a short cutting height.

Cutting the grass too high also has detrimental effects. Tall grass has a shaggy, unkempt, puffy look. The thatch layer may increase in thickness and cause problems. The shoots droop, and the tops may bend over. Leaf texture becomes coarser when the grass is cut to tall heights. Density decreases, and

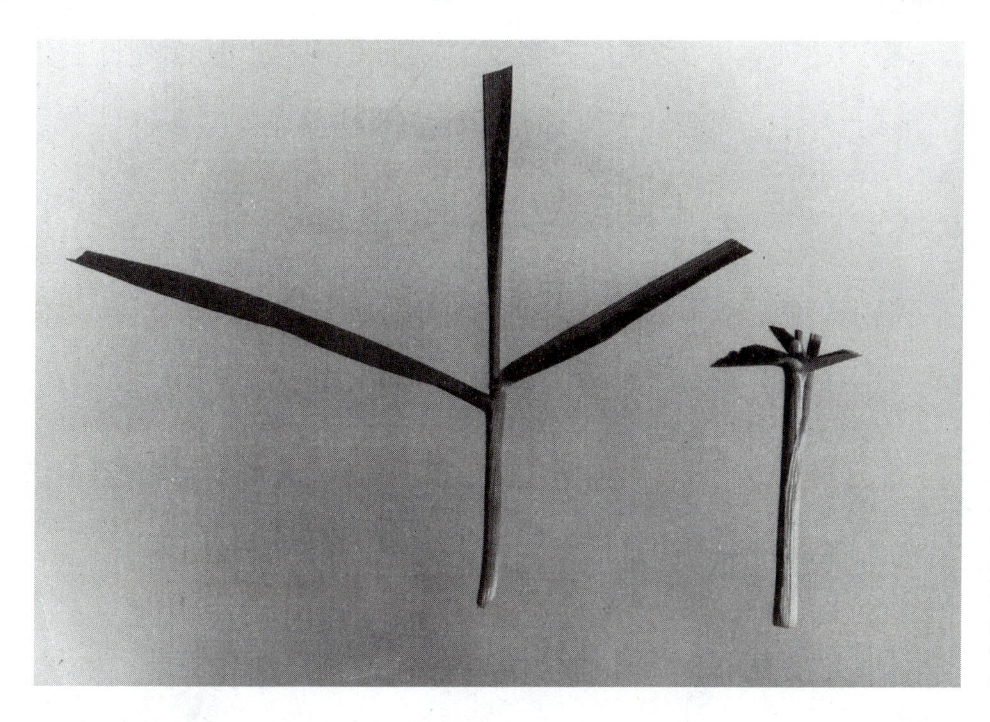

FIGURE 11-3
This tall fescue cultivar has an erect growth habit. The lower leaves occur relatively high above the ground. A 3-inch (7.6-centimeter) mowing height is ideal. A 2-inch (5-centimeter) mowing height (right) will seriously weaken the plants.

the turf is more open. The turf is less satisfactory aesthetically if the mower blade is raised too high.

Turf managers have generally preferred to mow grass to the lowest heights that the species or cultivar can tolerate because closer mowing results in a better-looking turf. However, closer mowing also results in a turf that is less tolerant of environmental and disease stresses. When grass is cut at the lower limit of its mowing tolerance range, it requires a higher level of maintenance and a more expert management program. For example, because shorter grass has a less extensive root system than grass that is mowed to a taller height, it requires more frequent fertilization and irrigation (Figure 11-4). The shallower roots of the grass that is cut lower have a decreased ability to obtain nutrients and water from the soil.

If the turfgrass manager is willing to give turf the extra care is needs to be maintained successfully at closer mowing heights, the results can be very satisfying. Bermudagrass and creeping bentgrass growing on golf greens exhibit superior quality even though they are mowed at the lowest heights in their tolerance range. However, a tremendous amount of maintenance is required to keep a green in healthy condition. Unfortunately, many homeowners cut their lawn at the lowest limit of its tolerance without providing an adequate level of maintenance.

When the grass is experiencing stress, it is best to raise the mowing height. During summer months, cool season grasses are normally cut higher to increase their heat and drought tolerance. Warm season grasses may be cut

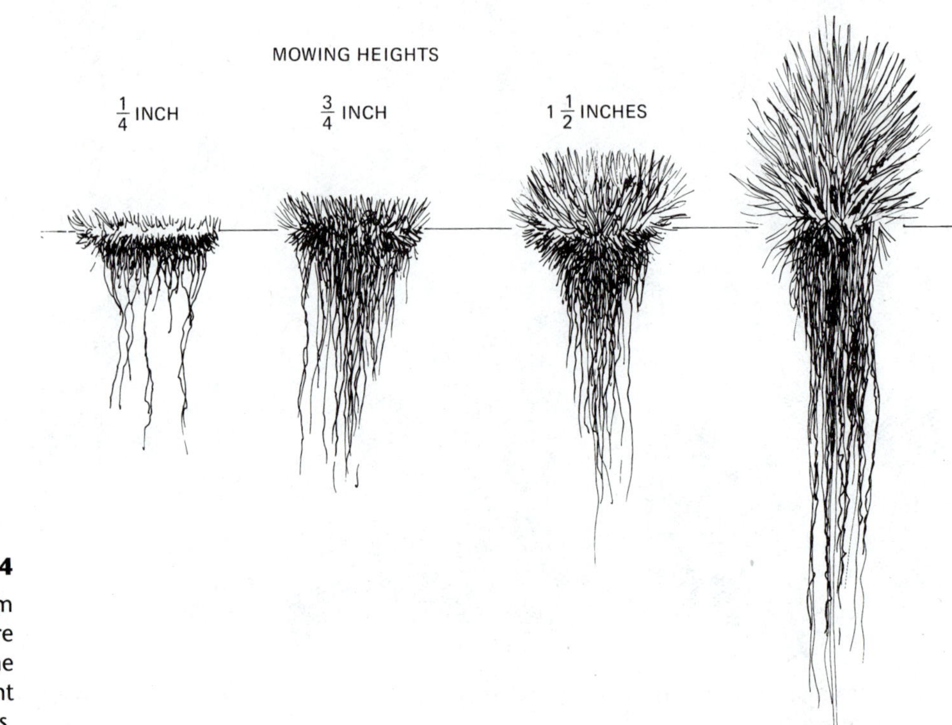

FIGURE 11-4
The root system becomes more extensive as the mowing height increases.

higher in the early and late parts of the growing season when colder temperatures occur. If the turf is recovering from injury due to insects, diseases, traffic, or other stresses, the mower blade should be raised. Turfgrass growing beneath trees is often in a weakened state because of the lack of sunlight and other problems. Mowing grass to a taller height enables it to adapt better to the shaded conditions.

Grass is occasionally mowed lower than its minimum height tolerance when it is dormant. Before the growing season begins, turf may be cut very short to remove dead leaf tissue and "clean up" the area. This close mowing also hastens spring green-up because shading of the soil surface is reduced following the removal of topgrowth. A greater amount of solar radiation reaches the soil surface, causing the soil to warm up more quickly. A certain degree of scalping may be permissible before the growing season when the grass is still dormant. However, it is important not to cut so low that the grass is seriously injured.

For each 0.125 inch (3.2 millimeters) the cutting height is raised, the photosynthetic area of the grass can be increased by as much as 300 ft^2 (28 m^2) per 1,000 ft^2 (93 m^2) of turf.

In many cases, the height to which the turfgrass is actually mowed differs from the height at which the mower blade is set. The mowing height is often adjusted and checked on a level surface such as a driveway or a bench in the equipment storage building. This is known as the bench setting. When the mower is operated on the turf, its wheels force down grass shoots and then ride on top of them. As a result, the mower is raised higher above the soil surface and the grass is cut at a taller height than the bench setting. In some cases, however, the mower cuts lower than the bench setting. This occurs when the wheels sink into the turf because of soft ground or a thick, spongy thatch layer.

MOWING FREQUENCY

Mowing frequency should not be determined by mowing when it is most convenient or on a set date such as every Saturday. How often the grass should be cut is determined by the growth rate of the grass. The growth rate depends upon environmental conditions, the degree of maintenance, and the species and cultivars that make up the turf.

Cool season lawn grasses may require mowing twice a week in the spring when temperatures are optimal and rainfall is abundant. However, in the summer, when weather conditions are unfavorable, mowing once every two weeks may be sufficient. Warm season grasses require frequent mowing in the summer, but the number of days between mowings increases at other times of the year because of cooler temperatures. Turf that is heavily fertilized and irrigated grows faster; it needs to be cut more often than turf that receives minimal maintenance. Some species such as centipedegrass or the fine fescues tend to have a slow growth rate and require less frequent mowing.

Turfgrass managers usually determine mowing frequency by using the one-third rule that was mentioned previously. The grass should be cut often enough so that not more than one-third of the topgrowth is removed at any one mowing (Figure 11-5). Cutting off more than one-third of the total leaf surface causes a drastic decrease in the plant's ability to photosynthesize because of the extensive loss of leaf area. The reduction in food production can result in the death of a large portion of the root system. All available carbohydrates are used to produce new shoot tissue, and there is not enough food remaining to sustain many of the roots. When one-third or less of the vertical shoot growth is removed per mowing, the impact on the root system is minimal.

To observe the one-third rule, the turfgrass manager should cut the grass when it is 50 percent higher than the desired mowing height (Figure 11-6). For example, if the blade is set at a 2-inch mowing height, the grass should be cut when it is 3 inches tall. The 1 inch of topgrowth that is removed is one-third of the total height. The lower a turf is cut, the more frequently it needs to be mowed. For example, if a certain species is cut to 1 inch, it is mowed when it reaches a height of 1.5 inches. If that species is maintained at a mowing height

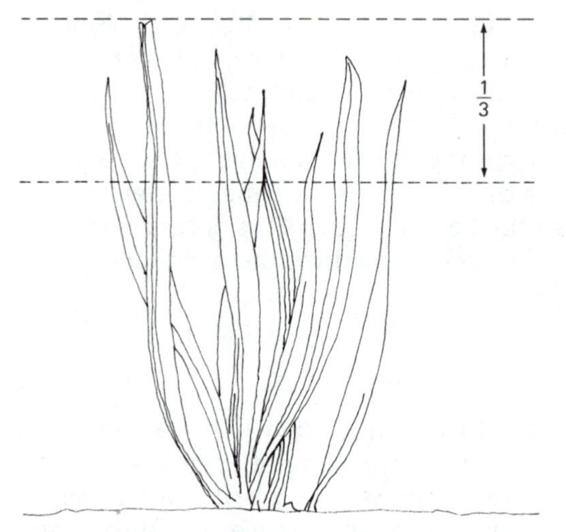

FIGURE 11-5
Approximately one-third of the topgrowth should be removed per mowing. Stated another way, leave twice as much leaf surface as you remove at any single mowing.

of 3 inches, it need not be cut until it is 4.5 inches tall. The plants, of course, will produce 0.5 inch of vertical growth much faster than they will increase in height 1.5 inches.

If the grass becomes excessively long, it should not be chopped down to the normal mowing height at one time. Such a severe scalping can stop root growth for a month or longer. Instead, the grass should be lowered to the desired height gradually by mowing at frequent intervals. For example, if the regular mowing height is 2 inches and the grass is 6 inches tall, it could be cut to 4 inches immediately, lowered to 3 inches after a few mowings at 4 inches, and then mowed at 2 inches a few weeks later. This gradual lowering requires more time and labor than a single mowing, but is usually worth the extra effort, especially if the grass is suffering any type of serious stress.

How seriously turf is weakened by violations of the one-third rule depends upon the health of the grass, the severity of the scalping, and the number of times that too much leaf tissue is removed. Significant injury is most likely

FIGURE 11-6
How to determine when to mow by following the one-third rule.

THE EASIEST METHOD OF DETERMINING MOWING FREQUENCY IS TO MOW THE GRASS WHEN IT IS 50% TALLER THAN THE HEIGHT TO WHICH IT IS TO BE CUT. THIS IS CALCULATED BY DIVIDING THE MOWING HEIGHT BY 2 AND THEN ADDING THIS LENGTH TO THE MOWING HEIGHT. GRASS THAT IS CUT TO 2 INCHES SHOULD BE MOWED WHEN IT IS 3 INCHES HIGH.

$$\frac{2 \text{ in}}{2} = 1 \text{ in}$$

$$2 \text{ in} + 1 \text{ in} = 3 \text{ in}$$

when the grass is suffering from some type of stress. If the turf is healthy and not more than 100 percent higher than (double) the mowing height, the effect may be insignificant. Repeated failures to observe the one-third rule often result in a gradual thinning of the stand and a dramatic reduction in turf quality.

On sites where turf quality is relatively unimportant, there is less concern about mowing the grass frequently enough to avoid removing more than one-third of the leaf tissue. Turfgrass growing beside highways may be cut only a few times during the growing season. Money is saved by mowing low-maintenance, utility areas infrequently.

Removing considerably less than one-third of the topgrowth at each mowing also leads to problems. Cutting the grass before it reaches the proper height results in a greater mowing frequency. The adverse effects that may result from mowing the grass more often than is necessary include decreased root, rhizome, and shoot growth, reduced food reserves, and increased succulence. The wounds occurring at the cut leaf tips are ideal sites for fungi and other pathogens to enter the plant. Disease problems can be encouraged by frequent mowing because of the increased amount of wounding. Mowing too often also results in unnecessary labor and expense.

TYPES OF MOWERS

Mowing equipment has increased tremendously in sophistication since 1830, when the first mechanical grass mower was developed (Figure 11-7). Today there are two basic types of mowers used to cut turfgrass—reel and rotary mowers.

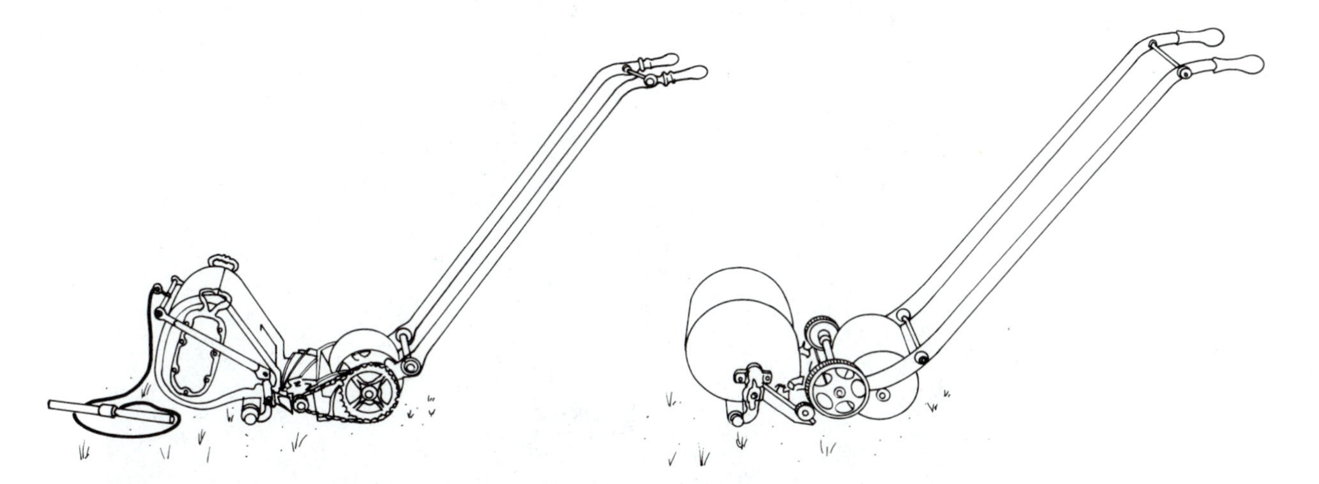

FIGURE 11-7 Two examples of early mowers. *(Courtesy of Brooklyn Botanical Garden)*

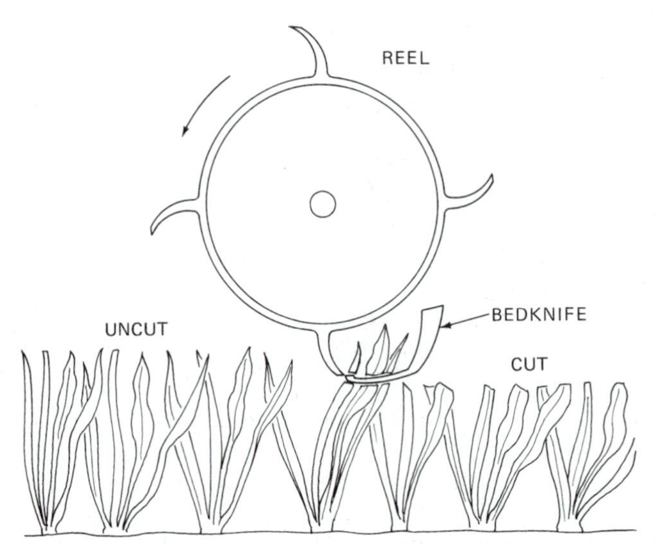

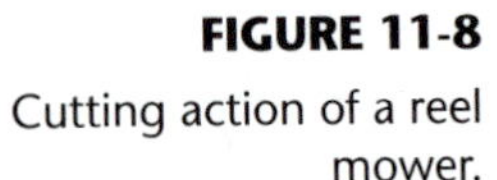

FIGURE 11-8
Cutting action of a reel mower.

Reel Mowers

Reel-type mowers have blades attached to a cylinder which is known as the reel. As the reel rotates, leaves are pushed against a sharp bedknife and cut off. The blades, as they turn, guide the leaves to the bedknife, which does the actual cutting (Figure 11-8). A basket is often attached to catch the clippings (Figure 11-9). When the cutting edge is sharp and the mower is properly ad-

FIGURE 11-9
Baskets can be used to collect the clippings. *(Used with permission of The Toro Company. "Toro" is a registered trademark of The Toro Company, Minneapolis, Minnesota.)*

FIGURE 11-10
Reel mowers produce superior mowing quality. *(Used with permission of The Toro Company. "Toro" is a registered trademark of the Toro Company, Minneapolis, Minnesota.)*

justed, the reel mower exhibits better mowing quality than the rotary mower. It produces a sharp, clean cut because the blade and bedknife create a scissor-type cutting action.

Reel mowers are the most commonly used type of mower on higher-quality turf areas because of the superiority of their cut. Golf course greens, tees, and fairways, better lawns, athletic fields, and sod farms are usually maintained with reel mowers (Figure 11-10). Some reel mowers can be set to cut grass as low as 0.125 inch (3.2 millimeters). They also offer the advantages of being safer and more fuel efficient than rotary mowers.

There are some disadvantages associated with reel mowers. They have difficulty cutting through seed stalks and cannot be used to mow some warm season species that produce tough seed heads. Reels often cannot mow tall grass that is higher than 4 to 6 inches (10.2 to 15.2 centimeters). Best mowing quality is obtained when a reel-type mower is operated on a relatively smooth surface.

Approximately 98 percent of all mowers sold are the rotary type. Homeowners generally prefer rotary mowers because reel mowers are often more expensive and require special maintenance. When reel blades and bedknives are nicked or worn, they must be ground and lapped. Professional turf managers use machines to perform the grinding and lapping operations (Figure 11-11). Most homeowners do not have access to this type of sharpening equipment or the skill to use it.

FIGURE 11-11
A lapping machine helps to keep the mower cutting properly between grindings.

The reels should be disengaged whenever the mower is driven across surfaces that are not covered with grass. The liquids in the grass leaves act as a lubricant and prevent the metal from heating and expanding as the blades strike the bedknife. If the reel is allowed to rotate when it is not cutting grass, excessive wear can occur.

Rotary Mowers

Rotary mowers have blades that rotate horizontally at high speeds. The sharpened edges at the ends of the blades cut off leaf tissue by impact (Figure 11-12). A rotary mower is able to cut grass because the blade spins so rapidly. It does not produce the sharp, clean-cut characteristic of a reel mower. Rather than cutting like a pair of scissors, its cutting action is more like that of a machete or scythe.

The mowing quality produced by a rotary mower is generally not good enough for higher-maintenance areas. However, if the blade is kept sharp, a rotary is quite satisfactory for use on most lawns and lower-maintenance areas (Figure 11-13). A sharp blade is essential. Grass cut with a dull blade is torn or shredded at the point of impact; the cut tip appears jagged and frayed (Figure 11-14). When the tissue is mutilated the wound heals more slowly. The result is greater water loss and an increased likelihood of disease problems. Repeated mowing with a dull blade seriously weakens the turf.

FIGURE 11-12

The rotary blade revolves at speeds as great as 200 miles per hour. *(Used with permission of The Toro Company. "Toro" is a registered trademark of The Toro Company, Minneapolis, Minnesota.)*

FIGURE 11-13

Rotary mowers such as this riding model are very popular for home lawns. A sharp blade is essential. *(Used with permission of The Toro Company. "Toro" is a registered trademark of the Toro Company, Minneapolis, Minnesota.)*

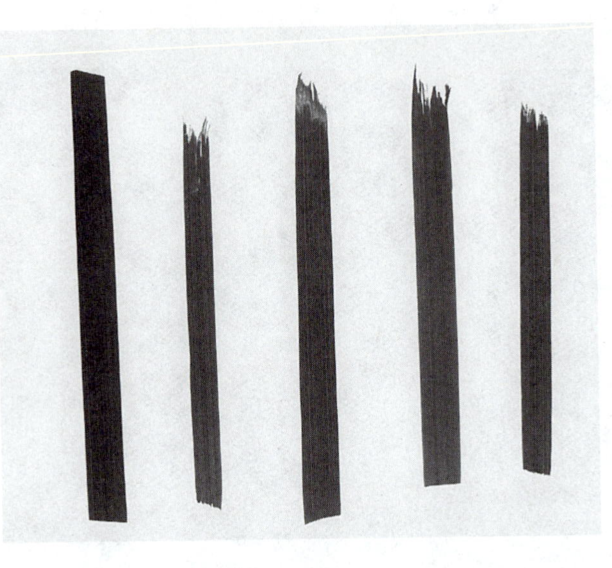

FIGURE 11-14
The leaf on the left was cut by a sharp blade. The others were cut by a dull blade.

When a rotary blade is dull or nicked it should be sharpened (Figure 11-15). A file or power grinder can be used. Excessively worn blades should be replaced; new blades are relatively inexpensive. The blade must be balanced before it is put back on the mower (Figure 11-16). The turf manager should examine the mower blade and cut leaf tips regularly to be sure that the blade is sharp.

FIGURE 11-15
The cutting edge of this blade is nicked and very dull.

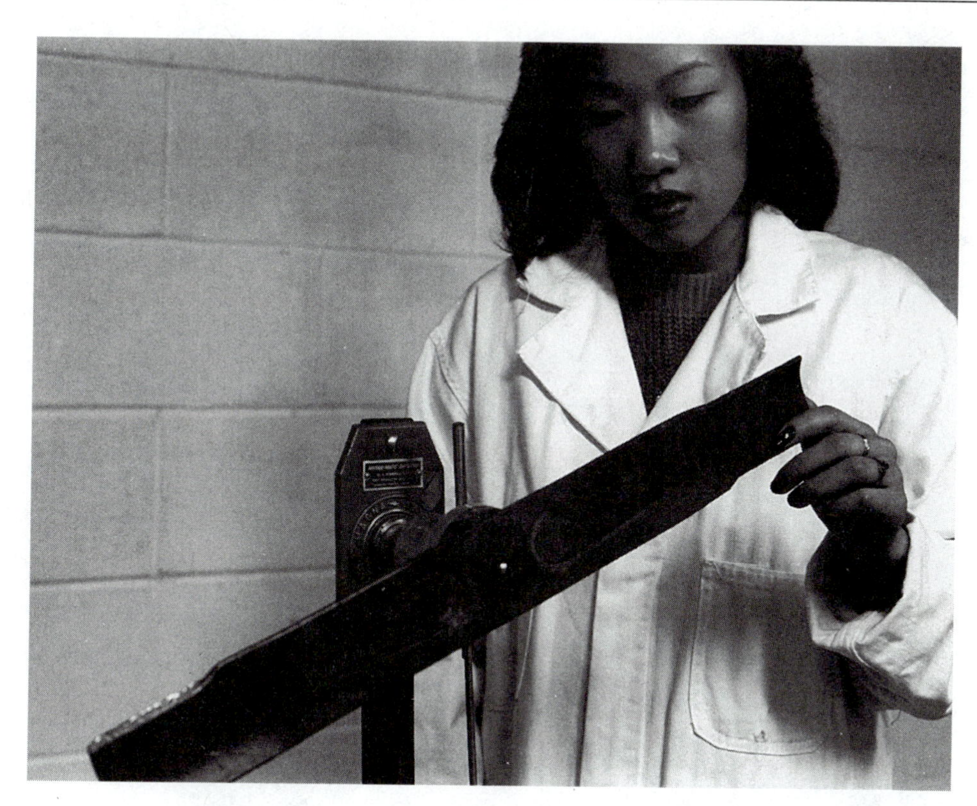

FIGURE 11-16
Rotary blade on a balancer.

There are other advantages to the rotary mower in addition to its simple maintenance requirements. Rotaries are more versatile than reel mowers. They can mow very tall grass and cut through tough seed stalks and weeds. The rotary is preferred for species such as bahiagrass, centipedegrass, and St. Augustinegrass because of their extensive seed head formation. It is also easier to trim around trees and buildings with a rotary mower than with a reel mower.

Rotary mowers have disadvantages as well. A serious accident can occur if a hand or foot is placed under the mowing deck (housing) while the blade is rotating. It is important to buy machines that have the best safety features to help prevent these types of injuries. It is also very important not to disconnect or in any way circumvent the safety devices. Rotary mowers can also be dangerous if the blade strikes a rock or stone. The object can be thrown from beneath the mower at the speed of a bullet. Rotaries are not used on turf that is mowed to a height lower than 1 inch (2.5 centimeters) because the mowing quality would be unsatisfactory. Reel mowers generally have two wheels and follow the contour of the surface more exactly than a rotary mower, which travels on four wheels. At close mowing heights, the rotary is more likely to scalp as it moves over small mounds or ridges (Figure 11-17).

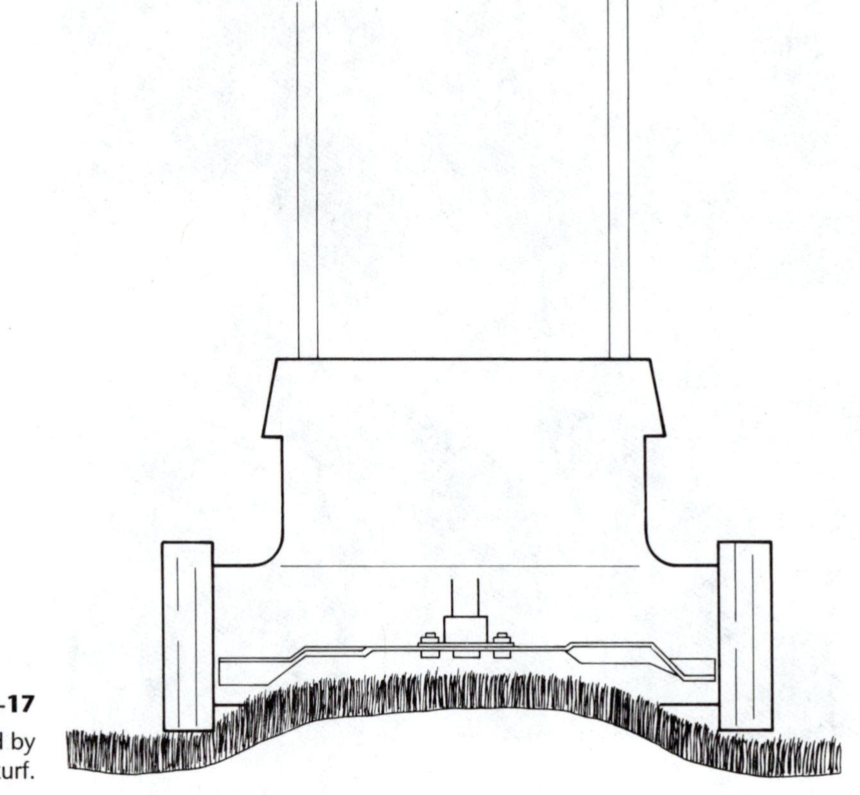

FIGURE 11-17
Scalping caused by high spots in the turf.

Flail Mower

A flail or hammer knife mower is occasionally used by turf managers. It has numerous, loose-hanging small knives, which are held out by centrifugal force as the shaft rotates at high speeds (Figure 11-18). The blades sever grass by impact. Flail mowers are used on low-maintenance, utility sites that are cut infrequently. They can mow grass that is 12 inches (0.3 meter) high. One disadvantage is the amount of time it takes to sharpen the many small blades on the machine. Mowing quality is inferior compared to a reel or rotary mower. Flail mowers do not throw objects they hit as rotary mowers do.

Selecting a Mower

The best type of mower for a particular turf area (Figure 11-19) depends upon many factors, including turf quality, mowing height, species and cultivar, safety considerations, and the sharpening equipment available. The mowing width of the mower is also important. For an average-sized home lawn, a small

FIGURE 11-18 A flail or hammer knife mower. *(Courtesy of Mott Corp.)*

SPECIES	TYPE OF MOWER
Bahiagrass	Rotary
Bermudagrass	
Common	Reel or rotary
Hybrids	Reel
Carpetgrass	Rotary
Centipedegrass	Reel or rotary
St. Augustinegrass	Rotary, primarily
Zoysiagrass	Reel
Creeping bentgrass	Reel
Colonial bentgrass	Reel
Fine fescue	Reel or rotary
Kentucky bluegrass	Reel or rotary
Perennial ryegrass	Reel or rotary
Tall fescue	Reel or rotary

FIGURE 11-19 The type of mower recommended for various turfgrass species. Certain cultivars may be exceptions.

mower with an 18- or 21-inch (0.46- or 0.53-meter) width is usually sufficient (Figure 11-20). Large areas such as golf courses, estates, parks, sod farms, athletic fields, school grounds, and cemeteries require large mowing machines. The mowing width of a rotary or reel mower is increased by cutting with several blades or reels (Figure 11-21). Some gang mowers have as many as nine reel units and can cut a swath over 20 feet (6.1 meters) wide (Figure 11-22).

FIGURE 11-20
A typical walk-behind rotary mower used on small to medium-sized lawns. (*Photo by Brian Yacur*)

FIGURE 11-21
A rotary mower with three blades *(Used with permission of The Toro Company. "Toro" is a registered trademark of The Toro Company, Minneapolis, Minnesota.)*

FIGURE 11-22

A nine-gang reel mower. *(Courtesy of Jacobsen Division of Textron, Inc.)*

The turf manager should purchase mowing equipment that is large enough for his or her needs. The initial amount of money saved by buying a smaller, cheaper machine may be small when compared to the cost in time and labor saved by purchasing a machine with a greater mowing capability. For example, it takes over two and a half hours to cut an acre (0.4 hectare) with a walk-behind mower that has an 18-inch (0.46-meter) cutting width. A 60-inch (1.5-meter) riding mower can mow the acre in approximately thirty minutes, while a gang mower with a 20-foot (6.1-meter) mowing width takes less than ten minutes.

Commercial-grade mowers, though more expensive than homeowner-type mowers, are better built and will last longer.

MOWING PATTERN

It is advisable to vary the mowing pattern. The grass should not be mowed in the same direction each time it is cut. Grass tends to lean or grow in the direction in which it is mowed. The horizontal orientation of grass foliage in one direction is called *grain*. Grain results in a less even cut, a streaked appearance, and poorer putting quality on golf greens. Alternating mowing directions reduces grain. A brush or groomer can be attached to the front of the mower to help lift leaves and stems.

FIGURE 11-23
The mower must not bump into trees. *(Used with permission of The Toro Company. "Toro" is a registered trademark of The Toro Company, Minneapolis, Minnesota.)*

Varying the mowing pattern prevents continual scalping of the same high spots and reduces compaction and rutting by mower wheels, which can develop if the wheels run over the same ground each time. Repeatedly turning the mower at the same places results in damage to the turf.

The operator must be very careful when mowing around trees (Figure 11-23). If the mower hits the trunk, the vascular tissue of the tree may be injured. Disease-causing microorganisms can enter the tree through mower wounds. Placing a mulching material such as wood chips at the base of the tree helps to prevent mower damage because the operator does not have to mow close to the trunk. Monofilament trimmers are good tools for trimming around trees (Figure 11-24).

CLIPPINGS

The pieces of grass tissue cut off by the mower are called clippings. They can be collected in a catcher or bag that is attached to the mower. If the clippings are short it is best not to remove them. Clippings are a valuable source of fertilizer nutrient, and the need for fertilization is decreased when they are left on the turf to decompose. As they decompose, the clippings return substantial quantities of nutrients to the soil. Clippings are often composed of 3 to 5 percent nitrogen on a dry weight basis. They also contain significant amounts of phos-

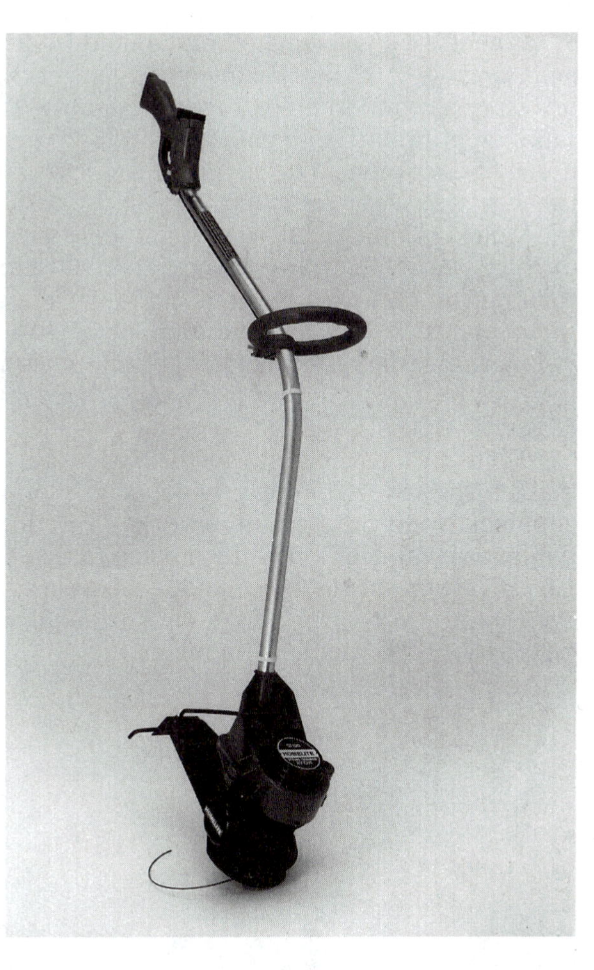

FIGURE 11-24

A nylon line or string trimmer. *(Courtesy of Homelite Division of Textron, Inc.)*

phorus and potassium. Research has shown that grass plants can obtain 25 to 40 percent of the nitrogen they require from recycled clippings.

Observing the one-third rule is advantageous because the mower removes small pieces of leaf and stem tissue. These short clippings fall close to the soil surface and do not cause aesthetic problems. They decompose rapidly, so their contribution to thatch accumulation is minimal.

Long clippings should be collected or raked up after mowing. These longer leaf pieces look unattractive on the surface of the turf. Large clumps can smother the grass or lead to disease problems because the grass beneath them remains damp. When turfgrass is diseased, clippings may be collected in an attempt to remove some of the disease-causing organisms that are living in the leaves. Clippings are not allowed to remain on golf greens even though the leaf pieces are quite short. The clippings detract from the superior appearance of the green and interfere with putting.

Whenever possible, clippings should be left on the turf. Time is wasted emptying the catcher or raking when the cut tissue is collected unnecessarily. It is especially important to leave the clippings for a few mowings after a fertilization application. Approximately 60 to 70 percent of the nutrients that are immediately available to the plants are found in the clippings the first three times the grass is cut after fertilization.

Mulching mowers are rotary mowers that are designed to cut leaf tissue into very small pieces. These pieces are small enough to fall into the turf rather than remain on top of the grass, and they will decompose rapidly because of their small size. Mulching mowers either do not have a discharge chute or have a chute that can be closed. The clippings are trapped beneath the mowing deck in the cutting chamber and are recut until they are finely chopped (Figure 11-25).

There are a number of walk-behind and riding mulching mowers on the market. They are becoming very popular in communities where the disposal of clippings is difficult because of local restriction or lack of dumping sites. A major problem with mulching mowers is that they have not performed satisfactorily when grass is wet or if long pieces of tissue have to be removed. However, improvements in design and in air flow in the cutting chamber are helping to overcome these problems.

FIGURE 11-25

Clippings are trapped under the deck of a mulching mower because it does not have a discharge chute. *(Used with permission of The Toro Company. "Toro" is a registered trademark of The Toro Company, Minneapolis, Minnesota.)*

MOWER SAFETY

The Consumer Product Safety Commission reports that each year in the United States approximately 80,000 people are injured by mowers. Of these accidents, 10,000 result in the amputation of toes or fingers. One must always be very careful when operating a mower. Obeying the following safety rules is essential.

- Never allow children near the mower.
- Always stop the engine when you leave a mower. The engine should be shut off even if you are going away from the mower only for a minute.
- Pick up any stones, sticks, or debris lying on the area that is to be mowed.
- Remove the ignition wire from the spark plug before touching a rotary blade. If the wire is not disconnected, the engine could start accidentally.
- Be sure that feet and hands are away from the blade before the engine is started. Start the mower on a level spot. Before starting a walk-behind rotary mower, place one foot firmly on top of the mower deck and be sure that the other foot is on solid ground, a safe distance away from the mower. Never place any part of the body near the blade when the mower is operating.
- Do not remove the bag that catches the grass clippings while the mower is running.
- Make certain that you know how to operate the mower before starting the engine. Be sure that you know how to stop the engine quickly in case of an emergency.
- Always push a rotary mower forward—never pull it backward toward you. There is too great a chance that it could be pulled back over a foot.
- Wear heavy, protective work boots or shoes when mowing. The safety-tipped type of shoe is best. Shoes should also have good traction to help prevent slipping.
- When using a walk-behind mower, mow across slopes—not up and down. A riding mower is often driven up and down a slope to reduce the risk of tipping over. Never operate a mower on a steep slope.
- If possible, do not mow wet grass. The operator is more likely to slip when walking on damp turf. Wet clippings also tend to stick together and clog the mower. Large clumps of cut tissue may remain on the turf after mowing.
- Never attempt to pour gasoline into the tank when the engine is hot. Allow the engine to cool off before refueling.
- Do not add fuel or oil to the mower when it is on turf. Spills will result in injury to the grass.
- Always pay careful attention when mowing. Operating a mower can be boring, but the chance of an accident increases significantly if you daydream or let your mind wander.
- Read the instruction manual before operating a mower. Additional important safety rules are listed there.

Mowers manufactured today are required to have certain safety features. For example the blades must stop immediately if the operator releases his or her grip from the handle or leaves the seat. This is an excellent safety feature because it helps to prevent people from accidentally placing their hands in the path of a mowing blade.

When the turf manager is deciding which type or model of mower to buy, safety considerations should have an important bearing on the decision. Also, it is important to check constantly to make sure that safety devices are working properly.

CHEMICAL MOWING

A large percentage of the turfgrass manager's maintenance budget is spent on mowing operations. Consequently, there is great interest in plant growth regulators (PGRs), chemicals that slow down the growth rate of grass and reduce the need for mowing. One type of PGR retards plant growth by inhibiting cell division, which results in less leaf elongation. A second type inhibits the production of gibberellic acid. Treated plants have shortened cells and are more compact. The best PGRs reduce shoot growth without significantly affecting root growth.

FIGURE 11-26
The grass on the right was treated with Embark® growth regulator. The tall grass on the left is untreated. *(Courtesy of Agricultural Products/3M Corp.)*

Plant growth regulators have been sprayed primarily on low-maintenance sites such as golf course roughs, roadsides, and difficult-to-mow areas such as steep banks and cemeteries, and around trees, fences, or ponds (Figure 11-26). Their use on fine turf has been limited because of their potential to reduce turfgrass quality. When growth is slowed plants become more susceptible to stress and less able to recover from injury. PGRs should not be applied to areas that receive much traffic. They may initially discolor the turf, and mixed stands of grasses can have an uneven appearance if one species is inhibited more than another.

The characteristics of PGRs have improved significantly in recent years. Some of the newer materials are used on lawns and golf course fairways. One product actually produces a darker green, denser-appearing turf. However, turf managers must be very careful when treating better-quality sites. It is important to experiment on small areas first and to start out using the lower recommended rates. An application of PGR generally slows growth for a four- to ten-week period. Its use may reduce the number of mowings that are necessary by 50 percent (Figure 11-27). The result is a decrease in maintenance costs and a reduction in the likelihood of mower accidents when hazardous mowing sites such as slopes are treated.

Another significant advantage of PGRs is the reduction in the volume of clippings. This is very helpful where clippings need to be removed from the site and disposal is a problem.

Plant growth regulators are primarily used at times of the year when the grass is growing rapidly. In the North, for example, half of the yearly top-

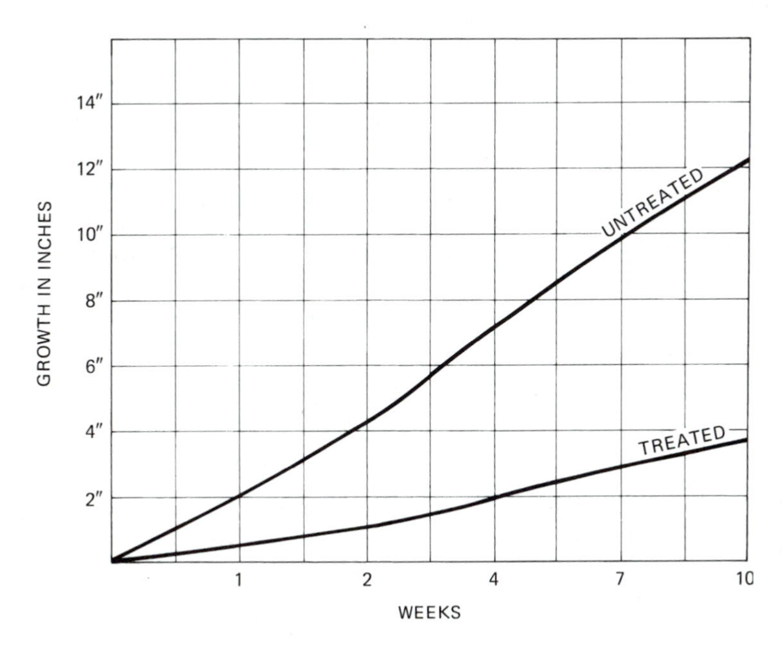

FIGURE 11-27

Average growth rate on untreated turf and turf treated with Embark® plant growth regulator. *(Courtesy of 3M Corp.)*

growth can occur in the spring. It is difficult to keep up with the mowing, especially if it rains a good deal. Turf quality can be greatly reduced if the grass gets too high and is scalped down or if heavy mowing equipment must be operated on wet turf and soils. A PGR can help to solve these problems.

There is also interest in using plant growth regulators to reduce water and fertilizer use. When its growth rate is slowed, a grass plant does not need as much irrigation or fertilization.

Many broadleaf weeds are not affected by PGRs used on turf. It may be necessary to apply a herbicide when a PGR is used to prevent the weeds from taking advantage of the grasses' lack of competiveness.

SELF-EVALUATION

1. The size of the turfgrass root system becomes ___________ if the mowing height is lowered.
2. Leaf width ___________ if the mowing height is raised.
3. Not more than ___________ of the vertical shoot growth should be removed at each mowing.
4. Extreme defoliation is referred to as ___________.
5. The actual mowing height may be higher or lower than the ___________.
6. Grass cut to a height of 0.5 inch (1.27 centimeters) should be mowed when it is ___________ inch(es) (___________ centimeter [s]) tall.
7. The best mowing quality is obtained with a sharpened, well-adjusted ___________ mower.
8. ___________ mowers are very popular because they are versatile and easy to sharpen.
9. The reel unit consists of blades and a ___________.
10. Ragged, shredded leaf tips indicate that the rotary mower blade is ___________.
11. When the turf is stressed because of unfavorable environmental conditions, the mowing height should be ___________.
12. Whenever possible, clippings should be left on the turf because they are a valuable source of ___________.
13. The horizontal orientation of shoots resulting from repeatedly mowing in the same direction is called. ___________.
14. Remove the ignition wire from the ___________ before touching a rotary mower blade.
15. Allow the engine of the mower to ___________ before refueling.
16. When using a PGR, a ___________ should also be applied.
17. What factors should the turfgrass manager consider when deciding what type of mowing equipment is most appropriate for his or her needs?
18. If you were a turf manager describe the mower safety program you would use to protect your employees.
19. Discuss the advantages of using plant growth regulators.

CHAPTER

12

Water and Irrigation

OBJECTIVES

After studying this chapter, the student should be able to

- Explain how water moves in the soil
- Describe why not all soil water is available for plant use
- Discuss irrigation needs
- List the signs that indicate when irrigation is needed
- Explain how the amount of water to be applied is determined
- Describe irrigation systems

INTRODUCTION

Water is essential for the survival of turfgrass and all other living organisms. There is serious concern in the United States about the quantity and quality of water that will be available in the future. If current predictions are correct, there could be a crisis in agriculture in coming years because of a lack of water. Some areas of the country have already experienced water use restrictions, and in many other areas people have been urged to conserve water on a voluntary basis. The turfgrass manager must understand the relationship between water and soil and plants so that he or she can use irrigation water effectively and efficiently.

TURFGRASS AND WATER

The living tissue of grass plants is 80 to 95 percent water. If the percentage drops to 60 percent, the plants die. Each plant cell is a container of water. When the cells are filled with water, the tissue is rigid and said to be turgid. If the cells lack water and lose their turgidity, they collapse. The leaf or stem droops and often shrivels up. It is said to be wilted.

Water has other functions in a turfgrass plant besides maintaining turgor pressure. The living protoplasm in cells depends on water for its existence. Water is necessary for photosynthesis and many other metabolic processes. The transport of nutrients and other compounds in a plant is a result of the movement of water in the vascular tissue. Water also protects the plant against potentially damaging temperature fluctuations.

Turfgrass plants obtain water from the soil. Root hairs absorb soil moisture. The water then enters the xylem and moves upward in the plant. This upward movement occurs because of transpirational pull created by the evaporation of water from the leaves. Water has a tendency to move from areas of higher water concentration to areas where the concentration is lower. If the soil contains sufficient moisture, water will move into plant roots. If the air surrounding a leaf is drier than the interior of the leaf, water moves from the leaves through open stomata into the atmosphere. The water is converted from a liquid into a gas (water vapor). The loss causes water to be pulled up from he roots to replace the water that has evaporated. The process of water movement from the soil into roots, up the xylem, and out of the leaves into the atmosphere is called *transpiration*.

The rate at which water evaporates from turfgrass leaves depends on the atmospheric conditions. The lower the relative humidity and the higher the temperature and level of solar radiation (sunlight), the greater the transpirational losses. Wind also contributes to an increased transpiration rate. On a hot, sunny, dry, windy day an acre of turfgrass can lose in excess of 10,000 gallons (38,000 liters) of water to the atmosphere.

As much as 90 percent of the water taken up by roots may be lost through the stomata. A plant needs 500 to 600 pounds (60 to 70 gallons) of water to produce 1 pound of dry matter. The amount of water required by turfgrass depends on many factors such as species and cultivar, rooting depth, weather, level of maintenance, intensity of traffic, soil type, and the desired turf quality.

When the roots are able to obtain enough moisture from the soil to replace the water vapor pulled from the leaves by the surrounding air, the turfgrass continues to grow normally. If transpiration losses exceed root absorption, changes occur in the grass plants. The effects of a water deficit include an increased rooting depth, fewer and thinner leaves, decreased leaf size and surface area, less tillering, reduced shoot elongation, and decreased turgidity and succulence. The rate of metabolic processes such as photosynthesis also slows.

Prolonged moisture stress results in wilting. Leaves droop, roll, or fold and turn a blue-green or grayish-green color. A rain or irrigation will restore turgidity after wilting first occurs. If soil moisture does not become available, the water content of cells may decrease to such an extent that the plant cannot recover. This is called the permanent wilting point. Usually, however, turfgrasses can escape serious injury or death by entering a dormant state. Most species, when confronted with prolonged droughty periods, slow their metabolism and become dormant. Shoot growth ceases, and the leaves may die. Buds in the crown, rhizomes, and stolons survive and will initiate new topgrowth after soil moisture is replenished by rain or irrigation.

Most perennial species are able to survive lengthy periods of moisture stress if the plants are mature, deep-rooted, and healthy (Figure 12-1). Annual species and plants that are weak and shallow-rooted are most likely to succumb to drought injury. Some species increase their ability to adapt to dry conditions by mechanisms such as storing water in roots or closing stomata sooner when moisture stress occurs. Structural modifications such as fewer stomata, increased cuticle thickness, the presence of hairs on leaf surfaces, and smaller cells contribute to greater drought hardiness. Fine fescues have narrow, rolled leaves that help to reduce transpirational losses because less leaf surface area is exposed to the air.

Drought-tolerant grasses normally can survive extended periods of dry weather without supplemental irrigation, but a temporary reduction in turf quality will occur. If brown, dormant turfgrass is unacceptable, the turf manager will have to water to maintain desirable quality. If the species or cultivars are not drought hardy, irrigation may be necessary to keep them alive during periods of moisture stress. Even drought-tolerant grasses can be endangered by a severe drought, and irrigation may be required to ensure their survival. Immature or seedling grasses must be watered regularly until they are well established and have developed an adequate root system.

EXCELLENT
Bahiagrass
Blue grama
Buffalograss
Common bermudagrass
Wheatgrass
Zoysiagrass
VERY GOOD
Fine fescue
Tall fescue
GOOD
Canada bluegrass
Hybrid bermudagrass
Kentucky bluegrass
FAIR
Perennial ryegrass
St. Augustinegrass
POOR
Annual bluegrass
Annual ryegrass
Carpetgrass
Centipedegrass
Colonial bentgrass
Creeping bentgrass
Rough bluegrass

FIGURE 12-1
Relative drought tolerances of turfgrass species. There can be significant variation between cultivars.

SOIL AND WATER

To understand how moisture becomes available for absorption by the roots of turfgrass plants, it is important to understand how water moves in the soil. The spaces between soil particles are called pores. The larger spaces are called macropores or aeration pores. The smaller spaces are called micropores or capillary pores. The larger pores normally contain air. Water drains down through these macropores too rapidly for plants to use. Water moves slowly through micropores because of their small size. They continue to hold moisture long after it has drained out of the macropores.

Soil texture has a major effect on water movement, storage, and availability, as explained in Chapter 5. Macropores are abundant in sandy soils, and consequently these coarse-textured soils are well drained but have a limited water holding capacity. Clayey soils contain a significantly larger percentage of micropores than sands and therefore drain more slowly. However, fine-textured soils hold more water because of their greater amount of surface area and total pore space. Loam soils exhibit intermediate drainage and water storage characteristics.

The speed at which a soil can absorb water is known as the *infiltration rate* or capacity. As much as 3 inches (7.6 centimeters) of water can enter a coarse sand in an hour because of the presence of a large number of macropores. The infiltration rate of a clay soil can be as little as 0.1 inch (0.25 centimeter) per hour because of the smaller number of macropores. Structure also has a major impact on how rapidly water can infiltrate a soil. A sandy loam soil with a granular structure may have an infiltration rate of more than 1 inch (2.5 centimeters) per hour, but the amount of water this soil can absorb in an hour drops off to 0.3 inch (0.76 centimeter) if it becomes heavily compacted. Compaction results in a loss of macropores because the soil particles are forced closer together.

Percolation is the term for the movement of water down through a soil. The percolation rate is generally fast in coarse-textured soils and slow in fine-textured soils. In a sandy loam, soil water can penetrate to a depth of 2 feet (0.6 meter) in 30 minutes, but it takes water 4 hours to reach the same depth in a clay loam. Drainage decreases as compaction increases.

Water-holding capacity is also directly related to soil texture. As a general rule, a clay soil can hold approximately twice as much water as a loam, and about four times more than sand. The storage capacities of soil textures can be compared by determining the amount of water required to wet them to a certain depth. Three inches (7.6 centimeters) of water are needed to wet a clayey soil 12 inches (0.3 meter) deep, 1.5 inches (3.8 centimeters) for a loam, and 0.75 inch (1.9 centimeters) for a sandy soil. The water-holding capacity of coarse- and fine-textured soils can also be compared by measuring how deep 1 inch (2.5 centimeters) of water will penetrate into these various soils (Figure 12-2). A coarse-textured soil requires a greater amount of water during the

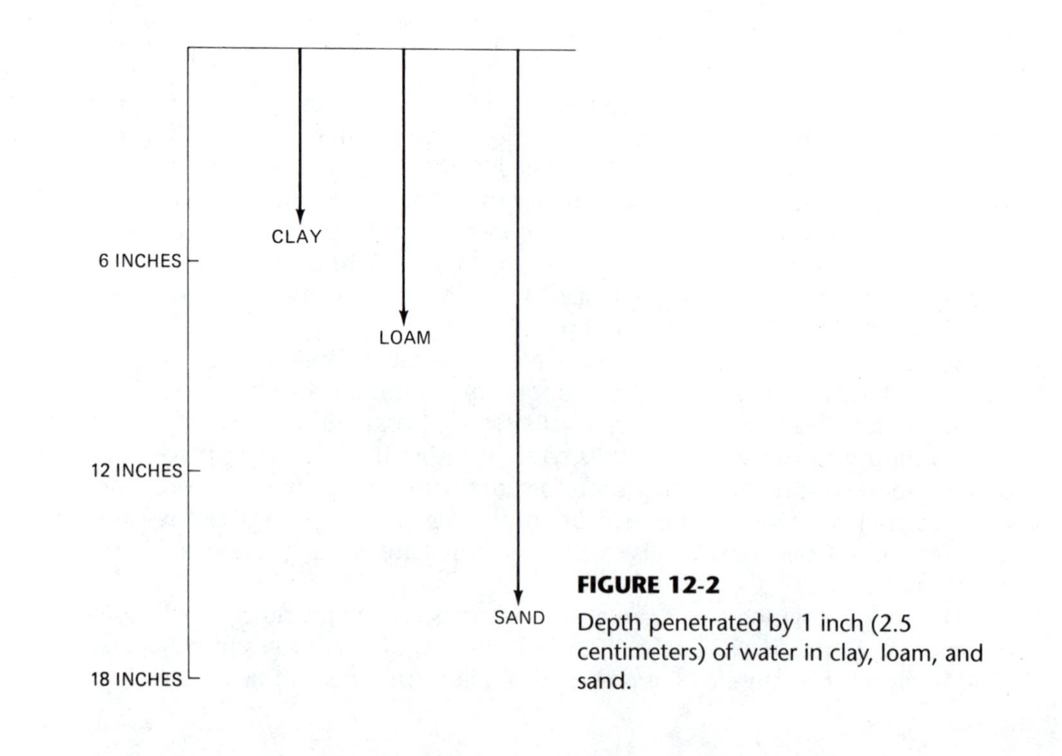

FIGURE 12-2

Depth penetrated by 1 inch (2.5 centimeters) of water in clay, loam, and sand.

growing season than a fine-textured soil because it loses more water through drainage and evaporation.

A sandy soil exhibits a high infiltration rate (unless compacted), rapid drainage, and a relatively low water-storage capacity. A low infiltration rate, slow drainage, and a relatively high water-holding capacity are associated with a clay soil. The behavior of water in a soil depends on the ratio of macropores to micropores in that soil. This ratio is primarily influenced by the texture of the soil, but structure also has a significant effect. Organic matter improves the structure and contributes to an increase in both the number of large pores in the soil and its water-holding ability (see Chapter 5).

It is helpful to understand what happens to water during and after a heavy rain or irrigation. If the infiltration and percolation rate of the soil is poor because of fine texture or compaction, runoff and puddling may occur. If the site is sloped, the water that does not enter the soil will move down the slope across the surface of the soil. This runoff is wasteful and may cause soil erosion. If the site is level or in a low area, standing water may accumulate above the soil surface. These puddles interfere with maintenance operations and other types of traffic, and can contribute to compaction. On hot, sunny days puddling can result in scald, which causes the grass to appear burned or scorched.

Following a heavy rain or irrigation, all of the pores in the upper soil may be filled with water. The soil is then said to be saturated. The soil remains saturated for only a brief period unless a restricted layer beneath the surface prevents drainage. The force of gravity soon causes the water in the large pores to move down through the soil. The water that drains from the macropores is called gravitational water. It moves down below the root zone too quickly to be used by plants. Macropores are also referred to as aeration pores because they contain air when the soil is not saturated.

After the percolation of gravitational water from the upper soil is completed, the soil is said to be at field capacity. Downward movement of water has virtually ceased at this point. The water that remains in the upper soil is found in the smaller pores as films of moisture around soil particles.

At field capacity, about one-half of the water in the root zone is available for plant use. To see why only half of the water is available, it is necessary to understand two attractive forces that affect water. *Cohesion* is the attraction between molecules of water. *Adhesion* is the attraction between water molecules and the surface of soil particles.

The attractive force between soil particles and water can be measured on a scale of soil moisture tension. The greater the attraction, the higher the soil moisture tension measurement. Areas of the soil that contain abundant moisture exert less of an attractive force on water than areas that contain smaller amounts of moisture. Water tends to move from areas of high moisture content to areas of low moisture content because of the greater moisture tension or attractive force exerted by soil locations containing small amounts of water.

When the soil is at field capacity, relatively thick films of water surround soil particles and fill the micropores. The water molecules in the film are held together by cohesion and bind to soil particles by adhesion. The molecules of water closest to the surface of the particle are held tightly, but the attractive force decreases as the distance from the particle increases. Plant roots can exert a great-enough moisture tension to absorb the outer layer of the film of water. However, as the soil dries the thickness of these films is reduced. Eventually the film may become so thin that the remaining water is held by the soil particles at a higher moisture tension than the roots can exert. This tightly bound water is unavailable for plant use.

At field capacity a soil contains the maximum amount of water available for plant use. The moisture content in a soil drops beneath field capacity because of transpiration and evaporation. Turfgrass roots remove large amounts of moisture from the upper regions of the soil profile, and much of this absorbed water is lost through the leaves via transpiration. A significant amount of water near the surface of the soil evaporates and is lost to the atmosphere in the form of water vapor. Evaporation losses, like transpiration losses, are greatest during periods of hot, dry, sunny weather. Wind also increases the rate of soil water evaporation.

The combined loss of water from the soil because of evaporation and from the plant because of transpiration is referred to as *evapotranspiration.* As much as 85 percent of soil moisture loss is caused by evapotranspiration. The amount of water available to turfgrass continues to decrease because of these losses until the soil moisture content is again increased as the result of rain or irrigation.

Capillarity is another concept that needs to be understood. Water in the smaller pores can be pulled upward in the soil to drier areas that exert a higher moisture tension. This upward movement of water in micropores is called *capillarity* or *capillary movement.* When the soil surface becomes dry, evaporation losses continue because capillary water is pulled up to the surface.

Capillarity is also involved in the root absorption process. When root cells exert a higher moisture tension than the soil particles holding moisture, water enters the roots. Water then moves by capillarity from the adjacent soil to the areas that have been partially depleted because of root uptake. It can then be absorbed by the roots. Capillary movement is slow, however, so roots obtain most of their water by growing to areas of greater moisture content rather than pulling water to them over long distances.

IRRIGATION

When rainfall is not sufficient to keep the grass alive or to maintain the desired turf quality, supplemental irrigation is necessary. The total amount of irrigation water required depends on several factors, including the grass species and cultivar, the soil type, the level of maintenance, the frequency and number of inches of rain, and weather conditions such as humidity and temperature. The amount of rainfall received varies greatly throughout the

TABLE 12-1 Average Monthly Precipitation at Sample Locations

	April	May	June	July	August	Sept.	Oct.	Annual
Atlanta	4.61	3.71	3.67	4.90	3.54	3.15	2.50	48.34
Dallas	4.72	4.85	3.27	1.80	2.36	3.25	3.18	35.94
Miami	3.60	6.12	9.00	6.91	6.72	8.74	8.18	59.80
Nashville	4.11	4.10	3.38	3.83	3.24	3.09	2.16	46.0
Richmond	2.77	3.42	3.52	5.63	5.06	3.58	2.94	42.59
Phoenix	0.43	0.12	0.08	1.00	0.93	0.71	0.44	7.62
Los Angeles	0.97	0.38	0.09	0.00	0.01	0.20	0.54	14.77
Denver	2.05	2.20	1.64	1.36	1.43	1.08	1.01	14.20
Salt Lake City	1.76	1.56	0.91	0.61	0.97	0.74	1.34	14.74
Albuquerque	0.68	0.68	0.68	1.46	1.25	0.94	0.73	8.40
Boston	3.77	3.34	3.48	2.88	3.66	3.46	3.14	42.77
Indianapolis	3.74	3.99	4.62	3.50	3.03	3.24	2.62	39.25
Columbus	3.49	4.00	4.16	3.93	2.86	2.65	2.11	36.67
Trenton	3.21	3.62	3.60	4.18	4.77	3.50	2.84	41.28
Albany	2.77	3.47	3.25	3.49	3.07	3.58	2.77	35.08
Boise	1.16	1.29	0.89	0.21	0.16	0.39	0.84	11.43
Minneapolis	1.85	3.19	4.00	3.27	3.18	2.43	1.59	24.78
Spokane	0.91	1.21	1.49	0.38	0.41	0.75	1.57	17.19
Sacramento	1.04	0.54	0.16	0.00	0.00	0.30	0.80	15.88
Houston	3.54	5.10	4.52	4.12	4.35	4.65	4.05	48.19
Helena	0.83	1.56	2.23	1.03	0.89	0.95	0.66	10.85
Seattle	2.15	1.58	1.43	0.66	0.81	1.83	3.50	36.11

United States (Table 12-1). In many regions the evapotranspiration rate exceeds rainfall during periods of the growing season, especially in the summer (Table 12-2). The reverse is true in the spring and fall, and irrigation may not be necessary. In drier areas of the country irrigation may be required throughout the growing season to ensure acceptable turf quality.

This variability of irrigation needs is illustrated by the amount of water used by turfgrass per week during the summer in different sections of the United States. In humid coastal areas less than 1 inch (2.5 centimeters) is used. In areas with moderate summer temperature and humidity 1 inch is the approximate amount used. Where hot, dry summers occur the amount increases to 2 inches (5.1 centimeters) or more. In arid desert areas turfgrass may use 3 inches (7.6 centimeters) of water per week.

TABLE 12-2 Average Monthly Rainfall/Evapotranspiration Rates at Sample Locations*

	Jan.	Feb.	Mar.	Apr.	May	June	July	Aug.	Sept.	Oct.	Nov.	Dec.
Chicago	1.9	1.6	1.9	1.3	–0.2	–1.8	–3.6	–2.7	–1.0	0.6	1.5	1.9
Dallas	1.4	1.3	–0.1	–0.2	–1.2	–4.6	–6.6	–6.4	–3.1	–0.8	0.6	1.4
Ft. Lauderdale	0.1	–0.3	–1.0	–1.3	–1.1	0.2	–1.2	–0.7	–3.1	3.1	–0.5	–0.2
New York	3.4	2.8	3.5	1.7	–0.2	–2.3	–3.1	–1.3	–0.2	1.1	3.0	3.3
Phoenix	0	–0.4	–1.8	–3.8	–6.3	–8.2	–9.1	–7.4	–5.9	–3.5	–1.3	–0.2
San Francisco	3.3	2.7	0.9	–1.2	–3.1	–4.3	–5.0	–4.5	–3.7	–1.9	0.2	3.0
Seattle	5.2	3.9	2.8	0.4	–1.4	–2.4	–4.1	–3.4	–1.1	2.3	4.8	5.8

*A negative number indicates a water deficit. For example, –1.8 is listed for Chicago in June. This means that on the average evapotranspiration exceeds rainfall by 1.8 inches during the month of June. This information is compiled for many cities in the United States in *Rainfall Evapotranspiration Data,* published by the Toro Company.

Determining When Grass Needs Water and Application Rates

Turfgrass managers use many different methods to tell when irrigation is necessary. A simple method is to water when the symptoms of moisture stress appear. Wilted grass turns a blue-green or grayish-green color. If footprints or tracks of machinery can be seen after walking or driving across the turf, the grass is experiencing moisture stress, as grass starts to wilt it loses turgor pressure and does not spring back up. This method is not appropriate for areas such as putting greens where superior turf quality must be maintained constantly. The method is also unacceptable for a heavily trafficked site because turfgrass is less wear tolerant when it is suffering from moisture stress.

Another way of determining when turfgrass should be watered is by examining the soil. A knife or soil probe (core sampler) can be used. If the soil is dry to a depth of 4 to 6 inches (10.2 to 15.2 centimeters), it is time to irrigate. Most of a grass plant's root system is located in the upper 4 to 6 inches of soil. Dry soil will appear a light color; moist soil is a darker color.

Some turf managers use an instrument called a tensionmeter to measure water needs. A porous ceramic cup on the bottom of the tensionmeter is connected to a piece of metal tubing. At the other end is a vacuum gauge that registers the tension at which water is held in the soil. The tensionmeter is filled with water and inserted in the soil (Figure 12-3). As the soil dries out, water is pulled from the porous cup, causing the gauge to indicate higher soil moisture tension. The turf manager can determine the proper timing of an irrigation application based on these measurements.

Gypsum, nylon, and fiberglass blocks containing electrodes can also be used to estimate the amount of available water in the soil. The percentage of available water is determined by the measurement of electrical resistance be-

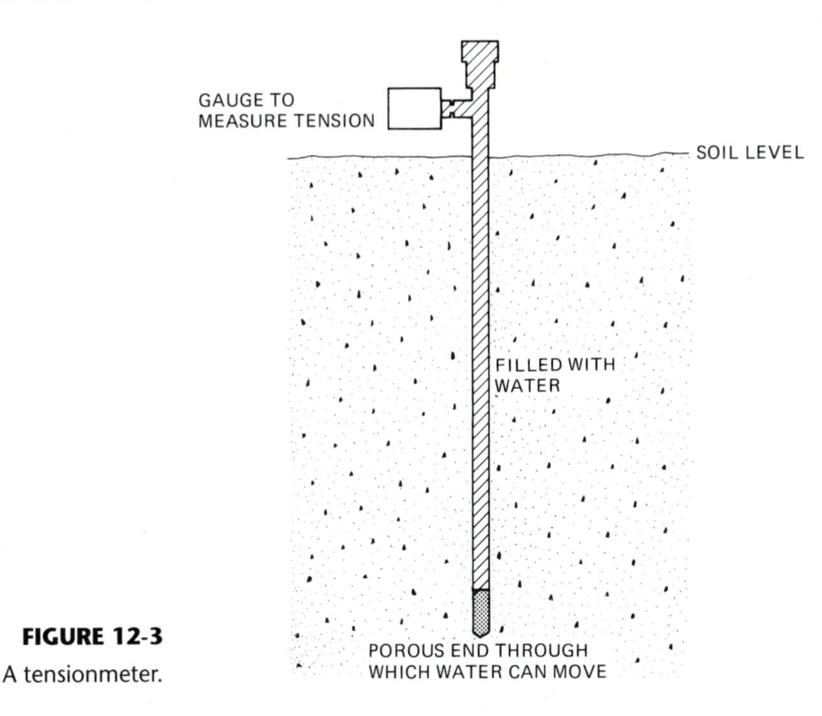

FIGURE 12-3

A tensionmeter.

tween electrodes in the block. The porous blocks are buried in the soil, and water moves in or out of them depending on the soil moisture tension. These blocks are attached to a moisture meter, which reads 100 percent at field capacity and 0 percent at the wilting point.

A rough estimate of water loss from the soil can be established by placing an evaporation pan filled with water in a sunny location. The amount of water that evaporates from the pan is similar to the amount lost from the turf because of evapotranspiration. A correlation between water loss from the evaporation pan and the appearance of moisture stress symptoms on turfgrass can be developed by the manager. Usually the amount of irrigation water applied is 75 to 85 percent of the depth of the water lost from the evaporation pan.

A rain gauge is a valuable tool which allows the manager to accurately measure the amount of rain that falls. Irrigation can be postponed or the rate decreased when significant rainfall occurs.

As a general rule, during drier periods of the growing season 1.0 to 1.5 inches (2.5 to 3.8 centimeters) of water should be applied per week to maintain green, actively growing turfgrass. In hotter, more arid areas 2 inches (5.1 centimeters) or more a week may be necessary (Figure 12-4).

A heavy irrigation once or twice a week is best. The total amount of water required per week can be applied in one application if the soil has the capability to store that much water in the root zone. Sandy soils should be irrigated twice a week. Half of the amount of water needed per week should be applied

To apply 1 inch of water per 1,000 ft^2 requires 620 gallons or 83 cubic feet of water.

1 acre inch = 27,154 gallons or 3,630 cubic feet

1 gallon = 8.33 pounds = 0.13 cubic feet = 3.785 liters

1 cubic foot = 7.48 gallons = 62.4 pounds = 28.3 liters

FIGURE 12-4

Some relevant weights and volumes of water.

every three or four days. If an inch or more is applied in one irrigation, much of the water may drain down below the roots and be wasted.

Daily watering is not usually recommended. If the soil surface is constantly moist, the roots remain near the surface. If the upper several inches of soil are allowed to dry out between irrigations, the roots are forced to grow down deeper in the soil to search for water (Figure 12-5). Shallow-rooted grass is weaker, more susceptible to stress and injury, and less recuperative than deep-rooted turf.

Irrigating too frequently also leads to greater disease and weed problems. When the surface soil is continually moist, weed seed germination will increase. However, there are circumstances when frequent and even daily irrigation is necessary. Grasses can become temporarily unable to support a deep root system because of problems such as spring root die-back of warm season species; root diseases; injury by grubs, nematodes, and other root feeders; or severe scalping. Cool season species may become shallow-rooted when summer temperatures are so high that the turf suffers from heat stress.

The turf manager should regularly check root depth with a core or profile sampler. A shallow root system resulting from some type of serious stress requires frequent shallow irrigation until the grass recovers.

Daily irrigation is also an acceptable practice on some high-maintenance sites such as putting greens. Superior quality is expected, and this may require frequent irrigation. Grasses growing on greens have a shallow root system because of close mowing. The shorter grass is cut, the shallower its root system. Because the upper few inches of soil can dry out quickly, the turf may wilt if it is not irrigated frequently. Wilted grass looks less attractive and is readily injured by traffic.

FIGURE 12-5

The plant on the left has a deep root system because the upper inches of soil are allowed to dry out between irrigations. The very shallow root system on the right is the result of daily irrigation.

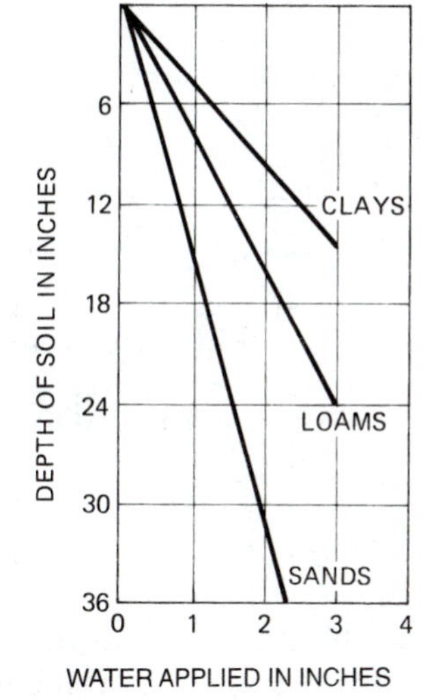

FIGURE 12-6

Depth of soil wet by the application of 1, 2, or 3 inches of water.

One way to tell when the proper amount of water has been applied is to examine the soil. Generally, sufficient water has been applied when the soil is wet to a depth just below the majority of the root system, which is usually 4 to 6 inches (10.2 to 15.2 centimeters). The amount of irrigation water applied can also be estimated by using Figure 12-6. If, for example, the manager wants to put down 1 inch of water and the grass is growing on soil with a clay texture, the soil should be wet to a depth of approximately 4 inches (10.2 centimeters). Once the manager has determined the amount of time it takes to moisten soil to the proper depth, he or she knows how long to water the area in the future.

Another method of determining the amount of water applied is to measure the amount of water each sprinkler delivers in a given time. At least four cans of the same diameter are placed at various distances from the sprinkler (Figure 12-7). After one hour all of the water collected is poured into one can, and the depth is measured. Then the depth in inches is divided by the number of cans to determine the application rate. If, for example, five cans were used and a total of 2.5 inches was collected, the application rate is 0.5 inch per hour. This is also called the *precipitation rate*. A turf manager with an inground irrigation system can accurately estimate precipitation rate by knowing system characteristics such as head spacing, nozzle size, and pressure.

Clayey or compacted soils and slopes can present a problem because of their low infiltration rate. If water moves into the soil at a slow rate, runoff may occur. To prevent this, the sprinkler should not be operated continually for a long period. Instead, water should be applied gradually by turning the sprinkler off and then on again several times. If, for example, 30 minutes of irrigation is required to apply the correct amount of water, three 10-minute applications spaced at least 1 hour apart would be recommended for an area

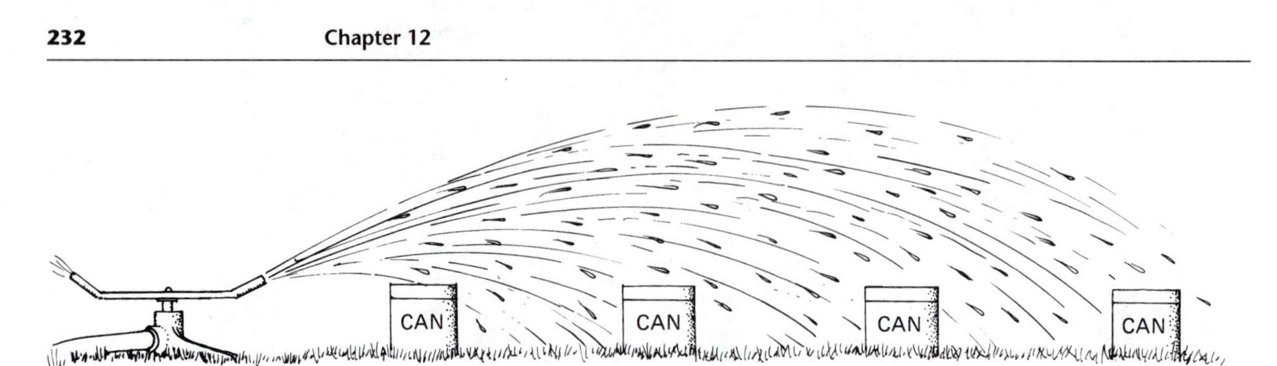

FIGURE 12-7 Place at least four containers at varying distances from the sprinkler.

exhibiting a low infiltration capacity. Increasing irrigation frequency also helps to solve the problem of poor infiltration. The manager may have to water twice a week at a half-rate each time, for instance, rather than supplying the full amount of water once a week.

Time of Day to Irrigate

An old adage among golf course superintendents is that "the best time of day to water is whenever it is convenient." Certain times of day, however, are preferable. The best time to irrigate is when there is little wind, high humidity, and lower temperatures. This substantially reduces water loss due to evaporation. The least amount of water loss occurs when turfgrass is irrigated at night or early in the morning. As much as 50 percent of the water applied during a midday irrigation may evaporate before it reaches the ground.

At night and in the early morning there is normally higher humidity and less wind than in late morning and the afternoon. On a typical summer day the relative humidity is 100 percent at 6 A.M., but drops to below 40 percent in the early afternoon. Little or no wind is preferred because air movement results in greater evaporation losses and poor water distribution. It is difficult to irrigate an area evenly when wind is blowing the water droplets in one direction. The higher temperatures and greater solar radiation in the middle of the day also increase evaporation losses.

Evening is the most convenient time to irrigate; however, disease problems are sometimes more prevalent if the grass is damp overnight. Fungi, the microorganisms that cause most turfgrass diseases, are better able to penetrate plant tissue when the leaves and stems are wet for long periods. For this reason some turf specialists recommend early-morning irrigation because the leaf surfaces will dry off quickly rather than remain moist all night.

Fungi are most likely to spread and infect when plant surfaces are moist for at least 12 continuous hours. If the grass is wet with dew most of the night, it is best to avoid irrigating at a time that would extend the wet period. It would be a mistake to water early in the evening before the dew forms or later in the morning when the grass is starting to dry off.

Many golf course superintendents water in the evening after the golfers are through because golfers begin playing again early in the morning. Putting greens are often sprayed regularly with fungicides, so the possibility of increased disease problems because of night irrigation is not a major concern. Another advantage of turning on the sprinklers at night is that the water will have more time to drain before golfers and equipment move onto the turf the next morning.

Midday irrigation is discouraged because of the amount of water that is wasted. It may also result in compaction problems if the turf is heavily trafficked during the day when the soil is wet. In areas where water use is restricted, irrigation may only be allowed at night. However, if water conservation practices are not necessary in the turf manager's region, irrigation in the middle of the day is acceptable.

Syringing

Syringing is the application of small amounts of water to turfgrass. Usually the sprinklers are turned on for only five minutes or less. The common reason for syringing is to cool the grass. This practice occurs primarily on golf greens. Turfgrasses growing on greens are very susceptible to heat injury because they are cut to such a low height. The short leaves shade the crown to only a limited degree. During hot summer days greens are syringed in the early afternoon to reduce the temperature of plant surfaces and the surrounding air. As the water evaporates from the leaf and stem surfaces, it has a cooling effect (Figure 12-8). Syringing is also performed to remove dew or frost from greens.

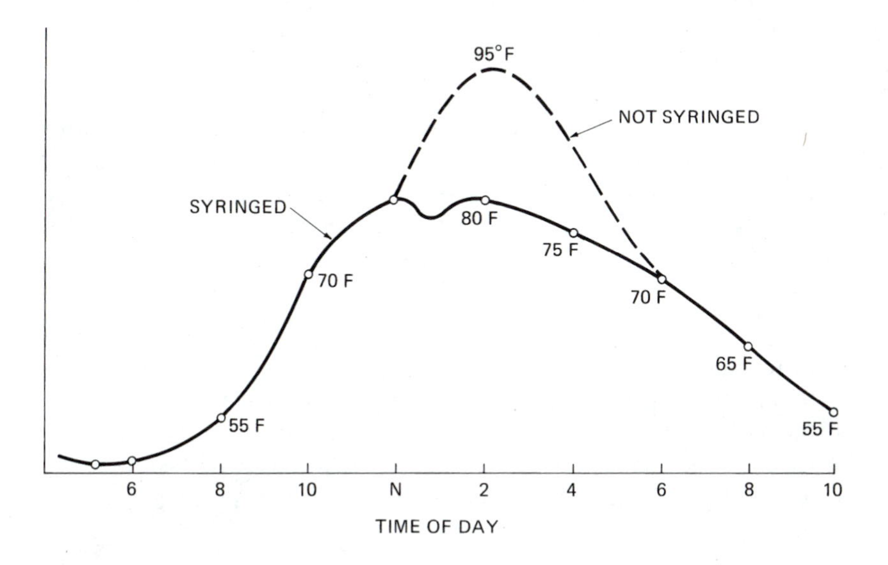

FIGURE 12-8

The cooling effect of syringing can be quite significant.

IRRIGATION SYSTEMS

An irrigation system is composed of the water source, a pump or pumps, water distribution lines (piping), valves, sprinkler heads, and, in the case of automatic systems, controllers. The controller turns on the sprinklers automatically at preset times by activating a remote control valve. When the valve is opened, water pressure forces the sprinkler heads up out of the ground and irrigation begins. When the preset irrigation time is completed, the valve closes and the sprinkler heads drop back into the ground. The majority of irrigation systems installed in recent years have been the automatic type.

The manual systems do not have an automatic controller. Instead, the water is turned on and off manually. The other major difference between the two systems is that in a manual system the sprinkler heads are not installed permanently in the ground but must be inserted into a water valve before irrigating and then be removed afterward.

Water Source

The water source for an irrigation system can be a river, stream, lake, reservoir, pond, well, or municipal water line. The water source must be able to provide sufficient quantities of water of acceptable quality throughout the growing season, even during droughty periods. The suitability of water for irrigation purposes depends on the type and concentration of substances in the water. High concentration of certain chemicals in irrigation water can result in serious injury to turfgrass. The total salt concentration (salinity) and the concentration of sodium specifically are important factors in determining water quality. The United States Department of Agriculture has developed a classification system for irrigation waters based on salinity and sodium hazards. Irrigation water should be tested regularly in areas where water quality is a problem.

In some regions of the United States droughts and population increases have resulted in water shortages. In parts of Arizona, California, Florida, and Texas, for example, there is concern about "wasting" valuable drinking water supplies for turfgrass irrigation. This concern has created and interest in using effluent water reclaimed from sewage for irrigation. Past safety concerns have been greatly reduced as a result technological advances in sewage reclamation.

Pumps

A pump or pumps are needed to pull water from the water source unless the irrigation system is small and is tapped directly into a city or town water main. The pumps that pull water from the source are called system-supply pumps. Booster pumps are installed in a pressurized line to increase pressure if the existing pressure in the system is insufficient. Both types of pump are generally centrifugal. Pressure is created by an impeller inside the pump housing which rotates and forces water through a discharge outlet.

Water Distribution Lines

Distribution lines are pipes that transport water throughout the system. Larger-diameter lines, called mains, carry water from the source, and then smaller lines branching from the mains lead to the sprinkler heads. Today virtually all the pipe used is plastic (Figure 12-9). Metal pipe made of galvanized steel, cast iron, or copper was used in the past, but has been largely replaced by plastics because they are cheaper, do not corrode or rust, and are lightweight and easier to work with.

There are basically two types of plastic material used for irrigation system pipe: polyvinyl chloride (PVC) and polyethylene (PE). PVC pipe is stronger and more durable than PE pipe and is far more popular. Pieces of PVC pipe can be easily joined by a solvent weld. A solvent-cement is applied to the outside of the ends of the two pieces of pipe and to the inside of the fitting (often a plastic sleeve) which will join the two. The solvent-cement chemically fuses the pipe ends and fitting together, producing a joint that is as strong as or stronger than the pipe itself. In cold climates water in the pipes could freeze and rupture the plastic. To prevent this the pipe must be installed deep enough in the soil to escape freezing or the water must be drained from the system for the winter.

FIGURE 12-9
Plastic pipe being installed with a trenching machine. *(Courtesy of Charles Machine Works, Inc.)*

Valves and Controllers

All irrigation systems have a series of valves which regulate the flow of water through the system. In an automatic system, remote control valves are operated by the controller.

The basic components of a controller are a clock, timers, and a series of terminals called stations (Figure 12-10). Each terminal (or station) is connected to one or more remote control valves by electric wire or hydraulic tubing. Each valve in turn operates one or several sprinkler heads. The area that a station controls is called a zone. The clock turns on a timer at preset times. The timer energizes a series of terminals, one after the other and in sequence. This automatic progression of energizing stations is called a cycle. As each station is energized, the zone it controls is irrigated. One zone is irrigated at a time because usually the capacity and pressure of the water supply are not great enough to satisfactorily operate all of the sprinkler heads in a system simultaneously.

There are two major types of remote control valve: electric and hydraulic. Electric valves are energized and opened when electric current flows from the station in the controller to the valve (Figure 12-11). The wire that carries the current is buried underground. Hydraulic valves are connected to the station by small-diameter plastic or copper hydraulic tubing. The valve opens or closes according to water pressure in the tubes. The remote control valves are buried in the ground near the sprinkler head or heads they control (Figure 12-12).

FIGURE 12-10

A controller. *(Used with permission of The Toro Company. "Toro" is a registered trademark of The Toro Company, Minneapolis, Minnesota.)*

FIGURE 12-11
An electric remote control valve. *(Courtesy of Weathermatic Division, Telsco Industries)*

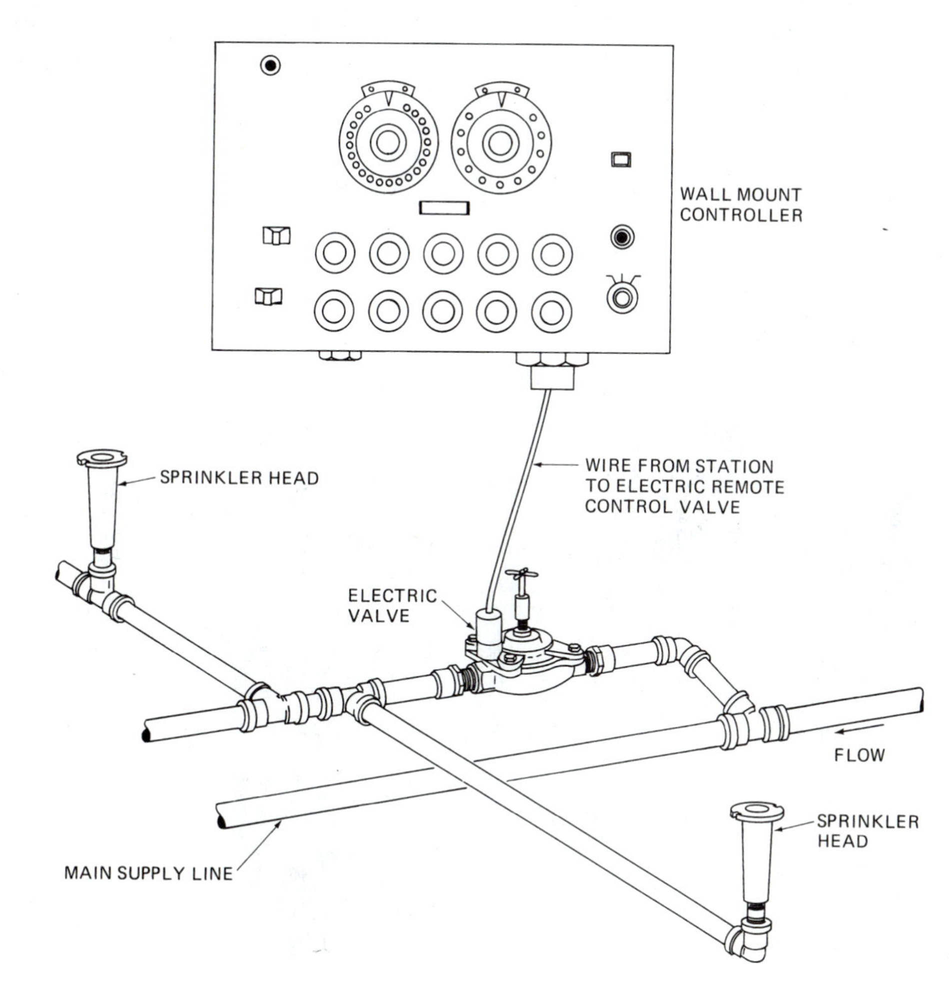

FIGURE 12-12
The station energizes the electric remote control valve, which opens, allowing water to flow to the sprinkler heads controlled by that valve.

Smaller irrigation systems have one controller, while systems at large facilities such as golf courses may have one central controller which programs and controls a series of satellite field controllers. The more sophisticated controllers have accessories such as sensory devices that shut off the system automatically if rain occurs or if the pressure in the system becomes too high or low. The controller can also be attached to tensionmeters which enable it to measure soil moisture levels. On golf courses and other large irrigated areas computer-controlled systems are common.

Sprinkler Heads

Sprinkler heads distribute water to the turf. The water is forced out through orifices (openings) in the sprinkler head nozzles. The heads used in an automatic system are called pop-ups because they are forced up out of the ground by water pressure when the valve to which they are connected opens (Figure 12-13). When the valve is closed the nozzle recedes into the body of the sprinkler. In some models the valve is built into the head. When not operating the sprinkler is completely concealed in the ground except for the cover plate.

There are two types of pop-up sprinkler heads—spray and rotary. Spray heads discharge a fine, uniform spray of water in all directions (Figure 12-14). They cover relatively small areas and are primarily used on lawns or other

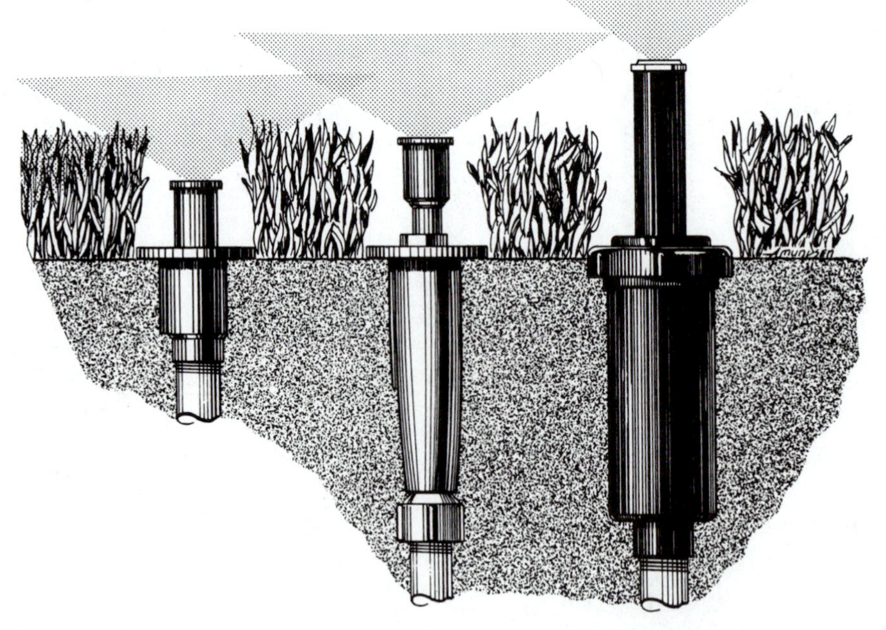

FIGURE 12-13
Three pop-up sprinklers. The one on the left is not suitable for this situation because the nozzle does not rise above the grass. *(Courtesy of Weathermatic Division, Telsco Industries)*

small turf areas. The circular distribution pattern is normally 16 to 30 feet (4.9 to 9.1 meters) in diameter.

Rotary heads are also called rotating stream heads because a stream of water is released through one or several nozzles that rotate as the head operates (Figure 12-15). They are more widely used than spray heads because a rotary head can irrigate a larger area (50 to 200 feet in diameter). Consequently, fewer heads are needed, and so a rotary sprinkler system often costs less than half as much to install as a spray head system. However, spray heads apply water at a much faster rate and provide a more uniform distribution than the rotary type.

Rotary sprinklers have different types of drive mechanisms that make the nozzles rotate. The most common is the impact-drive type, which has a spring-loaded drive arm that causes the nozzle assembly to rotate (Figure 12-16). A stream of water from the nozzle deflects the drive arm sideways, and the spring pulls the arm back into the path of the stream. As the drive arm completes each swing it impacts against the nozzle assembly and rotates it slightly. The gear-driven type works differently; water spins a rotor inside it, which in turn drives a gear that causes the head to rotate (Figure 12-17).

Quick-coupler sprinkler heads are used in manual systems. A quick-coupler valve is permanently installed at each sprinkler location. Water pressure and a spring hold the valve closed when not in use. When a quick-coupler sprinkler is inserted in the valve and turned the value is opened. Normally quick-coupler sprinklers are the impact-drive rotary type. After the irrigation is completed the heads are removed from the valves. Some quick-coupler valves may be installed in an automatic system as a backup in case of problems with the regular system. A hose connection can also be screwed into a quick-coupler valve.

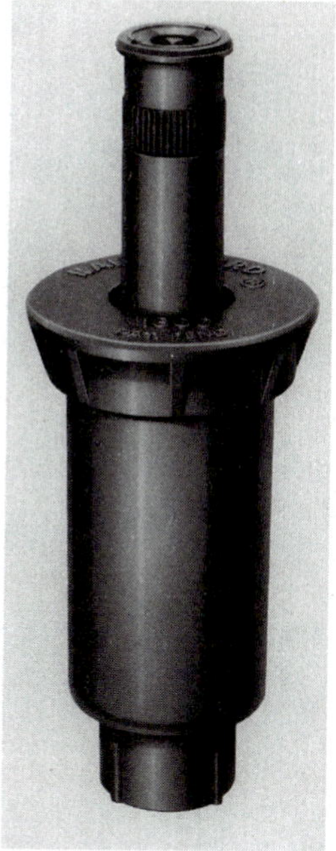

FIGURE 12-14

A spray head. *(Courtesy of Rain Bird Sprinkler Mfg. Corp.)*

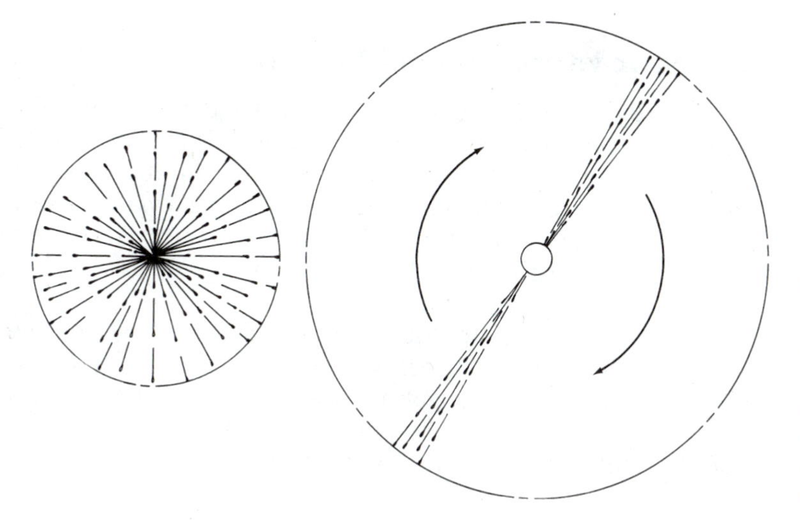

FIGURE 12-15

A spray head releases water in all directions simultaneously (left), while a rotary head (right) usually releases one to two streams of water as it rotates.

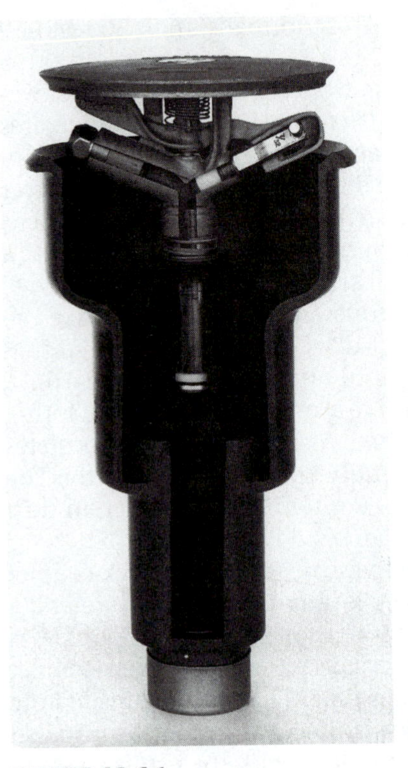

FIGURE 12-16

An impact-drive rotary type pop-up sprinkler head. *(Courtesy of Royal Coach/Buckner)*

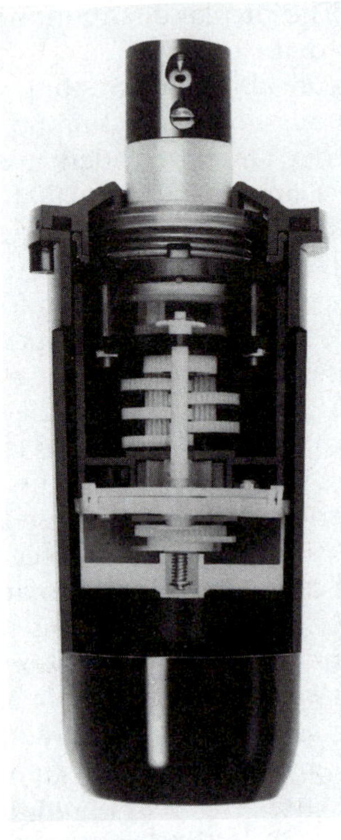

FIGURE 12-17

A view of the inner mechanisms of a gear-driven rotary-type pop-up sprinkler head. *(Courtesy of Weathermatic Division, Telsco Industries)*

Automatic versus Manual Systems

The only advantage to a manual system is its cheaper installation cost. An automatic system is very expensive initially, but this large installation cost is offset by a major reduction in labor costs. An automatic system is more efficient than a manual system and uses less water, so water costs are also lower. Another advantage of an automatic system is that it gives the manager complete control over irrigation. He or she programs the controller and is not dependent on an irrigator.

Automatic systems are not totally labor-free. They require maintenance, constant checking for malfunctions, and repairs when problems occur.

Some facilities have irrigation systems that are part automatic and part manual. Systems that are a combination of both types are called semiautomatic. Examples are those that have pop-up sprinklers and manual control valves or quick-coupler heads operated by an automatic controller. A golf course may have an automatic system for greens and a manual system on fairways. A semiautomatic system is installed as a compromise when money is not available for a fully automated one. The system may be converted to a completely automatic system later.

An automatic system is far superior to other types of systems because of labor savings, water use efficiency, and manager control of irrigation practices.

Irrigation System Design

Irrigation systems should be designed by experienced professionals because there are many considerations involved. The number of zones and sprinkler heads per zone depends on the volume of water available and water pressure. The spacing of heads depends on their coverage, which is determined by factors such as water pressure, nozzle size, size of the orifice (the opening in the nozzle), and wind. Heads are normally spaced closer together when wind is a problem.

Sprinkler-head distribution patterns must overlap. As the radius of sprinkler coverage increases, water is distributed over an increasingly large area. Consequently, as the distance of the turf from the head increases, the amount of water it receives decreases (Figure 12-18).

Overlapping sprinkler patterns help to equalize water distribution. Normally the system is designed with 70 to 100 percent overlap. Equilateral triangular spacing provides the most uniform coverage, but square patterns are also used (Figure 12-19). Spray heads are generally installed 20 feet (6 meters) apart, while the best spacing between rotary heads is usually 65 to 70 feet (20 to 21.3 meters) (Figure 12-20).

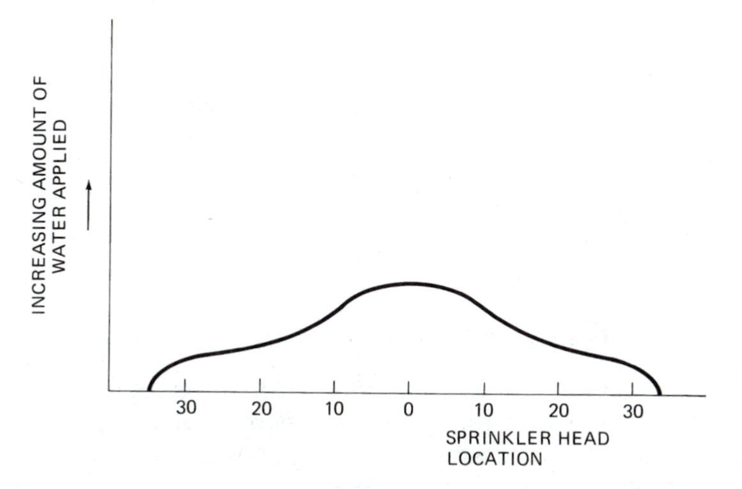

FIGURE 12-18

Areas farther away from the sprinkler head receive less water than areas close to it.

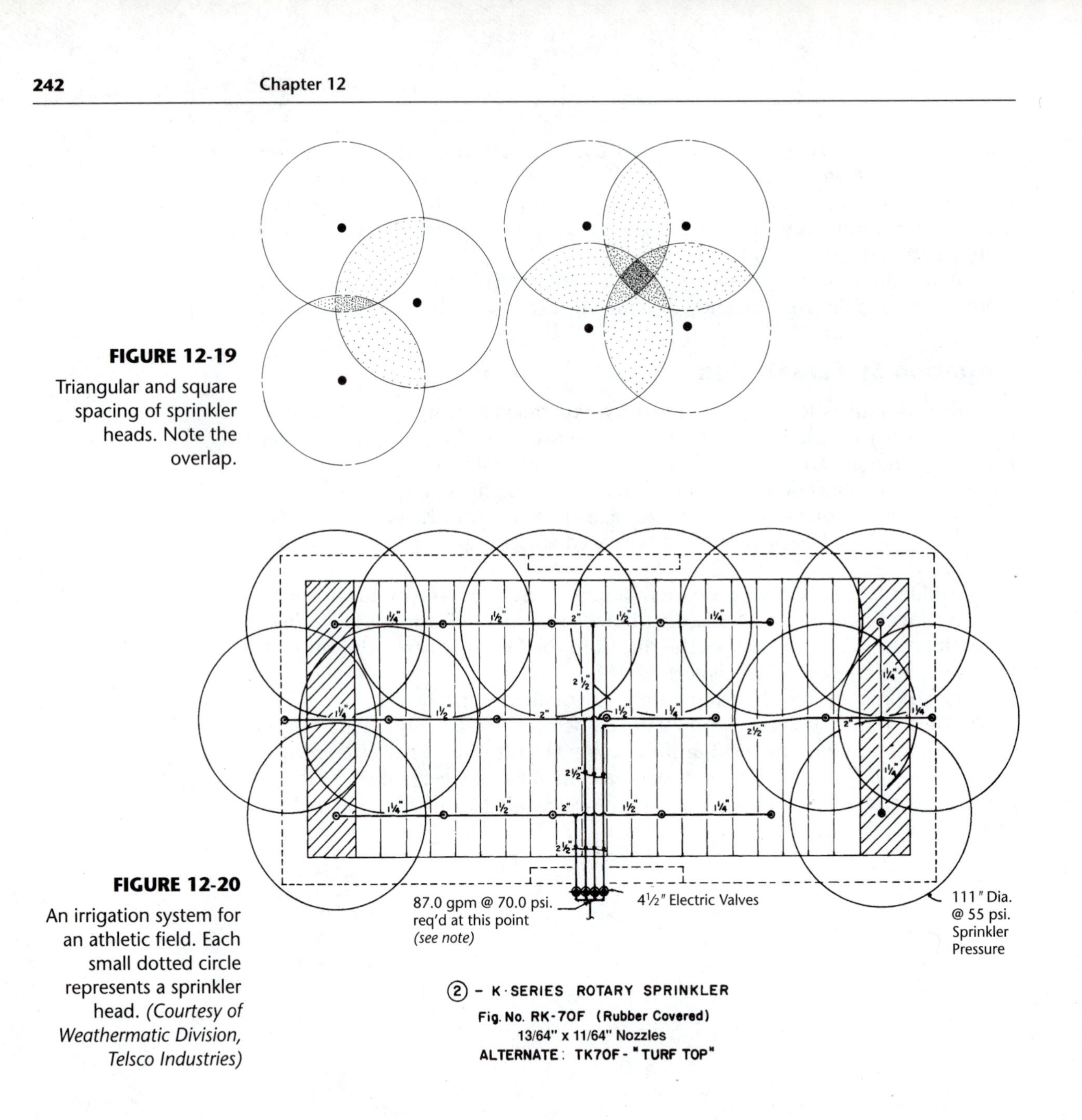

FIGURE 12-19

Triangular and square spacing of sprinkler heads. Note the overlap.

FIGURE 12-20

An irrigation system for an athletic field. Each small dotted circle represents a sprinkler head. *(Courtesy of Weathermatic Division, Telsco Industries)*

Other Types of Irrigation Systems

Subirrigation, as the name implies, involves underground irrigation. Small plastic or ceramic tubes that have narrow slits or pinhole-sized openings are buried in the root zone. The irrigation water moves through these tiny holes into the soil. A subirrigation system conserves water because atmospheric

FIGURE 12-21

A type of mobile, aboveground irrigation system. *(Courtesy of Northrup King Co.)*

evaporation losses are not a factor when the water is distributed. Sheets of plastic may be placed beneath the root zone to prevent water from percolating deeper into the soil. Subirrigation is becoming popular in areas where water availability is a problem.

The system shown in Figure 12-21 is used on large, open turf areas such as seed production fields and sod farms. The wheels turn slowly and move the irrigation pipe and sprinkler heads across the field.

Many residential lawns are watered by connecting a hose to an outside faucet and periodically moving the hose and attached sprinkler. A traveling sprinkler is a bit more sophisticated. It is attached to a hose, and when the water is turned on the pressure causes a gear mechanism to slowly move the sprinkler across the turf (Figure 12-22).

FIGURE 12-22

A traveling sprinkler. *(Courtesy of Royal Coach/Buckner)*

LIMITING WATER USE

Various techniques can be incorporated into the turfgrass management program to decrease the amount of water used for irrigation purposes. The following practices help to conserve water.

1. Raise the mower blade 0.5 to 1 inch (1.3 to 2.5 centimeters). Taller grass has a deeper root system. Because soil dries out from the surface down, the roots will have a better chance of finding water deep in the soil. Longer leaves result in higher transpiration losses because of the increased surface area, but the advantages of a deeper root system outweigh this disadvantage. The longer leaves will shade the soil surface and reduce evaporation losses. Shading also protects the crown from high-temperature injury. Cool season grasses may be unable to produce deep roots during very hot weather. In this case it is better to lower the cutting height temporarily to reduce transpiration losses

2. Reduce mowing frequency. Significant moisture loss occurs through mower wounds. The more often the grass is cut, the longer the period during which mower wounds are open. Be sure to use a sharp blade. Mowing with a dull blade causes jagged wounds that take longer to heal.

3. Apply lower rates of nitrogen fertilizer during droughty years. Grass stimulated by high rates of nitrogen grows fast and needs more water. Leaves become lush and succulent and are more prone to wilting. Use turf fertilizers that contain ample amounts of phosphorus and potassium. These two nutrients have been shown to increase the drought tolerance of grasses.

4. Remove thatch with a vertical mower when the thatch layer becomes too thick. A deep layer causes shallow rooting and slows water movement into the soil.

5. Core-cultivate compacted soils. Compaction reduces the infiltration rate and causes runoff. A core cultivator or aerator makes holes in the soil which allow water to penetrate more easily.

6. Use herbicides sparingly. Some herbicides injure turfgrass roots.

7. When establishing new turfgrass, plant drought-tolerant species and cultivars. Improve the water-holding capacity of a sandy soil before planting by adding organic matter or other amendments. Water-absorbent ploymers are new products that show promise. They can absorb 50 to 500 times their weight in water and will release the water for plant use as the soil dries.

8. Before irrigating, check the weather forecast to see if it will rain soon.

In drier parts of the country strict water conservation measures are being mandated. The turf manager may not be able to irrigate as much as he or she would like. Even where water use is not restricted, the increasing cost of water may necessitate less irrigation. Such water limitations will not have a catastrophic effect on the turfgrass industry. Expectations may have to change, however. People may have to learn to be satisfied with turfgrass that has a slightly lower quality than they expected in the past. Plant breeders will develop new cultivars that exhibit greater drought tolerance, and this will help to compensate for reduced irrigation.

WETTING AGENTS

Wetting agents reduce the tendency of water to form drops and reduce the tension between water and solids or other liquids. Their application to turf increases the wetting capacity of water in the soil. Some soils are difficult to wet—they are said to be hydrophobic. Organic matter in the soil may repel water, resulting in poor wetting and dry spots.

Localized dry spots can occur when fungi grow on soil particles and coat them with a waxlike material. This phenomenon can be observed by pulling a plug of the soil and placing drops of water on it. The water will not penetrate the soil. This problem is most common with sand particles.

Thatch is also hydrophobic. It can impede water movement into the soil and is very hard to wet.

When wetting agents are present water penetrates rapidly into the soil and wets the thatch. There is a more uniform wetting of the root zone, and drainage is improved.

Another benefit is water conservation. Wetting agents help water to infiltrate into hard-to-wet soils and on slopes. Less water is wasted because of evaporation and runoff.

The primary use of wetting agents is on hydrophobic soils. They can be sprayed or spread on problem areas or applied through the irrigation system. In some cases their use may result in improved drainage in layered soils.

SELF-EVALUATION

1. The combined loss of water from plants and soil is called ___________.
2. ___________-rooted grasses tend to be more drought tolerant.
3. Soil ___________ and ___________ have a major influence on infiltration and percolation rates.
4. At field capacity, gravitational water has drained from the ___________.
5. A tensionmeter measures soil moisture ___________.
6. ___________ irrigation is not usually recommended.
7. Reducing the temperature on a golf green by applying a small amount of water is called ___________.
8. How can the turf manager tell when irrigation is necessary?
9. An automatic irrigation system has ___________ control valves and ___________ sprinklers.
10. ___________ heads can irrigate a larger area than ___________ heads.
11. The size of the root system can be increased by what simple change in the maintenance program?
12. Why are automatic irrigation systems superior to manual systems?
13. When should turfgrass be irrigated everyday?
14. Discuss ways to reduce irrigation needs.

CHAPTER

13

Pesticides

OBJECTIVES

After studying this chapter, the student should be able to

- Discuss the different types of pesticides and formulations available to the turf manager
- Explain the process of pesticide applicator certification mandated by the Environmental Protection Agency
- Describe the fate of pesticides in the environment
- Discuss pesticide toxicity and methods of measuring toxicity
- State why it is important to read a pesticide label and follow the directions on the label
- List the protective clothing that must be worn when handling various pesticides
- Identify the symptoms of pesticide poisoning
- Describe safe methods of pesticide storage

INTRODUCTION

A turfgrass pest is defined as any organism that causes a decrease in turf quality. These pests include fungi and other microorganisms which cause turf diseases, insect and nematode species that feed on turfgrass (Figure 13-1), and weeds. A pesticide is a toxic chemical that destroys pests. Pesticides are very controversial because of problems that can result from their misuse. It is very important that turf managers be knowledgable about pesticides so they can protect themselves and the environment if they use them. Methods of reducing pesticide use are explained in Chapter 17, which discusses integrated pest management techniques.

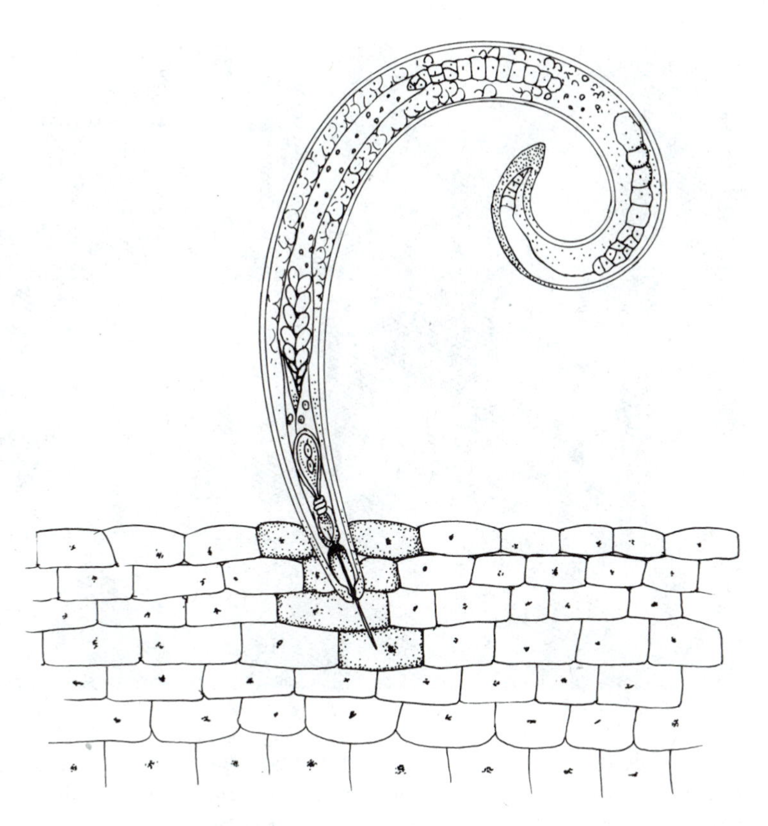

FIGURE 13-1
A nematode feeding on the roots of a turfgrass plant.

TYPES OF PESTICIDES

A number of different types of pesticides are used to control various pests. Insecticides are chemicals used to control insects. Most insecticides kill the insect pest by disrupting its nervous system. These toxic chemicals enter the insect's body when it comes into contact with the insecticide. A second type of insecticide acts as a stomach poison and must be ingested (swallowed) by the insect to be effective. Some insecticides work both ways. Insecticides are often the most dangerous type of pesticide because humans and insects have certain biological similarities. A chemical that is lethal to insects can also be injurious to humans because of these similarities. Some insecticides will kill mites, but many do not. Miticides are pesticides that are developed specifically for mite control.

Fungicides control fungi, microorganisms that can cause plant diseases (Figure 13-2). Bactericides kill bacteria, another type of microorganism that can incite plant diseases. Antibiotics are substances produced by microorganisms which inhibit or destroy other microorganisms. Some of these chemical compounds are sold commercially and can be used to control plant diseases.

FIGURE 13-2
Applying a fungicide to turf to prevent diseases. *(Used with permission of The Toro Company. "Toro" is a registered trademark of The Toro Company, Minneapolis, Minnesota.)*

Herbicides control unwanted plants. These chemicals are also called weed killers. Nonselective herbicides are toxic to all plants; selective herbicides kill some plants, but do not seriously affect others. Herbicides usually control either broadleaf plants or grassy plants. Preemergence herbicides are applied before a weed species appears in the turf. They kill the seedlings before their emergence from the soil. Postemergence herbicides are used after a weed has appeared in the turf area.

Nematicides control nematodes, tiny wormlike animals which usually live in the soil and feed on plant roots. Rodenticides control rats, mice, and other rodent pests.

Systemic pesticides enter a plant and spread internally by moving through the vascular system. They are absorbed by the roots or leaves. Most herbicides are systemics. Some insecticides and fungicides also work systemically. Systemics protect turfgrass for a longer period than nonsystemics because once they enter the plant they will not be washed away by rain or irrigation water and are not subject to decomposition by sunlight or microorganisms.

One advantage of systemic fungicides is their ability to control disease-causing organisms already established inside the plant (Figure 13-3). Systemic insecticides are beneficial because they primarily kill the destructive insects feeding on the plant rather than indiscriminately poisoning all insects. It is important to save as many beneficial insects as possible. They will help to keep insect pests under control. Systemics can also move into and protect plant tissue produced after the pesticide application.

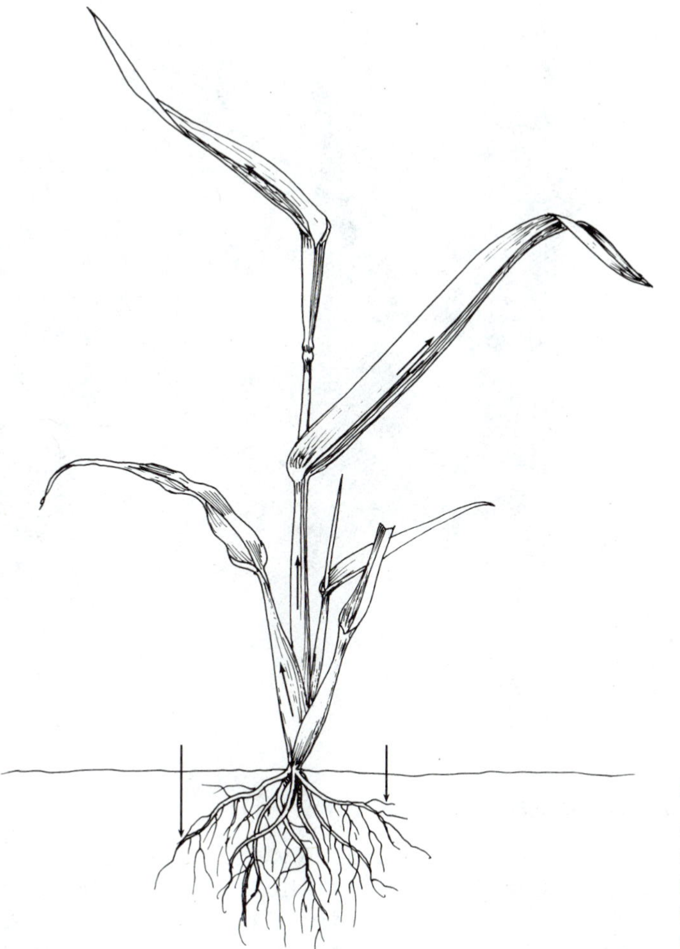

FIGURE 13-3
A systemic fungicide enters the roots of the turfgrass plant and is then translocated to the topgrowth.

PESTICIDE FORMULATIONS

Toxic chemicals used for pest control are often unsafe to handle or difficult to apply in their pure form. The manufacturer must mix or dilute the toxic active ingredient, called the *toxicant,* with other materials to make it safe and easy to spread evenly over the area to be treated. The toxicant is formulated in liquid or dry carriers by combining it with materials such as solvents, wetting agents, granules, or powders. The final product is called the *pesticide formulation.* All formulations used on turfgrass, except for granulars, are mixed with water and applied with a sprayer. Fertilizer spreaders are used to apply granular formulations. A single pesticide is often sold in several different formulations. The formulations described here are those commonly used by turfgrass managers. The list will change as pesticide chemists develop new and better formulations.

FIGURE 13-4 Two granular pesticides.

Granular Formulations (G)

Granular formulations (abbreviated as G) are dry materials that are easily spread using a seeder or a fertilizer spreader. The pesticide is coated on or absorbed by particles of clay, vermiculite, corn cobs, walnut shells, fertilizer granules, or other solid carriers (Figure 13-4). Most granular formulations have a low concentration of pesticide. The toxicant is released by moisture. When the granules become wet the pesticide is washed off or out of the carrier. Granulars are normally used to treat the soil or thatch. Most do not stick to foliage; liquid sprays adhere to leaves better than granules. Systemic pesticides that enter the turfgrass plant by root absorption are often applied in granular form. In some products the toxicant is mixed with fertilizer nutrients.

The remainder of the formulations to be discussed are mixed with water and applied as a liquid. For the pesticide to be sprayed evenly over the turf area the chemical must be diluted uniformly in water. The pesticide must be concentrated equally throughout the mixture. The pure form of the toxicant usually does not mix properly with water. Various chemical additives must be combined with the toxicant to achieve a uniform mixture. Some of these formulations dissolve in water and form a solution. Once it is in solution, the pesticide remains equally dispersed in the water and will not settle out of the mixture. Other formulations produce a mixture called a suspension. A pesticide in a suspension remains uniformly concentrated when the mixture is agitated. If the mixture is not in motion the pesticide will settle out. The turbulence necessary for agitation can be provided by the movement of a mechanical paddle located in the sprayer tank. Another method of agitation involves the recirculation of the mixture from the pump back into the tank. The pesticide mixture streams through a nozzle into the tank, keeping the tank contents in constant motion whenever the sprayer is operating.

Emulsifiable Concentrates (EC or E)

The toxicant in an emulsifiable concentrate is oil soluble but not water soluble. It is dissolved in an oil solvent such as xylene. An emulsifying agent is added to the oil-pesticide solution to enable the pesticide-containing oil to mix with water. The result is a relatively stable, milky suspension called an emulsion. Tiny oil globules are dispersed uniformly throughout the mixture. Some agitation may be necessary to maintain this uniform dispersal. Emulsifiable concentrates usually contain a high concentration of active ingredient. They may occasionally injure plants if misused because the oil solvent can be phytotoxic.

Wettable Powders (WP or W)

These preparations are sold in a dry, flourlike, powdered form. The toxicant exhibits low solubility in water. It is combined with a filler such as clay or talc and a wetting agent. The wetting agent allows the pesticide to be dispersed as a suspension in water. The mixture must be continually agitated to prevent the wettable powder from settling out (Figure 13-5). A wettable powder can be washed off foliage more rapidly than an emulsifiable concentrate.

Flowables (FL or F)

Flowables, like wettable powders, are pesticides manufactured as solid materials only. However, a flowable is sold in a liquid form. The very finely ground powder is suspended in a small amount of water or oil. The liquid paste, or slurry, that is formulated is easier to mix with water in a spray tank than a wettable powder because the pesticide in a flowable is already fully wetted. Flowables require moderate agitation to maintain a uniform tank mix.

Dry Flowables (DF)

A dry flowable holds the active ingredient in small granule-sized particles. These particles are easily measured and mixed, and there is less inhalation hazard than with a wettable powder. Some dry flowables are called water-dispersible granules (WDG).

FIGURE 13-5
Before a wettable powder is added to a spray tank it should first be mixed with water in a bucket. After vigorous stirring a slurry will form. The slurry is then poured into the tank.

Soluble Powders (SP)

These dry powders are highly soluble in water. The pesticide dissolves in water and forms a stable solution. No agitation is required after the soluble powder is mixed with water in the spray tank.

Soluble Liquids (SL)

These liquid preparations are also highly water soluble and form a stable solution in the spray tank that does not require agitation (Table 13-1).

Other Formulations

A few other formulations are occasionally used by the turf manager. Salt concentrates are formulated by producing a chemical reaction that changes a

TABLE 13-1 Pesticide Application Rates*

	Amount per Acre AI/A (lb)				
Formulation	**1**	**2**	**4**	**5**	**10**
1G	100 lb	200 lb	400 lb	500 lb	1000 lb
2G	50 lb	100 lb	200 lb	250 lb	500 lb
5G	20 lb	40 lb	80 lb	100 lb	200 lb
10G	10 lb	20 lb	40 lb	50 lb	100 lb
1EC or FL	1 gal	2 gal	4 gal	5 gal	10 gal
2EC or FL	2 qt	1 gal	2 gal	2.5 gal	5 gal
4EC or FL	1 qt	2 qt	1 gal	1.25 gal	2.5 gal
50WP or SP	2 lb	4 lb	8 lb	10 lb	20 lb
75WP or SP	1 lb 5 oz	2 lb 11 oz	5 lb 5 oz	6 lb 11 oz	13 lb 5 oz
80WP or SP	1 lb 4 oz	2 lb 8 oz	5 lb	6 lb 4 oz	12 lb 8 oz

	Amount per 1,000 ft² AI/A (lb)				
Formulation	**1**	**2**	**4**	**5**	**10**
1G	2 lb 5 oz	4 lb 10 oz	9 lb 3 oz	11 lb 8 oz	23 lb
2G	1 lb 2 oz	2 lb 5 oz	4 lb 10 oz	5 lb 13 oz	11 lb 8 oz
5G	7 oz	15 oz	1 lb 14 oz	2 lb 5 oz	4 lb 0 oz
10G	4 oz	8 oz	15 oz	1 lb 3 oz	2 lb 5 oz
1EC or FL	3 oz	6 oz	12 oz	15 oz	29 oz
2EC or FL	1.5 oz	3 oz	6 oz	7.3 oz	15 oz
4EC or FL	.75 oz	1.5 oz	3 oz	3.7 oz	7.5 oz
50WP or SP	.7 oz	1.5 oz	3 oz	3.7 oz	7.3 oz
75WP or SP	.5 oz	1 oz	2 oz	2.5 oz	4.9 oz
80WP or SP	.5 oz	.9 oz	1.8 oz	2.3 oz	4.6 oz

*These rates may be stated as pounds of active ingredient (AI) per acre or 1,000 ft². The number preceding the letter abbreviation of the formulation tells how much active ingredient is in the formulated material. For example, a 2EC or 2FL contains 2 pounds of active ingredient per gallon. A 50WP or 50SP is 50% active ingredient; that is, a 4-pound bag would have 2 pounds of active ingredient. Granular formulations have low concentrations of active ingredient. A 5G material has 5% active ingredient.

The table lists the amounts of formulated material that are necessary to apply the correct rate of active ingredient. 100 pounds of a 1G-formulated material has to be applied per acre if the recommended application rate is 1 pound of active ingredient per acre.

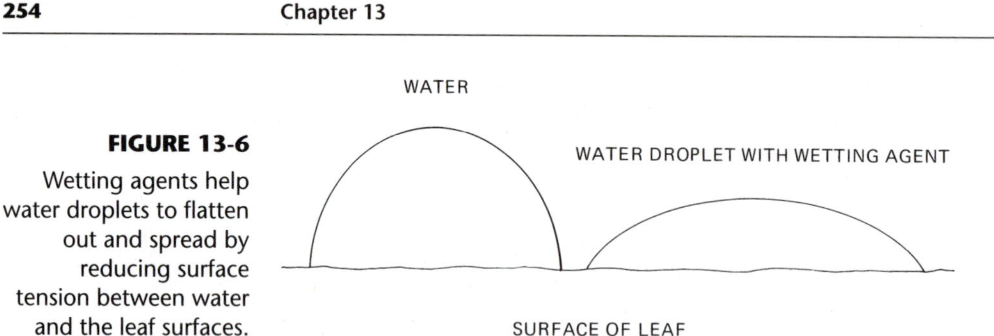

FIGURE 13-6
Wetting agents help water droplets to flatten out and spread by reducing surface tension between water and the leaf surfaces.

water-insoluble pesticide into a salt. The resulting salt form of the pesticide is highly water soluble. The widely used broadleaf weed killer 2,4-D is manufactured in a salt concentrate formulation. Fumigants are pesticides that form a gas when applied. Turf managers may fumigate a soil before establishment to kill pests such as nematodes. Poison baits are a mixture of pesticide and a food that is attractive to pests such as slugs, mole crickets, ants, grasshoppers, or yellowjackets. The pest eats the bait and is poisoned.

Adjuvants are chemical additives that improve the performance of a pesticide. Two types of adjuvants that allow the toxicant to mix with water have already been mentioned—wetting and emulsifying agents. Three other types commonly added to pesticide formulations are stickers, extenders, and spreaders. Stickers increase the initial adherence of the pesticide to foliage and enable it to remain there for a longer period of time. Without stickers rain or irrigation water may wash the pesticide off leaves too quickly for it to be effective. Extenders act like a sun screen, protecting the chemical from photodegradation by ultraviolet light. Spreaders help the toxicant to spread evenly over treated surfaces. A pesticide combined with a spreader will form a thin film and cover more leaf surface area, Figure 13-6.

THE FATE OF PESTICIDES

To be effective against turfgrass pests, the pesticide must be applied to the target zone in the turf where the pest is found. It must then retain its toxicity for a long enough period to control the pest. Drift, the movement of spray droplets by wind, may result in an insufficient concentration of pesticide reaching the target zone. Drift not only reduces the uniformity of spray coverage, it can also lead to problems in the surrounding environment. This is especially true if a herbicide drifts to desirable plants nearby and damages them. Pesticides normally should not be sprayed when the wind speed is greater than a few miles per hour. Early morning and evening are usually less windy than other times of the day.

Water can also remove the pesticide from the target zone. The chemical may be washed off treated foliage. Pesticide runoff occurs when water moves the chemical across the surface of the turf. Runoff is most common on slopes. Leaching is a process that occurs when rain or irrigation water carries the pesticide downward in the soil. The result is a lower concentration of pesticide near the soil surface. Some leaching is desirable if the pest inhabits the soil or if the pesticide is a systemic that must be taken in by plant roots. Too much leaching, however, not only moves the chemical below the target zone, it may lead to contamination of water supplies if the pesticide is carried down to the groundwater.

Volatilization is another process by which the pesticide can move from the target zone. When a chemical volatilizes, it changes into a gas and diffuses into the atmosphere (Figure 13-7). This loss results in reduced pest control and is a serious problem if herbicides in a gaseous state are carried by the wind to sensitive plants. Volatilization is most common at higher temperatures.

Even when a pesticide remains in the target zone, it eventually loses its ability to kill pests. Some pesticides retain their toxicity for only a short time. Chemicals that have a brief residual effectiveness are said to be nonpersistent. Pesticides that retain their toxicity for a long period are said to be persistent. The long-lasting pesticides can be desirable, since fewer treatments may be required because of their persistence. However, pesticides that linger too long in the environment can build up in the food chain and cause other environmental problems.

There are several reasons why pesticide residues (deposits) lose their toxicity. Chemicals adhering to foliage are broken down by sunlight. This process is called *photodecomposition*. Adsorption and microbial degradation occur in the soil and thatch. Adsorption is a process by which pesticides bind tightly to soil particles or thatch and are inactivated. The rate of adsorption increases in soils with high clay or organic matter contents. Microbial degradation is the breakdown of pesticides by microorganisms which use certain components of the

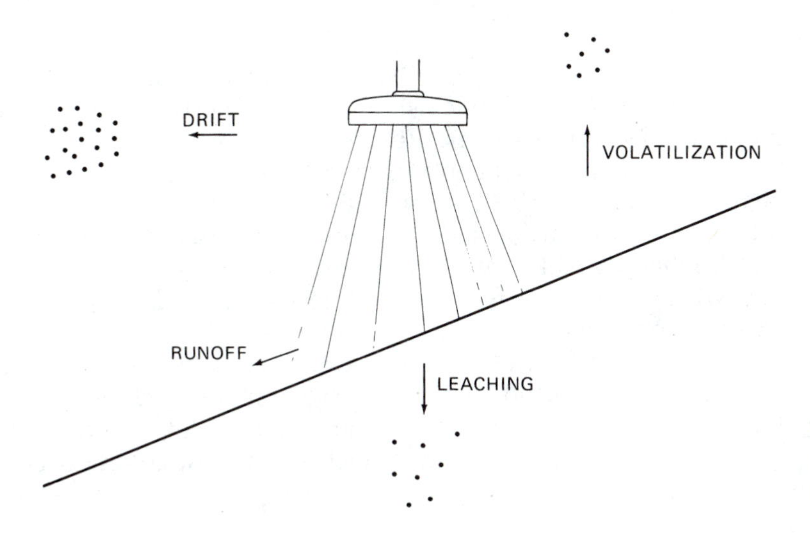

FIGURE 13-7

Movement of pesticides from the target zone.

chemicals as a food source. Decomposition by microorganisms is the major reason for the disappearance of pesticides from the environment.

Various chemical reactions can occur that result in the inactivation of these toxic chemicals. For example, the pesticide can be transformed by hydrolysis when it reacts with water in the soil. Some pesticides are affected by pH and will begin to break down in the spray tank if mixed with water that has an incompatible pH. Most pesticides should not be kept in a sprayer overnight because the mixture may become unstable.

The possible contamination of groundwater by pesticides is a serious concern. A number of pesticides, including some used on turf, have been found in groundwater. It is important to understand why this can occur.

Certain characteristics of a pesticide can be used to predict whether it is likely to move down to the water table. Chemicals that have a long half-life may reach groundwater before they are degraded. Some are so persistent that they are still found in groundwater years after their use was discontinued. Pesticides that are very water soluble can readily move downward in the soil. The degree to which a pesticide is adsorbed to soil particles or thatch is significant. Some are held tightly and do not move.

A pesticide that is used frequently and at high application rates is more likely to cause problems. Soil texture and structure have a significant impact on pesticide movement. Leaching is more common in well-drained soils. Other important factors include the amount of irrigation and the distance from the soil surface to the groundwater.

There are also major concerns about pesticide runoff into reservoirs, lakes, and rivers. Protection of drinking water has become a top priority of government agencies and environmental groups. Further restrictions on pesticide use will occur, especially in areas with sandy soils.

PESTICIDE TOXICITY

Pesticides can be hazardous to both the environment and people. Some can kill or seriously injure people, while others are fairly safe. All can be dangerous if misused. Even relatively nontoxic pesticides may cause skin irritations.

To poison or injure people or animals, the pesticide must be on or in the victim's body. Thus, the applicator should never allow the chemical to come into contact with his or her body. Pesticides enter the body by three routes: oral contact, dermal contact, and inhalation (Figure 13-8). Oral contact occurs when a pesticide is swallowed accidentally, usually by children. Oral entry often results in serious injury. Dermal contact occurs when a pesticide contacts the skin. Many chemicals can be absorbed through the skin. Pesticides are readily absorbed through skin on the neck, groin, feet, armpits, hands, and wrists. Contact with the eyes or a cut or scrape also results in rapid pesticide entry. Dermal contact is the most common entrance route. Inhalation occurs when a person breathes in a pesticide. Gaseous vapors or

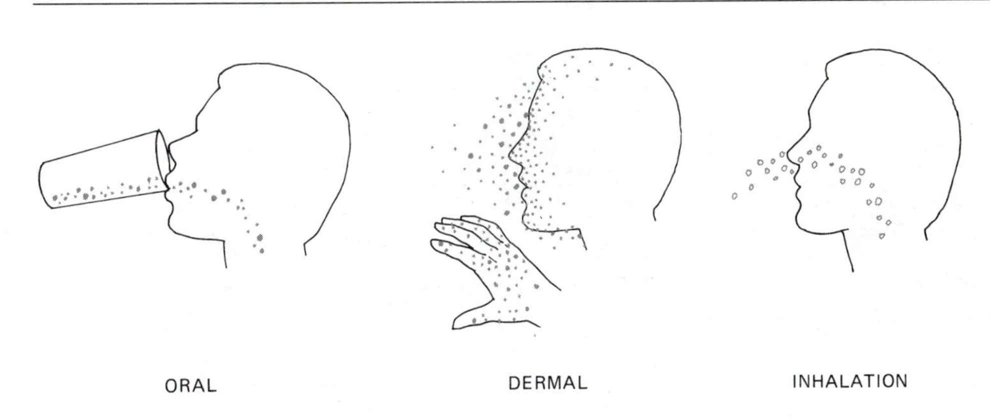

FIGURE 13-8
Ways pesticides enter the body.

very fine dry particles of pesticide in the air can be inhaled through the mouth or nose.

Types of Toxicity

Acute Toxicity is a measure of how poisonous a pesticide is in a single exposure. The acute toxicity of a pesticide indicates how great the immediate danger is to a person after exposure to the chemical. *Chronic toxicity* is a measure of how poisonous a pesticide is after repeated exposures over a period of time. Some pesticides accumulate in the body and can eventually cause cancer or other serious problems. A person may handle pesticides for many years before the physiological disorders associated with chronic toxicity become apparent.

Measuring Acute Toxicity

Acute toxicity is relatively easy to measure. Test animals such as rats or mice are exposed to various concentrations of a pesticide, and their death rate is observed. Acute toxicity is measured by determining an LD_{50} value. The LD stands for "lethal dose"—the amount of pesticide required to cause death. The 50 means that 50 percent of the test animals are killed by this dose. The LD_{50} value is expressed in milligrams of pesticide per kilogram of test animal body weight (mg/kg). This metric measurement is the same as parts per million (ppm). If the LD_{50} of a pesticide is 250, this means that a dose of 250 milligrams of the chemical per kilogram of body weight (or 250 parts of pesticide for every one million parts of body weight) results in the death of 50 percent of the test animal population.

The lower the LD_{50} number, the more poisonous the pesticide. A chemical with an LD_{50} of 10 is 100 times more toxic than a substance with an LD_{50} of 1,000. LD_{50} values are established for both the acute oral and acute dermal toxicity of a pesticide. Acute inhalation toxicity is measured by LC_{50} (lethal concentration) values. The LC_{50} is expressed in milligrams of pesticide per liter

TABLE 13-2 Signal Words and Categories of Acute Toxicity

Category	Signal word required on label	Categories: LD_{50} Oral (mg/kg)	LD_{50} Dermal (mg/kg)	LC_{50} Inhalation* (mg/l)	Probable Oral Lethal Dose for 150-lb (68-kg) person
I Highly toxic	DANGER (skull and cross-bones) POISON	0–50	0–200	0–2.0	Few drops to 1 teaspoon
II Moderately toxic	WARNING	Over 50 to 500	Over 200 to 2,000	Over 2.0 to 20	Over 1 teaspoon to 1 ounce
III Slightly toxic	CAUTION	Over 500 to 5,000	Over 2,000 to 20,000	Over 20 to 200	Over 1 ounce to 1 pint or 1 pound
IV Relatively nontoxic	CAUTION	Over 5,000	Over 20,000	Over 200	Over 1 pint or 1 pound

*Values depend on exposure time.

of air. A liter is a metric measurement of volume approximately equal to one quart. The lower the LC_{50} number, the more poisonous the pesticide.

Label Warning Statements

To alert pesticide users to the acute toxicity of a pesticide, a signal word must appear on the label on the pesticide container. Pesticide products are categorized as highly toxic, moderately toxic, slightly toxic, or relatively nontoxic. This decision is based on both the acute toxicity of the active ingredient and its concentration in the formulated material. For example, a product containing a small amount of highly toxic active ingredient may be less toxic than a product containing a high concentration of lower-toxicity active ingredient. The category in which the pesticide is placed is indicated by the signal word on the label (Table 13-2). All pesticides need to be handled safely; however, those identified by the signal words DANGER-POISON must be handled with extreme care (Figure 13-9).

Measuring Chronic Toxicity

Chronic toxicity is the result of repeated exposure to pesticides. It is usually impossible to predict the rate of accumulation of a pesticide in the body and its eventual effect on the health of the individual. There is no standard measure like LD_{50} for chronic toxicity. The symptoms of chronic toxicity may take years to develop, and laboratory test animals seldom live long enough for its effects to be determined. Some pesticides are known to be carcinogens, but

PACKAGE SIZE(S):
4-1 Gallon Containers
5 Gallon Container
30 Gallon Drum
55 Gallon Drum

LANNATE® L

INSECTICIDE
WATER SOLUBLE LIQUID
1 GALLON CONTAINS 1.8 LBS. METHOMYL

RESTRICTED USE PESTICIDE
For retail sale to and use only by Certified Applicators or persons under their direct supervision and only for those uses covered by the Certified Applicator's certification.

ACTIVE INGREDIENT:
Methomyl
S-methyl-N-[(methylcarbamoyl)oxy] thioacetimidate **24%**
INERT INGREDIENTS **76%**
U.S. Pats. 3,576,834, 3,639,633 & 3,862,316 EPA Reg. No. 352-370-AA

Keep out of reach of children.

DANGER ☠ POISON

STATEMENT OF PRACTICAL TREATMENT
If swallowed, call a physician or Poison Control Center. Drink 1 or 2 glasses of water and induce vomiting by touching back of throat with finger. **Do not induce vomiting or give anything by mouth to an unconsious person.**
If inhaled, remove from exposure and have patient lie down and keep quiet. If patient is not breathing, start artificial respiration immediate. **Never give anything by mouth to an unconscious person.**
In case of contact, wash skin with plenty of soap and water, for eyes flush with water for 15 minutes and get medical attention; remove and wash contaminated clothing before re-use.
ATROPINE IS AN ANTIDOTE—Seek medical attention at once in all cases of suspected poisoning.
If warning symptoms appear (see **Warning Symptoms**), get medical attention.
DO NOT SUBJECT TO TEMPERATURES BELOW 32°F.

FIGURE 13-9
A highly toxic pesticide.

information such as how many exposures to a chemical result in cancer is difficult to establish because other factors influence this disease as well.

One type of chronic toxicity can be measured accurately by blood tests. Organophosphates and carbamates, two groups of chemicals used to control insect, mite, and nematode pests, disrupt the nervous system by inhibiting an enzyme called cholinesterase. This enzyme breaks down acetylcholine, a chemical involved in the transmission of nerve impulses. If acetylcholine accumulates in the body, serious nervous system disorders can occur. Constant exposure to organophosphates or carbamates can result in a decline in cholinesterase activity and a buildup of acetylcholine. Applicators who use these chemicals regularly should arrange for blood tests to determine cholinesterase levels. If the level is too low, the applicator should avoid further exposure to these pesticides until the body regenerates sufficient cholinesterase.

PESTICIDE APPLICATOR CERTIFICATION

In the 1970s the Environmental Protection Agency (EPA) became increasingly involved in the regulation of pesticide use because of the potential threat of pesticides to the environment. This federal agency set standards for the handling and use of pesticides. States were required to develop and administer pesticide applicator certification programs. To be allowed to purchase certain hazardous pesticides, an applicator must first pass a written test that requires a basic knowledge of pesticides and their use. Restricted-use pesticides have a high

FIGURE 13-10
The turf manager must keep records of all pesticide applications. *(Courtesy of Broyhill Co.)*

acute toxicity or pose a significant environmental risk because of their persistence or tendency to leach. A certified pesticide applicator must be recertified every few years. This is usually accomplished by attending workshops, conferences, or other activities at which the applicator receives pesticide training.

Federal and state agencies require that applicators keep records of pesticide applications. These records include information such as the name of the pesticide and the formulation used, the date and location of the application, and the pest controlled (Figure 13-10). The applicator must also have special pesticide insurance in case of an accident. The turfgrass manager must be aware of all federal and state pesticide laws. Failure to obey these regulations can result in fines and other penalties.

THE PESTICIDE LABEL

Pesticide manufacturers are required by law to list certain information on the label attached to the pesticide container. Each statement on the label is reviewed and approved by the EPA. The chemical company obtains this information by conducting years of research costing millions of dollars. To ensure safe and proper use of a pesticide, the turfgrass manager must read, understand, and follow all directions stated on the label (Figure 13-11).

The following information is found on a pesticide label:

1. Name and address of the manufacturer.
2. Trade (product) name of the pesticide. The same chemical may be sold under several different trade names.

TERSAN® 75
TURF FUNGICIDE

ACTIVE INGREDIENT:
Thiram (Tetramethylthiuram disulfide) **75%**
INERT INGREDIENTS **25%**
EPA Reg. No. 352-160

Keep out of reach of children.

CAUTION! MAY IRRITATE EYES, NOSE, THROAT, AND SKIN. HARMFUL IF INHALED OR SWALLOWED.
• Do not get in eyes, on skin, on clothing. • Do not inhale dust or spray. • Wash thoroughly after handling.
The consumption of alcoholic beverages increases the possibility of harm.
In case of contact, flush skin or eyes with plenty of water; for eyes, get medical attention.

6-79

IMPORTANT

Do not permit animals to feed or graze on treated areas. Do not use grass clippings from treated areas for food or feed purposes.

This product is toxic to fish. Keep out of lakes, streams, or ponds. Do not contaminate water by cleaning of equipment or disposal of wastes.

NOTICE OF WARRANTY

Du Pont warrants that this product conforms to the chemical description on the label thereof and is reasonably fit for purposes stated on such label only when used in accordance with the directions under normal use conditions. It is impossible to eliminate all risks inherently associated with the use of this product. Crop injury, ineffectiveness or other unintended consequences may result because of such factors as weather conditions, presence of other materials, or the manner of use or application, all of which are beyond the control of Du Pont. In no case shall Du Pont be liable for consequential, special or indirect damages resulting from the use or handling of this product. All such risks shall be assumed by the Buyer. DU PONT MAKES NO WARRANTIES OF MERCHANTABILITY OR FITNESS FOR A PARTICULAR PURPOSE NOR ANY OTHER EXPRESS OR IMPLIED WARRANTY EXCEPT AS STATED ABOVE.

DIRECTIONS

"Tersan" 75 should be used only in accordance with recommendations on this label, or in separate published Du Pont recommendations available through local dealers. It is a violation of Federal Law to use this product in a manner inconsistent with its labeling.

Turf — Du Pont "Tersan" 75 Turf Fungicide used in conjunction with good turf management will usually prevent and control large brown patch and dollar spot on golf greens, grass tennis courts, lawns and other fine turfs. Make regular spray applications at intervals of 7 to 10 days** beginning before the earliest probable appearance of the diseases and continuing throughout the season.

When such a preventive spray schedule is not followed, the turf should be inspected each morning and at the first appearance of brown patch or dollar spot, the curative rate of "Tersan" 75 should be applied. When treating golf greens, always treat aprons and approaches.

Treatment	Rate per 1000 sq. ft.	Rate per 5000 sq. ft.	This Pkg. Treats
Preventive	3 ozs. or 2/3 cupful	1 lb.	3 Greens††
Curative	4-1/2 ozs. or 1 cupful	1-1/2 lbs.	2 Greens††

††Based on an area of 5000 sq. ft. per green.

For the Control of Snow Mold — Use 8 ozs. "Tersan" 75 per 1000 sq. ft. applied as late in the Fall as possible, prior to snowfall. Follow with a Spring application of 3 ozs. per 1000 sq. ft. applied as soon as the snow melts.

To Prepare Sprays — Use only enough water to spray or sprinkle the area thoroughly (usually 5 to 10 gals. per 1000 sq. ft.). For high pressure sprayers, use from 200 to 300 lbs. pressure and 5 gals. of water per 1000 sq. ft. applied as a fine spray. "Tersan" 75 is insoluble in water and will tend to settle out; constant agitation is necessary.

NOTE: Do not water or mow grass after treatment until the fungicide is thoroughly dry, preferably 24 hours after application.

**When conditions are unusually favorable for the development of disease or the disease is active, reduce the interval between sprays to 3 to 5 days and use the rate for curative treatment.

Sweet Potato Sprouts — "Tersan" 75 is generally effective for control of stem rot and scurf. Dip the roots of the sprouts for ½ minute in a suspension made by mixing 1 lb. "Tersan" 75 in 7½ gals. of water. Agitate suspension frequently. Plant sprouts promptly after treatment.

Gladiolus Bulbs — As a storage or preplanting treatment, "Tersan" 75 usually prevents basal rot and decay. Treat any time after drying. Coat bulbs well with "Tersan" 75 by shaking in paper bag or similar device, or by dipping in a suspension made by mixing 1 lb. "Tersan" 75 in 8 gals. of water. Agitate suspension frequently. If bulbs are to be stored, spread out to dry.

Storage and Disposal — Do not contaminate water, food or feed by storage or disposal. Do not re-use container. Bury empty container, or product that cannot be used, in a safe place away from water supplies, or dispose of by alternative procedures recommended by Federal, State or local authorities. Open dumping is prohibited.

FIGURE 13-11
Pesticide label information. The entire label is not included in this photograph.

3. List of active ingredients. The percentage, common name, and chemical name of all active ingredients (the toxic chemical that kills the pest) must be listed. The common name is the official name of the pesticide accepted by the EPA. The chemical name is the chemical formula of the toxicant.
4. Type of pesticide (insecticide, fungicide, herbicide, etc.).
5. Type of formulation (wettable powder, emulsifiable concentrate, granular, etc.).
6. EPA registration number.
7. Storage and disposal precautions. These precautions follow federal and state regulations.
8. Hazard statements. All precautions that are necessary for the safe use of the pesticide must be stated. Dangers to people and to the environment must also be stated. Typical warnings are "Avoid contact with skin, eyes, and clothing" or "Keep out of lakes, streams, and ponds." Hazard state-

FIGURE 13-12 The warning "Keep out of the reach of children" must be prominently displayed on a pesticide container because a large percentage of pesticide accidents involve children. *(Courtesy of Stanford Seed Co.)*

ACTIVE INGREDIENTS:
N-butyl-N-ethyl-alpha, alpha, alpha-trifluoro-2, 6-dinitro-p-toluidine* 0.92%
INERT INGREDIENTS 99.08%
TOTAL 100.00%
***Balfin™ – The trademark for Elanco Products Company Benefin**

KEEP OUT OF THE REACH OF CHILDREN
CAUTION
See Back Panel for Additional Precautions

EPA Reg. No. 9198-40-12428 **EPA Est. No. 9198-OH-1**

ments include information such as antidotes that will counteract the effects of the poison and protective clothing that should be worn. All pesticide labels must state the appropriate signal word and the warning "Keep out of reach of children" (Figure 13-12).

9. Directions for use. The specific crops and pests for which the pesticide is registered are listed. It is a violation of the law to use the pesticide on crops and pests that are not listed on the label. The directions also explain how to apply the pesticide. Mixing instructions, rates, the type of equipment to use, and proper timing of the application are included in these directions.
10. Net contents.

PESTICIDE SAFETY

Safety precautions are essential when handling or applying a pesticide. The following is a list of important safety precautions.

- Read the label carefully before opening the container. All directions stated on the label must be followed exactly. It is illegal to use a pesticide in any manner that is inconsistent with the label.
- Avoid contact with the pesticide. This is especially important when mixing a pesticide because the chemical is in a more concentrated form before it is diluted in the spray tank.
- Wear the protective clothing that is indicated on the label. Protective clothing such as elbow-length rubber gloves, respirator, face shield, goggles, wide-brimmed waterproof hat, rubber boots, and waterproof coveralls prevent the absorption of pesticide into the body (Figure 13-13). Wash protective clothes after each use.
- Have clean water and detergent available. Any pesticide that contacts the applicator's skin should be washed off immediately. Change contaminated clothing immediately. Shower thoroughly after applying a pesticide.
- Never eat, smoke, or drink any liquids while handling pesticides.

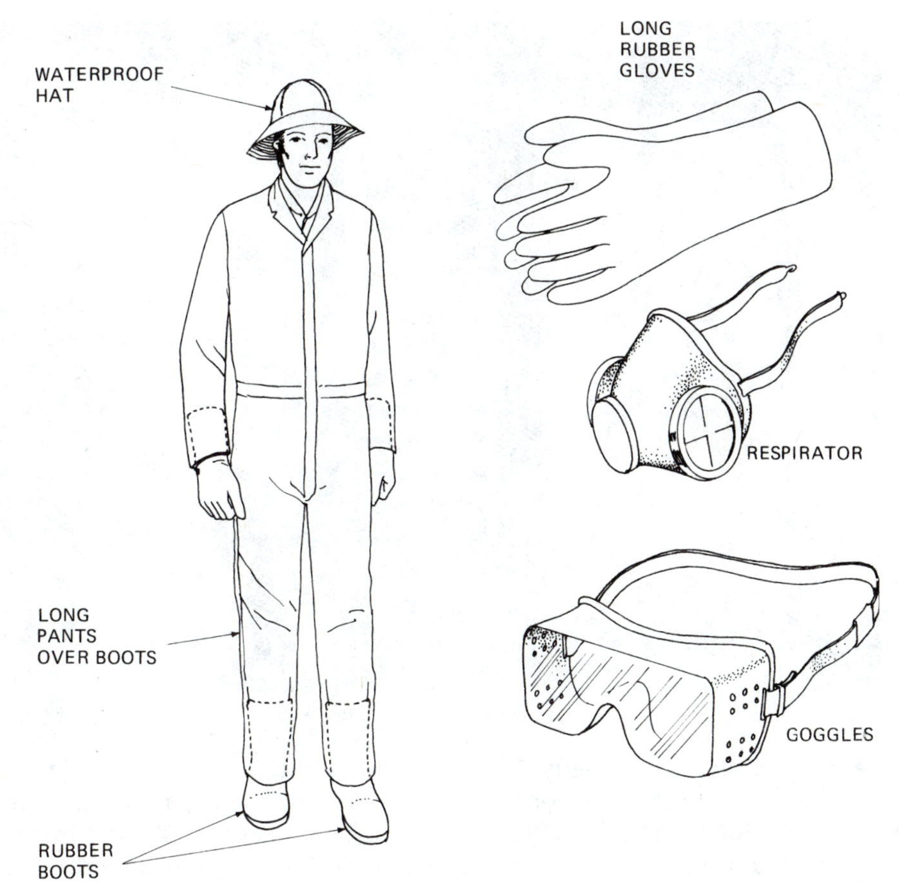

FIGURE 13-13
Protective clothing.

- Do not use a highly toxic pesticide if a safer one will do the job.
- Do not increase the recommended pesticide rate. Many applicators think that they can achieve better pest control by doubling the application rate. This is a hazardous practice and it does not result in improved control. Do not attempt to save money by decreasing the rate. Lower rates will be less effective than the recommended rate.
- Always mix just enough pesticide to cover the area to be sprayed. Large amounts of leftover spray are difficult to dispose of properly. Small amounts of unused spray can be applied to an untreated area where the pesticide will not contaminate or damage the environment.
- Be as certain as possible that correct weather conditions will occur during and after a pesticide application. Wind can cause drift. A heavy rain may wash away the chemical. The temperature has an effect on some pesticides.
- Rinse out empty containers at least three times. Do not dispose of them carelessly. Be aware of state regulations concerning pesticide container disposal.

FIGURE 13-14
Pesticides must be kept locked up, and the storage site must be identified by a warning sign.

- Store pesticides properly. Keep them locked up, away from children and animals. The locked door should be posted with a warning sign that says "DANGER PESTICIDES" (Figure 13-14). Do not store pesticides in a building that houses people or animals. Do not store them near food or animal feed. Herbicides should be kept separate from other pesticides. Always store pesticides in their original containers. The storage building or room should be dry and have ventilation. Most pesticides have a shelf life of at least two years if they are not exposed to freezing or high temperatures and are kept tightly sealed and dry. However, it is best to use up pesticides as soon as possible.
- Clean up any pesticide spills immediately. Absorptive clay, activated charcoal, pet litter, vermiculite, or sawdust can be used to soak up liquid spills. Any major spills should be reported to the appropriate authorities immediately.
- Report the location of stored pesticides to the local fire department. Many pesticides can give off poisonous fumes if heated or burned. Fire department personnel will need to wear respirators if a fire occurs. Advance notice will alert the fire department to these potential problems.
- Be sure that all employees who work with pesticides are aware of the necessary safety precautions. The employer has a legal responsibility to train workers in handling pesticides safely.

PESTICIDE POISONING

The vast majority of people who work with pesticides never experience a serious mishap. If the applicator is careful and uses common sense, pesticide accidents are unlikely. However, when a person comes into contact with a pesticide, the following steps should be taken immediately after the exposure occurs:

1. Remove the contaminated clothing.
2. Wash off the pesticide thoroughly with a soap or detergent and plenty of water. A shower is the most effective method of removing a pesticide from the skin. If a shower is not available, drench the exposed area with a hose or under a faucet. The faster the toxic chemical is washed off, the less chance of injury.
3. Telephone a doctor immediately if poisoning has occurred. Read the label to the doctor and follow his or her instructions precisely. Every area has a hospital with a poison control center that specializes in treating poisoning victims. If the local doctor is not knowledgeable about pesticide poisoning or is unavailable, call the nearest poison control center. The telephone numbers of the physician and the poison control center must be posted in a location where they are readily available in case of an emergency. The Chemical Manufacturers Association will also provide emergency information when a pesticide accident occurs. This information can be obtained 24 hours a day by calling the toll-free number 800-424-9300.

   Do not hesitate to seek medical help if pesticide poisoning is suspected. The symptoms of pesticide poisoning include headache, fatigue, dizziness, restlessness, nervousness, nausea, excessive perspiration, diarrhea, loss of appetite, loss of weight, moodiness, abnormal thirst, fever, increased breathing rate, vomiting, lack of muscle coordination, uncontrollable muscle twitches, pinpoint pupils, convulsions, fainting, blurred vision, inability to breathe, and skin, eye, nose, or throat irritations.

Remember that important first aid information is stated on the pesticide label.

PESTICIDE APPLICATION EQUIPMENT

Most pesticides are applied with sprayers, but spreaders are also used. The pesticide application equipment must be properly calibrated, or the application may be ineffective, injurious to the turf, or unsafe. Unfortunately, inaccurate calibration is a common problem. Correct application techniques are essential (Figure 13-15). Sprayers and their calibration are discussed in Appendix B; spreaders are discussed in Appendix C.

FIGURE 13-15
Marking the edges of the treated area with foam so that the applicator knows where she has already sprayed. *(Courtesy of Richway Industries)*

SELF-EVALUATION

1. Pesticides that are absorbed into a plant are called ___________.
2. An emulsifiable concentrate is a type of pesticide ___________.
3. Granular pesticides are applied with a ___________.
4. To keep a wettable powder uniformly dispersed in water, the mixture has to be ___________.
5. Stickers and spreaders are examples of ___________.
6. ___________ is the movement of spray droplets by wind.
7. Pesticides that retain their toxicity for a long time are said to be ___________.
8. Pesticides enter the body orally, through the skin, or by ___________.
9. Acute toxicity is a measure of how poisonous a pesticide is in a ___________ exposure.
10. ___________ toxicity is the result of repeated exposure to pesticides over time.
11. The lower the LD_{50}, the ___________ poisonous the pesticide.
12. A highly toxic pesticide is identified with a skull and crossbones and the signal words ___________.
13. Some pesticides disrupt the nervous system by inhibiting the enzyme ___________.
14. If pesticide poisoning occurs, important first aid information can be found stated on the ___________.
15. Rinse out empty pesticide containers at least ___________ times.
16. Report the location of stored pesticides to the local ___________.
17. What telephone numbers should be posted in case of a pesticide emergency?
18. Discuss how to use pesticides safely.
19. What problems can result if pesticides are used incorrectly?

CHAPTER

14

Weeds

OBJECTIVES

After studying this chapter, the student should be able to

- Explain why a correct turfgrass maintenance program results in less weed competition
- List some of the more important weed species
- Identify the different types of herbicides
- Discuss the methods of controlling annual grasses, perennial grassy weeds, and broadleaf weeds

INTRODUCTION

Any plant that is growing where it is not wanted can be called a weed. Coarse-leafed tall fescue is desirable in roadside areas, but is considered a weed on a better-quality lawn. Creeping bentgrass can be a serious problem on a lawn, but is the perfect choice for a golf course green (Figure 14-1). Many plants such as dandelions and crabgrass are described as weeds regardless of the type of turf in which they are growing.

Broadleaf weeds are very noticeable in a turf area because their appearance is much different from that of grasses. Their broader leaves ruin the uniform appearance that is expected of a quality turf (Figure 14-2). These weeds have a particularly disruptive aesthetic effect if they produce flowers and seed heads. Certain grass species are considered weeds because they destroy the uniformity of the turf. Their leaf texture may be coarser than that of the desirable turfgrasses. They may grow in unattractive clumps, have poor color, form unsightly seed stalks, or have a different growth habit (Figure 14-3). Annual grasses such as crabgrass and goosegrass and the annual subspecies of *Poa annua* die each year, leaving dead areas in the turf.

Weeds also cause problems by competing with turfgrass for sunlight, nutrients, and moisture. Weeds can spread rapidly in a turf by means of seeds, rhizomes, stolons, and various underground storage organs such as bulbs and tubers (Figure 14-4). For example, the appearance of a few broadleaf plantain

FIGURE 14-1
Patches of creeping bentgrass disrupt the uniformity of a Kentucky bluegrass lawn.

FIGURE 14-2
Plantains are a common nuisance on many lawns. *(Photo by Brian Yacur)*

FIGURE 14-3
A clumpy, coarse-textured grass is not compatible with these desirable lawn grasses.

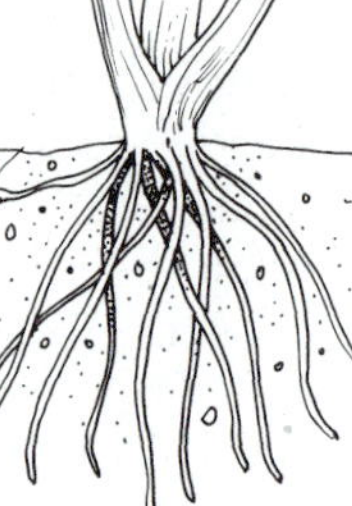

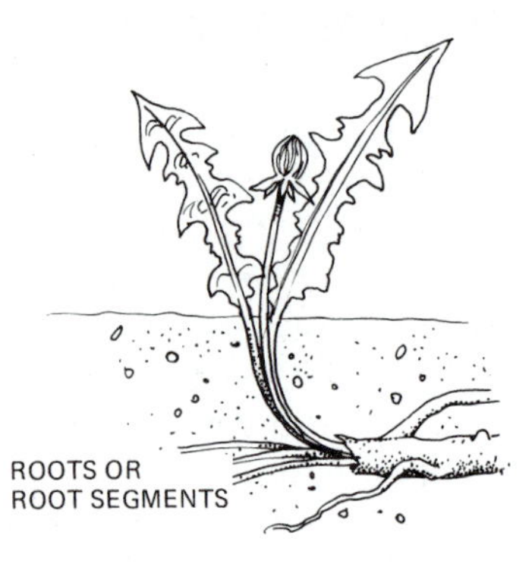

BULBS

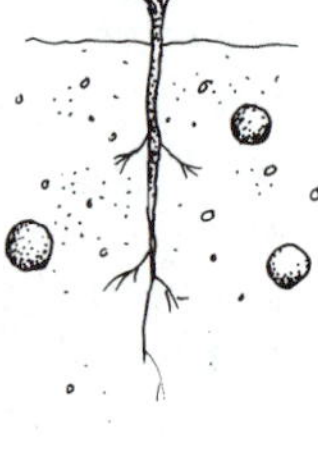

FIGURE 14-4.
How weeds spread.

COMMON NAME	SCIENTIFIC NAME	SEEDS PER PLANT
Annual bluegrass	*Poa annua*	2,000
Barnyardgrass	*Echinochloa crusgalli*	7,000
Common chickweed	*Stellaria media*	15,000
Curly dock	*Rumex crispus*	30,000
Dandelion	*Taraxacum officinale*	12,000
Green foxtail	*Setaria viridis*	34,000
Knotweed	*Polygonum aviculare*	6,500
Broadleaf plantain	*Plantago major*	36,000
Veronica or purslane speedwell	*Veronica peregrina*	2,900

FIGURE 14-5

Approximate number of seeds produced by single plants of certain weed species.

plants in a turf may not be considered a serious problem, but each plant can produce thousands of seeds (Figure 14-5). If bare spots are present in the turf, some of these seeds will successfully germinate and in a short time numerous plantains may appear.

Weed control implies chemical control to many people. However, the best weed control strategy is to prevent weed populations from appearing in the turf at all. This can be accomplished by intelligent establishment and maintenance. Weeds become a problem when turfgrasses either lose or do not achieve the proper density. Weeds gain a foothold and flourish on bare areas not covered by turfgrass. If the grasses are adaptable to the site and the maintenance operations are performed correctly, the turf should be thick enough to choke out most weed problems.

The presence of a sizable weed population indicates that the turf is too open or thin and that the management program needs improvement. Practices such as fertilization, irrigation, mowing, and disease and insect control should be reevaluated. Weaknesses in the program should be identified and corrected. It may be necessary to reseed or overseed with turfgrass species and cultivars that are better adapted to the site. Properly maintained, well-adapted turfgrasses are able to outcompete weeds. Weed problems are normally minimal when turf density is high (Figure 14-6).

By identifying the weed species that predominate, the turf manager can sometimes discover why the weeds outcompete the desirable grasses. For example, the presence of annual bluegrass, pearlwort, moss, sedges, rushes, or alligatorweed as the major weed problem suggests that the site is too wet. Cutting back irrigation or improving drainage may reduce the weed population. Goosegrass, birdsfoot trefoil, prostrate spurge, black medic, or yellow woodsorrel populations indicate droughty conditions. Increased irrigation will make turfgrasses more aggressive. When weeds such as clovers, common speedwell, birdsfoot trefoil, or hawkweed are predominant low fertility is often

FIGURE 14-6
A turf composed of adaptable species and cultivars that are maintained properly should remain relatively weed-free. *(Courtesy of O. M. Scott and Sons Company)*

the problem. Applying more nitrogen will cause the desirable grasses to be more competitive.

Large populations of shallow-rooted weeds may be the result of significant soil compaction. Annual bluegrass, broadleaf plantain, prostrate spurge, knotweed, goosegrass, and corn speedwell are examples of weeds that do well on compacted soils. Core cultivation will help the turfgrasses by allowing deeper rooting. Mowing too close leads to an invasion of low-growing weeds such as annual bluegrass, chickweeds, some of the speedwells, and moss, which are better adapted to low mowing heights than most of the turfgrasses. Raising the cutting height will strengthen the desirable grasses and encourage their spread. Taller weeds such as burdock, teasel, and bull thistle can be partially controlled by lowering the cutting height.

TYPES OF WEEDS

The characteristics of turfgrass weed species vary tremendously, but all have the ability to persist even though the grass is cut regularly. Generally, they survive mowing because of a low growth habit. Weed species are either broadleaf (dicotyledon) or grassy type (monocotyledon). They can be further divided into annuals, biennials, and perennials.

Annuals complete their life cycles in one year or less. Summer annuals, such as crabgrass, germinate in the spring and die in the fall with the onset of colder temperatures. The seeds they produce during the growing season remain in the soil over the winter and germinate the following spring when soil temperatures increase. Winter annuals, such as common chickweed and annual bluegrass, germinate in late summer or early fall, overwinter in a dormant state, and continue to grow the following spring. Death occurs during the summer after seed is produced.

Biennials, such as wild carrot, burdock, and bull thistle, live for two years. *Perennial* weeds, like most turfgrasses, live for longer than two years. Simple perennials reproduce by seed. Common examples are dandelions, plantains, and chicory. Creeping perennials reproduce by seed, but can also spread by means of rhizomes, stolons, and underground storage organs such as bulblets. Examples of creeping perennials are quackgrass, nutsedge, and white clover.

Selecting the proper herbicide and application time for controlling a weed species is determined by the life cycle of the weed and whether it is a broadleaf or grassy type. Generally weeds fall into three control groups—annual grasses, perennial grasses, and broadleaf weeds. This classification is important to the turf manager because each group is controlled differently (Figure 14-7).

ANNUAL GRASSES	PERENNIAL GRASSES	BROADLEAF WEEDS		
Annual bluegrass	Bentgrass	Creeping beggarweed	Wild garlic	Pigweed
Barnyardgrass	Bermudagrass	Field bindweed	Hawkweed	Pineappleweed
Hairy crabgrass	Dallisgrass	Bittercress	Healall	Broadleaf plantain
Smooth crabgrass	Tall fescue	Burdock	Henbit	Buckhorn plantain
Green foxtail	Johnsongrass	Creeping buttercup	Ground ivy	Common purslane
Yellow foxtail	Kikuyugrass	Carpetweed	Knapweed	Shepherdspurse
Goosegrass	Knotgrass	Wild carrot	Knotweed	Red sorrel
Fall panicum	Nimblewill	Common chickweed	Lambsquarters	Speedwells (spp. *Veronica*)
Rescuegrass	Purple nutsedge*	Mouse-ear chickweed	Lespedeza	Prostrate spurge
Sandbur	Yellow nutsedge*	Chicory	Mallow	Spotted spurge
Six-weeks fescue	Quackgrass	Cinquefoil	Black medic	Spurweed
	Smutgrass	Hop clover	Mugwort	Wild strawberry
	Torpedograss	White clover	Wild mustard	Thistles
	Velvetgrass	Cranesbill	Wild onion	Prostrate vervain
		English daisy	Birdseye pearlwort	Wild violet
		Oxeye daisy	Field pennycress	Yellow woodsorrel
		Dandelion	Pennywort	Common Yarrow
		Dichondra	Parsley piert	Yellow rocket
		Curly dock		

*Actually a sedge, not a true grass.

FIGURE 14-7 Examples of annual grasses, perennial grasses, and broadleaf plants found as weeds in turfgrass.

FIGURE 14-8
The area on top was treated with a herbicide and is weed-free. The untreated area on the bottom is infested with weeds and many dandelion seed heads are visible. *(Courtesy of O. M. Scott and Sons Company)*

HERBICIDES

Herbicides are chemicals that control weeds. They can be selective, killing one type of plant but not injuring another. 2,4-D is a widely used herbicide that controls a number of broadleaf weed species but will not adversely affect mature turfgrass if applied properly. Nonselective herbicides, also called total vegetation killers, control both broadleaf and grass plants.

Some herbicides are applied to the soil before the time when the weeds are expected to appear in the turf. These *preemergence* herbicides form a toxic chemical barrier near the soil surface and destroy seedlings before they emerge. This type of herbicide is primarily used to control annual grasses. Mature, established grasses are not killed by preemergence herbicides because their roots are beneath the chemical barrier. *Postemergence* herbicides are applied after the weeds have appeared above the soil surface. This type is used primarily against broadleaf weeds and perennial grasses (Figure 14-8).

The vast majority of herbicides used by the turf manager are systemic. They enter the weed through its roots or leaves and are translocated throughout the plant by the vascular tissue. Eventually the systemic herbicide kills all parts of the plant. A few herbicides occasionally used by turf specialists are the nonsystemic or contact type. They kill only the parts of the plant to which they are applied. The turf manager must be aware of the characteristics of herbicides (Figure 14-9).

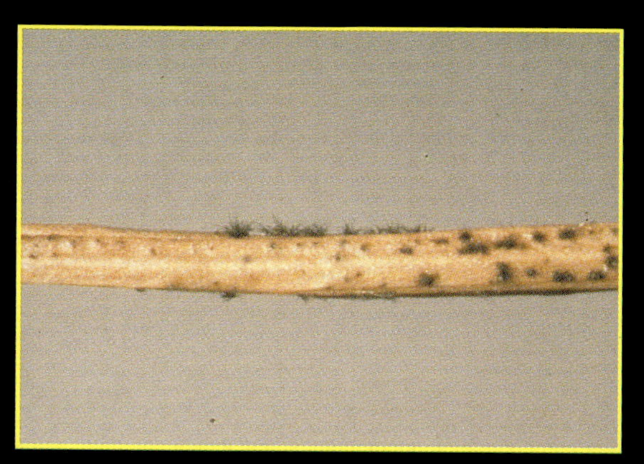

PLATE 1 Anthracnose reproductive structures (acervuli) on a rhizome. *(Courtesy NYSTA)*

PLATE 2 Brown patch on a putting green. *(Courtesy NYSTA)*

PLATE 3 Brown patch on a tall fescue lawn. *(Courtesy NYSTA)*

PLATE 4 Dollar spot on a bentgrass putting green. *(Courtesy NYSTA)*

PLATE 5 Mycelium produced by the dollar spot pathogen. *(Courtesy NYSTA)*

PLATE 6 Fairy ring with mushrooms. *(Courtesy NYSTA)*

PLATE 7 Gray leaf spot on St. Augustinegrass. *(Courtesy William Knoop)*

PLATE 8 Leaf spot and crown rot symptoms. *(Courtesy NYSTA)*

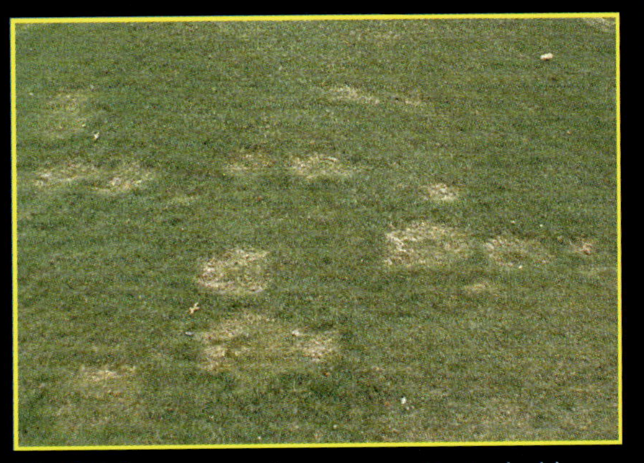

PLATE 9 Necrotic ringspot on a Kentucky bluegrass lawn. *(Courtesy NYSTA)*

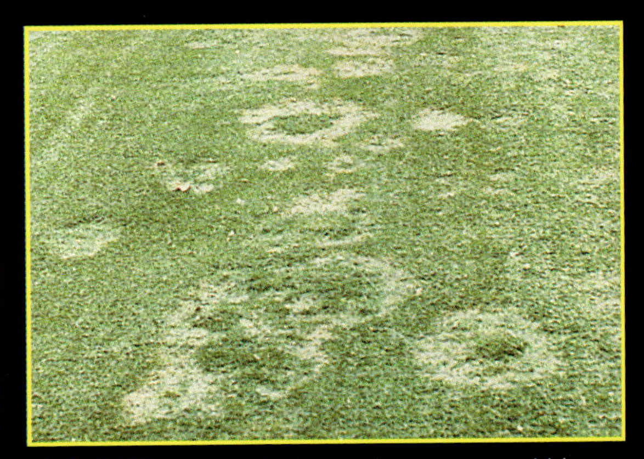

PLATE 10 Necrotic ringspot on an annual bluegrass fairway. *(Courtesy NYSTA)*

PLATE 11 Nematode injury on a putting green. *(Courtesy NYSTA)*

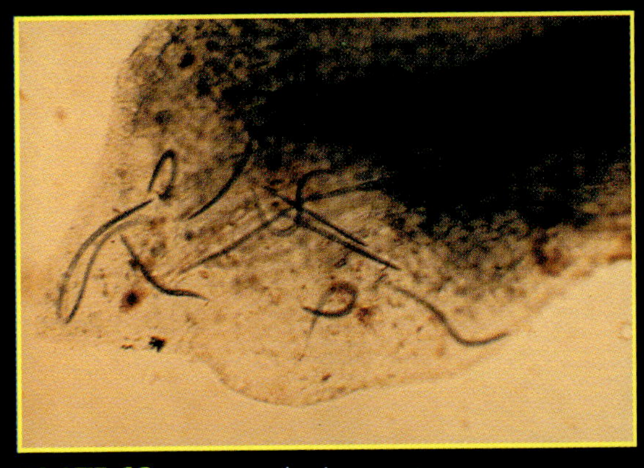

PLATE 12 Nematodes in a root. *(Courtesy William Knoop)*

PLATE 13 Powdery mildew. *(Courtesy NYSTA)*

PLATE 14 Pythium blight spread by a mower. *(Courtesy NYSTA)*

PLATE 15 Mycelium produced by the Pythium blight pathogen. *(Courtesy NYSTA)*

PLATE 16 Oospores of the Pythium root rot pathogen. *(Courtesy NYSTA)*

PLATE 17 Red thread. *(Courtesy NYSTA)*

PLATE 18 Rust. *(Courtesy NYSTA)*

PLATE 19 Red and black reproductive structures of the rust pathogen. *(Courtesy NYSTA)*

PLATE 20 St. Augustine decline symptoms. *(Courtesy William Knoop)*

PLATE 21 Slime mold. *(Courtesy NYSTA)*

PLATE 22 Gray snow mold. *(Courtesy NYSTA)*

PLATE 23 Sclerotia of gray snow mold pathogen. *(Courtesy NYSTA)*

PLATE 24 Pink snow mold. *(Courtesy NYSTA)*

PLATE 25 Spring dead spot on bermudagrass. *(Courtesy William Knoop)*

PLATE 26 Striped smut. *(Courtesy NYSTA)*

PLATE 27 Summer patch on an annual bluegrass fairway. *(Courtesy NYSTA)*

PLATE 28 Root rot symptoms on Kentucky bluegrass caused by the summer patch pathogen. *(Courtesy NYSTA)*

PLATE 29 Take-all patch on a bentgrass putting green. *(Courtesy NYSTA)*

PLATE 30 Yellow patch; also called cold weather or winter brown patch. *(Courtesy NYSTA)*

PLATE 31 Typical white grub. *(Courtesy NYSTA)*

PLATE 32 Life stages of a European chafer – egg, three larval instars, beginning and true resting stage (pupa), and adult. *(Courtesy NYSTA)*

PLATE 33 Injured sod is easily picked up exposing the grubs. *(Courtesy NYSTA)*

PLATE 34 Japanese beetle adults. *(Courtesy NYSTA)*

PLATE 35 V-shaped arrangement of spines on the raster of a Japanese beetle grub. *(Courtesy NYSTA)*

PLATE 36 Bluegrass billbugs adults. The two darker ones are mature adults. *(Courtesy NYSTA)*

PLATE 37 Bluegrass billbug larvae. *(Courtesy NYSTA)*

PLATE 38 Annual bluegrass weevil adult, pupa, and larva. *(Courtesy NYSTA)*

PLATE 39 Annual bluegrass weevil larva burrowing into Poa annua crown. *(Courtesy NYSTA)*

PLATE 40 Black turfgrass ataenius larva, pupa, callow (young), and mature adult. *(Courtesy NYSTA)*

PLATE 41 One of the lawn moth species. *(Courtesy NYSTA)*

PLATE 42 One of the sod webworm species. *(Courtesy NYSTA)*

PLATE 43 Hairy chinch bug adults. Note that the female is larger. *(Courtesy NYSTA)*

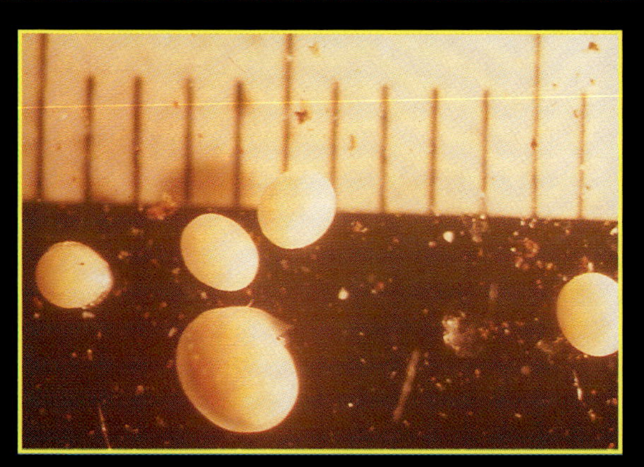

PLATE 44 Ground pearls. *(Courtesy William Knoop)*

PLATE 45 Bermudagrass mites. *(Courtesy William Knoop)*

PLATE 46 Bermudagrass mite injury. *(Courtesy William Knoop)*

PLATE 47 Normal bermudagrass plant (left). Plant damaged by bermudagrass mites (right). *(Courtesy William Knoop)*

PLATE 48 One species of mole cricket. *(Courtesy Brian Yacur)*

COMMON NAME	TRADE NAME	TIME OF APPLICATION PR = Preemergence PO = Postemergence	TYPE OF WEED CONTROLLED AND OTHER COMMENTS
Asulam	Asulox	PO	Primarily used to control annual grasses in St. Augustinegrass turf
Atrazine	Aatrex, Atrazine	PR, PO	Annual grasses and broadleaf weeds in St. Augustinegrass, centipedegrass, and zoysiagrass
Benefin	Balan	PR	Annual grasses
Bensulide	Betasan, Bensumec, Pre-San, Weedgrass Preventer, Lescosan	PR	Annual grasses
Bentazon	Basagran, Lescogran	PO	Nutsedge
Chlorsulfuron	TFC	PO	Primarily used for tall fescue control
2,4-D	Many	PO	Broadleaf weeds
DCPA	Dacthal	PR	Annual grasses, also postemergence for speedwell (veronica)
Dicamba	Banvel, K-O-G Weed Control	PO	Broadleaf weeds
Diclofop-methyl	Illoxan	PO	Goosegrass control in bermudagrass
Diquat	Diquat	PO	Annual weed control in dormant bermudagrass
Dithiopyr	Dimension	PR, PO	Preemergent for annual grasses with some postemergent activity if applied before grass has tillered
Ethofumesate	Prograss	PR, PO	Annual bluegrass, primarily as a postemergent
Fenoxaprop	Acclaim	PO	Annual grasses
Glufosinate-ammonium	Finale	PO	Nonselective
Glyphosate	Roundup	PO	Nonselective, kills all types of plants including perennial grasses
Imazaquin	Image	PR, PO	Some grasses, sedges, and broadleaf weeds in warm season turf
Isoxaben	Gallery	PR	Preemergence control of some broadleaf weeds, limited control of annual grasses

FIGURE 14-9 Characteristics of commonly used herbicides. The pesticide label must be read to obtain more complete information.

(continues)

COMMON NAME	TRADE NAME	TIME OF APPLICATION PR = Preemergence PO = Postemergence	TYPE OF WEED CONTROLLED AND OTHER COMMENTS
MCPA	MCPA	PO	Broadleaf weeds
Mecoprop	MCPP and several others	PO	Broadleaf weeds
Metolachlor	Pennant	PR	Some annual grasses and sedges in warm season turf
Metribuzin	Sencor	PO	Annual grasses and broadleaf weeds in bermudagrass
Napropamide	Devrinol	PR	Annual grasses in warm season turf
Organic arsenicals (DSMA, MSMA, etc,)	Several	PO	Annual grasses
Oxadiazon	Ronstar	PR	Annual grasses
Pendimethalin	PRE-M, Weedgrass Control	PR	Annual grasses
Prodiamine	Barricade	PR	Annual grasses and annual broadleaf weeds
Pronamide	Kerb	PR, PO	Primarily annual bluegrass in bermudagrass
Sethoxydim	Vantage	PO	Annual and perennial grasses in centipedegrass
Siduron	Tupersan	PR	Annual grasses in seedbeds
Simazine	Princep	PR, PO	Annual grasses and broadleaf weeds in St. Augustinegrass, centipedegrass, zoysiagrass, and some bermudagrass cultivars
Triclopyr	Turflon	PO	Broadleaf weeds
Some Combinations			
Benefin and trifluralin	Team	PR	Annual grasses
2,4-D and dichlorprop and dicamba	Super Trimec	PO	Broadleaf weeds
2,4-D and MCPP and dicamba	Many	PO	Broadleaf weeds
2,4-D and triclopyr	Turflon II	PO	Broadleaf weeds
MCPA and MCPP and dichlorprop	Triamine II	PO	Broadleaf weeds
Oxadiazon and bensulide	Goosegrass/ Crabgrass Control	PR	Annual grasses
Triclopyr and clopyralid	Confront	PO	Broadleaf weeds

FIGURE 14-9 *(continued)*

Soil sterilants are nonselective chemicals that retain their toxicity for long periods. The sterilant kills any sprouting seeds or plant parts that begin to grow in the soil where it is located. Sterilants are used in parking lots, on paths, under fences, and other areas where any vegetation is undesirable.

Fumigants can be applied to the soil to control weeds, seeds, nematodes, disease-causing microorganisms, and insects. However, because of pesticide regulations and expense they are generally used only on specialty turf areas such as golf greens.

All chemicals suggested for use in the following discussions should be applied in accordance with the directions given by the manufacturer on the container label. Any deviation from the label instructions is a violation of the law. The turf manager should check the current status of a pesticide being considered to be certain that its use has not been restricted by federal, state, or local regulatory agencies. Some pesticides registered by the Environmental Protection Agency are not registered for use in all states.

The turf manager must also be sure that the herbicide is not toxic to the desirable turfgrasses. Dicamba, for example, is safe to use on bahiagrass, bermudagrass, carpetgrass, and zoysiagrass, but will injure St. Augustinegrass (Table 14-1). Varietal differences can occur. Meyer zoysiagrass is tolerant to MSMA, but the cultivar Emerald is sensitive to it. Closely mowed turfgrass is often more susceptible to herbicide injury. Stress can reduce tolerance. Cool

TABLE 14-1 Tolerance of Established Warm Season Turfgrasses to Some Postemergence Herbicides*

Herbicide	Bahiagrass	Bermudagrass	Carpetgrass	Centipedegrass	St. Augustinegrass	Zoysiagrass
Atrazine	NR, t	D, S	NR, T	T	T	t
Bentazon	T	T	NR, T	T	T	T
2, 4-D	T	T	t	t	S	T
Dicamba	T	T	T	t	S	T
DSMA, MSMA	NR, S	T	NR, S	NR, S	NR, S	t
Fenoxaprop	NR, S	NR, S	NR	NR, S	NR, S	T
Imazaquin	NR, S	T	NR, t	T	T	T
Mecoprop	T	T	t	t	t	T
Metribuzin	NR, t	T	NR, S	NR, S	NR, S	NR, S
Sethoxydim	NR, S	NR, S	NR, t	T	NR, S	NR,t

*D = use when turf is dormant; NR = not registered for use on this turfgrass species; S = sensitive, herbicide will injure this turfgrass species; t = species shows intermediate tolerance to this herbicide, injury may occur, especially at higher rates; T = species tolerant to this herbicide at labeled rates. The pesticide user must always check the product level for species and cultivar tolerance information. Some cultivars may be less tolerant than others. Closely mowed or stressed turf is often more susceptible to injury.

season grasses are more sensitive to herbicides during the hottest, driest parts of the summer, while warm season species are more sensitive during spring green-up or droughty periods.

When warm season grasses are dormant in the winter, they can be safely treated with some herbicides that would cause severe damage if applied when the grasses were actively growing. In some southern states diquat is used for winter annual weed control in dormant bermudagrass. The pesticide would injure actively growing bermudagrass.

Essential information about the use of a herbicide is found on the product label. Any questions that remain after reading the label should be referred to the Cooperative Extension Service or to the product manufacturer. Most companies have toll-free 800 numbers.

It is important to consider the efficacy of a herbicide. How well will it work against the weed species prevalent in the turf area? The site should be surveyed to determine which weed species are present and which are most common. Obviously, the herbicide or herbicides selected must be able to successfully control the weed population growing on the site. Herbicides do not work equally well against all weed species. Bensulide is very effective against crabgrass, but its control of another annual grass, goosegrass, is relatively poor.

When selecting a herbicide, turf managers should also try to use the material that is least toxic and most unlikely to cause environmental problems. For larger areas, cost is a factor. Some studies have suggested that 2,4-D is a possible carcinogen. However, the chemical has been widely used, partly because of its effectiveness, but also because it is very inexpensive.

CONTROLLING ANNUAL GRASSES

Annual grasses such as crabgrass, goosegrass, foxtail, barnyardgrass, annual bluegrass, fall panicum, and field sandbur are often controlled with preemergence herbicides. Commonly used preemergence chemicals include benefin, bensulide, DCPA, dithiopyr, metolachlor, oxadiazon, pendimethalin, prodiamine, and siduron. Most applications occur in the spring and are aimed at summer annuals such as crabgrass and goosegrass.

The toxic chemical barrier formed at the soil surface by the preemergent herbicides usually lasts 6 to 12 weeks, depending upon which chemical is used. Eventually microorganisms break down the herbicide.

Proper timing of an application is very important. These materials must be put down before seed germination. With the exception of dithiopyr, they have little effect if applied after weed emergence (Figure 14-10). However, it the chemical is put down too early, it may lose its effectiveness before the peak germinating period ends.

Preemergence herbicides should be applied one to two weeks before seed germination. In some areas a traditional method for crabgrass control is to time the application with the full bloom of forsythia. Another method is to

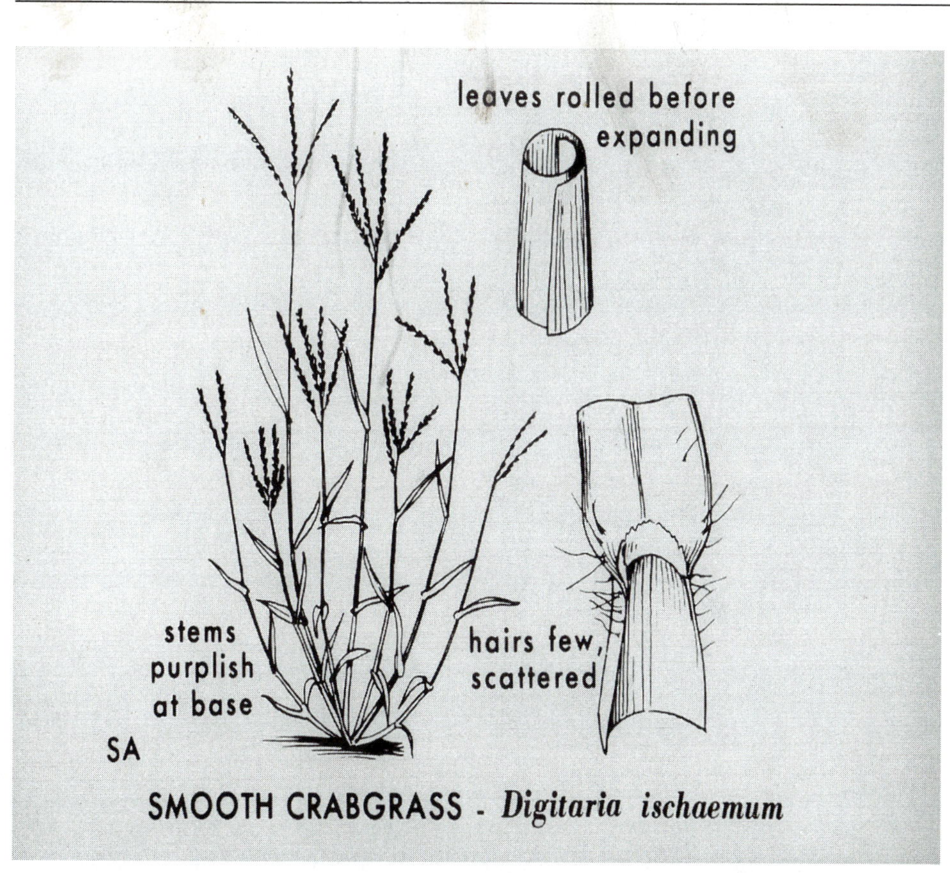

FIGURE 14-10
Crabgrass seeds overwinter in the soil and germinate in the spring. Preemergence herbicides must be applied before the seedlings emerge. *(Courtesy of New York State Turfgrass Association)*

apply the herbicide after the soil temperature at a 2-inch (5.1-centimeter) depth has been 55°F (13°C) for three or four consecutive days. Some turf managers wait until the mean air temperature is in the 55°–60°F (13°–15.5°C) range for two weeks. Goosegrass germinates a few weeks after crabgrass.

Liquid and granular formulations are equally effective (Figure 14-11). In southern states two or three applications per growing season are usually necessary. If good control is achieved one year, it is often possible to skip a year before retreating with a preemergence herbicide.

Postemergent control is also possible. Fenoxaprop is effective against summer annual grasses in cool season turf and zoysiagrass. It works best when weeds are in an early growth stage. Dithiopyr will kill annual grass plants that have not yet tillered. It can be used in most warm season and cool season turf and also acts as a preemergent.

The organic arsenicals, MSMA and DSMA, are labeled for use against emerged annual grasses. Two or more applications are required for adequate control, and they can be phytotoxic to desirable grasses. Cool season species are especially susceptible to injury.

FIGURE 14-11
A granular formulation combined with fertilizer. Granular preemergence herbicides are popular with homeowners because they can be applied with a spreader. *(Courtesy of Stanford Seed Co.)*

Postemergence applications of metribuzin alone or in combination with MSMA provide good control of goosegrass in bermudagrass turf. Diclofop-methyl is used for the same purpose. Asulam is applied for postemergence control of goosegrass and crabgrass in St. Augustinegrass turf. Sethoxydim controls annual grasses in centipedegrass.

It is very important to check the label carefully before applying herbicides. Tolerances to herbicides can vary significantly between turfgrass species.

Preemergent applications for winter annual grasses occur in late summer or early fall.

Annual Bluegrass Control

Annual bluegrass (*Poa annua*) is a winter annual that can be very difficult to control. Because of its prolific seeding ability it is so competitive that large populations of annual bluegrass can be common on higher-maintenance turf sites (Figure 14-12). Complete control of annual bluegrass with preemergence herbicides is not possible.

As explained in Chapter 4, both annual and perennial types of annual bluegrass exist. The annual subspecies is prevalent in warm season areas and produces seed in the spring which germinates in the late summer. Application

FIGURE 14-12

Annual bluegrass is constantly producing seed. *(Courtesy of New York State Turfgrass Association)*

of a preemergent such as bensulide or prodiamine before germination provides some control.

In bermudagrass, pronamide can be used for preemergent and postemergent annual bluegrass control. Atrazine, diquat, or glyphosate can be applied to dormant bermudagrass in the winter to kill annual bluegrass plants.

Annual bluegrass control is more difficult in cool season turf. The perennial biotype usually predominates, and its seed germinates throughout the season. *Poa annua* is dormant in the winter when the desirable cool season species are dormant, so it is not possible to kill the weed with a nonselective herbicide such as glyphosate, as in the South.

The annual bluegrass problem is greatest on golf courses. Preemergent herbicide applications in the spring and fall can help. However, successful control requires both chemical and cultural programs. Flurprimidol and paclobutrazol are growth regulators that suppress *Poa annua* by reducing its growth and competitive ability. The desirable grasses continue to grow and spread. The conversion can be speeded up even more by overseeding with turfgrass seed.

Ethofumesate provides very good control of annual bluegrass when applied as a postemergent. It is labeled for use in perennial ryegrass, creeping bentgrass, and Kentucky bluegrass turf. Because it kills *Poa annua*, dead spots will result when the chemical is applied to turf areas that have large patches of the weed. For this reason ethofumesate gives the best results when the turf is less than 20 percent annual bluegrass. Overseeding helps to fill in the areas where *Poa annua* has died.

Some golf course superintendents, when confronted with large fairway areas composed primarily of annual bluegrass, will spray with the nonselective herbicide glyphosate and kill everything. The fairway will then be reseeded. However, annual bluegrass will almost certainly dominate again in the near future if proper cultural techniques are not implemented.

The perennial biotype of *Poa annua* is very competitive in wet areas. Improving drainage and avoiding overwatering is important. Core cultivation is helpful because annual bluegrass tolerates compaction better than the desirable turfgrasses. Reducing nitrogen fertilization and changing application times may also help. For example, an early-spring nitrogen application benefits the annual bluegrass because it starts to grow sooner than most desirable grasses in the spring.

Lightweight mowing has resulted in a significant decrease in *Poa annua* populations. The lighter mowers cause less compaction and wear on fairways. Picking up the clippings in the mower baskets is also important. This may be because seeds are removed or because the clippings supply nitrogen, which favors annual bluegrass. Some theories hold that the clippings promote disease or decomposition products that injure the turfgrasses. Whenever turf is weakened or killed there is a good chance that annual bluegrass will appear in the bare spot because in the soil there are as many as 20,000 of its seeds per square foot (0.09 m^2).

There is interest in using a bacterium for annual bluegrass control. The organism, *Xanthomonas campestris* pv. *poannua* causes a disease that can be quite destructive. The pathogen works best in the southern United States because the weather is warmer and the annual subspecies is more susceptible than the perennial type.

Some golf course superintendents decide to coexist with their annual bluegrass. This is often because past control efforts have failed. Annual bluegrass performs most satisfactorily where temperature extremes do not occur. The grass is very susceptible to winter injury. In the spring and summer annual bluegrass must be protected from the annual bluegrass weevil, the black turfgrass ataenius, and diseases such as anthracnose and summer patch.

CONTROLLING PERENNIAL GRASSES

Perennial grassy weeds can cause serious problems because they are so similar to turfgrasses that it is difficult to develop selective herbicides which will control them and not kill the desirable grass. This is especially true in cool season turf. Some experimental chemicals look promising. Presently a nonselective herbicide such as glyphosate is commonly used in to spot-treat the patches of grassy weeds. The chemical is normally applied with a hand sprayer (Figure 14-13). The turf manager must be careful not to spray glyphosate on the adjacent desirable turfgrass.

FIGURE 14-13
Spot-treating perennial grassy weeds with glyphosate. *(Courtesy of H. D. Hudson Manufacturing Co.)*

Glyphosate is inactivated in the soil because it is bound tightly to soil particles. Consequently, it has no effect on seed. The areas that contained perennial grassy weeds can be reseeded shortly after a glyphosate application. A waiting period of seven days is necessary to ensure that the chemical has been translocated to the roots of the weeds. One selective material, chlorsulfuron, will control tall fescue in Kentucky bluegrass, fine fescue, bahiagrass, and bermudagrass lawns. However, it is usually spot-applied because it can discolor and suppress the growth of desirable grasses.

In warm season turf there are some opportunities to selectively control perennial grassy weeds. For example, ethofumesate will suppress common bermudagrass in St. Augustinegrass turf, and sethoxydim can be used to remove bahiagrass from centipedegrass. Fenoxaprop will suppress bermudagrass growing in zoysiagrass turf, and repeat applications of MSMA will help to control bahiagrass that is infesting bermudagrass. Glyphosate can also be used to spot-treat undesirable perennial grasses.

Nutsedge, or nutgrass as it is sometimes called, is a sedge and not a true grass. However, it is a perennial monocot and is normally grouped with perennial grasses in discussions of weed control. Organic arsenicals were used for years to eradicate nutsedge, but they were not totally effective and could be phytotoxic to the turf. Bentazon offers better control of yellow nutsedge with minimal phytotoxicity problems. Imazaquin is used for purple nutsedge control in warm season turf, with the exception of bahiagrass.

CONTROLLING BROADLEAF WEEDS

The majority of herbicide applications are aimed at broadleaf weeds. Fortunately, most broadleaf weeds are relatively easy to control in turf because they are dicots and can be killed with herbicides that do not injure the turfgrass. One of the reasons broadleaf weed killers are selective is that dicots have their major meristematic area exposed at the top of the plant, while the crown of grasses is located at the base of the plant and is more protected from herbicides.

Selective herbicides such as 2,4-D, mecoprop, dichlorprop, dicamba, triclopyr, and clopyralid are commonly used to control broadleaf weeds. They are applied directly to the leaves.

A combination of at least two broadleaf herbicides is best. Using only one is less effective because some weeds growing in the turf may be resistant to the chemical (Table 14-2). For example, mecoprop alone controls only about 25 percent of the broadleaf weed species common in turf, but when mixed with 2,4-D it controls 65 percent. A mixture ensures better broad-spectrum weed control.

Broadleaf weed killers must be used carefully because they can injure nearby trees, shrubs, flowers, fruits, and vegetables if they come into contact with these plants. Drift is a common problem. The turf manager should spray

TABLE 14-2 Advantages of Using a Herbicide Mixtures because of Resistance of Certain Weed Species

	Herbicide			
Weed	**2,4-D**	**Mecoprop**	**Dicamba**	**2,4-D + Mecoprop Dicamba**
Common chickweed	R*	I	S	S
Black medic	R	I	S	S
Purslane	I	R	S	S
Red sorrel	I	R	S	S
Broadleaf plantain	S	I	R	S
Buckhorn plantain	S	I	R	S
Knotweed	R	I	S	S
Spurweed	I	S	I	S
White clover	I	S	S	S
Field pennycress	S	I	S	S

*R = resistant; S = susceptible; I = intermediate susceptibility—applications may have to be repeated for satisfactory control.

FIGURE 14-14
If broadleaf weed killers drift from the target area, serious problems can result because desirable broadleaf plants are often nearby. (Photo by Brian Yacur)

these chemicals only when there is little or no wind (Figure 14-14). Amine formulations of 2,4-D are usually preferred because ester formulations are more volatile. Dicamba may leach through the soil, so it should not be applied above the root zone of trees and shrubs unless the rate is sufficiently low. It should not be used at all on turf areas that are directly above the roots of very sensitive ornamental plants.

Postemergence herbicides such as 2,4-D, dicamba, and mecoprop kill broadleaf weeds by entering the plant and disrupting its metabolic processes. The herbicide must remain on the foliage for several hours or even a full day to allow lethal quantities of the chemical to penetrate the leaves. Rain should not be expected for 24 hours after the application. The turf manager should wait a few days before mowing the turf. Dicamba enters plants through both roots and leaves.

Granular herbicides should be applied when the weed foliage is moist. These dry materials stick better if leaf surfaces are wet. More herbicide is absorbed by the weed if there is good adhesion. Granular broadleaf weed killers should be spread after a rain or irrigation, or early in the morning when the foliage is wet with dew.

If performed properly, spraying, usually provides better control than granular applications. Best control with either occurs when application coincides with vigorous weed growth. When the broadleaf weeds are actively growing, the herbicides are more effective because of greater absorption and

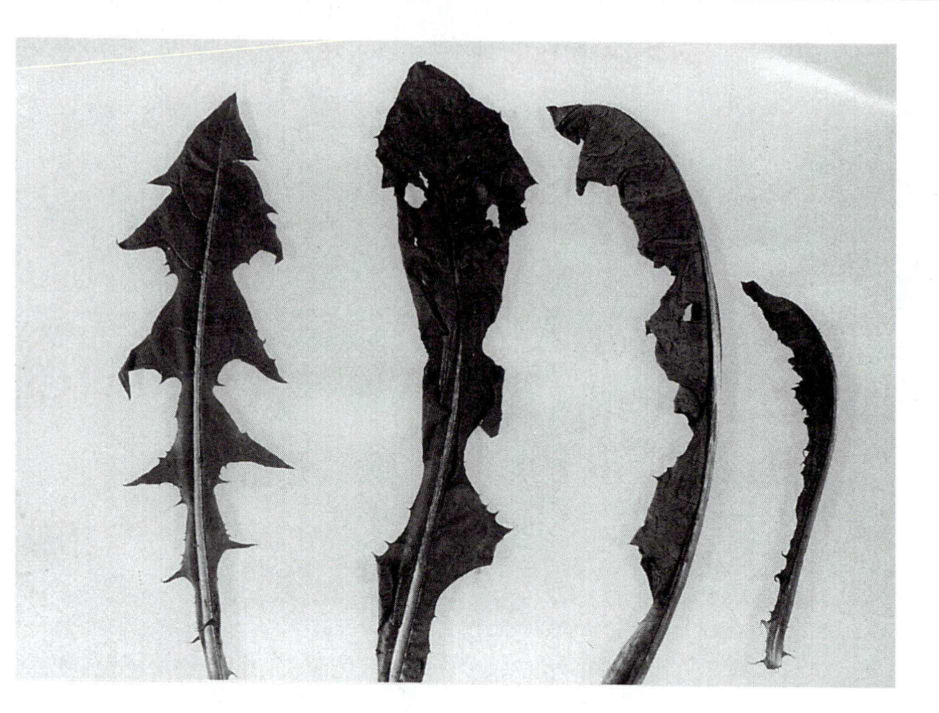

FIGURE 14-15
The healthy dandelion leaf on the left is untreated. The other three have been sprayed with a herbicide. The leaves curl (middle two), then desiccate (far right), and eventually disappear from the turf.

translocation (Figure 14-15). The materials work especially well if the weeds are young and succulent. However, the plants should not be so small that it is difficult to contact their leaf surfaces with the herbicide. The turf manager should not mow the site for a few days before application. Cutting the weeds reduces the size of their leaves and the likelihood that the herbicide will land on them.

Seedling grasses are susceptible to injury from broadleaf weed killers. Application should be delayed until the new turf has been mowed two or three times. After an application has occurred the turf manager should wait at least three or four weeks before seeding.

Even mature grasses can be injured if the turf is suffering from heat or drought stress. Warm season turfgrasses are susceptible to injury if treated in the spring before they have fully recovered from the winter.

Other postemergent herbicides such as metribuzin, metsulfuron, pronamide, atrazine, and simazine are used in warm season turf. Tolerances to these materials vary greatly among warm season turfgrass species, so it is very important to read the label thoroughly. St. Augustinegrass and centipedegrass are sensitive to the phenoxy herbicides 2,4-D, meloprop, and dichlorprop.

Knotweed, lambsquarter, purslane, spurge, and yellow woodsorrel are common summer annual broadleaf weeds. Henbit and common chickweed are winter annual broadleaf weeds often found in turf. Preemergence herbicides applied to control annual grasses, will also kill some annual broadleaf weeds as they germinate. Perennial broadleaf weeds also produce seeds, and they will be affected. Isoxaben is a preemergent herbicide that is used primarily for broadleaf weed control.

RULES FOR APPLYING HERBICIDES TO CONTROL BROADLEAF WEEDS

1. Broadleaf herbicides generally work best on young, succulent plants.
2. The weeds should be actively growing at the time of application. Best control occurs when the air temperature is 65°–85°F (18°–29°C). At higher temperatures leaf stomata may be closed, reducing herbicide movement into the leaves. Herbicides may volatilize (change to a gas) and drift when temperatures are greater than 85°F. Phytotoxicity to grass is also more likely.
3. Soil moisture levels should be adequate to ensure active growth, good translocation of the herbicide, and satisfactory root uptake if the material is absorbed by the roots. Irrigate before the application if the soil is dry. Avoid applications during long, excessively dry periods. Weed control is poor at times of prolonged moisture stress, and herbicides can be damaging to turfgrass roots.
4. Do not mow the turf right before the application. This ensures that there is sufficient leaf tissue on the weed for the herbicide to contact.
5. Avoid drift. If it is windy, do not spray.
6. Use low volumes of water when spraying. One gallon of water per 1,000 ft^2 is a common recommended application rate.
7. Weed leaf surfaces should be moist if a granular material is used. Keep traffic off the treated area for at least eight hours.
8. Rain or irrigation should not occur for at least 8 hours and preferably 24 hours after the application.
9. Do not mow until two days after the application. Mowing immediately will result in a removal of the herbicide before it can start to work.
10. Wait at least two weeks before retreatment. Do not be impatient—it may take one to four weeks for the weeds to die.
11. Grass seedlings are very sensitive to herbicides. It is best to delay applications until the newly planted area has been mowed two or three times. If an application is necessary sooner, in some instances it is possible to apply the regular broadleaf weed killers at half the normal rate.
12. Do not apply herbicides to newly installed sod until it is firmly rooted.
13. Do not seed a treated area until at least one month after application.
14. Do not use grass clippings on gardens or around shrubs until the third or fourth mowing after application of the herbicide.

Broadleaf weed control in warm season species is performed anytime weeds are actively growing. For cool season turf, control occurs in the spring, late summer/early fall, or late fall. Fall applications are generally preferable to spring treatments. Fall broadleaf weed control works very well on cool season turf because the turfgrasses are more likely to fill in bare spots vacated by the weeds then and the weeds are transporting a large amount of food to the roots for winter storage. There is excellent movement of herbicides into the weeds' roots, which results in better weed control. Another advantage of a fall application is that any accidental injury to desirable broadleaf plants will be less important because most will soon drop their leaves or die anyway.

Proper timing is critical when broadleaf species that are difficult to kill with herbicides are present in the turf. Applications should occur at the specific time of year when the weed is most susceptible. When large weed populations or difficult-to-kill weeds are present, repeat applications may be necessary.

Grasses under stress can be injured by broadleaf weed killers they would normally tolerate. Excessively high application rates can result in injury to healthy grass. Proper application techniques are essential (see Appendices B and C). Herbicide residues can remain in a sprayer even after it has been carefully cleaned. Serious injury can occur if the same sprayer is used to apply other pesticides to plants that are sensitive to the herbicide.

SELF-EVALUATION

1. The presence of a large weed population indicates that changes should be made in the ___________.
2. ___________ annuals germinate in the late summer or early fall.
3. ___________ perennials reproduce by seed, rhizomes, stolons, and underground storage organs.
4. ___________ herbicides kill all types of plants.
5. Annual grasses are primarily controlled with ___________ herbicides.
6. Broadleaf herbicides in a granular form work best when the leaf surfaces of weeds are ___________.
7. Herbicides should be sprayed when there is little or no ___________.
8. Discuss the differences between the two subspecies of *Poa annua*.
9. The presence of which specific weeds indicates that the site is too wet?
10. Discuss the rules to follow for the proper application of a broadleaf herbicide.
11. What type of cultural techniques can result in lower annual grassy weed populations?

CHAPTER

15

Insects

OBJECTIVES

After studying this chapter, the student should be able to

- Diagnose turfgrass injury caused by insects
- Describe the ways in which insect pests injure turfgrass
- Identify the insect species that are serious pests of turfgrass
- Discuss the life cycles and characteristics of major injurious insects
- Explain how insect pests are controlled

INTRODUCTION

Insects are small animals that have three pairs of legs. Adult insects always have six legs, but immature forms may not. Insect bodies are divided into distinct segments. The three major sections are the head, thorax, and abdomen (Figure 15-1).

Insects that feed on turfgrass have two different types of mouthparts. Some, such as grubs, caterpillars, and maggots, chew plant tissue. Their jawlike mouthparts tear, chew, and grind grass shoots and roots. Others, such as chinch bugs, scales, and aphids, have piercing-sucking mouthparts. They pierce plant tissue with their beaks and then suck juices from the stems, leaves, or roots (Figure 15-2). Both types of feeding can result in serious injury to turfgrass plants.

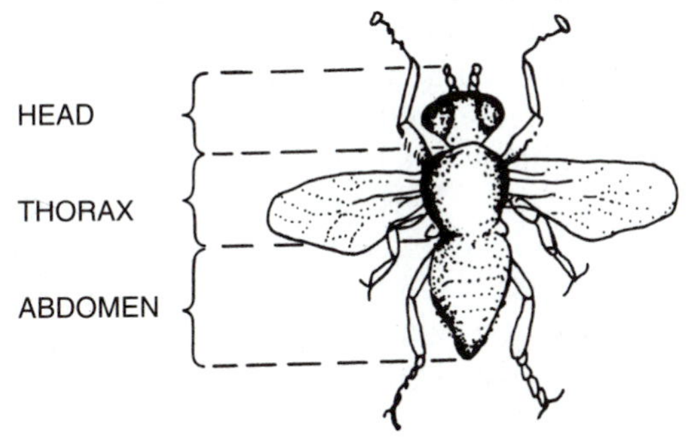

FIGURE 15-1
The three major body parts of a frit fly.

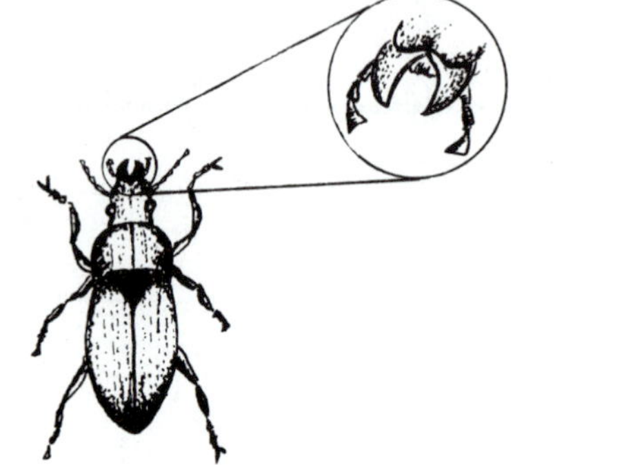

FIGURE 15-2
Typical chewing and piercing-sucking mouthparts.

Insects have their skeletons on the outside of their bodies. This protective shell is called an *exoskeleton.* As insects feed and grow in size, the relatively rigid exoskeleton must be shed periodically or growth is restricted. Insects use chemicals to dissolve their old exoskeletons and then produce new, larger ones. This process is called *molting.* Each period between a molt is called an *instar.*

The insect life cycle begins with an egg that is deposited on or near a food source, in this case a turfgrass plant. The egg hatches, and an immature insect emerges and begins to feed. This young insect is in its first instar. As it grows larger, the immature insect molts, enters the second instar, and continues feeding and molting until it becomes an adult. A female adult mates with a male and then lays eggs, beginning a new generation.

As insects progress through their life cycles, visible changes occur externally. Growth that results in changes in shape or form is called *metamorphosis.* Insects such as beetles, moths, and flies change tremendously in external appearance as they develop from egg to adult. They are said to undergo complete or complex metamorphosis. An immature beetle, for example, looks much different from an adult (Figure 15-3). Other insects such as chinch bugs, mole crickets, grasshoppers, and aphids do not change greatly in shape or form as they grow. They are said to undergo gradual or simple metamorphosis. An adult chinch bug, even though it is larger and has fully developed wings, appears similar to its immature version (Figure 15-4).

The immature stages of insect species that undergo gradual metamorphosis are called *nymphs.* Fully grown, last-instar nymphs molt and become adults. Three stages occur in the life cycles of these insects: egg, several nymphal instars, and adult. Immature stages of insect species that experience complete metamorphosis are called *larvae.* Fully grown, last-instar larvae are unable to simply molt and change into adults because the two stages are so different. Instead, the larva enters an inactive, nonfeeding stage during which it changes greatly in appearance and is transformed into an adult. This process is referred to as *pupation.* Four stages occur in the life cycles of these insects: egg; several larval instars, pupa, and adult (see Figure 15-3). The larva of a beetle is a *grub,* the larva of a moth or butterfly is a *caterpillar,* and the larva of a fly is a *maggot.*

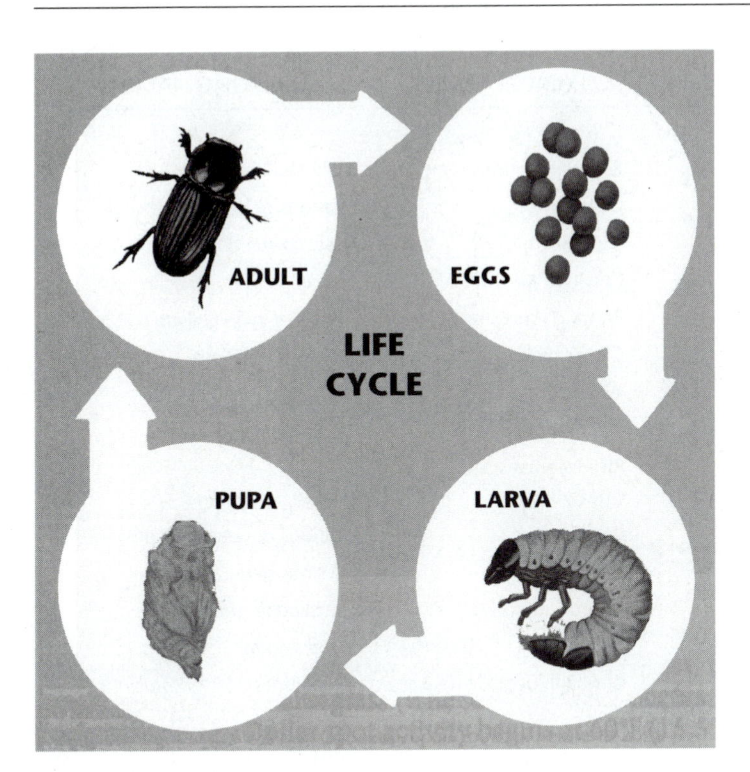

FIGURE 15-3

The life cycle of a beetle is an example of complete metamorphosis. *(Courtesy of TUCO Agricultural Chemicals, Division of Upjohn Co.)*

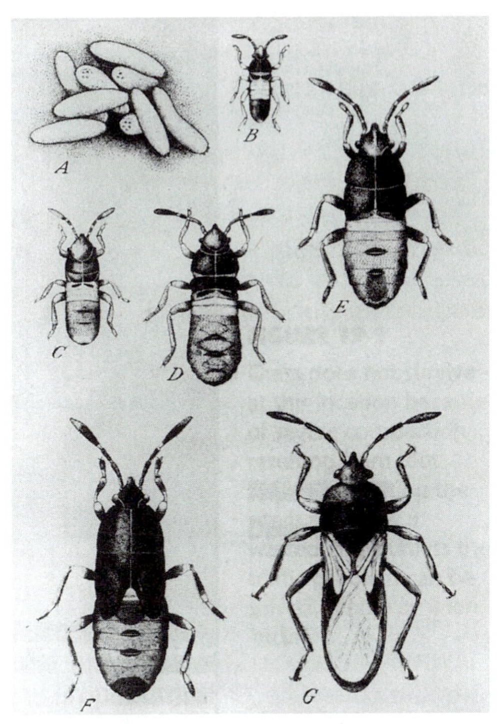

FIGURE 15-4

The life cycle of a chinch bug is an example of simple metamorphosis. It begins with the egg (A), followed by five immature stages (B–F), and finally the adult (G). *(Courtesy U.S. of Department of Agriculture)*

How seriously turfgrass is injured by insects depends primarily upon the size of the insect population present in the turf area. Each individual insect consumes only a small amount of food because of its small size. The more insects that are feeding, the greater the injury to the turf.

The size of an insect population is influenced by several factors. Some insects have rapid reproduction rates, and large numbers can build up in a relatively short time. An insecticide application can kill many insects and decrease the population significantly. Weather has the greatest effect on insect populations. Weather conditions that are unfavorable result in smaller populations. For example, wet springs and summers contribute to a sharp reduction in the number of chinch bugs present in a lawn. In moist conditions a fungus disease destroys many of them. Most insects can complete their life cycles and lay eggs at a faster rate at high temperatures. The longer that temperatures are at optimal levels for insect development, the larger the populations.

COMMON NAME	SCIENTIFIC NAME	COMMON NAME	SCIENTIFIC NAME
Grubs			
Japanese beetle	*Popillia japonica*	Sod webworms	*Crambus* spp.
May beetle (June bug)	*Phyllophaga* spp.	Cutworms	Various genera and species
Southern masked chafer	*Cyclocephala immaculata*	Armyworm	*Pseudaletia unipuncta*
Northern masked chafer	*Cyclocephala borealis*	Fall armyworm	*Spodoptera frugiperda*
European chafer	*Amphimallon majalis*	Bermudagrass mite	*Eriophes cynodoniensis*
Green June beetle	*Cotinus nitida*	Clover mite	*Bryobia praetiosa*
Asiatic garden beetle	*Maladera castanea*	Winter grain mite	*Penthaleus major*
Oriental beetle	*Anomala orientalis*	Bermudagrass scale	*Odonaspis ruthae*
Black turfgrass ataenius	*Ataenius spretulus*	Rhodesgrass scale (mealybug)	*Antonina graminis*
Bluegrass billbug	*Sphenophorus parvulus*	Leafhoppers	Various genera and species
Hunting billbug	*Sphenophorus venatus vestitus*	Grasshoppers	Various genera and species
Phoenix billbug	*Sphenophorus phoeniciensis*	Greenbug	*Schizaphis graminum*
Tawny mole cricket	*Scapteriscus vicinus*	Frit fly	*Oscinella frit*
Southern mole cricket	*Scapteriscus acletus*	Annual bluegrass weevil	*Hyperodes* near *anthracinus*
Wireworms	Various genera and species	Ants	Various genera and species
Ground pearls	*Margarodes* spp.	Bees and wasps	Various genera and species
Hairy chinch bug	*Blissus leucopterus hirtus*	Periodical cicada	*Magicicada septendecim*
Southern chinch	*Blissus insularis*		

FIGURE 15-5 The major turfgrass insect pests in the United States.

In most regions of the United States insects overwinter in a resting state. They usually stop feeding sometime in the fall in response to the colder temperatures. When warmer weather returns in the spring, they become active and begin feeding again. In the warmest areas of the South insects may continue to feed throughout the winter.

The following species are the primary insect pests of turfgrass in the United States (Figure 15-5).

ROOT-FEEDING INSECTS

Grubs

Grubs are the larvae of beetles. The grubs of several beetle species, including the Japanese beetle, the May (June) beetle, the European chafer, the Asiatic

FIGURE 15-6
A typical grub.
(Courtesy of TUCO Agricultural Chemicals, Division of Upjohn Co.)

garden beetle, and the oriental beetle, live in the soil and feed on turfgrass roots. These immature beetles are from 0.5 inch (1.3 centimeters) to 1.5 inches (3.8 centimeters) in length when fully grown, depending upon the species. The soft, wormlike larvae are white to grayish in color, have hard brown heads, and have six distinct legs. They are often found curled in a C-shaped position in the soil (Figure 15-6).

Each of the species can be identified by examining the arrangement of the hairs or spines on the grub's *raster,* the terminal segment. A hand lens is necessary to distinguish the rastrel pattern (Figure 15-7). Species identification is important because some of these beetles have different life cycles. Insecticide efficacy can vary between species. The turf manager is able to apply the best control product when the grubs are most vulnerable if he or she knows the life cycle of the species attacking the turfgrass.

The majority of the beetle species that produce grubs which injure turfgrass have one-year life cycles. May beetles are an exception, completing their life cycles in two to four years. Japanese beetles are a typical example of

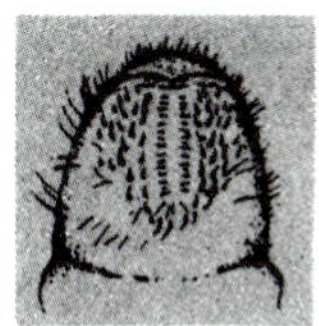

MAY BEETLE

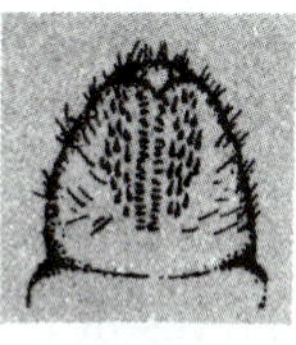

EUROPEAN CHAFER

ASIATIC GARDEN BEETLE

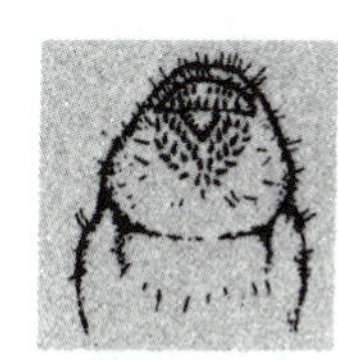

JAPANESE BEETLE

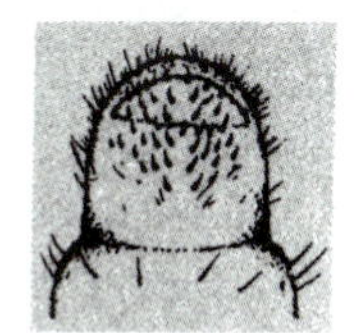

NORTHERN MASKED CHAFER

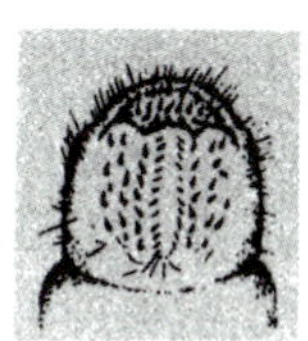

ORIENTAL BEETLE

FIGURE 15-7 Rastrel patterns of common white grubs.

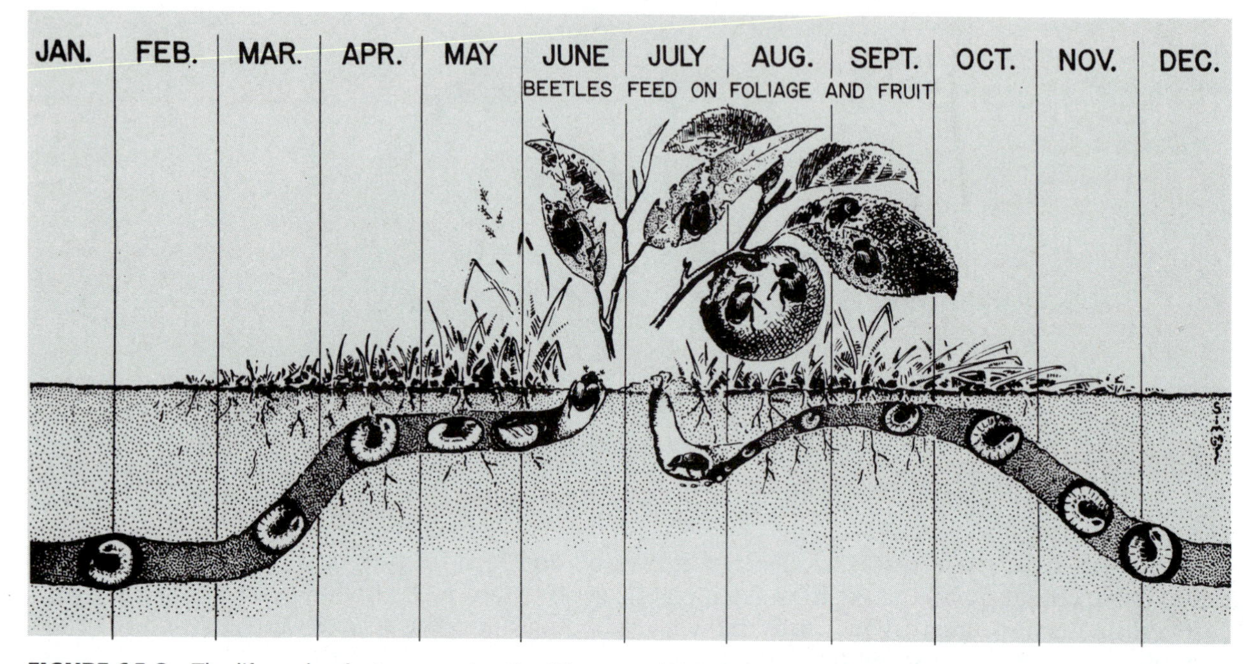

FIGURE 15-8 The life cycle of a Japanese beetle. *(Courtesy of U.S. Department of Agriculture)*

the species that require one year to complete their life cycle (Figure 15-8). They spend at least ten months of the year in the soil. In the early summer the grubs pupate and become adults, leaving the soil to fly around and find mates. After copulation the adult females lay their eggs in the soil beneath grass sod. In July or August the eggs hatch, and the young larvae begin to feed on turfgrass roots. As the grubs feed, they molt and grow progressively larger. This rapid growth continues through the fall until soil temperatures become cold enough that the grubs stop feeding. They then tunnel down deeper in the soil and overwinter. In the spring, when the soil warms, the grubs move back up near the soil surface and resume feeding. In the late spring they reach full size and enter the pupal stage. After pupation the adults emerge from the soil, mate, lay eggs, and the cycle starts over.

Severe summer droughts have a major effect on grub populations. When the soil is extremely dry, many eggs desiccate and are destroyed. Grub problems are often minimal after a location has experienced droughty weather for several years, but the populations increase again when the soil is moist during midsummer. Unusually cool summer weather can result in a decrease in mating and egg laying.

Various beetle grubs attack many turfgrass species, but the adults do not feed on turfgrass. The leaves of trees and shrubs serve as their major food source. Japanese beetle adults can feed on more than three hundred different plants.

FIGURE 15-9
Grubs are exposed when the injured sod is pulled back. *(Courtesy of Art Wick, Lesco Products)*

Japanese beetles are found throughout much of the United States, but are most common east of the Mississippi River. Asiatic garden beetles and oriental beetles are primarily a problem in the Northeast. The European chafer is most serious in New York and neighboring states. Masked chafers and May beetles are widely distributed in the United States. The green June beetle is a pest mainly in the South.

Grubs have chewing mouthparts and are capable of causing extensive damage to the turfgrass root system. As few as three grubs per square foot (0.09 m^2) may be enough to seriously injure the grass. There is a direct correlation between the size of the grub and the amount of injury it causes. Populations of 40 to 80 grubs per square foot (0.09 m^2) of turf are not unusual.

When turfgrass suffers severe root loss because of grubs, the plants wilt, turn yellow or brown, and eventually die. Irregular patches of injured, dying turf appear. Grub damage can be diagnosed by examining the affected turf, which can be easily lifted from the soil if grubs are the problem. Grub-infested turf can be pulled up readily (Figure 15-9), because the roots are cut by the chewing mouthparts of the insects—the turf is no longer strongly anchored in the soil when grubs chew through the roots. Lifting the sod exposes the grubs. They are usually located in the top 3 inches (7.6 centimeters) of soil.

A quick survey for potential grub problems can be accomplished by cutting into the turf with a spade and examining the root zone. A golf green cup cutter can also be used to examine the soil. When the grub population averages more than three per square foot (0.09 m^2), insecticide treatment may be necessary. This decision is based on the species present, the health of the grass, irrigation capability, and other factors. A number of animals feed on grubs. The presence of moles or large flocks of birds feeding on turf areas is a good indication that grubs are in the soil. Skunks, raccoons, and other mammals will rip up the turf when searching for grubs.

Grub damage is most evident during the spring (April and May) and fall (September to November). Insecticides are the primary method of control. Insecticide applications are most effective in July and August when the young larvae have recently hatched out. The grubs become less susceptible to insecticides as they mature. Frequent irrigation may also help the turf to recover. Daily watering will reduce the moisture stress caused by root injury and promote the growth of new roots.

The black turfgrass ataenius (BTA) is another beetle that produces root-feeding grubs. This species is known to occur in at least forty states, but it is far less of a problem then the other grub species previously discussed. The injury caused has been primarily to annual bluegrass *(Poa annua)* on golf course fairways in northern states. Bentgrass and Kentucky bluegrass may also be attacked.

The adults, which are 0.04 inch (1 centimeter) long, overwinter along the edge of fairways in the soil or under tree leaves or other debris on the ground. In the spring the adults emerge from these protective sites. Eggs are laid in the soil or thatch beneath turfgrass in May or early June. The eggs hatch, and the grubs begin to feed on roots. The first instar larvae are so tiny that they are difficult to see with the naked eye. After feeding, molting, and growing for approximately a month, the larvae reach maturity and tunnel down into the soil to pupate in late June and July. A BTA larva has two distinctive pads on the tip of its abdomen. The adults appear in July and early August.

In most areas a second generation occurs. The adults lay eggs in July and August. The second-generation larvae feed throughout the remainder of the summer and pupate in September. In September and October the second-generation adults leave the fairways and seek shelter for the winter.

The first wilting symptoms appear in mid-June and continue throughout the summer. Annual bluegrass is most severely injured. Because of the small size of the larvae, large populations are necessary to cause serious root-feeding problems. More than 30 to 40 BTA grubs per square foot (0.09 m^2) are required for significant damage to occur, unless the turf has been weakened by other stresses.

Billbug grubs feed on turfgrass roots, crowns, and stems. The adult beetles, members of the weevil family, have their chewing mouthparts at the tip of a snout or bill. The larvae are less than 0.5 inch (1.3 centimeters) long and legless, with thick white bodies and brownish heads (Figure 15-10). The Phoenix

FIGURE 15-10
Billbug larva and adult.

billbug attacks bermudagrass, the hunting billbug feeds on zoysiagrass, and the bluegrass billbug is primarily a problem on Kentucky bluegrass.

Generally these species overwinter as adults, hibernating in the turf or nearby protected areas. In the spring the adult females lay eggs in the lower part of the turfgrass stem. After hatching, the larvae feed inside the stem and then feed on the crown. Eventually the grubs move down into the soil and attack roots and rhizomes. They change into pupae at the end of the summer or in early autumn, and the adults emerge from the soil in September or October. There is usually only one generation a year.

Injured turfgrass wilts and turns brown during the summer. Damaged plants are easily pulled up, especially if the grubs have chewed off the stems at the crown. The presence of a fine, white, sawdustlike material, known as *frass,* at the feeding sites on the turfgrass plant is evidence of billbug problems. Observing billbug adults on sidewalks, driveways, and other paved areas adjacent to the turf in the spring and fall is another method of detecting potential problems. Insecticides are applied either in the spring to prevent the adults from laying eggs, or in June to kill the newly hatched larvae.

Mole Crickets

Mole crickets are a species of cricket that lives in the soil and feeds on grass roots, stolons and rhizomes, insects, and earthworms. They are a pest of many turfgrasses in the southeastern United States and Texas. Bahiagrass and bermudagrass receive the most damage. The southern and the tawny mole cricket are the two species that cause the greatest problems. Mole crickets are light brown, covered with velvety hairs, and have short, spadelike legs that are adapted for digging and tunneling through the soil (Figure 15-11). The adults are approximately 1.5 inches (3.8 centimeters) long. The mole cricket undergoes simple metamorphosis. Immatures are called nymphs.

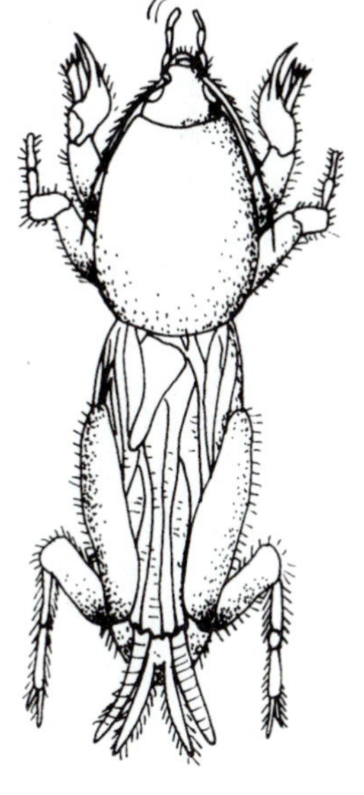

FIGURE 15-11
Mole cricket.

Eggs are laid in the soil in the spring. The nymphs hatch out in May and look like the adults, except that they are smaller and lack fully developed wings. The nymphs live in the soil the remainder of the year and normally become adults the following spring. Occasionally nymphs change into adults in the fall. There is one generation per year.

Turfgrass may wilt and dry out because of mole crickets' root feeding and tunneling. As they burrow through the soil, plants can be uprooted. Damage can be severe, especially on newly seeded or sprigged turf. Mole crickets often burrow 6 inches (15.2 centimeters) or deeper into the soil. They are not found in heavy soils. At night, especially after a rain or irrigation, the come up near the soil surface to feed.

Insecticide applications give best control in June when the nymphs are small and relatively easy to kill. Spreading bait containing an insecticide on the turf is also an effective method of control. There is a great deal of interest in using natural enemies such as parasitic nematodes and flies for control. This will be discussed in Chapter 17.

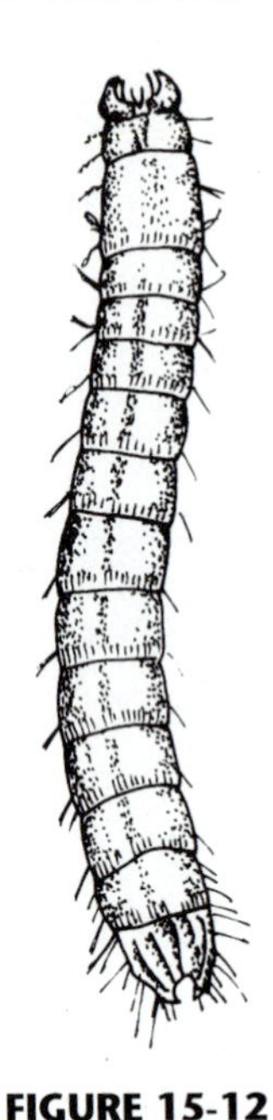

FIGURE 15-12 Wireworm.

Wireworms

Wireworms are slender, shiny, smooth, hard-shelled, wormlike larvae. They are pale yellow to dark brown and feel tough or wirelike, not soft like most beetle larvae (Figure 15-12). Their length is usually 0.5 to 1 inch (1.3 to 2.5 centimeters). The adult is called a click beetle.

Wireworms live in the soil and feed on roots. They are common in turfgrass soils, but their numbers are seldom large enough to cause noticeable injury. Problems are most like to occur in low, wet areas.

Ground Pearls

Ground pearls are scale insects found in the southern United States. They secrete a waxy material that covers and protects their bodies like a shell. This process is common to all scale insects. The shells are spherical and look like tiny, yellowish-purple pearls. The insects are usually around 0.125 inch (3 millimeters) in diameter, but they can be as small as a grain of sand. Ground pearls have piercing-sucking mouthparts and suck juices from turfgrass roots. They attach themselves to the roots by penetrating the tissue with their needlelike beaks (Figure 15-13).

Ground pearls feed on the roots of centipedegrass, bermudagrass, zoysiagrass, and St. Augustinegrass. Their life cycle is not fully understood, but it is usually completed in one year. The adult females leave their protective shells and deposit eggs in the soil. The eggs hatch during the summer. The newly hatched nymphs, called crawlers, insert their beaks in roots and then secrete the pearllike shell around their bodies.

Chemical control of ground pearls has been difficult because they may feed as deep as 10 inches (25 centimeters) below the soil surface. They injure the plants by removing liquids containing carbohydrates and minerals from

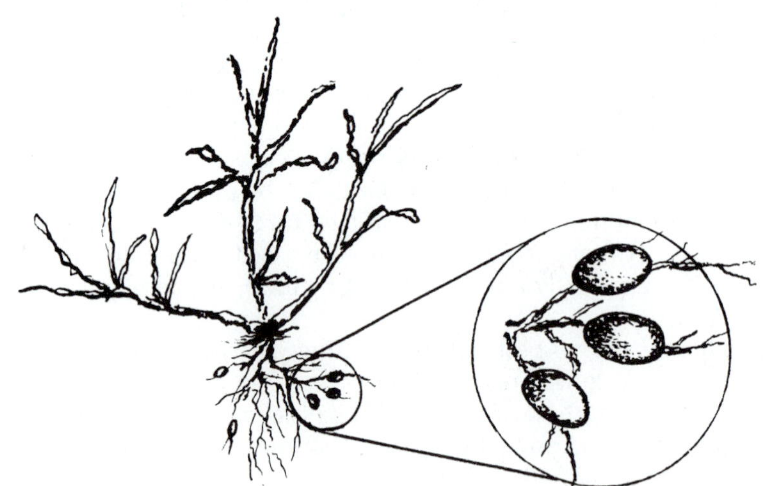

FIGURE 15-13 Ground pearls.

the roots. The effect of this feeding damage can be reduced by fertilizing, watering, and mowing properly to keep the grass healthy as possible.

LEAF AND STEM FEEDERS

Chinch Bugs

There are two species of chinch bugs that are important turfgrass pests in the United States. The hairy chinch bug attacks cool season grasses, and the southern chinch bug feeds on warm season grasses. The latter species is a serious threat only to St. Augustinegrass. Chinch bugs are primarily a problem in eastern and midwestern states. They injure plants by sucking juices from the shoots and by injecting a toxic salivary fluid into the tissue. This fluid damages the xylem, the water-conducting tissues, causing the plant to wilt and turn brown. If the weather is hot and dry, the affected turfgrass will die.

The number of generations that occur in a year depends upon temperature. The warmer the temperature, the faster the chinch bug completes its life cycle. In the northernmost areas of the United States there are only one or two generations a year. In southern Florida there can be as many as seven. Chinch bugs may continue to feed and be active throughout the winter in warmer climates. In colder regions they overwinter as adults in tall grass or other protected sites, and begin egg laying in the spring.

Eggs are deposited on or in lower leaf sheaths, stolons, or in the thatch. Each female can lay over 200 eggs. The first instar nymphs that hatch from the eggs are the size of a pinhead, 0.05 inch (1.3 millimeters) long. They are bright red, wingless, and have a white band across their backs. There are usually five nymphal instars (refer to Figure 15-4). As these immatures molt and grow larger, their color changes to orange, then brown, and finally to black in the last instar. All of the nymphs have the white stripe, although it is partially hidden on the final-instar nymph. The adults are approximately 0.2 inch (5 millimeters) long and are black with white, folded wings (Figure 15-14). The white wings have black markings.

Chinch bugs are commonly found in sunny areas. Heavily infested, sunny sites may contain over 200 chinch bugs per square foot (0.09 m^2). Their damage becomes most apparent during droughty periods in the summer (Figure 15-15). People often mistake chinch bug injury for dormancy resulting from moisture stress because they expect the grass to turn yellow or brown during hot, dry weather. They become aware of the problem when the dead grass does not green up again after rain or irrigation. This misdiagnosis is a serious error because turf injury can be easily prevented by an insecticide application if the chinch bug infestation is discovered in its early stages.

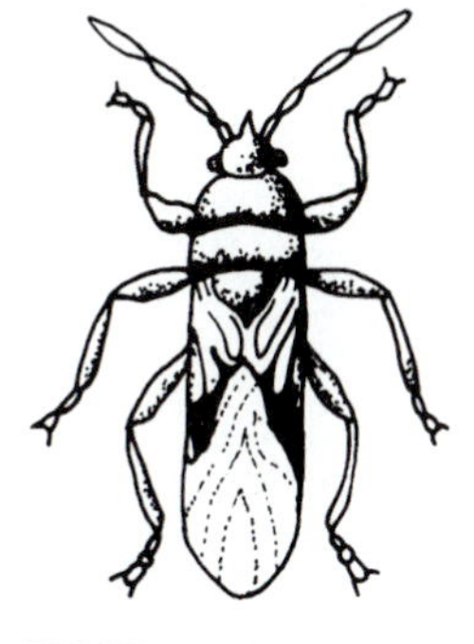

FIGURE 15-14
Adult chinch bug.

When suspicious yellow spots appear in a turf area during the summer, the turf manager should push aside the grass and examine the surface of the soil or thatch for chinch bugs. A careful examination is necessary because the insects

FIGURE 15-15
Chinch bug damage.
(Courtesy of Art Wick, Lesco Products)

may be difficult to see. They are small and will attempt to hide when disturbed. A simple method of detection is to remove both ends of a coffee can or similar large can, cut off or file one rim to produce a sharp edge, and twist and push the can 1 to 2 inches (2.5 to 5.1 centimeters) into the soil. It should be worked into the soil at the edge of a declining spot, preferably where the grass is just beginning to turn yellow. The can is then filled with water, and within five or ten minutes any chinch bugs present will float to the surface. It may be necessary to add more water to keep the water level above the grass. Flooding a small area and then covering it with a piece of white cloth also works. Chinch bugs will crawl up to the top of the leaf blades and then cling to the underside of the cloth.

Significant chinch bug populations, usually 20 or more per square foot, require an insecticide application. Early detection and control protect the turf from serious injury. Insecticide applications do not completely eradicate the pest, but do result in a significant decrease in its population. In prolonged periods of wet, rainy weather, a fungus disease infects and kills chinch bugs. An insect predator, the big-eyed bug, feeds on chinch bugs. Floratam, a St. Augustinegrass cultivar, exhibits resistance to this pest. Perennial ryegrass, tall fescue, and fine fescue cultivars high in endophytes are also resistant.

Sod Webworms

Sod webworms are actually caterpillars, the larvae of moths. There are several species of sod webworms that injure turfgrass. The caterpillars chew the leaves

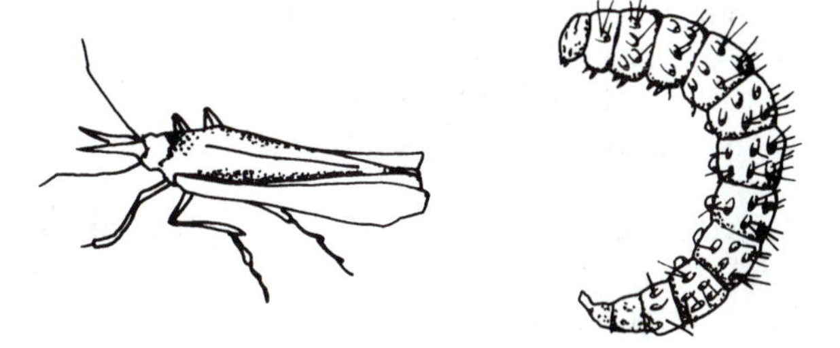

FIGURE 15-16
Lawn moth (adult) and sod webworm (larva).

of turfgrass plants with their mouthparts, but the adult moths do not feed on turf. The adults are usually referred to as lawn moths because they are often seen laying their eggs on lawns.

The moths are grayish-white to brown in color and 1 inch (2.5 centimeters) or less in length. They have two distinguishing features. The adults are sometimes called snout moths because of the snoutlike projection that extends from the front of their heads. When they are not flying, lawn moths fold their wings around their bodies (Figure 15-16). Most moths, when at rest, keep their wings spread.

The caterpillars have dark brown heads. Their size varies with each instar, but the full-grown larvae are approximately 0.75 inch (1.9 centimeters) long. The sod webworm's body is tan, green, or gray depending upon the species. Most have black or dark brown circular spots scattered over their bodies.

In northern areas there are usually two generations per year, while in southern climates the number is greater. This pest normally overwinters in the larval stage. In the spring the caterpillar pupates and becomes an adult. Female moths fly over the turf at dusk and drop their eggs randomly. During the day the moths hide in the grass or in shrubs and trees near the turf area. The caterpillars hide in the thatch during the day and feed on the grass foliage at night. As they grow larger, they construct silk-lined tunnels in the thatch.

The sod webworm injures grass plants by chewing off leaves. If the population is small, the damage is usually minor. Plants can often produce new leaf tissue and recover. However, the intense feeding of a large sod webworm population may result in a significant loss of turfgrass quality. This is especially true during periods of dry weather. Golf greens and other closely mowed turfs are more readily injured than areas where the grass is cut to a taller height.

If flocks of birds return frequently to feed on a turf area, it may indicate an infestation of sod webworms or other larvae. Pencil-sized pecking holes made by birds when they are searching for larvae are another possible sign of insect problems. If close examination of irregular brown areas in a turf reveals that leaves have been chewed off, the turf manager should look for caterpillars in the thatch. During the day they are normally found curled up near or on the soil surface. The presence of green pellets of excrement, called frass, is also evidence of sod webworm feeding.

Sod webworm and other caterpillars can be detected by pouring household detergent on any suspicious brown spots in the turf. The detergent acts as an irritant and brings the insects to the surface, where they can be seen more easily. A common mixture used is 1 ounce (30 milliliters) of liquid detergent per gallon of water.

Most people become aware of potential sod webworm problems when they observe numerous lawn moths flying over the turf in the early evening. The moths may also be seen during the day because they fly out of the grass when disturbed. The presence of adults does not mean that damage is inevitable. It is best to check for larvae before applying an insecticide. Turfgrass can usually tolerate up to 10 sod webworms per square foot (0.09 m^2), unless it is cut very short.

The sod webworm population builds up as the growing season progresses, and turf injury is often greatest late in the summer. Damage is most likely on well-managed turf because the sod webworms are attracted to higher-quality, vigorous grass. If insecticide treatment becomes necessary, an evening application is best because this is when the caterpillars emerge from the thatch to feed.

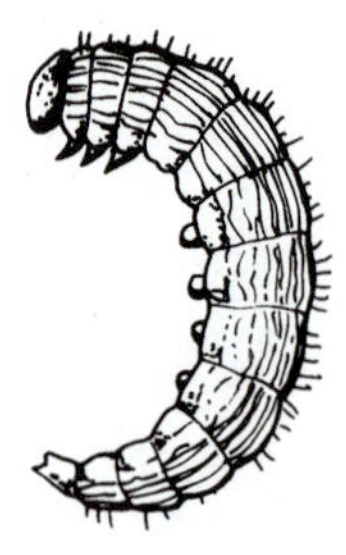

FIGURE 15-17 Cutworm.

Cutworms

Cutworms are caterpillars with dark brown heads and blackish, gray or brown bodies. They may have stripes or spots, depending upon the species. Mature larvae are 1.5 to 2 inches (3.8 to 5.1 centimeters) long (Figure 15-17). Cutworms feed at night on foliage and larger instars chew off turfgrass shoots near the bottom of the stem. Usually they are only a minor problem, but golf greens and new seedings may occasionally be seriously injured. Even a small amount of damage is unacceptable on a green.

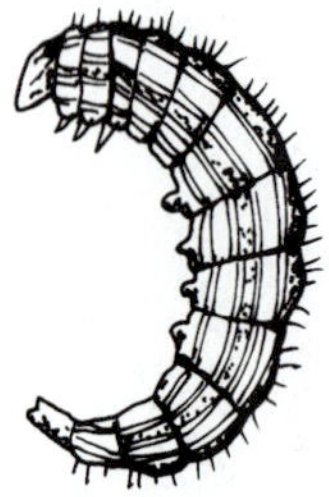

Armyworms

Armyworms are caterpillars that are 1.5 to 2 inches (3.8 to 5.1 centimeters) long when fully grown. Their color varies from green to gray, and they have stripes (Figure 15-18). The caterpillars feed on grass leaves at night, but unlike cutworms and sod webworms they often do not hide during the day. Large numbers of armyworms are usually found feeding together. This "army" devours one turf area and then crawls to another site at night. Armyworm problems are most common in the South.

FIGURE 15-18 Armyworm.

Mites

Mites are very similar to insects, but they have four pairs of legs rather than three. Mites are extremely small, and a hand lens or microscope is needed to examine them (Figure 15-19). Enormous populations can build up in a short time because of their rapid reproduction rate. Some species complete their life cycles in a week. Hundreds of mites may be found feeding on a single grass leaf or stem. Leaves become blotched or spotted as mites suck the juices from the shoots. Continuous feeding by large numbers of mites results in the death of the leaves.

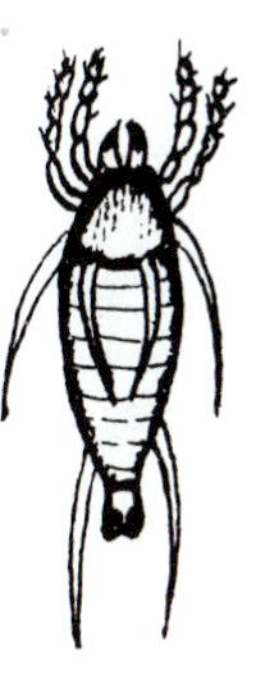

FIGURE 15-19

Bermudagrass mite.

The bermudagrass mite is a serious pest of bermudagrass in southeastern and southwestern states. It is very tiny, only 0.01 inch (0.25 millimeter) long, and feeds under the leaf sheaths. A microscope is necessary to identify these yellowish-white, wormlike pests. Because they are difficult to see, their presence is often diagnosed by the characteristic injury they cause to turfgrass. Initially, the leaves of infested plants curl abnormally and turn light green. As the feeding continues, leaf tissue swells, stem internodes become shortened, and the leaves are crowded closely together on the stem. The injured plants have a clumpy or tufted appearance.

The clover mite and the winter grain mite are two other species that occasionally cause injury to turfgrass. The winter grain mite has an unusual life cycle because it feeds in the winter but not during the summer. Neither of these species is an important a pest as the bermudagrass mite.

Some insecticides will be effective against mites because they are close relatives of insects. Miticides, pesticides developed specifically to kill mites, can also be used. Chemical control may be necessary when mite populations are very large and the grass is suffering from moisture stress.

Scales

Scales are small insects that secrete a waxy material which serves as a protective covering. Ground pearls, scale insects that feed on roots, have already been discussed. Two other scale species are also pests of warm season turfgrasses. They suck plant juices from grass leaves and stems.

The bermudagrass scale, as its name implies, attacks bermudagrass. It is oval or circular, 0.04 to 0.07 inch (1 to 1.7 millimeters) in diameter, and covered by a hard, white, waxy shell. The rhodesgrass mealybug, which is also called the rhodesgrass scale, feeds on bermudagrass, St. Augustinegrass, and centipedegrass. Its dark, spherical body, covered with a mass of white cottony material, is 0.125 inch (3.2 millimeters) or less in diameter.

When eggs hatch, the first-instar nymph, called a crawler, is mobile and finds a place to feed on the stem or leaf. It then secretes its protective covering and does not move again. Scale infestations are easy to diagnose because of the insect's immobility and distinctive appearance.

Leafhoppers

Leafhoppers are small, wedge-shaped insects that readily fly or hop around (Figure 15-20). The adults are normally 0.5 inch (1.3 centimeters) or less in length. The immature nymphs closely resemble the adults, but are smaller and lack wings. Leafhoppers are common, but do not often cause serious injury to turfgrass. Occasionally large numbers may migrate into a lawn and damage

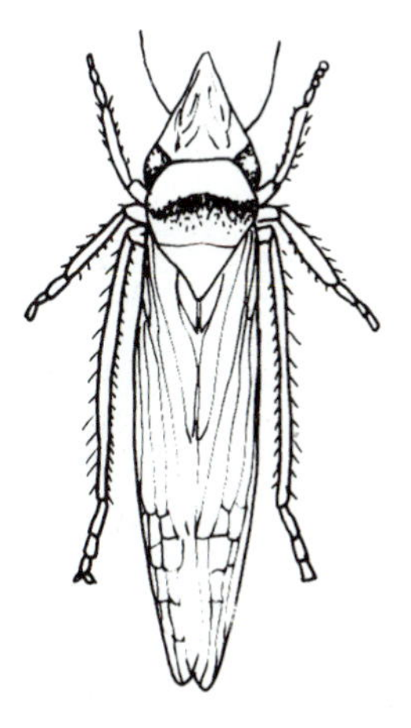

FIGURE 15-20

Leafhopper.

plants by sucking out fluids from the shoots. Usually they move on to another area in a few days. Most serious injury occurs in the prairie states.

Grasshoppers

Grasshoppers are chewing insects that commonly feed on range and pasture grasses, as well as grasses growing in waste areas. They are seldom a threat to well-maintained turf unless their numbers are quite large and food is scarce.

Aphids

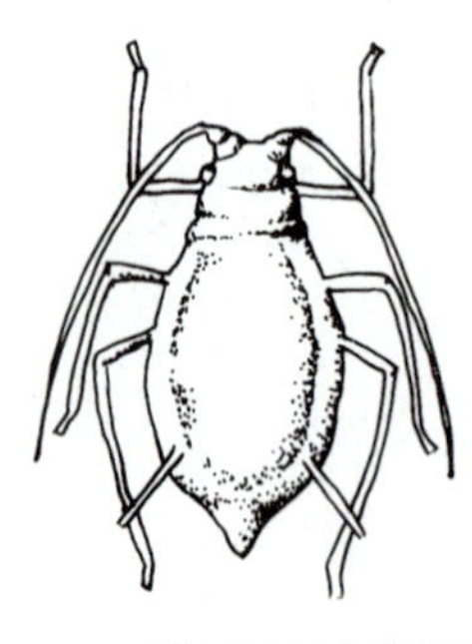

FIGURE 15-21 Aphid.

Aphids are small, piercing-sucking insects that are common pests of many plant species. However, until recently they were not considered a serious problem on turfgrass. In the 1970s one species of aphid began to cause extensive injury to Kentucky bluegrass on lawns in some midwestern and northeastern states. These aphids are called greenbugs because the nymphs and adults have soft, pear-shaped bodies that are light green in color. The antennae, eyes, and tips of the legs are dark in color. These aphids are usually wingless, approximately 0.06 inch (1.6 millimeters) long, and have two tubes extending from their abdomens (Figure 15-21).

Greenbugs can become a serious problem because of their rapid reproduction rate. Females do not need to mate with a male. They give birth to living young during the spring and summer rather than laying eggs. At optimum temperatures, a greenbug is able to develop into an adult and begin reproducing a week after it is born. Populations build up very quickly, and thousands of these aphids may be found per square foot (0.09 m^2).

Greenbugs injure turfgrass plants by inserting their mouthparts into leaf tissue and sucking sap from the phloem. Further damage is caused by salivary fluids which destroy the leaf tissue surrounding the area that is pierced by the insect's beak. Infested grass blades turn yellow, then orange, and finally brown when they die. Injury is most common during summer months, especially on turf areas shaded by trees. Greenbugs are thought to overwinter in the egg stage.

Diagnosis is simple because of the large numbers of greenbugs present on leaves when turfgrass injury occurs. Several insecticide applications are often necessary to control this pest. Many aphids will be killed by the insecticide, but some will survive. The greenbug's rapid reproduction rate enables the population to rebuild quickly unless insecticide treatment is repeated. Moreover, it has exhibited resistance to some insecticides.

Frit Flies

Frit flies are tiny, shiny black flies approximately 0.06 inch (1.6 millimeters) long (see Figure 15-1). The larvae, called maggots, tunnel into grass stems near the soil surface and feed on the tissue with their chewing mouthparts. During hot, dry weather the infested shoots are killed. The maggots are small, legless, yellowish-white in color, and have two black hooks on their heads. These hooks are chewing jaws and can be seen with a hand lens.

The maggots overwinter in grass stems, and in the spring, after pupation, the adult flies lay eggs on turfgrass leaves. The newly hatched larvae tunnel into stems and begin to feed. Several generations occur each year. The principal hosts are bentgrass, annual bluegrass, and Kentucky bluegrass. Frit flies are common throughout the United States, but generally are a minor problem on turfgrass.

The adult flies are attracted to white objects. This curious habit aids in diagnosis. If present, they will immediately appear on a piece of white paper or cloth that is placed on the turf. Frit flies will hover around and land on a golf ball lying on an infested green.

Annual Bluegrass Weevil

The annual bluegrass weevil, which feeds on annual bluegrass, is also referred to as the hyperodes weevil. This insect is a serious pest primarily on golf course fairways and collars around greens that contain significant annual bluegrass populations. It is a problem in the northeastern United States. The adult weevils are black, have snouts on their heads, and are less than 0.2 inch (5.1 millimeters) in length. The larvae are also small and have soft, white, legless bodies (Figure 15-22). Their heads are brown.

The annual bluegrass weevil overwinters as an adult in protected sites under fallen tree leaves or needles or in tall grass in the roughs adjacent to infested fairways. In the spring the adults crawl to the turf areas populated with annual bluegrass and lay eggs in the leaf sheaths. The larvae feed in stems until the final instar, which feeds on the crown of plants. Injury becomes apparent by late spring. Mature larvae pupate in the soil near the surface. There are usually two generations a year.

Insecticides are applied before the adult has laid its eggs in annual bluegrass stems. Treatments usually occur during the period when forsythia and then flowering dogwood bloom because this is when the adults move from winter hibernation sites.

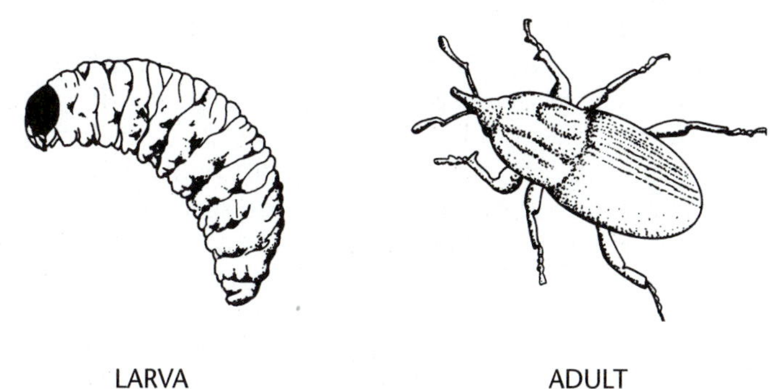

FIGURE 15-22
Annual bluegrass weevil—larva and adult.

OTHER INSECT PESTS

Other insects occasionally injure turf, although they do not feed on turfgrass plants. Several species of ants build underground nests in turf areas. As the ants excavate soil, grass plants close to the mounts are buried or desiccated. Anthills also disrupt surface uniformity. Some bee and wasp species nest in turf as well, and people may be stung when they walk near the nests.

Periodical cicadas may be a pest in certain years. The immature nymphs live in the soil and feed on tree roots. After 13 or 17 years the last-instar nymphs emerge from the soil, become adults, and mate, and the females lay their eggs in tree twigs. When the eggs hatch, the small, antlike nymphs drop to the ground, enter the soil through cracks, and then remain in the soil or 13 or 17 years. Damage to turf is because of the emergence holes made by the full-grown nymphs when they leave the soil. As many as 50,000 cicadas may emerge from the soil beneath a large tree.

The European crane fly is found in southern British Columbia and in western Washington state and Oregon. The adult looks like a large mosquito, and the immature is called a leatherjacket. The larvae feed on turfgrass roots, crowns, and leaves.

CONTROL

When weather conditions are favorable for the growth of insect pest populations, injury to turfgrass may occur. If insect populations are large enough to damage a turf area, the turfgrass manager may consider applying an insecticide (Figure 15-23). There are some alternative control measures, but for the present, insecticides are still the primary method of managing turfgrass insect pests. In the future many effective insect control strategies not requiring the use of traditional insecticides will be available to the turfgrass manager.

Insecticides can be a very valuable tool if used wisely. Safe handling requires both common sense and following the direction on the label. The turf manager must read and obey all label instructions and precautions. When a turf manager has any questions regarding pesticide use, he or she should contact the Cooperative Extension Service for assistance.

Usually insecticides should not be applied until an insect problem occurs. Most of the insecticides commonly used on turfgrass in recent years retain their ability to kill insects for less than two weeks after application. They do not persist long enough to protect the turf from future problems. If injury is not apparent, it is often a waste of time and money to treat the turf areas with these materials because of their brief residual effect. It is also environmentally unsound to apply insecticides unnecessarily.

Until the mid-1970s chlordane, an insecticide belonging to a group of chemicals called chlorinated hydrocarbons, was widely used on turfgrass. It had a lengthy residuality, often killing insects for five years or more after it was

COMMON NAME	TRADE NAME
Acephate	Orthene
Azadirachtin	Turplex Bioinsecticide
Bacillus thuringiensis kurstaki	Dipel
	MVP
	Steward
	Worm-ender
Bendiocarb	Turcam
Bifenthrin	Talstar
Carbaryl	Sevin and many others
Chlorpyrifos	Dursban and many others
Cyfluthrin	Tempo
Diazinon	Diazinon
Dicofol	Kelthane
Ethion	Ethion
Ethoprop	Mocap
Fenamiphos	Nemacur
Fluvalinate	Mavrik
Fonofos	Crusade
	Mainstay
Imidacloprid	Merit
Isofenphos	Oftanol
Isazofos	Triumph
Lambda-cyhalothrin	Battle
	Scimitar
Lindane	Lindane
Malathion	Malathion
Permethrin	Torpedo and others
Steinernema carpocapsae	Exhibit
Trichlorfon	Dylox
	Proxol

FIGURE 15-23 Common and trade names of pesticides used to control insect and mite pests on turfgrass in 1994.

applied. However, chlordane and many other chlorinated hydrocarbons persisted for so long that they caused environmental problems, and their use was banned. The insecticides that replaced them, the organophosphates, carbamates, and pyrethroids, are relatively short-lived. They are usually effective against insects for 3 to 14 days. One of the organophosphate chemicals,

FIGURE 15-24
The key to insect control is early detection. *(Photo by Brian Yacur)*

isofenphos, sold under the trade name Oftanol, retains its toxicity to insects for several weeks or longer.

The key to effective insect control is early diagnosis and treatment of problems. Preventing losses to insects requires constant vigilance on the part of the turf manager, who must continually look for insect pests or signs of their presence. Any yellow grass or declining turf must be examined immediately (Figure 15-24). Early detection enables the manager to apply an insecticide before significant turf injury occurs. It is important to remember that the presence of a pest does not mean that significant injury will occur. What will happen depends primarily on how many pests are present. It takes a certain size population to injure the grass. This is called a threshold and is usually expressed as the number of insects per square foot (0.09 m^2). For example, treatment is usually recommended when 5 to 10 white grubs are found per square foot in a moisture-stressed turf. However, the threshold may rise to 15 or more if the turf can be irrigated frequently to compensate for root injury. Threshold population numbers are being compiled for many insect pests.

Insect problems often become serious in the summer because this is when the populations of many species increase in size to injurious levels. Grass is also weakened by heat and drought during summer months and is more likely to succumb to insect injury. A common mistake is to assume that damage resulting from insect feeding is merely summer dormancy caused by hot, dry weather. The symptoms exhibited in both cases may be identical because grass injured by insects experiences moisture stress owing to the loss of fluids, roots,

FIGURE 15-25
Ignoring insect problems can result in severe injury to the turf area. *(Courtesy of New York State Turfgrass Association)*

or the disruption of vascular tissue (Figure 15-25). This misdiagnosis can be avoided by carefully searching the affected turf for insect pests.

Understanding the life cycles of insect pests is essential. This knowledge allows the turf manager to anticipate when problems are likely to occur. Familiarity with life cycles also aids the manager in preparing a spray schedule. Insecticides should be used whenever injury becomes apparent; however, the proper timing of applications helps to ensure effective control. As a general rule, insects are more susceptible to insecticides during their early instars. Susceptibility may decrease as larvae or nymphs mature. Best control often occurs when a treatment is timed to coincide with egg hatching.

There is another advantage to treating the younger immatures. The earlier instars are usually too small to cause extensive damage. The insecticide will not eradicate the entire pest population, but many will be killed. This prevents the majority of the insects from growing large enough to cause serious feeding injury.

The turf manager should be aware that some of the insects will survive an insecticide application. The survivors will not be numerous enough to present an immediate threat to the turf, but when they reproduce the population may build up to injurious levels. The manager should expect and watch for the recurrence of insect problems in the future. Retreatment may eventually become necessary.

To be killed, the insect must come into contact with the insecticide. This contact occurs in several ways. The chemical may be applied directly on the insect. In many cases the pest contacts the insecticide as it moves around in treated grass, thatch, or soil. The insect may also ingest the toxic material if it feeds on tissue covered with or containing insecticide.

Insect pests that live and feed in the soil are the most difficult type to control. It is not possible to apply the insecticide directly to the insect. The chemical must move through the turf and thatch and penetrate the soil to be effective. It takes longer for insecticides to come into contact with soil-inhabiting insects. This can be a problem, especially if the material is tied up in the thatch layer. The shorter-term insecticides may lose much of their toxicity by the time they reach grubs or other soil insects.

The movement of insecticides into the soil is accelerated by applying 0.5 inch (1.3 centimeters) of water after treatment. The water leaches the chemical down to the soil. When irrigation is not possible, the turf manager should try to apply the material right before or even during a heavy rain. Sprays should be watered in immediately after application. If irrigation is delayed, the liquid will adhere to the foliage. It may not readily wash off and can be degraded by ultraviolet light. Granular insecticides will not stick to the foliage if the leaves are dry.

Watering before treatment can also be helpful. Insecticides move more readily through the thatch when it is wet. Using large volumes of water when spraying helps to wash liquid materials off the foliage and through the thatch.

Soil inhabitants such as white grubs prefer to feed at the thatch-soil interface. However, if the surface soil is dry they will stay deeper. Most insecticides do not penetrate very deep into the soil. Heavy irrigation before or after treatment will keep the grubs near the surface, where they are most likely to come into contact with the insecticide.

Insects that feed on shoots are easier to control than soil-inhabiting insects. Sprays work very quickly, killing many insects in the first 48 hours after application. Treatment in the late afternoon or early evening is ideal because most of the pests that attack the shoots feed at night. An application late in the day ensures maximum contact. The liquid material will either directly contact the insects or dry on and stick to the foliage and the surface of the thatch. Insects that are not killed immediately will come into contact with the insecticide as they move around and feed on the leaves and stems. Sprays directed at shoot-feeding insects are not usually watered in after application because the insecticide would be removed from the target area. Rain should not occur immediately after the application. The turf manager should also avoid mowing the grass for a few days.

The pH of the water used to dissolve or suspend the insecticide can have an impact on the performance of the chemical. Some insecticides have a greatly reduced period of effectiveness when mixed with an alkaline carrier. For example, trichlorfon is stable for over 80 hours in a spray tank with water that has a pH of 6.0, but it loses its ability to kill insects in minutes when the water has a pH of 9.0. Acidifying agents can be added to the water if it has a high pH and the insecticide is sensitive to alkalinity. This information can be found on the pesticide label.

Granular insecticides do not usually kill shoot-feeding insects as quickly as liquid sprays. Granular materials consist of inert carriers such as clay, vermiculite, or corn cob particles that are coated with insecticide. The granules release the insecticide when they absorb moisture. A light irrigation following application hastens the leaching of insecticide from the carrier. Granular materials usually have longer residual activity than sprays because the liquids adhere to the foliage and are broken down by light. Most of the granules fall to the surface of the thatch layer and are less exposed to sunlight. Insects that feed on stems and leaves come into contact with the granular insecticide as they crawl over or through the thatch.

Turf managers must remember to read the label on the insecticide container carefully before using the material. The Cooperative Extension Service or the manufacturer should be consulted if the manager has any questions concerning the proper use of the insecticide.

Well-managed, healthy turfgrass has a better chance of surviving an insect infestation than poorly maintained grass. Strong, vigorous grass plants are more likely to recover from insect damage. Plants that are weak and nonrecuperative are more likely to be severely injured by insect feeding. Proper fertilization, mowing, and watering are an important part of an effective insect control program.

A method of biological control has been available for many years for use against Japanese beetle grubs. *Bacillus popilliae* is a bacterium that causes a disease which kills only Japanese beetle grubs. A powder or granular formulation containing spores of this organism can be purchased and spread on a turf area. As the grub feeds in the soil, it ingests these spores and becomes infected. The disease is called milky spore disease because the grub's blood turns a white, milky color. Several years are required before the grub population is controlled, but once the bacteria are well established in the soil the disease may continue to kill Japanese beetle grubs for many years. Milky spore disease products are used on low-value turf areas because of their slow initial control of grubs. They are most effective when soil temperatures are greater than 70°F (21°C). Some turfgrass entomologists are skeptical about the effectiveness of milky spore disease.

Bacillus thuringiensis var. *kurstaki* is a bacterium that is marketed for the control of caterpillars such as sod webworms. It produces a toxic protein that damages the caterpillar's stomach. When a larva ingests a lethal dose it stops feeding within the hour and dies in several days.

The use of nematodes to control white grubs and other insect pests is generating tremendous interest. Species of the genera *Steinernema* and *Heterorhabditis* enter insects' bodies and release bacteria that kill the insects. Commercial products are on the market that can be applied with a pesticide sprayer. Nematodes dry out very easily, so they should be sprayed on moist turf and watered in immediately.

Beneficial insects are being investigated for their value as control agents. A fly has been imported from South America to help control the mole cricket.

The immature fly, called a maggot, feeds on the pest. A wasp has been released to parasitize rhodesgrass mealybugs in Texas.

Researchers are experimenting with insect growth regulators which can be used to adversely affect insect behavior. One type of hormone prevents immatures from becoming adults. Alternatives to insecticides and integrated pest management techniques are discussed in greater detail in Chapter 17.

Another area of great interest is the development of resistant cultivars. Some varieties presently available contain compounds that are toxic or distasteful to insect pests, but the majority are susceptible to insect injury. In the future resistant varieties will become more common. Some of the perennial ryegrass, fine fescue, and tall fescue cultivars contain endophytes, fungi that live inside the plants. These fungi produce toxic chemicals which discourage feeding by surface-feeding insects. Unfortunately, endophyte levels in the roots are very low, so endophytic varieties are not resistant to insects that feed on roots.

SELF-EVALUATION

1. The larvae of beetle species that live in the soil and feed on turfgrass roots are called ___________.
2. The period between molts is referred to as an ___________.
3. Turfgrass insect pests have two types of mouthparts—chewing and ___________.
4. The ___________ is the immature stage of the law moth.
5. Insects that secrete a waxy protective covering are called ___________.
6. The greenbug is actually an ___________.
7. Chinch bug injury is often mistaken for ___________.
8. How many grass species serve as the food source for the annual bluegrass weevil?
9. The larva of a frit fly is called a ___________.
10. ___________ have eight legs and are closely related to insects.
11. The front legs of mole crickets are adapted for ___________.
12. Before using an insecticide, the turf manager must read the ___________.
13. An insecticide that is directed against soil insects should be ___________ after it has been applied.
14. A turf manager attempting to control shoot-feeding insects should not ___________ the grass for a few days following an insecticide application.
15. Discuss why an insecticide application might fail to result in satisfactory control of an insect pest.
16. Discuss the significance of thresholds.
17. Why are white grubs sometimes harder to control than chinch bugs?

CHAPTER

16

Turfgrass Diseases

OBJECTIVES

After studying this chapter, the student should be able to

- Explain what causes turf diseases
- Understand how diseases are identified
- Describe common turfgrass diseases
- Explain how to prevent and control turfgrass diseases
- Discuss nematodes and their control

INTRODUCTION

Plant diseases are disorders caused by microorganisms such as fungi, bacteria, and viruses. The study of plant disease is called plant pathology, and microorganisms that cause plant abnormalities are referred to as *pathogens*. Almost all common turfgrass diseases are caused by fungi (Figure 16-1).

FIGURE 16-1
A fungal disease (snow mold) on a lawn. (*Courtesy of New York State Turfgrass Association)*

FUNGAL PATHOGENS

Description

Fungi are microscopic organisms that cannot produce their own food because they lack chlorophyll. Fungi that cause plant disease obtain their food by stealing it from higher, green plants such as turfgrass. Some fungi, called saprophytes, feed on dead organic matter; others, called parasites, feed on living hosts. Most fungi that cause diseases are able to obtain their nutrition from both living and dead plant material. They can live as saprophytes or parasites. Two turf pathogens, *Erysiphe* and *Puccinia,* can obtain their food only from living hosts and are called obligate parasites.

Fungi lack roots and conductive tissue and therefore have difficulty obtaining and transporting water. Because of this limitation fungi are usually inactive during periods of dry weather. Disease problems are most likely to begin when there is a film of moisture on plant surfaces for several hours or more. Severe disease injury can occur during prolonged periods of high humidity, rain, fog, or heavy dew. Abundant moisture because of excessive irrigation also favors diseases.

The body of a fungus is composed of threadlike filaments known as *hyphae* (Figure 16-2). A mass of hyphae is referred to as the *mycelium (mycelia,* plural), which has several functions. The mycelium is a feeding structure, absorbing nutrients from plant cells. Mycelia can also become dormant, allowing the fungus to survive adverse conditions such as cold temperatures or a drought.

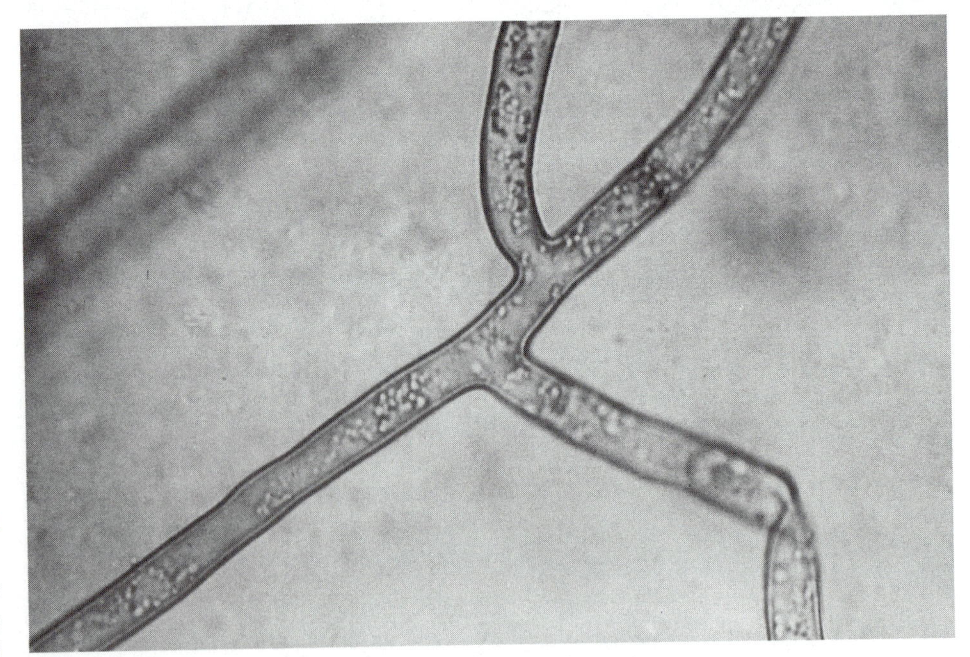

FIGURE 16-2 Fungus hyphae. *(Courtesy of New York State Turfgrass Association)*

When the fungus is established inside the host (the plant being attacked), specialized strands of hyphae grow back out to the surface of the leaf, stem, or root. They produce the spores responsible for spreading the disease. The tiny, seedlike spores are commonly carried to healthy plants by wind, splashing rain drops, and lawn mowers. The spores then germinate, and the fungus invades the healthy plant if there is a film of moisture on the plant surface. Hyphae can also spread the disease when the turf is wet by growing from plant to plant. The hyphae and spores of some pathogens infect roots.

Life Cycle

To identify and control a disease successfully, the turfgrass manager must understand the life cycle of the pathogen that causes the disease. The life cycle of the fungus responsible for brown patch disease illustrates the yearly disease cycle caused by a typical turfgrass pathogen.

Brown patch is a common disease problem in many regions of the United States and occurs on all of the major turfgrass species (Figure 16-3). The scientific name (Genus, species) of the causal organism is *Rhizoctonia solani.*

This fungus overwinters in the form of *sclerotia*—hard, thick-walled resting masses of mycelium that are able to survive cold temperatures. The sclerotia are found either on the surface of the soil or embedded in plant tissue. In the spring, when soil temperatures reach the 60s (15–20°C), new mycelium grows

FIGURE 16-3

Brown patch disease. *(Courtesy of New York State Turfgrass Association)*

from the sclerotia. Significant disease injury does not occur until the temperatures are high enough to cause heat stress and weaken the grass.

Hyphae usually enter grass leaves through open stomata or mowing wounds. Most fungi also have the ability to chemically dissolve or mechanically force their way into plant tissue.

If weather conditions favorable to *Rhizoctonia solani* and unfavorable to the turf persist, the mycelium continues to invade host cells and feed on their nutrients. The result is dead or dying brown patches of grass. When the temperatures and humidity necessary for injury are no longer present, the disease temporarily disappears. However, it can reappear throughout the growing season whenever the optimum environmental conditions recur, because the pathogen is always present.

Severity of a Disease

For a disease to occur there must be a pathogen present, favorable environmental conditions, and a susceptible plant. The severity of the disease depends on a number of factors. Some pathogens are said to be virulent—they have the ability to destroy turfgrass. Other pathogens do not normally kill plants.

Host plants vary in their susceptibility to a disease. Some are relatively resistant and sustain only minor injury, while others are highly susceptible and seriously injured. The degree of susceptibility is also affected by the health of the plant. If the grass is weak, it becomes more susceptible and is less recuperative than a vigorous plant.

Environmental conditions have a great impact on disease severity. If weather conditions optimal for the growth of a pathogen occur for a long period, disease problems will be increased. The disease can also be worse when weather extremes result in a weakened turf. Weather changes unfavorable to the pathogen can stop the disease.

DISEASE IDENTIFICATION

The turf manager must be able to identify a disease accurately because the correct method of control often differs for each disease. Identification is accomplished by observing symptoms, signs, weather conditions, and the host species.

Symptoms

The symptom is the external reaction of the plant to the disease. The disease causes a distinctive change in the appearance of the host. Different diseases cause different symptoms.

Common symptoms are leaf spots. If a spore lands on the foliage, germinates, and penetrates the leaf, a dead spot will appear where the fungus has fed. These spots often have different shapes, sizes, and colors. For example,

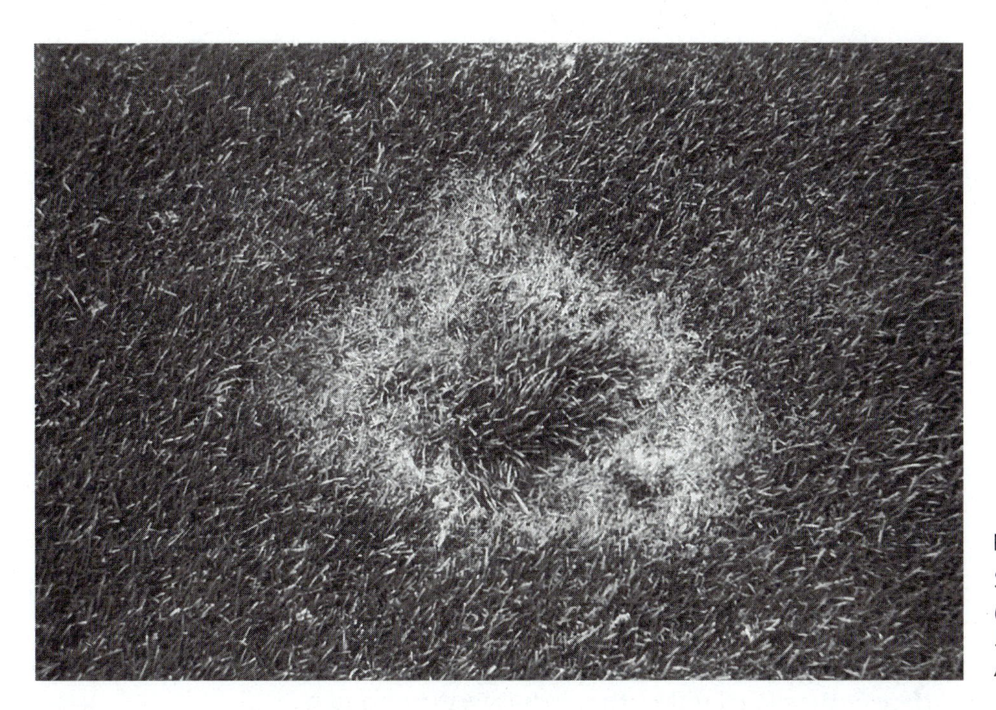

FIGURE 16-4
Summer patch.
(Courtesy of New York State Turfgrass Association)

gray spots on the leaves of St. Augustinegrass indicate the presence of *Pyricularia grisea. Drechslera* and *Bipolaris* fungi cause purple to black spots on Kentucky bluegrass.

The overall appearance of diseased grass is also important when identifying diseases. Older patches of necrotic ring spot and summer patch diseases often exhibit a distinctive pattern—light brown dead patches with green areas in the center (Figure 16-4). This frog-eye appearance occurs when the grass in the center recovers. Dollar spot disease is first noticed when circular, tan spots roughly the size of a silver dollar appear in a lawn or on a golf course green.

To the untrained observer all diseases may look alike. Turfgrass managers must carefully inspect diseased areas, often getting down on their hands and knees to observe the grass closely. If root diseases are suspected, plugs of soil must be removed and examined.

Signs

A sign is any observable part of the pathogen. Although pathogens are very small, it is sometimes possible to see them without a microscope. The fungus causing powdery mildew disease produces abundant white mycelium and spores visible to the naked eye. A magnifying glass is very helpful when searching for signs.

The mycelium dries out very readily, so the best time to look for signs is early in the morning, when the relative humidity is high and dew is still on

FIGURE 16-5
Whitish mycelium is visible in this photograph. *(Courtesy of Lofts Seed, Inc.)*

the grass (Figure 16-5). Later in the day the mycelium is not found on the surface of the plants unless the grass is wet from a rain.

Seeing the fungus is often the most important step toward identifying a disease. For example, *Microdochium nivale* (formerly *Fusarium nivale*) has pink mycelia, and the disease it causes is called pink snow mold. Stripe smut disease is easy to identify because masses of spores on the leaf look like black streaks (Figure 16-6).

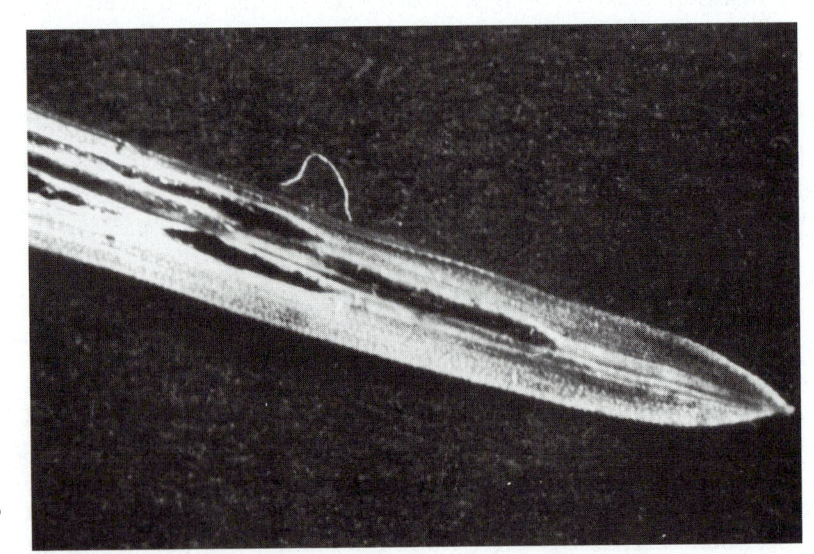

FIGURE 16-6
Strip smut.

Weather Conditions

Every pathogen has specific optimum environmental conditions during which it is most active. Although the majority of fungi need abundant moisture to grow, temperature requirements often vary. Severe anthracnose disease problems are most likely to occur, for example, when air temperature is 80°F (26°C) or higher. Gray snow mold (Typhula blight) is a cool-temperature disease prevalent at 32° – 40°F (0° – 4°C).

Temperature and humidity requirements can be used to predict disease occurrence. A serious outbreak of Pythium blight can occur when there are at least 14 hours of relative humidity above 90 percent and it is hot. Specifically, during that 14-hour period the maximum temperature must be greater than 82°F (28°C) and the minimum temperature at least 68°F (20°C).

Instruments are available that collect weather data to predict the potential occurrence of Pythium blight and several other diseases. Data collected include relative humidity, air and soil temperature, leaf wetness, soil moisture, and amount of rainfall. The instrument uses this information to determine if a fungicide application is necessary.

Host

Determination of the grass species being attacked contributes to successful disease diagnosis. St. Augustine decline commonly attacks only one host, St. Augustinegrass. Take-all patch only occurs on the bentgrasses. Unfortunately, some pathogens have a wide host range. A disease such as brown patch is found on many grass species.

Identification Summary

When confronted with a disease, the turf manager, like a detective, examines the clues. For example:

A disease occurs on perennial ryegrass during hot weather when the temperature is in the 85°–95°F (29°–35°C) range. The affected grass feels greasy and appears water-soaked. White, cottony-looking mycelium is observed. The manager compares his or her list of observations with the disease identification information found in the next section of this chapter or in the turfgrass disease books listed in the bibliography. The disease is identified as Pythium blight. Several chemical companies that manufacture fungicides provide booklets that contain excellent color photographs of turfgrass diseases.

Disease diagnosis can sometimes be difficult because diseases do not always look or behave the way they are supposed to. Fungi have a tremendous potential for genetic variability and change because of their rapid reproduction rate. Consequently, new subspecies or strains can arise that have characteristics somewhat different from those generally exhibited by the species.

If a disease problem is difficult to diagnose, the turf manager should con-

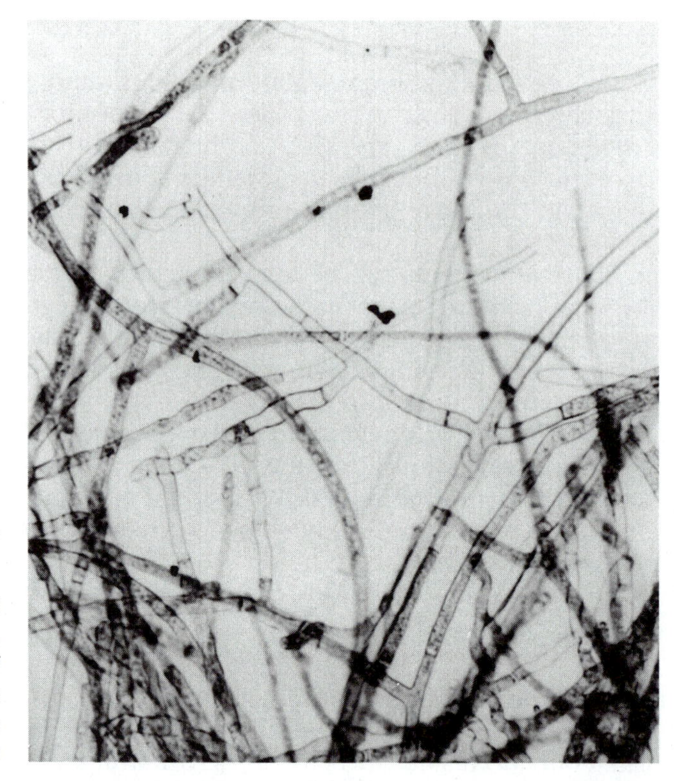

FIGURE 16-7 Right-angle branching of *Rhizoctonia solani* (brown patch pathogen). *(Courtesy of Patricia Sanders and the Pennsylvania Turfgrass Council)*

sult a plant pathologist or turf specialist. A diseased grass sample (including roots and leaves) is collected and immediately taken or sent by priority mail or delivery service to the nearest plant disease diagnostic lab. The sample must be fresh and include all of the stages of the disease on the entire grass plant. Information concerning proper sampling and handling techniques should be obtained from the Cooperative Extension Service.

The pathologist will examine the specimen through a microscope. Under high magnification the pathologist can observe fungal structures clearly. *Rhizoctonia solani,* for example, has hyphae that branch at right angles (Figure 16-7). Snow mold disease caused by *Microdochium* can be identified by the presence of boomerang-shaped spores (Figure 16-8). Some turf managers are equipped to perform microscopic identification themselves.

Common Turfgrass Diseases

Some scientific names may change because the reclassification of pathogens is common.

Anthracnose *(Colletotrichum graminicola).* Anthracnose is a foliage disease that is primarily a problem on fine fescues, perennial ryegrass, and annual

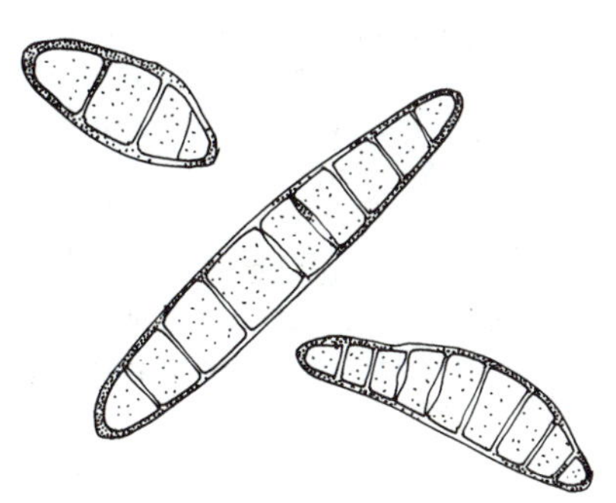

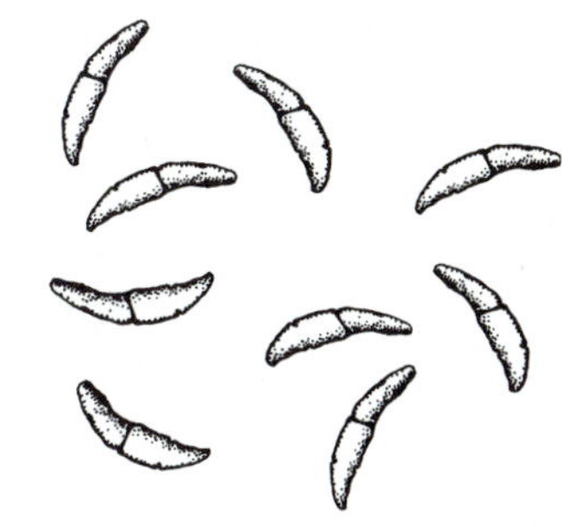

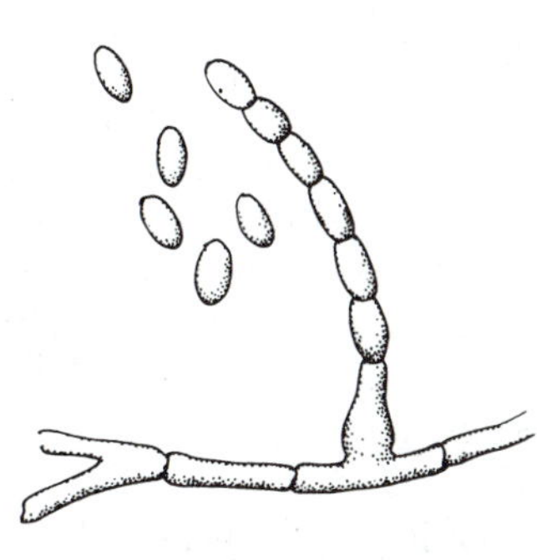

FIGURE 16-8
Some spores produced by turfgrass pathogens.

bluegrass, though it can also occur on bermudagrass, Kentucky bluegrass, creeping bentgrass, and centipedegrass. The disease is most severe when periods of high humidity and temperatures greater than 80°F (22°C) occur. Its symptoms are reddish-brown leaf lesions and irregular-shaped patches of infected grass from a few inches to several feet in diameter. The patches are initially yellow and then turn to a bronze color. Eventually, numerous black fruiting bodies (acervuli) appear on infected leaves. These reproductive structures are covered with black spines and are easily seen with the naked eye. Anthracnose development is favored when the grass is weakened by drought and low fertility. Balanced fertilization and irrigation help to prevent the disease. Reducing compaction and improving soil drainage can also be helpful.

A less common disease caused by *Colletotrichum* spp. results in a crown rot. The crown appears dark because it is filled with black acervuli. Infected plants die. Crown-rotting anthracnose is a serious problem on annual bluegrass and creeping bentgrass.

Bermudagrass Decline *(Gaeumannomyces graminis).* Bermudagrass decline occurs on bermudagrass in the southeastern United States in the late summer and fall. It is most severe on golf course putting greens. The symptoms are irregularly shaped yellow patches several inches to a few feet in diameter. Eventually the patches become dark brown. The roots are short, black, and rotted. The stolons and rhizomes also become dark colored.

Maintenance practices that promote root growth help to compensate for root losses caused by the disease. Raising the mowing height, regular core cultivation, light, daily irrigation, and fertilization are recommended. Applying an acidifying fertilizer such as ammonium sulfate can inhibit the pathogen.

Brown Patch *(Rhizoctonia solani).* Brown patch occurs on all the major turfgrass species, with the bentgrasses, perennial ryegrass, St. Augustinegrass, and annual bluegrass usually most seriously affected. Disease activity begins at 60°F (15.5°C), but is greatest in the 80°–90°F (27°–32°C) range. Also called Rhizoctonia blight, brown patch is a summer disease of cool season grasses, but is more common on warm season grasses in the spring and fall. Roughly circular, light brown patches a few inches to several feet across appear in the turf. When the grass is wet a "smoke ring" of grayish-black mycelia may be observed around the edge of the patch. Leaves, stems, crowns, and roots may be infected. Excessive thatch, heavy nitrogen fertilization, and poor drainage increase the likelihood of disease occurrence. Removing dew from putting greens early in the morning helps to decrease problems.

Rhizoctonia cerealis causes yellow patch, also called cool weather brown patch. It occurs during cool, rainy periods from fall to spring on closely mowed sites such as golf greens and tees. The patches are yellow or light brown.

Copper Spot *(Gloeocerospora sorghi).* Copper spot is primarily a problem on bentgrasses. Warm, wet weather favors development. Initial symptoms are small reddish lesions on leaves. The spots increase in size and coalesce (grow together), and pink or copper-colored spore masses may appear on the leaves. The overall symptom is small copper or pink patches a few inches in diameter.

Damping-Off and Seedling Diseases. Species of *Pythium, Rhizoctonia, Bipolaris, Drechslera,* and *Fusarium* can cause seed rots and seedling blights. Both diseases are also referred to a damping-off. Seedlings can be very susceptible to disease, especially if overwatered or planted in a poorly drained seedbed. Excessive seed rates can result in an overcrowded stand of weak seedlings, which are prone to disease problems. Seeds can be coated with a fungicide to help prevent seed rot, and seedlings can be sprayed if necessary.

Dollar Spot *(Sclerotinia homoeocarpa).* Though dollar spot is traditionally attributed to *Sclerotinia homoeocarpa,* many pathologists believe that the pathogen is actually in the genus *Lanzia* or *Moellerodiscus.* Dollar spot is very

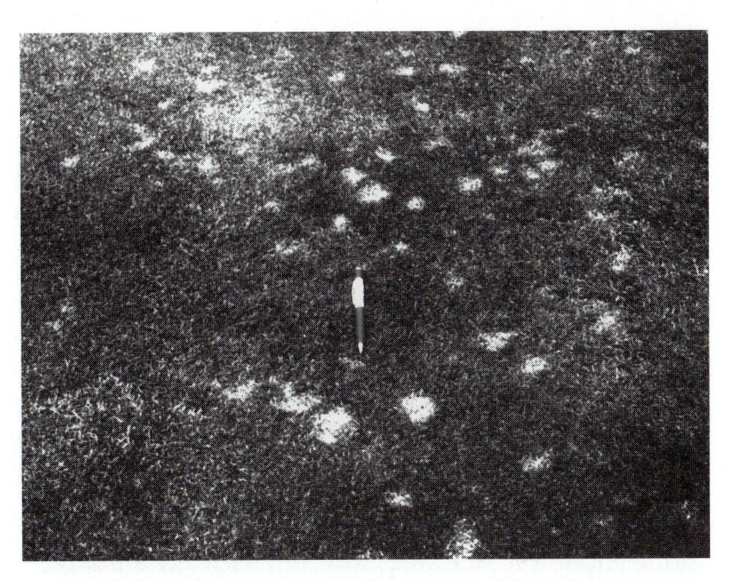

FIGURE 16-9
Dollar spot.

common and can cause severe injury to the bentgrasses, fine fescues, bermudagrass, and annual bluegrass. The disease also occurs on the other major turfgrass species. Dollar spot activity begins at 60°F (15.5°C) and is optimum at 70°–80°F (21°–27°C). It is thought that there may also be a second strain that occurs at warmer temperatures. The first symptoms are straw-colored bands on the leaves. On all hosts except annual bluegrass these spots have reddish-brown borders or margins. As a result of this leaf damage, small bleached spots appear in the turf. On closely mowed turf the spots are first readily observable when they are the size and roughly the shape of a silver dollar (Figure 16-9). On turf cut higher the spots are often larger and less regular in shape. If the disease remains unchecked, the numerous spots will coalesce and overlap each other. Early in the morning white, cobwebby mycelium may be seen on the surface of infected tissue. Predisposing factors include low soil moisture and fertility and excessive thatch. Removing dew as early as possible in the morning is helpful.

Fairy Ring. (Several species of Fungi in the class Basidiomycetes). Fairy rings are partial or complete circular bands of grass that are darker green and faster growing than the remainder of the turf. The organisms that cause fairy rings feed on organic matter in the soil. One of the decomposition products released as the organic material decays is nitrogen. The grass plants growing directly above the spots where the fungus mycelium is feeding absorb some of this nitrogen. During wet, rainy times of the year mushrooms may appear in the circular band. Though grass is not usually injured by fairy rings, it can be killed if the fungi secrete toxic compounds or if the layer of mycelia becomes so thick that water cannot penetrate to the roots.

Gray Leaf Spot *(Pyricularia grisea)*. Gray leaf spot is primarily associated with St. Augustinegrass, though it can occur on other warm season species and perennial ryegrass. Optimum temperature range for disease development is 70°–90°F (21°–32°C). Initial symptoms are brownish-gray leaf spots with darker margins. Numerous spots may occur, resulting in the death of infected leaves. Excessive nitrogen, drought, and soil compaction are contributing factors. The disease is most severe on newly established grass.

Helminthosporium Diseases *(Drechslera* and *Bipolaris* spp.*)*. A number of common diseases caused by *Drechslera* and *Bipolaris* spp. are still referred to by many turf managers as Helminthosporium diseases because for many years the pathogens were classified as *Helminthosporium* spp. Most of the diseases caused by these pathogens are leaf spots, but blights and crown and root rots also occur (Figure 16-10). The crown and root rot stage can be very severe and is called melting-out. Generally, leaf spot symptoms are common during wet periods in the spring and fall, while crowns and roots are attacked later in the spring and in the summer. However, there are a number of fungal species involved, and one causes leaf spots during warmer weather. The Helminthosporium diseases occur primarily on cool season grasses, but bermudagrass and zoysiagrass can be susceptible as well. The leaf spots have gray to tan centers and purplish to brown margins. High nitrogen or too little nitrogen fertility, excessive irrigation, close mowing, and poor drainage favor disease development. Light, daily irrigation can reduce severity.

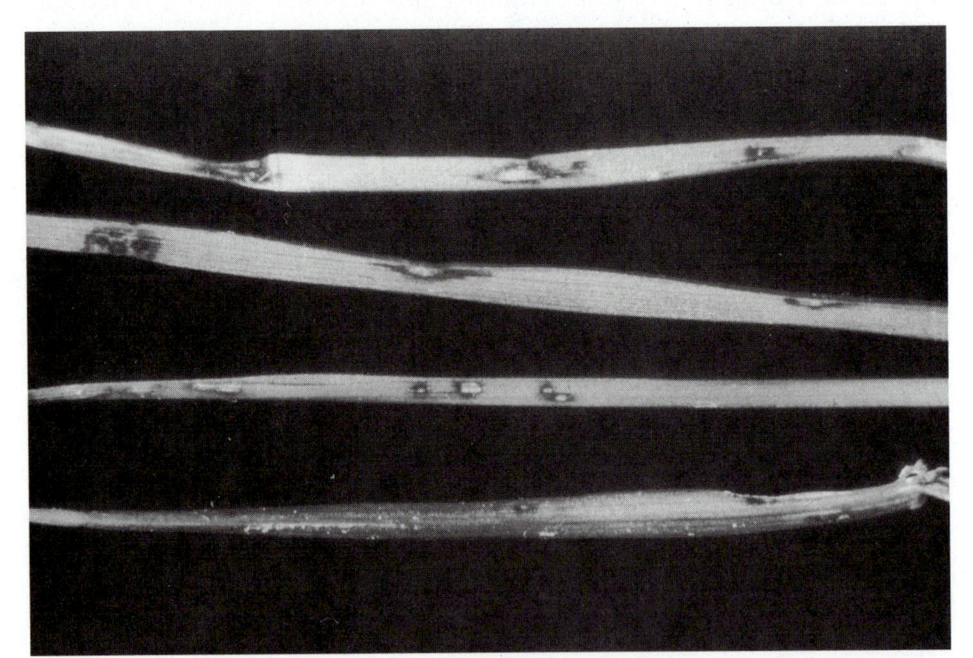

FIGURE 16-10
Leaf spot disease. *(Courtesy of New York State Turfgrass Association)*

Necrotic Ring Spot *(Leptosphaeria korrae).* Necrotic ring spot occurs on the bluegrasses and fine fescue. It is the most common patch disease on Kentucky bluegrass. The turf manager first becomes aware of the disease during late spring and summer, when small patches less than a foot in diameter appear. The above-ground symptoms are the result of root injury that occurred earlier in the spring and in the previous fall. The roots and rhizomes turn black, and dark brown hyphae can sometimes be seen growing on them. When hot, dry weather occurs the infected plants wilt because of the root losses.

Older patches often have a healthy green center surrounded by a ring of dead brown grass. This is known as a "frog eye." Some plants survive and grow back after the fungus stops feeding in the center of the patch.

Prevention of drought stress is critical in fighting necrotic ring spot. Light, daily irrigation may be necessary to ensure the survival of plants with a reduced root system. The application of a slow-release nitrogen fertilizer helps to stimulate recovery. Organic fertilizers or composts have been used to increase populations of microorganisms that suppress *L. korrae*. Damaged turf can be overseeded with perennial ryegrass or more resistant Kentucky bluegrass cultivars.

Pink Patch *(Limonomyces roseipellis).* Pink patch is similar to red thread disease, but is not as injurious. Hosts include the bentgrasses, annual bluegrass, fine fescues, perennial ryegrass, and bermudagrass. Small patches appear during cool, wet weather. Pink mycelium is observed when the infected grass is wet. This disease spreads very slowly. Turfgrass experiencing insufficient nitrogen fertilization is especially susceptible.

Powdery Mildew *(Erysiphe graminis).* Powdery mildew is a disease that is most troublesome on Kentucky bluegrass, but it may occur on fine fescue and bermudagrass as well. It is easy to identify because of the presence of whitish mycelia and spores on the surface of the leaves. Cooler temperatures, 55°–70°F (12°–21°C), and low light intensities are necessary for its development. The pathogen is inhibited by sunlight, so the disease is commonly found in the shade. Theoretically powdery mildew should not be too injurious because *Erysiphe graminis* is an obligate parasite—it can only feed on living tissue and therefore does not normally kill healthy plants. However, it can be quite injurious in the shade, when the grass is already weakened because of the lack of sunlight and other stresses.

Pythium Blight *(Pythium* spp.*).* There is probably no other turfgrass disease that can devastate a turf area as rapidly as Pythium blight. Whole stands can be destroyed in less than twenty-four hours when environmental conditions are optimum. High temperatures, 80°–95°F (26°–35°C), and cloudy, wet, rainy weather favor disease development. The disease can also be destructive during cooler weather. All cool season grasses and bermudagrass are attacked, but the bentgrasses, annual bluegrass, and perennial ryegrass exhibit the greatest susceptibility. When humidity is high the infected plants have a dark-colored, slimy appearance; thus the disease is sometimes called grease spot. Another name is cottony blight because early in the morning a cottony-looking gray to white mycelium may be seen on leaf tissue. As the leaves die they become reddish-brown and then finally straw-colored. Pythium blight is most severe in association with excessive nitrogen applications and poor drainage and air circulation.

Pythium spp. also cause crown and root rots. These diseases are most common on golf greens, attacking both creeping bentgrass and annual bluegrass. Root losses can be significant. When examining severely infected plants with a microscope it is often possible to see the round resting spores characteristic of *Pythium* inside the roots.

Red Thread *(Laetisaria fuciformis).* Red thread is most active at temperatures of 60°–70°F (15°–20°C). Fine fescue, perennial ryegrass, bentgrasses, annual bluegrass, and Kentucky bluegrass are suscepts. Irregularly shaped patches of tan grass often have a pink to reddish cast because of pink or red mycelium on the leaves. Red strands of mycelium arising from leaf tips may be highly visible and give the disease its name. Inadequate nitrogen fertilization can be a contributing factor.

Rust *(Puccinia* spp.*).* Different species of *Puccinia* attack the leaves, stems, and crowns of Kentucky bluegrass, perennial ryegrass, tall fescue, bermudagrass, and zoysiagrass. Rust occurs over a wide temperature range. The first symptoms are light yellow spots on the infected tissue. Soon rust-colored (orangish-red to orangish-brown) fruiting bodies burst through the cuticle. These spore-containing pustules are easily seen and are the chief diagnostic characteristic. Turf growing under infertile, shaded, droughty conditions is most susceptible.

Slime Mold *(Physarum* spp., *Fuligo* spp., and *Mucilago crustacea).* Slime molds produce different-colored reproductive structures on turfgrass during cool, wet weather. This fungal growth is very noticeable and appears ominous. However, these fungi do not attack the plants. Normally they live in the soil and feed on microorganisms and decaying organic matter. Slime molds grow up on grass leaves and stems to reproduce because their spores travel further when released from an elevated surface.

Inexperienced turf managers may mistake the sudden appearance of large amounts of fungal growth as a serious threat. Control is not necessary, though the fruiting bodies can be hosed or brushed off the grass. Slime molds disappear when the leaves dry.

Smut *(Ustilago* and *Urocystis* spp.*).* Smut usually causes significant problems only on Kentucky bluegrass and creeping bentgrass. Its optimum temperature range is 50°–65°F (10°–18°C). Long yellow streaks appear on infected leaves; then they turn gray, and eventually epidermal cells and the cuticle rupture to expose black spore masses. Smut is often called stripe smut because of the black streaks on the leaves. Infected leaves split and curl from the tip downward. Excessive nitrogen in the summer and drought significantly increase smut's severity.

Snow Mold *(Typhula* spp. and *Microdochium nivale).* The snow mold diseases, as their name implies, appear when temperatures are relatively cold. Optimum temperature range is 32°–45°F (0°–7°C) and the diseases occur in late fall, winter, and early spring. Snow mold is most common in the colder regions of the cool season zone. All cool season species can serve as hosts, with injury usually most severe on the bentgrasses.

Microdochium nivale causes Microdochium patch which is also called pink snow mold. The disease can occur even when the temperature is in the high 50s, though the grass is more resistant than it is at lower temperatures. Microdochium patch can develop with or without a snow cover. During cool, wet weather circular reddish brown spots a few inches to a foot in diameter appear. When there has been a snow cover they can be larger, and pink mycelium is visible at the edges of the patches.

Typhula incarnata and *T. ishikariensis* causes Typhula blight, which many people refer to as gray snow mold because of the presence of grayish white mycelium. Typhula blight requires a snow cover and is most severe when the turf is covered for more than 90 consecutive days.

Both types of snow mold are most injurious on turf that is lush due to fertilization late in the season. It is important to apply nitrogen early enough in the fall so that the grass is able to harden off before winter. Poor drainage contributes to greater disease incidence. Allowing the grass to go into the winter too high is also a problem. Tall grass gets matted down and is very slow to dry out, creating a damp microclimate ideal for pathogens.

Spring Dead Spot *(Gaeumannomyces graminis, Leptosphaeria* spp., and *Ophiosphaerella herpotricha).* Brown, circular patches of dead bermudagrass, a few inches to a few feet in diameter, appear in spring after dormancy ends (Figure 16-11). Closer inspection shows that a root, crown, and stolon rot has occurred. A black discoloration is visible on the affected tissue. In late spring the dead spots may have green centers because of weed invasion.

There has been some confusion about the exact causes of this disease, but it is now believed that several fungi attack the turf during the previous growing season. The weakened bermudagrass then succumbs to cold weather stress. The longer the dormancy period and the colder the winter temperatures, the greater the severity of the disease. Spring dead spot is most injurious on intensively managed turf in the northern areas where bermudagrass is grown.

A thick thatch layer, heavy nitrogen fertilization in the fall, and poor drainage make the problem worse. The use of potassium and soil acidifying fertilizers such as ammonium sulfate and ammonium chloride can help to reduce spring dead spot severity.

FIGURE 16-11
Spring dead spot. *(Courtesy of William Knoop)*

Summer Patch *(Magnaporthe poae)*. Summer patch occurs on Kentucky bluegrass, annual bluegrass, and fine fescues. Yellow or brown patches 6 inches (15 cm) to a few feet in diameter appear during warm weather in the summer. Examination of roots shows root rot symptoms. In mixed stand of grass "frog eyes" can result because resistant species are not injured and stay green.

The initial root infection occurs in the spring, but the pathogen remains in the outer cortical tissue and causes minimal injury at this time. However, when soil temperatures are greater than 70°F (21°C) and heavy rain or irrigation forces oxygen out of the root zone, the fungus moves deeper into the root. The higher temperatures and oxygen depletion weaken roots to the extent that *Magnaporthe* can invade the vascular tissue. The result is a disruption of water and nutrient uptake and less movement of food from the leaves to the roots.

Low mowing heights, soil compaction, and poor drainage contribute to summer patch problems. Fertilization at moderate rates with a slow-release nitrogen source is very helpful. Applications should occur throughout the summer if summer patch is a threat. Light daily irrigation, usually in the afternoon, is the best program for turf with a weakened root system. Light watering will also cool the grass, but will not deplete oxygen from the soil.

Take-All Patch *(Gaeumannomyces graminis* var. *avenae)*. Take-all patch occurs on bentgrasses and is especially severe on newly established creeping bentgrass greens. The patches start out several inches in diameter, but may grow over the next three or four years to a few feet across. Usually the disease declines after several years, probably because of the buildup of microorganisms that suppress the pathogen. This helps to explain why the problem is worse on relatively sterile, sandy media and on soils that have been fumigated. They would initially have very few antagonistic microorganisms to compete with *Gaeumannomyces*.

The pathogen attacks roots and crowns. Black hyphae can sometimes be seen on the roots. The root injury primarily occurs in the fall and spring during periods of cool, wet weather. The disease is most severe on turfgrasses growing in soils with a pH above 6.5.

It is best not to lime where take-all patch is a problem. The use of alkaline irrigation water or topdressing materials should also be avoided. Applying acidifying fertilizers such as ammonium chloride or ammonium sulfate can reduce disease pressure. Core cultivation to improve drainage is helpful.

Yellow Tuft *(Sclerophthora macrospora)*. Yellow tuft, also known as downy mildew, occurs on all of the cool season grasses during cool, wet weather. The primary symptom is small yellow spots 0.5 to 3 inches (13 to 76 millimeters) in diameter, which are actually dense clusters of yellow shoots arising out of a single bud. White mycelium may be seen on leaf surfaces. Application of iron sulfate may green up the turf and hide the symptoms. Although unsightly, yellow tuft usually does not cause permanent injury to turf. Occasionally infected plants may die during periods of heat, drought, or winter stress because of their reduced root system. Problems are greatest on poorly drained, wet sites.

DISEASE CONTROL

Resistant Varieties

Some varieties of grass are more likely to be damaged by a disease than others. These susceptible cultivars are incapable of fighting off a disease attack. Resistant varieties, however, are able to survive an attack without suffering any serious injury (Figure 16-12).

There are different reasons why certain cultivars can withstand or repulse a disease invasion. Some produce chemicals toxic to the fungus. Others lack a nutrient that the fungus needs in order to survive.

When establishing new turf plantings, the turfgrass manager can avoid many future disease problems by selecting grass species and varieties that are resistant to the diseases common to his or her area. This information is obtained from sources such as extension bulletins, seed company representatives, horticultural magazines, and turfgrass specialists at the state university.

As time passes a variety may lose its resistance to a disease. Fungi change genetically by sexual reproduction and mutation. These changes may eventually enable them to overcome the mechanism or process that made the variety resistant. Merion Kentucky bluegrass was once immune to Drechslera melting-out, but today the cultivar is susceptible to this disease.

FIGURE 16-12
A disease-susceptible cultivar surrounded by resistant cultivars. *(Courtesy of New York State Turfgrass Association)*

This potential problem can be solved by growing more than one resistant variety or by planting a mixture of species. If one cultivar loses its resistance to a disease, the entire turf is not ruined. The other varieties will spread and fill in the damaged areas. This practice is avoided when planting multiple varieties would result in serious compatibility problems or an unacceptable lack of uniformity.

Cultural Techniques

Proper turf management techniques help to prevent diseases. Healthy, vigorous grass is better able to fight off a disease attack. Weak, poorly maintained turf cannot defend itself against a pathogen. Even a disease that is not usually severe can kill unhealthy grass.

Turf fertilized with too much nitrogen and not enough potassium becomes lush and succulent. Grass in this condition is very susceptible to diseases. Moderate nitrogen fertilization rates and the use of a well-balanced fertilizer containing at least a third as much potassium (K_2O) as nitrogen help to prevent this problem from occurring.

Low fertility also contributes to disease susceptibility. Dollar spot, red thread, and rust diseases are most damaging on turf deficient in nitrogen. Starving grass is not strong enough to resist them.

Natural organic fertilizers and composts have been shown to cause a significant reduction in disease problems. The organic materials are derived from animal and plant matter and serve as a food source for fungi, bacteria, and other microorganisms. Some of the natural organic fertilizers already contain beneficial organisms. Applying these materials results in a large, diverse popu-

FIGURE 16-13
Compost applied for disease control.

lation of microflora which can prevent a pathogen population from becoming high enough to cause serious plant injury (Figure 16-13).

Some of these antagonistic microorganisms give off chemicals that suppress pathogens. Others compete with pathogens for dead organic matter or feed on the pathogens themselves. During the decay process the organic material may release acids or other compounds that inhibit pathogens.

There is much research occurring in this area. Pathologists are trying to determine which organic sources do the best job of suppressing each pathogen and what application rates and frequencies are most effective.

Improper mowing practices are a major reason for disease problems. A dull mower blade tears and shreds grass plants. The shock weakens the grass, and the ragged wound heals very slowly. The wound offers a fungus an open door through which it can enter the plant.

Turf that is scalped—cut too short—is seriously weakened. By the time it recovers, the pathogen can be well established inside the plant. Higher cutting heights usually result in stronger more resistant turfgrass.

Excessive irrigation contributes to disease problems. Watering late in the day or at night is normally avoided unless there is no other convenient time to irrigate. Most fungal spores must be surrounded by a film of moisture for a number of hours before they can germinate and infect. Mycelial growth from plant to plant is encouraged when leaves remain wet. Grass is often wet some of the night because of dew. Watering early in the evening extends the length of time that the turf is wet and stimulates pathogens to germinate and grow at the beginning of the wet period.

Many turfgrass specialists recommend irrigating right before sunrise. This washes dew and guttation fluid from leaf surfaces. Guttation fluid exudes from the leaf tips and is rich in nutrients. It serves as a food source for foliar pathogens as they spread. Early-morning irrigation breaks drops of water it doesn't wash off into smaller droplets that dry up more quickly. Some turf managers use long flexible poles or drag hoses to knock dew off the grass in the morning.

Another possibility is a light daily irrigation, preferably in the early to mid-afternoon. Though there is some disagreement about this practice, it will relieve heat stress at the hottest part of the day by cooling the grass. During the warmest, driest times of the year the roots of cool season grasses can be quite shallow, so it makes sense to irrigate frequently at light rates if the root system is near the surface. Grass injured by disease often has a reduced root system. Light daily watering will help the grass to survive. Keeping the soil near the surface and thatch moist may also allow large populations of antagonistic microorganisms to develop.

Thatch is a layer of organic matter, produced by grass plants, that is above the soil surface. A thick layer of thatch weakens the turf. Reducing the thatch layer by core cultivation or topdressing may help to strengthen the turf and make it less vulnerable to injury by root diseases.

Removing or reducing any environmental stress affecting the grass decreases the likelihood of serious disease injury. Alleviating stress problems caused by shade or heavy traffic, for example, increases the disease resistance of the turf. It is important to use cultivars that are disease resistant, but they must also be tolerant of the other environmental conditions found at the growing site. If a cultivar is not adaptable it will be weak and may lose its normal disease resistance.

Reducing excessive moisture on the turf is helpful. If a site is frequently wet, improving air circulation and soil drainage often results in fewer disease problems.

Fungicides

Contact nonsystemic fungicides are applied to turf foliage and prevent pathogens from penetrating the plant. They form a protective barrier on the surface of the plant (Figure 16-14). Complete, uniform spray coverage is necessary for the chemical barrier to be effective. They do not protect roots.

Contact fungicides do not destroy a pathogen already inside the plant. Their function is to prevent diseases from occurring initially. However, if a foliar disease is present, their application will stop it from spreading to healthy plants.

A contact fungicide is effective for a relatively short period, usually a few days to two weeks. Rain washes the chemical off plant surfaces, and sunlight may break it down. Adjuvants can be mixed with the fungicide to make it stick longer or to protect it from ultraviolet light. Mowing the grass also removes the protective barrier.

Systemic fungicides offer longer-term protection, normally three to four weeks (Figure 16-15). They are absorbed by the roots or leaves and become inter-

FIGURE 16-14

The effectiveness of fungicides is seen in this picture. The area that was accidentally missed during the fungicide application has been injured by gray snow mold.

nal protectors. Though some of the material stays on the outside and acts as a surface protectant, the fungicide that enters the plant is not affected by rain. By being translocated throughout the plant tissue the fungicides are also able to stop a pathogen that is living inside the grass. Unlike contact fungicides, systemics can "cure" a diseased plant if applied at early stages of infection.

Systemic fungicides generally inhibit the growth of a fungus rather than kill it. Most systemics are taken into the plant by the roots. Irrigation after application helps to wash the chemical down to the root zone. Systemics also protect new growth that occurs because the roots continue to absorb the fungicide from the soil.

Fungicide treatment is expensive because of the cost of the chemical materials and the amount of time it takes to apply them. The turf manager must decide if this expense is justified.

Fungicides should be applied when a disease is causing significant injury to turfgrass. Minor damage does not usually require treatment unless a continuation of weather conditions optimum for the growth of the pathogen is predicted. A high-quality turf area may be sprayed at the first sign of disease activity if even minor injury is unacceptable. The use of resistant turfgrass varieties decreases the need for fungicide applications.

	COMMON NAME	SOME TRADE NAMES
Contact type (nonsystemic)	Anilazine	Dyrene
	Captan	Captan
	Chloroneb (has some systemic activity)	Teremic SP, Scotts Fungicide V
	Chlorothalonil	Daconil 2787
	Etridiazol (ethazole)	Koban, Terrazole
	Iprodione (has some systemic activity)	Chipco 26019, Scotts Fungicide X
	Mancozeb	Fore, Dithane, Formec 80
	Maneb	Manzate
	Oxycarboxin	Plantvax
	PCNB (quintozene)	Terraclor, Turfcide, Scotts F + F II
	Thiram	Tersan 75, Spotrete, Thiramad, Thiram
Systemics	Benomyl	Tersan 1991, Benomyl
	Fenarimol	Rubigan
	Flutolanil	ProStar
	Fosetyl-Al	Aliette
	Metalaxyl	Subdue, Apron, Scotts Pythium Control
	Propamocarb	Banol
	Propiconazole	Banner
	Thiophanate-methyl	Fungo, Spot Kleen, Scotts Systemic, Clearys 3336
	Triadimefon	Bayleton, Scotts Fungicide VII
	Vinclozolin	Vorlan, Touche, Curalan
Combination products	Thiophanate + thiram	Bromosan
	Fenarimol + chlorthalonil	TwoSome
	Metalaxyl + maneb	Pace
	Thiophanate-methyl + chloroneb	Scotts Fungicide IX
	Thiophanate-methyl + iprodione	Scotts Fluid Fungicide
	Thiophanate-methyl + mancozeb	Duosan
	Thiopanate-methyl + thiram	Bromosan
	Triadimefon + metalaxyl	Scotts Fluid Fungicide II
	Triadimefon + thiram	Scotts Fluid Fungicide III

FIGURE 16-15

Common turfgrass fungicides.

Preventive fungicide programs are practiced at many golf courses. Fungicides are applied to greens before diseases occur at times of year when weather conditions stress the turf or favor the activity of common pathogens. This is because susceptibility is increased by the effects of traffic and close daily mowing. Consequently, because greens are predisposed to disease and even minor injury is unacceptable, preventive applications of fungicides are often necessary. Tees are usually treated in a similar manner.

Turf specialists of the Cooperative Extension Service can help recommend the correct fungicide to control a disease when the turf manager is not certain about which chemical should be used. They in turn may contact a turfgrass pathologist for advice. If the disease can be correctly identified, the manager can select the appropriate fungicide by consulting extension bulletins which list the materials that control each disease. The label on a fungicide container also lists the diseases for which the product is effective.

When the manager is not positive about the identity of a disease, he or she can select a broad-spectrum fungicide that will control all of the logical possibilities, or apply two fungicides at the same application. If a fungicide does not seem to be stopping the disease within a few days after application, the manager can try another chemical. It is important to avoid overusing fungicides. The use of a chemical against one disease can sometimes lead to increased severity of another disease in the future. Applying the same systemic fungicide continually can cause a strain of the pathogen resistant to that chemical to become dominant. Systemics are said to be single-site fungicides because they usually control fungi by interrupting one specific metabolic process. It is easier for pathogens to become resistant to a chemical that has only one mode of action.

To avoid resistance, it is wise to alternate materials. It does no good to rotate one systemic with another in the same chemical family (benomyl and thiophanate, for example) because they have the same mode of action. There is disagreement about the best way to avoid developing resistance to a fungicide. Some turf specialists recommend mixing a systemic and a nonsystemic and applying them together or spraying a systemic and then using a nonsystemic at next application.

Mixing two systemic fungicides from different chemical families sometimes works. Metalaxyl is used to control Pythium blight, but resistance has occurred. Mixing metalaxyl with fosetyl al or propamocarb, or with both, significantly reduces the problem.

OTHER TYPES OF PATHOGENS

St. Augustine Decline Virus

One economically important turfgrass disease, St. Augustine decline (SAD), is caused by a virus. Viruses are extremely tiny pathogens that do not feed on plants in the traditional sense; they do not consume cell contents, but instead cause injury by forcing plant cells to produce more viral particles. When the cells are induced to replicate viruses, normal plant metabolism is disrupted. A typical symptom of viral infection is chlorosis—yellowing due to the lack of chlorophyll. Viruses are so extremely small that an infected cell may contain as many as 10,000,000 virus particles.

SAD is caused by the St. Augustine decline virus. St. Augustinegrass is a major turfgrass species in the southern United States, and SAD is its major disease problem. The disease is spread mechanically, primarily by mowing equipment. The initial symptom of SAD is a small amount of yellow mottling on the leaves. Gradually the number of chlorotic cells increases until photosynthesis becomes severely limited because of the lack of chlorophyll. Usually the grass dies within several years of the first appearance of decline symptoms.

St. Augustinegrass will inevitably die once it is infected by SAD. The decline can be slowed by proper fertilization. The decline is more rapid, however, when the grass is unhealthy or weak because of low fertility, poor maintenance, shade, drought, or other stress factors. Because the pathogen is a virus, there are no chemicals that will control SAD. The best hope for the future is the development and release of resistant cultivars. A few resistant St. Augustinegrass cultivars are already available. One of these, Floratam, also shows moderate resistance to chinch bug, but exhibits some undesirable characteristics such as low density, poor cold tolerance and wear tolerance, slow spring green-up, coarse texture, and lack of shade tolerance. Raleigh, Seville, and Floralawn have some resistance of SAD and are more cold tolerant than Floratam.

Bacterial Wilt

A bacterium, *Xanthomonas campestris,* causes a wilt disease on some vegetatively propagated creeping bentgrass cultivars such as Toronto (C-15) and Cohansey. The disease is associated with lengthy periods of rainfall. The bacteria fill and block the xylem, causing plants to wilt. The best solution is to replace susceptible varieties with resistant ones.

One type of *Xanthomonas campestris* has been found infecting annual bluegrass. There is interest in using *Xanthomonas* to control this species.

Nematodes

Nematodes are small animals found in the soil. They are nonsegmented worms generally ranging in length from 1/100 to 1/8 inch (0.25 to 3.2 millimeters). Sometimes called eelworms, nematodes are transparent when seen through the microscope (Figure 16-16). They feed by means of a spearlike structure called a stylet (Figure 16-17). The plant-parasitic nematode punctures plant cells with its stylet and withdraws the cell contents through it.

Plant-parasitic nematodes fall into two categories—ectoparasites and endoparasites (Figure 16-18). The majority of nematodes that attack turfgrass are ectoparasites. They insert their stylets into root cells, but their bodies

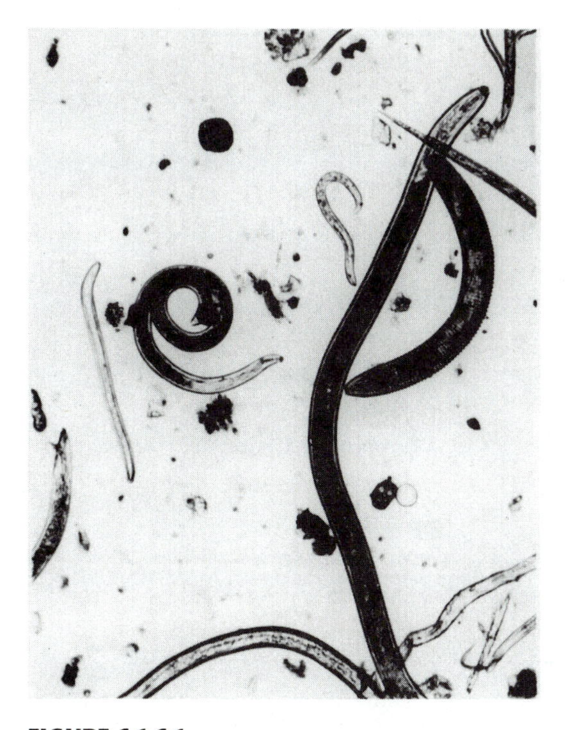

FIGURE 16-16

Different nematode species. *(Courtesy of Patricia Sanders and the Pennsylvania Turfgrass Council)*

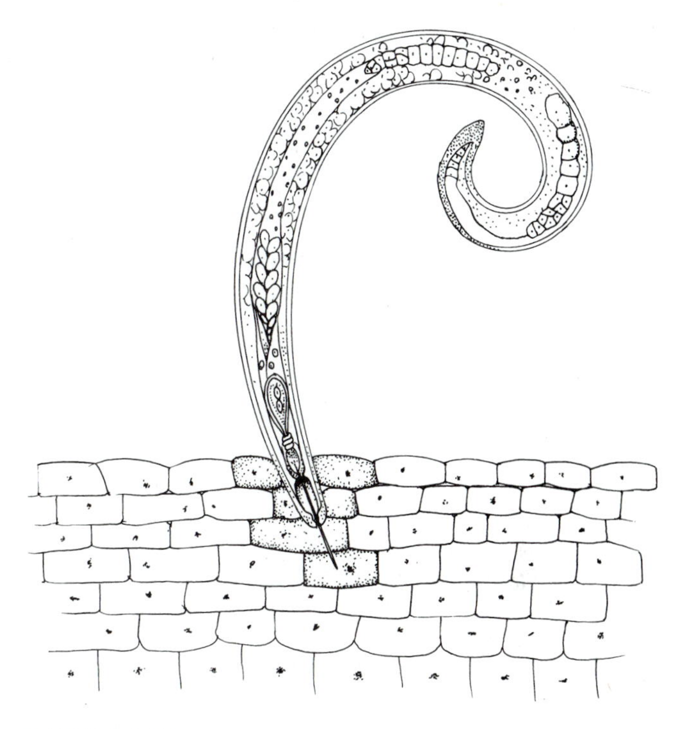

FIGURE 16-17

A nematode feeding on the roots of a turfgrass plant.

remain outside of the root. Endoparasitic nematodes actually enter the root and feed inside.

Nematodes injure turfgrass roots by feeding on root cells. They also secrete digestive substances that often cause abnormal enlargement of root cells. Damaged roots have necrotic lesions (dead tissue) and may have swellings or knots. Because the size of the root system and its ability to obtain water are reduced, the first aboveground symptom is wilting. Yellowing of the leaves, slow growth, and a gradual thinning of the turf are also common symptoms.

Nematodes usually do not directly kill the grass, but can seriously weaken it. The wounds caused by nematode feeding allow pathogenic fungi easy entry into the roots. Nematodes can even spread fungal diseases when they attack healthy roots after feeding on diseased root tissue. The interaction between nematodes and fungi combined with other stress factors such as drought and high temperature can result in severe injury.

Nematodes normally move less than 1 foot (0.3 meter) in the soil so they spread very slowly. Introduction into a turf area is usually the result of human activities. They can be brought in on vegetative plant material such as sod,

	COMMON NAME	SCIENTIFIC NAME
Endoparasitic		
	Cyst nematode	*Heterodera* spp.
	Lesion nematode	*Pratylenchus* spp.
	Root-knot nematode	*Meloidogyne* spp.
Ectoparasitic		
	Awl nematode	*Dolichodorus* spp.
	Dagger nematode	*Xiphinema* spp.
	Lance nematode	*Hoplolaimus* spp.
	Pin nematode	*Paratylenchus* spp.
	Ring nematode	*Macroposthonia* spp.
	Sheath nematode	*Hemicycliophora* spp.
	Spiral nematode	*Helicotylenchus* spp.
	Sting nematode	*Belonolaimus* spp.
	Stubby-root nematode	*Trichodorus* spp.
	Stunt nematode	*Tylenchorhynchus* spp.

FIGURE 16-18

Nematodes that affect turfgrass.

sprigs, or plugs, in infested soil adhering to equipment, in soil used for root zone modification or topdressing, or even in irrigation water.

Nematodes are generally a much greater problem in the southern United States than in the North. Nematode populations are more likely to increase to injurious levels in the South because of warm soil temperatures and a longer growing season. A typical nematode species might complete its life cycle in two weeks when the soil temperature is very warm, but six weeks or more may be necessary for completion if the soil temperatures are cool. Female nematodes produce from 500 to 2,000 eggs.

When a nematode problem is suspected, a soil sample should be sent to a diagnostic laboratory. A nematologist will interpret the laboratory estimates of nematode populations. Whether nematodes are a serious problem depends on the nematode species present, their populations, and other environmental conditions. For example, a population of 1,000 ring nematodes per 100 cubic centimeters of soil may be required to cause noticeable injury, but as few as 10 sting nematodes in the same volume of soil can be a large enough population to cause significant problems.

Proper sampling and handling techniques are essential. The sample must be kept cool and moist because nematodes are very perishable. The turf manager should check with the Cooperative Extension Service to learn the correct sampling procedure. In some states extension agencies sell inexpensive nematode sample kits.

Because nematodes destroy roots and disrupt water uptake, nematode injury is most severe during dry weather and on sandy, droughty soils. Irrigation can help the turfgrass to recover. Often nematode injury can be reduced to tolerable levels by removing any other stresses that also weaken the grass. Controlling other diseases and insect pests and correcting problems such as improper watering or fertilization, poor mowing practices, and soil compaction may minimize the stress caused by nematodes. Some cultivars show resistance to individual nematode species. Bahiagrass is the warm season grass that is least likely to have nematode problems.

Fumigants, chemicals that become gaseous when applied to the soil, can be used for nematode control before planting. These materials are phytotoxic so they cannot be applied to areas where turf already exists. The nonfumigant nematicides can be applied after planting. These chemicals are effective against some nematode species but not others, unlike the fumigants which are broad spectrum. Fumigants provide better control than nonfumigants. If the nematode problem is serious, it may be best to remove the turf, fumigate, and then replant. When nonfumigants are used the turf should be aerified or core-cultivated before the application to improve the penetration of the nematicide into the soil. Heavy irrigation is necessary to water it in.

Nematicide treatments are very expensive, and the chemicals are potentially hazardous to the user. The turf manager should apply nematicides only after a diagnostic lab has confirmed that a serious nematode problem exists.

SELF-EVALUATION

1. Most turfgrass diseases are caused by ___________.
2. A plant that is diseased is said to be a ___________.
3. Fungal pathogens reproduce by producing ___________.
4. The body of a fungus is composed of numerous threadlike filaments called ___________.
5. Disease problems are most common during periods of high ___________.
6. ___________ fungicides are effective for a longer period than the contact type.
7. An important turfgrass disease caused by a viral pathogen is ___________.
8. How do nematodes damage turfgrass plants?
9. What observations enable the manager to distinguish between Typhula snow mold and Microdochium snow mold?
10. The first step in a disease prevention program is to plant ___________ cultivars.
11. Which warm season turfgrass is least likely to be injured by nematodes?
12. What factors affect the severity of a disease?
13. Discuss how to identify a disease.

CHAPTER

17

Integrated Pest Management

OBJECTIVES

After studying this chapter, the student should be able to

- Understand the importance of an integrated pest management program
- Discuss the components of an IPM program
- Describe monitoring techniques
- Discuss biological control
- Understand how to develop an IPM program

INTRODUCTION

The controversy surrounding pesticides is well known. Improper pesticide use can be hazardous to both the applicator and the environment. Turfgrasses can be injured by applying too much, or the wrong, pesticide. Some of the chemicals kill earthworms, natural enemies of turf pests, and other beneficial organisms.

Pesticides may also be very expensive because of the cost of the materials and the labor and time required to apply them. Insurance premiums can be quite high for businesses using pesticides.

The public perception of pesticides continues to be largely negative. Increasingly, golfers, homeowners, and other people who use turf are expressing concerns about pesticide applications. Moreover, pesticide regulations are making it more difficult and costly to apply pesticides.

The turfgrass manager needs to reduce the use of toxic pesticides whenever possible. To accomplish this goal, he or she should develop an integrated pest management program.

WHAT IS INTEGRATED PEST MANAGEMENT?

Integrated pest management, abbreviated as IPM, is a program that integrates many different concepts and techniques to reduce a pest population or minimize its impact. Cultural, biological, and chemical controls are used. An important part of an IPM program is looking for pests. A scout determines which pests are present and monitors their population.

In the past pesticides were applied in anticipation of a problem or when a few pests were seen. In an IPM program treatment is not considered necessary until the pest population is high enough to cause unacceptable damage.

IPM is an approach that employs multiple tools to suppress pests rather than simply resorting to traditional pesticides every time a problem occurs. By carefully monitoring pest populations the turf manager can determine whether control is necessary and if past control efforts were successful.

An effective integrated pest management program requires the turf manager to be knowledgeable about turfgrass culture, turf pests, and all of the possible control measures (Figure 17-1). It is a much more sophisticated approach than the control programs of the past, where chemical sprays were the automatic response to the sighting of a single insect or weed. Education is an integral part of an IPM program.

An IPM program should be cost effective and site specific. Conditions vary greatly from one turf site to another, so the program must be developed with local conditions in mind.

IPM does not mean that pesticides are never used. Situations arise when there are no other effective alternatives. However, the pesticide selected should be the one that is least toxic to the applicator and the environment. It is applied in a safe manner at the correct rate when the pest is in a controllable stage.

COMPONENTS OF AN IPM PROGRAM

Setting Objectives

The first step when developing an IPM program is to set objectives or establish goals. In most cases the primary objective is to reduce pesticide use. An equally important objective is to improve the turf so that it is more tolerant of pest injury and better able to recover from pest problems.

The turf manager needs to discuss the proposed program with all of the involved parties. The golf course superintendent confers with the green committee, and the lawn care company representative talks to his or her customers. Everyone who will be affected by the programs should have input.

One of the crucial decisions to be made is how much damage will be acceptable. Generally, if there is zero tolerance for injury, such as on a putting

A. IDENTIFICATION OF THE PEST AND THE INJURY IT CAUSES TO TURF

1. What does the insect look like?
2. On what types of sites and on which turfgrasses does the injury occur?
3. What part of the plant is attacked?
4. What does the injury look like?
5. At what time of year is the turf injured by the pest?

B. LIFE CYCLE

1. Which is the injurious stage (instar 1, 2, 3, etc., adult)?
2. At what time of the year does the injurious stage occur?
3. Does this stage live in the soil, thatch, or on the surface?
4. Which stage is easiest to control?
5. When does this controllable stage occur?
6. What environmental conditions (rain, drought, heat, etc.) favor the insect and which ones result in a population decline?

C. MONITORING

1. Which monitoring techniques works best for this insect—checking a square foot of soil, flotation in a can, pheromone trap, irritating drench to make pest come to the surface, etc.
2. How many insects per square foot can the turf tolerate without showing injury?
3. How much damage is acceptable to the customer, golfer, owner, etc.?
4. Are there other stresses present that will cause the turf to be less tolerate of the insect injury?
5. Will the grass recuperate quickly from the injury?

FIGURE 17-1 Types of information a turfgrass manager needs to know about an insect pest in a monitoring program.

green, the amount of pesticide use is higher than in a rough or on a lawn, where some injury can be tolerated. This is called an aesthetic threshold, or sometimes an economic or client threshold (Figure 17-2). How much damage will the customer, club membership, or owner accept?

The aesthetic threshold in turn sets the action threshold, also known as the control threshold. The action threshold is the point at which the manager must take some kind of action to avoid reaching unacceptable levels of injury. For example, the manager may treat when she or he finds 10 first instar per square foot (0.09 m^2) because she or he knows this number will result in unacceptable injury when the grubs get bigger and feed more.

Turfgrass thresholds are hard to set because they are a subjective, aesthetic decision. The degree of injury that is tolerable varies significantly between people and is also dependent on the type of turf area (Figure 17-3).

IPM programs have the best chance of success when there is general agree-

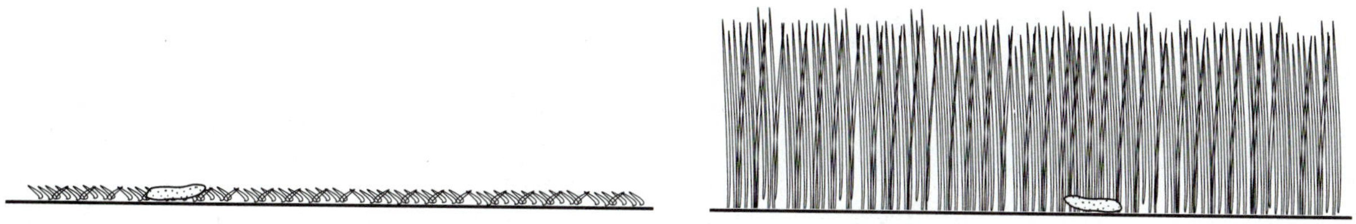

FIGURE 17-2 The sod webworm can cause visible injury on a putting green (left), but in taller grass the damage would not be noticed. On a lawn there is so much leaf tissue that the small amount eaten by the sod webworm is insignificant.

ment that pesticide use reduction is a desirable goal. At the beginning of the program mistakes may be made, and more injury than people are used to may occur. Alternative controls can also be more expensive than traditional pesticides. Consequently, the turf manager needs the support and understanding of all the parties affected by the program.

In some cases IPM programs are mandated by government agencies, school boards, or other organizations. It seems inevitable that eventually everyone who uses pesticides will be required to develop an IPM program.

FIGURE 17-3 Beauty is in the eye of the beholder. Some people may be upset by a small number of pests, while other clients are not concerned.

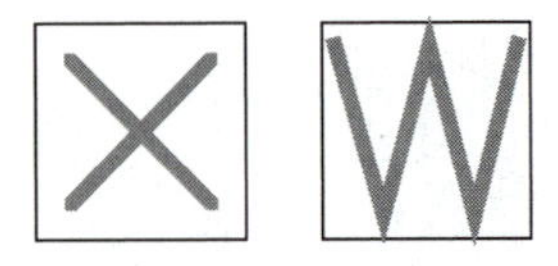

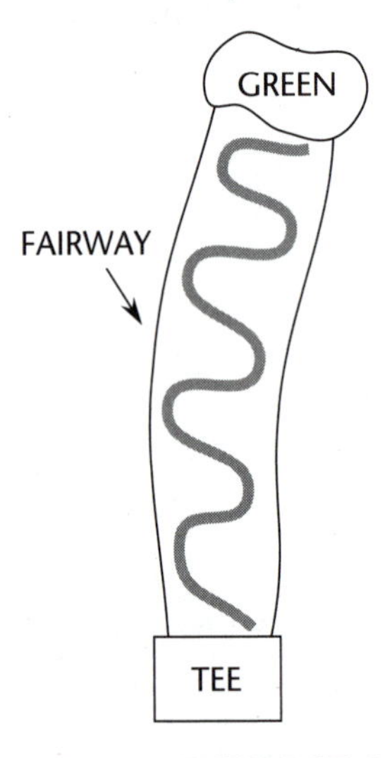

FIGURE 17-4

Typical patterns used by a scout. If the scout sees evidence of damage, he or she will check the spot even if it is not in the pattern.

Monitoring

The turf manager must constantly monitor pest populations. This is hardly new, as agricultural workers have always been instructed to look for pests. But in an IPM program this monitoring is performed in a more methodical, thorough way. Generally, the person who is monitoring does so regularly. Golf courses are scouted at least once a week. The scout checks all 18 greens, tees, and fairways.

The pattern is important. When monitoring an area, the object is to look at enough turf to be able to make an accurate appraisal of pest populations. However, to keep the program cost effective the scout cannot take the time to check every square foot. Consequently, patterns are selected that allow the scout to visually inspect the area in a thorough but time-efficient manner (Figure 17-4).

The scout needs to carry a hand lens, knife, hand trowel or soil sampler, plastic vials or bags to hold specimens collected, a notebook for keeping records, pest identification books if necessary, and anything else that is helpful. He or she has to be well trained in pest identification and monitoring techniques. The success of the IPM program depends on his or her conscientiousness and skills.

An important part of monitoring is recordkeeping. The scout must note any pest populations or problems of significance. Different types of forms have been developed to record this information (Figure 17-5). Besides inspecting for insects, diseases, and weeds, the scout also notes other types of problems such as a leaking sprinkler head, compaction, or a dry spot.

It is usually the turf manager's job to evaluate the information and make a decision about the necessity for action. The scout's report allows the manager to continually know what's happening on the golf course, athletic field, or lawn. The manager has to judge whether the pest population is or will become large enough to cause unacceptable damage or aesthetics.

The advantages of scouting are obvious. By having a regular report on turf and pest conditions there should be no unpleasant surprises such as a sudden loss of grass. The manager can intervene before serious problems occur. Unnecessary pesticide applications can also be avoided. A further benefit is the opportunity to improve the maintenance programs because cultural problems are noted as well as pest concerns.

To be most effective, monitoring should occur at least once a week. This is a problem for lawn care companies, since generally lawn care specialists visit their clients' properties only once a month, unless they are also mowing the turf, and the properties need to be scouted more frequently than this. Thus, an IPM program requires more visits, which will increase the cost of lawn care for the customer. Many people respond positively because they want a safer, more environmentally sound lawn care program. Other customers do not like the greater expense, especially when they see the scout walking around apparently doing nothing. Communication is very important. The customer must understand that even though the lawn care company employee is not spraying, he is performing a valuable service.

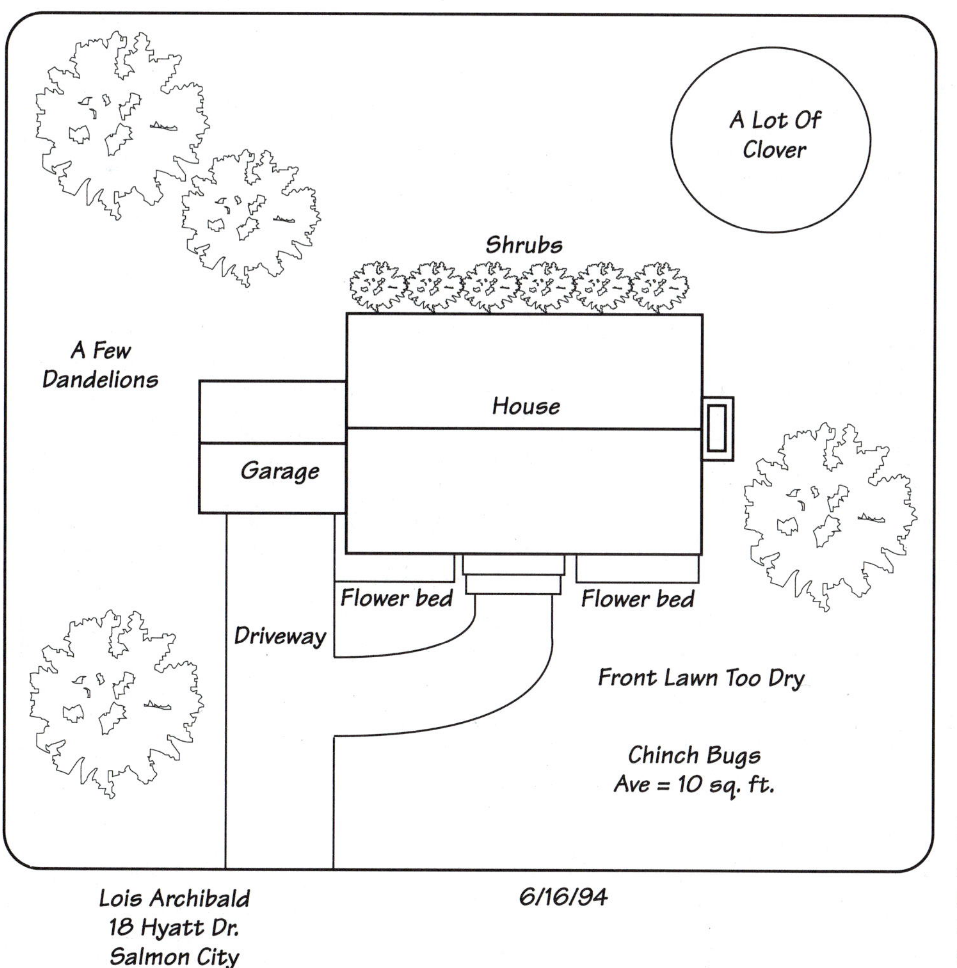

FIGURE 17-5

A map-type recordkeeping form used by a lawn care company.

The information collected by the scout can be reported in many different ways. An example of a form used on a golf course is shown in Figure 17-6. Insects are normally counted per square foot. Weeds can be counted, or the percentage of the turf area they cover can be estimated. Damage from insects, nematodes, or diseases can be reported as low, moderate, or high, or they can be quantified more exactly on a scale of 1 to 5 or 1 to 10 or as a percentage of the total area. After a pesticide application the scouting report can be used to evaluate the efficacy of the material.

At the end of the year the information collected is included in an annual report which will help the manager anticipate and prepare for future problems (Figure 17-7).

Though visual examination is the most common monitoring technique, there are other tools that can be used. A weather station is helpful because it

Course ____________________

Hole _____

GOLF COURSE IPM SCOUTING REPORT

	LOCATION	Non-Inft.	Diseases	Weeds	Insects	Samp#	COMMENTS
DATE	TEES						
	B/W/R L C R						
	B/W/R L C R						
	B/W/R L C R						
	B/W/R L C R						
	B/W/R L C R						
	B/W/R L C R						
	B/W/R L C R						
DATE	FAIRWAY						
	yds. L C R						
	yds. L C R						
	yds. L C R						
	yds. L C R						
	yds. L C R						
	yds. L C R						
	yds. L C R						
DATE	GREEN						
	fr/back L C R						
	fr/back L C R						
	fr/back L C R						
	fr/back L C R						
	fr/back L C R						
	fr/back L C R						
	fr/back L C R						

FIGURE 17-6 Golf course hole record form. *(Courtesy of Cornell University.)*

can measure air and soil temperature, leaf wetness, relative humidity, soil moisture, and rainfall. Some of the weather instruments have the ability to use this information to predict the likelihood of a disease problem (Figure 17-8). They will alert the manager if the weather conditions are appropriate for several diseases, and state if a fungicide spray is recommended.

Weather stations also record the accumulation of degree days, which is valuable in predicting the appearance of various stages in an insect's life cycle. Temperature has a major effect on insects' activities and the progression of the life cycle. Most insects are relatively inactive around 50°F (10°C), so 50°F or a temperature close to it is normally used as the base temperature. Averaging the daily maximum and minimum temperatures and subtracting the base temperature from this number calculates the degree-day accumulation.

For example, if the daily high is 90°F, the daily low is 60°F, and the base is 50°F, 25 degree days will be accumulated. (90° + 60° = 150°/2 = 75° average – 50° = 25 degree days.)

Degree-day information can be obtained from the Cooperative Extension Service if the turf manager does not have access to a weather station. Degree-

YEAR —

WEEDS	LOCATIONS	CONTROL METHOD
crabgrass	in bare spot by goals	rototilled, resodded
DISEASES	**LOCATIONS**	**CONTROL METHOD**
leaf spot	south end	was irrigating too much—cut back on irrigation
		slit seeded in resistant varieties
INSECTS	**LOCATIONS**	**CONTROL METHOD**
none seen		no control necessary
CULTURAL PROBLEMS	**LOCATIONS**	**CORRECTION**
overwatering	all	purchased moisture sensors for field
not mowing frequently enough in summer	all	talked to maintenance supervisor—he's made fields the top mowing priority

FIGURE 17-7

Typical annual report form for an athletic field.

FIGURE 17-8

Envirocaster monitoring station. *(Courtesy of Neogen Corporation)*

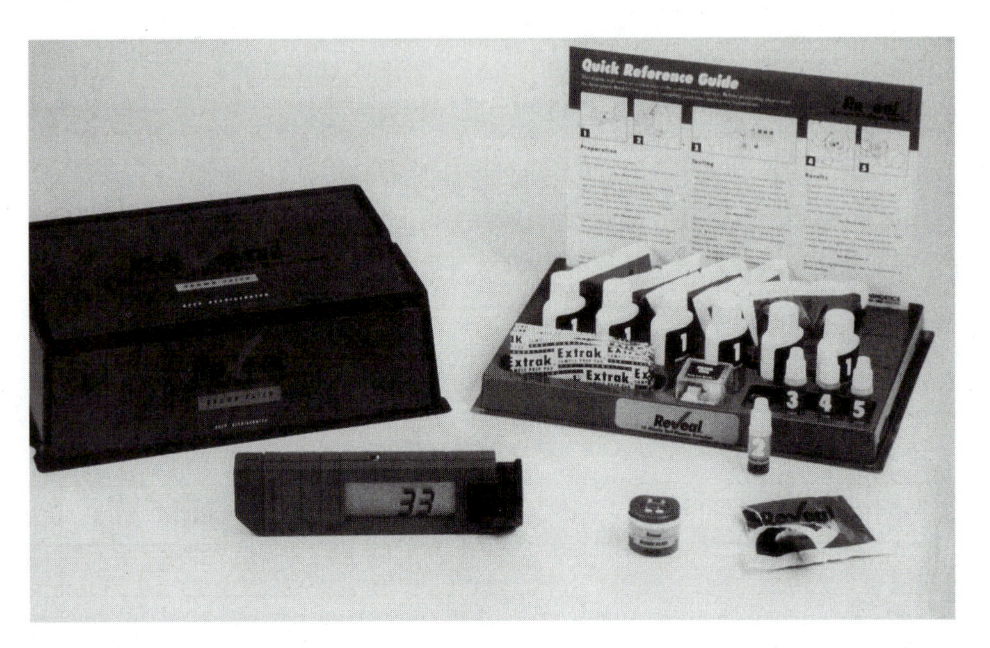

FIGURE 17-9
Immunoassay kit.
(Courtesy of Neogen Corporation)

day data are very helpful because they tell the scout when to be on the lookout for various insect pests and which stage to expect. For example, northern masked chafer adults begin to emerge from the soil at 930 degree days (base = 50°F), while southern masked chafers adults start to appear 1,085 degree days. The black turfgrass ataenius begins to lay its first generation of eggs when the degree-day accumulation is in the 210–300 (base = 55°F) range. The hatching of first-generation chinch bug eggs is completed at 240 degree days (base = 58°F) where there are two generations per season. The second-generation egg hatch is completed at 1,550 degree days.

This calculation is much more accurate than calendar date, since the dates will change from year to year because of differences in temperature that season. There is also interest in correlating degree days with weed and disease appearance.

Pathologists are attempting to develop immunoassays for plant diseases. The test kits have a disease detector with antibodies that are specific for a pathogen (Figure 17-9). If the grass sample contains the pathogen, the antibodies bind to the pathogen and a color change occurs. The darker the color, the greater the amount of pathogen that is present. These kits not only identify the disease, but can also predict the potential severity of the disease by indicating the size of the pathogen population in infected plants. The kits should be valuable tools in the future.

Pheromone traps can be used to monitor moth species whose larvae feed on turfgrass. The pheromone is a chemical attractant that lures the moths into a simple trap from which they cannot escape.

Cultural Techniques Affecting Pest Management

As mentioned in previous chapters, cultural techniques have a major impact on pest problems. Turfgrass that is established and maintained correctly will be less affected by pests and will recuperate quicker if injured. For example, healthy, vigorous grass produces a dense turf that resists weed invasion. When the turf is sparse and bare spots are present, weeds can gain a foothold and flourish (Figure 17-10). Poor management practices such as inadequate fertilization and cutting the grass too short are common reasons for a thin turf. Improving the maintenance program should increase the competiveness of the grass and result in a denser, less weedy turf and reduced herbicide use.

Correctly performing mowing, fertilization, irrigation and other cultural practices is an essential part of an IPM program. Proper establishment is also critical, including the selection of species and cultivars that are well adapted to the site and exhibit resistance to diseases common in the area. Because of endophytes it is also possible to select some turfgrasses that resist surface-feeding insects. New insect- and disease-resistant varieties can be slit-seeded into older, pest prone-turfgrass stands.

The cultural component is so important that some agriculturalists prefer the acronym *ICM* (integrated crop management) to *IPM*. Arborists have chosen the term *plant health care* to describe their management program because they believe that good cultural techniques are the most important part of any pest management strategy. There is a growing movement among turf professionals to use the expression *turfgrass management system* instead of *integrated pest management* for the same reason. TMS considers other potential environmental problems besides pesticides. Fertilizer pollution of groundwater and overuse of irrigation water are also addressed in this management program.

FIGURE 17-10

Fertilization helps to produce a dense, weed-resistant turf.

Whatever terminology is used, it is impossible to reduce pesticide use without a strong cultural program.

Biological Control

There are a number of living organisms that can be employed to help suppress pest populations. These biological agents are generally nontoxic to higher animals and have few if any negative environmental effects. Biocontrol is very attractive for these reasons, and a tremendous amount of research is presently devoted to finding new and more effective organisms.

Bacillus popillae, a bacterium, causes milky disease in Japanese beetle grubs. Another bacterium, *Bacillus thuringiensis kurstaki*, is sold commercially for caterpillar control. Other organisms being investigated include two fungi, *Beauvaria*, which kills chinch bugs, and *Metarhizium*, which causes a fatal disease in white grubs. *Xanthomonas campestris* var. *poannua*, a bacterium, has been effective in controlling annual bluegrass in some studies.

Insect-attacking nematodes in the genera *Steinernema* and *Heterorhabditis* have shown promise against turfgrass insect pests. These beneficial nematodes enter the insect and release bacteria which quickly kill the pest (Figure 17-11). Because nematodes are aquatic animals that live in the soil water, they are very susceptible to desiccation. They must be applied to moist turf and soil and be thoroughly watered in. Insect control using nematodes is sometimes unsuccessful. This is usually because the nematodes were weak or dead before the application or were allowed to dry out when sprayed on the turf. Some of the species are not aggressive enough in searching for insect pests.

Despite these problems, nematodes have great potential as control agents.

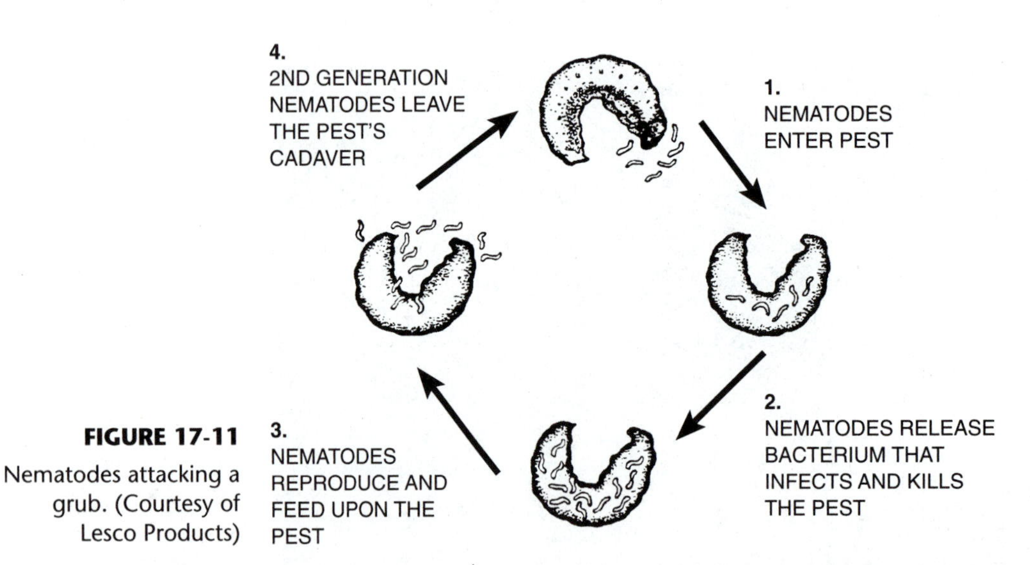

FIGURE 17-11 Nematodes attacking a grub. (Courtesy of Lesco Products)

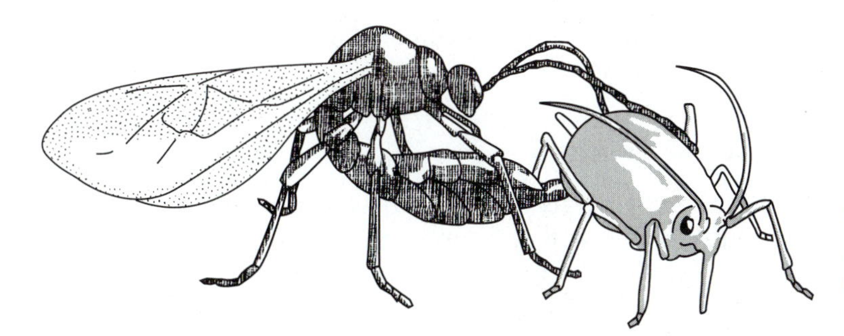

FIGURE 17-12
A wasp parasitoid laying an egg inside an aphid.

Researchers will find more effective species, and companies will learn how to formulate them so they will remain viable during storage and shipment. Turf managers will eventually become accustomed to working with living pesticides.

Endophytes have already been discussed in some detail in Chapter 15. They are fungi that live in turfgrass plants but cause no detrimental effect. Instead, they actually give plant insect resistance because they give off chemicals that repel insect pests. They also make the plant more tolerant of drought, heat, and other stresses.

Endophytic fungi in turn receive their nutrition from plants, are protected inside the plant, and are disseminated in the plant's seed. Presently there are endophytic cultivars of perennial ryegrass, fine fescue, and tall fescue. Major efforts are under way to establish these beneficial fungi in other turfgrass species. Because endophyte levels are highest in the leaves they are effective in preventing injury by surface-feeding insects such as sod webworms, aphids, chinch bugs, and billbugs.

Insect pests have many natural enemies such as ants, spiders, mites, ground beetles, wasps, and flies (Figure 17-12). Big-eyed bugs are predators of chinch bugs. A parasitic fly whose larvae feed on mole crickets has been used to help control this southern pest. One of the positive results of reducing insecticide use is an increase in the beneficial insect population. Unfortunately, most insecticides kill both "good" and "bad" insects.

Certain microorganisms, such as species of the fungi *Trichoderma* and *Gliocladium* and the bacteria *Pseudomonas* and *Enterobacter*, suppress turfgrass pathogens. They are said to be "antagonistic" to pathogens. In some cases the antagonist feeds on the pathogen, but many give off chemicals that inhibit the pathogens or compete in other ways. The presence of these types of beneficial microorganisms results in lower populations of disease-causing fungi.

Adding compost or natural organic fertilizers to turf will build up antagonist populations because they serve as a food source for these organisms. In some cases the organic material already contains pathogen suppressive microorganisms. Some field studies have shown that the application of composts and organic fertilizers results in disease control comparable or superior to fungicides.

Much more research is necessary in this area. Certain composts work better than others, and the age of the material is very important. There is also interest in finding ways to formulate antagonists and sell them individually as microbial fungicides.

Chemical Control

Alternative chemicals are being developed. The neem tree, which is widely distributed in tropical and subtropical regions of the world, produces an oil in its seeds which has remarkable properties. It repels insects and acts as an antifeedant (prevents insects from feeding) and as an insect growth regulator. Insect growth regulators (IGRs) mimic hormones in the insect and disrupt some vital function. Most IGRs prevent insects from successfully changing into the adult stage.

Neem oil contains many insecticidal chemicals. Azadirachtin is the one presently being used in commercially available neem products (Figure 17-13). Neem oil extracts are so nontoxic that in some countries they are used in toothpaste. They also degrade quickly and do not appear to contribute to any chronic health problems.

Other insect growth regulator compounds are being investigated. Insects produce juvenile hormones to prevent them from becoming adult before they have attained the proper size. When they are ready to become adults, the hormone is no longer produced and the transition occurs. If an insect is treated with its own juvenile hormone at this time, it will die without reproducing. Researchers are also studying hormones that could disrupt mating and egg laying.

FIGURE 17-13
A commercially available neem product. *(Courtesy of O. M. Scott and Sons Company)*

Even when traditional, more toxic pesticides are the only choice, there is an integrated pest management approach to their use. The least toxic chemical should be selected. It must be applied at the correct rate in a safe manner. If applied improperly, a control failure may occur, resulting in the need for another application.

The greatest problem applicators have is usually calibration. Many people who use pesticides apply too high a rate. Calibration is discussed in Appendices B and C.

The turf manager must understand all the characteristics of the pesticide selected. He or she must know enough about the life cycle of the pest to apply the material at the time when it will have the greatest effect.

Education

The key to a successful IPM program is knowledge. The people involved in its implementation must understand how to grow healthy plants, the life cycle and characteristics of the pests they encounter, how to monitor correctly, and what alternatives are available when a pest problem requires action. It is obvious that the most knowledgeable turf managers will have the most successful IPM programs.

The turf manager can obtain this knowledge by taking college courses, attending educational meetings, reading turf books and trade magazines, and talking with other turf specialists. Even is the manager is skilled, the program may fail if his or her employees are not properly trained.

PUTTING IT ALL TOGETHER—TWO EXAMPLES

The following two examples present a brief description of techniques associated with IPM programs.

Hairy Chinch Bugs in Lawns

Traditionally, hairy chinch bugs have been controlled by lawn care companies with "wall-to-wall" applications of insecticides. The response to any history of chinch bug injury in a town or neighborhood is often regular insecticide treatments of all lawns in this area every year. The size of the chinch bug population is ignored. IPM specialists seriously question the necessity and appropriateness of this approach.

Studies have shown that chinch bug numbers are largest on dry, sunny sites where there is a significant thatch layer and a large population of fine fescues. Chinch bug problems are much less likely on sites that are newly established, shaded, wet, have little thatch, and that are primarily tall fescue, Kentucky bluegrass, or endophytic fine fescues. It is obvious that every lawn in

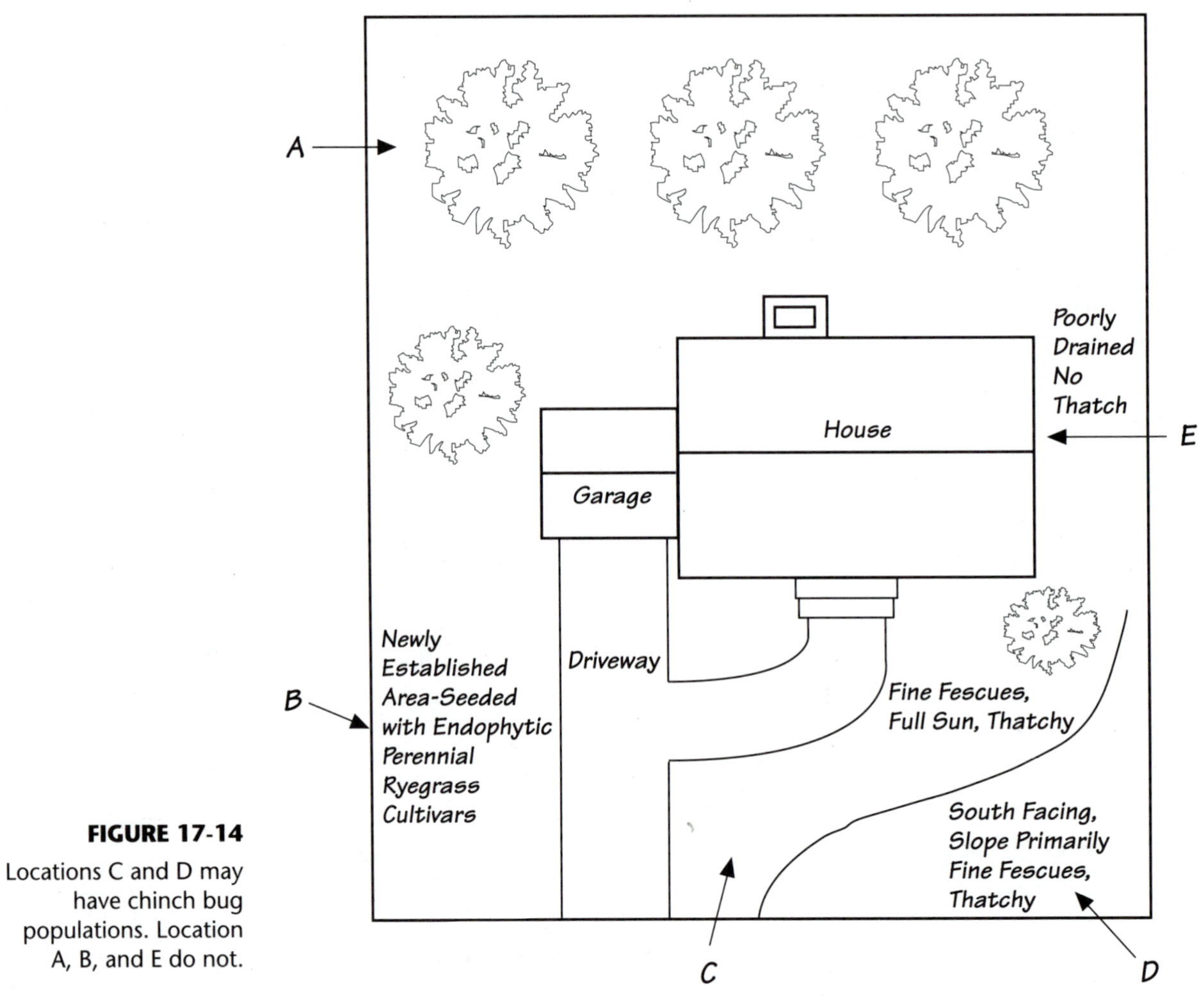

FIGURE 17-14
Locations C and D may have chinch bug populations. Location A, B, and E do not.

a town does not have the same potential for chinch bug problems. In fact, an individual lawn may have some areas that are prone to injury and others that are not (Figure 17-14).

A lawn care manager can accurately predict which lawns and parts of lawns are likely to have trouble. Even if a site is ideal for chinch bugs and has a history of injury, this does not mean that large pest populations are inevitable. Weather conditions have a great impact on the size of the population. Many of the adult chinch bugs die during the winter. This is especially true if there is no snow cover. Egg mortality can be quite high. During prolonged wet periods the disease caused by *Beauvaria bassiana* can devastate chinch bug populations. Natural enemies such as big-eyed bugs may reduce the numbers significantly.

The only way to know for certain whether a turf area will have problems is to sample it. The common method is to use a can or cylinder with both ends cut out. It is forced 1 to 2 inches (2.5 to 5.1 centimeters) deep into the soil and then filled with water to a level higher than the grass leaves. All chinch bugs in the turf will float up to the top in 10 minutes. This should be repeated in several places where chinch bugs are likely to be found. The most accurate count occurs when it is hot and sunny and the foliage is dry prior the water being poured in the cylinder.

Calculating the area of the end of the cylinder allows the scout to determine how many chinch bugs there are per square foot. Rather than assuming that chinch bugs will be a problem, the scout has numbers that will tell her whether some type of action is necessary. The threshold is normally 20 to 30 per square foot (0.09 m^2).

Monitoring has the greatest value when it is performed after most of the first-generation eggs have hatched. In regions where there is only one generation a year, this is at about 1,650 degree days (base = 45°F). Scouting should occur after the eggs of each generation have hatched. Sites prone to chinch bug injury are monitored several times, even if there are only one or two generations per season, because chinch bugs may move in from nearby areas.

No action is necessary if the populations are below the threshold. If the threshold is reached on certain parts of a lawn, those parts can be spot-treated. Even in warm, dry seasons, when chinch bug pressure is great, the turf manager knows that certain areas (wet, shaded, low thatch, etc.) will probably not require treatment.

If an insecticide must be used, the one that is least harmful to the applicator and the environment should be selected. Some entomologists recommend treating adults in the spring, when they first become active. This will minimize the effect on beneficial insects. Neem oil products are being used with some success. There is optimism that parasitic nematodes may eventually provide good control of chinch bugs.

Cultural practices can be quite helpful in reducing chinch bug populations. Thatch reduction and overseeding with endophytic varieties are common recommendations. Also recommended is increased irrigation, which lessens the moisture stress experienced by the grass. Frequent light irrigations to keep the surface moist will stimulate *Beauvaria* and result in more diseased chinch bugs. One solution is to plant more trees and increase the amount of shade.

White Grubs on Golf Courses

The white grubs discussed in this section are the larvae of the Japanese beetle, the European chafer, the northern and southern masked chafers, the Asiatic garden beetle, and the oriental beetle. They are more difficult to kill than chinch bugs and other surface-feeding insects because they are in the soil.

Many grub control failures occur because the pests do not come into contact with the insecticide. The result may be several applications and claims that the grubs are resistant to the chemicals. In fact, the reason for many failures is poor application timing. The ideal time to treat is when the grubs are in

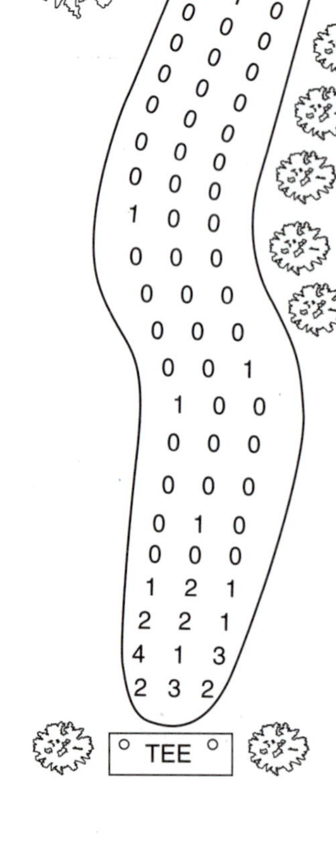

FIGURE 17-15

The number (1, 2, etc.) = the number of grubs per sample. Multiplying by 10 = the number per square foot. The results show that the area near the tee and green need to be treated, but much of the fairway has no grubs.

the first instar and still small, and when they are close to the surface feeding at the thatch/soil interface.

The date when eggs hatch and first instars appear varies from year to year because of weather differences. In one season eggs may hatch late because the soil is cool, and in another they may hatch early because of warm, moist conditions. If the soil remains very dry, eggs may desiccate and not hatch at all. As with chinch bugs, the key to a white grub IPM program is monitoring.

A scout must sample turf areas regularly until he or she discovers that the first instar grubs are near the surface. If the numbers exceed the threshold (usually 6 to 10 per square foot), treatment at this time has the greatest chance of success. Watering before or after the application will encourage the insects to feed near the surface. Irrigation will also help to wash liquid sprays off the foliage and into the soil.

On a golf course fairway the scout may take three or four samples across the fairway in rows 50 feet (15 meters) apart (Figure 17-15). The samples are pulled with a cup cutter, which has an area of 0.1 square foot. Multiplying the number of grubs found per plug by 10 converts to the number of grubs per square foot.

In many cases it is unnecessary to treat the entire fairway. Areas with grub populations at the threshold levels can be spot-treated. This type of monitoring often results in a 50 to 80 percent reduction in insecticide use. Some years environmental conditions are so unfavorable for white grub species that insecticides don't need to be used at all.

When insecticides are needed it is important to fully understand their characteristics. Repeat applications can be avoided if the turf manager selects the best material for the circumstances and applies it correctly. If a significant thatch layer is present, for example, chlorpyrifos would be a poor choice because it tends to be bound in the thatch. Trichlorfon however, is very soluble and will readily move down through a thatch layer to the grubs.

The use of nematodes for white grub control is an alternative to traditional pesticides. Their performance is steadily improving with the release of more effective species and the development of improved formulations and application techniques.

Cultural practices that affect the size and vigor of the root system are important. Plants with a root system that is large and recuperative will be able to tolerate more root injury than plants with a root system that is small and unable to replace lost root tissue.

Proper fertilization, mowing and irrigation will help to minimize the impact of white grub injury. Core cultivation is very important. Some turf managers apply biostimulants such as seaweed extracts to increase rooting.

The major effect of grub feeding is moisture stress due to root loss. Irrigating one or more times daily can help to compensate for the reduced root system. Frequent irrigation may keep turf alive and satisfactory for fairway use when the grub population far exceeds normal thresholds.

CONCLUSION

Integrated pest management is a relatively new concept in the turfgrass industry. Turf managers who have implemented IPM have experienced a reduction in the use of traditional pesticides. The future development and commercial release of insect growth regulators and biological control agents such as fungi, bacteria, insects, and nematodes will greatly strengthen IPM programs.

SELF-EVALUATION

1. What is an action threshold?
2. How often are golf courses scouted?
3. How many degree days are accumulated when the daily maximum temperature is 87°F, the minimum daily temperature is 52°F, and the base temperature is 50°F?
4. What is a pheromone?
5. What is an endophyte?
6. Why do nematode applications sometimes fail to control pests?
7. What is an antagonist?
8. What is an IGR?
9. Which sites would be most likely to have high populations of hairy chinch bugs?
10. How can the turf manager help to compensate for the loss of roots due to white grub feeding?
11. Develop an IPM program for a turf insect pest that is common in your area.
12. Develop and IPM program for a turfgrass disease.
13. Discuss why some people prefer the terms *integrated plant management* or *turfgrass management system* rather than *integrated pest management*.

CHAPTER
18

Other Turfgrass Problems

OBJECTIVES

After studying this chapter, the student should be able to

- Discuss the unfavorable growing conditions that occur in the shade
- Explain why certain shade-tolerant turfgrass species and cultivars are able to adapt to moderate or partial shade
- List the maintenance practices that help turfgrass plants survive on shaded sites
- Discuss the causes of soil compaction
- Describe the problems resulting from soil compaction
- List methods of preventing and alleviating compaction
- Explain why a thick layer of thatch is undesirable
- Describe the reasons for thatch buildup
- Explain how thatch is reduced and controlled

INTRODUCTION

Most of the serious problems encountered by turfgrass have been explained in previous chapters. Three important problems remain to be discussed—shade, compaction, and thatch. Shade is one of the major reasons for turfgrass decline. It is difficult to maintain satisfactory turf on a shaded site. Grass plants growing in the shade may be unable to survive because of low light intensities and other stress factors. Compaction occurs when soil particles are pressed close together and results in a decrease of larger pore spaces. Air, water, and fertilizer cannot readily enter a compacted soil. This unfavorable soil condition is caused by intense traffic on the turfgrass. Thatch is defined as the layer of partially decomposed or undecomposed organic matter formed above the soil surface. A thick layer of thatch exhibits many undesirable characteristics and reduces turf quality. These three problems and their solutions will be discussed in this chapter.

SHADE

Shade is one of the most common reasons for the deterioration of turf. It is estimated that as much as 25 percent of the turfgrass grown in the United States is shaded to some extent by trees. Grass and trees are highly desirable in a landscape, and it is inevitable that both are grown together. Unfortunately, they are somewhat incompatible. It is often difficult to maintain turf under trees.

Problems in the Shade

Tree leaves block sunlight and prevent it from reaching turf. Grass growing beneath tree species with dense foliage, such as maples, beeches, oaks, lindens, and hemlocks, may receive only 5 percent of the amount of sunlight that grass in nonshaded locations intercepts (Figure 18-1). The light that does filter through the tree canopy is of poorer quality for photosynthesis.

Low light intensities restrict the rate of photosynthesis. A typical turf that is heavily shaded is weak and sparse because photosynthesis is severely limited (Figure 18-2). Shoot and root growth are reduced, and the plants are unable to store adequate amounts of carbohydrate reserves. The likelihood of injury from stresses such as cold, heat, and drought increases when the plants have insufficient supplies of stored food. Grass growing in the shade is also easily damaged because it has succulent, tender foliage and shallow roots.

FIGURE 18-1

These front yards are heavily shaded.

FIGURE 18-2
Weak turf and bare spots are common in dense shade.

Light exclusion is not the only problem caused by trees. The roots of shallow-rooted tree species such as elms, willows, maples, beeches, and cottonwoods may interfere with grass development. Trees compete with the turf for nutrients and water. Grass plants may suffer from moisture and nutrient deficiencies because trees require large amounts of water and the same fertilizer elements that the turfgrass needs. The blocking of the sun's rays by tree foliage results in cooler air and soil temperatures beneath the tree. Turfgrass growth is slower at these lower temperatures. The roots of certain tree species may even exude toxic substances which can injure turfgrass.

Environmental conditions in the shade are ideal for the development of disease organisms. Pathogens are a serious threat when the surfaces of plants are wet. Turf growing under a tree is more likely to be covered with moisture than turf located on sunny sites. This is because the relative humidity is higher in the shade. Air movement and the heat of the sun have a drying effect. Trees hinder drying by reducing air circulation and sunlight.

This damp microclimate is not the only reason for increased disease problems in the shade. The pathogen that causes powdery mildew disease is inhibited by sunlight and is a greater problem at low light intensities. Grass plants situated under trees are more disease susceptible because of their weakened condition. They are less able to resist an attack by a disease than healthy, vigorous plants. This disease susceptibility is also a result of the succulent leaf tissue produced by grass in the shade. The thinner cell walls are more easily penetrated by fungi.

Growing Turfgrass in the Shade

The denser the shade, the more difficult it is for turfgrass to survive beneath trees. How dense the shade is depends on the tree species, the number of trees, and the distance between them. Conifers such as pines, spruces, and firs indi-

FIGURE 18-3

Many tree species cause only a light shading effect when a single tree is growing alone. Planted in groups, these same species may create serious shade problems.

vidually cause less shading than most deciduous trees. This is because most pines have an open canopy while spruces and firs have a narrow canopy. However, when planted in groups conifers will cause a shading problem (Figure 18-3). Some deciduous species that produce a light shading are poplar, locust, ash, birch, ginkgo, silk tree, Kentucky coffee tree, and the Japanese pagoda.

It is difficult to grow quality turf on heavily shaded sites where the grass receives less than four hours of full sunlight. When tree species with a dense canopy are planted close together it may be impossible to maintain a decent stand of turfgrass. Adequate amounts of sunlight may reach the grass if the trees are spaced far enough apart. People who are planning a landscape must consider the shading effect of trees 10 or 20 years after planting when the trees become large.

Some turfgrasses are more shade tolerant than others. Certain species and varieties can perform satisfactorily in partial or moderate shade, although no grass grows well in heavy shade. Several factors contribute to shade tolerance. These grasses are able to adapt to low light intensities because they use light more efficiently or require less light than intolerant species and varieties. They have to be resistant to diseases that are common in the shade, such as powdery mildew. Drought tolerance can be important because of the competition with tree roots for soil moisture.

St. Augustinegrass, zoysiagrass, centipedegrass, and fine fescue exhibit shade tolerance. Certain Kentucky bluegrass cultivars such as Glade, Eclipse, Chateau, and Bristol can adapt to semishaded conditions. However, the majority of the Kentucky bluegrass varieties respond poorly to low light intensities. Rough bluegrass *(Poa trivialis)* can tolerate wet, shady sites. Tall fescue has good shade tolerance when grown in the South. These species and cultivars are the best choices when turf is to be established beneath trees.

Even shade-tolerant grasses perform best when grown in sunny locations. They have a better chance of adapting to shaded conditions than other grasses, but special care is necessary to ensure their survival in the shade. Certain maintenance practices are recommended to reduce the stress experienced by turf situated under trees.

Soil testing is necessary to determine if soil pH and fertility levels are correct. The competition for plant food between tree and grass roots may result in nutrient deficiencies. Unfortunately, attempting to satisfy the nutritional needs of both trees and grass by applying a heavy rate of fertilizer to the turf often has disastrous results. As the nitrogen moves down through the soil it will encounter the grass roots first. Large amounts of nitrogen will be taken up by the turfgrass, and the plants become lush and succulent. The thinner cell walls cause the grass to be more disease susceptible and less wear resistant. The nitrogen also stimulates a greater rate of vegetative growth. This is a problem because rapid growth results in a depletion of carbohydrate reserves and the need for increased photosynthesis to replace the stored food that is used. Grass exposed to low light intensities may not be able to photosynthesize enough to replenish its carbohydrate supplies.

Turfgrass located in the shade, because of its slower growth rate, requires less fertilizer than grass receiving full sunlight. Light rates of fertilizer are usually sufficient (Table 18-1). Trees should be fertilized separately if the application occurs when the turf is actively growing. This is accomplished by inserting a tree root feeder into the soil beneath the majority of the turfgrass roots (Figure 18-4). Surface application with a lawn fertilizer spreader is an easy and effective method of fertilizing trees. However, when turf covers the tree roots, surface applications should occur in the late fall or winter, when the grass is not growing and the soil is not frozen. The turf should be watered heavily enough to move the fertilizer off the surface and into the soil. The tree can be fertilized without injuring the grass.

Shaded turfgrass should be cut to a taller height than nonshaded grass. As a general rule, the mower blade should be raised 1 inch (2.5 centimeters) in shaded areas. A mowing height of 2.5 to 3 inches (6.4 to 7.6 centimeters) or

TABLE 18-1 Maximum Annual Nitrogen Requirements of Turfgrass Growing in Shade

Species	Nitrogen per 1,000 ft^2 (lbs)
St. Augustinegrass	2–3
Centipedegrass	1–2
Zoysiagrass	2.5–3.5
Tall fescue	2–3
Fine fescue	1.5–2.5
Kentucky bluegrass	2–3
Poa trivialis	2–3
Annual bluegrass	2.5–3.5
Bentgrass	2.5–3.5

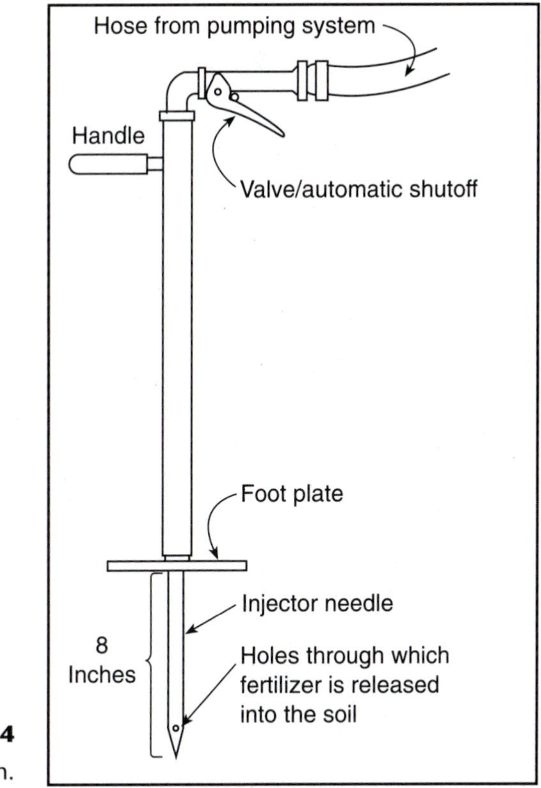

FIGURE 18-4
A tree root feeder or fertilizer gun.

more is best. Allowing grass beneath trees to grow taller is advantageous for three reasons. Grass plants exposed to low light intensities tend to grow very erect, as if they were reaching for the sunlight. This upright growth habit requires a higher than normal cutting height to avoid removing too much leaf tissue per mowing. Plants with longer leaves can also photosynthesize more because they have greater leaf surface area and more chlorophyll than shorter grass. This increased photosynthetic area helps to compensate for the reduced amount of sunlight intercepted by grass growing in the shade. A third advantage is that a higher cutting height results in a deeper root system, which allows the grass to compete better with tree roots for nutrients and water. Stopping to raise the mower blade when cutting shaded areas is only a minor inconvenience and has a very beneficial effect on the turf.

Moisture stress may occur because of the dual use of water by trees and grass. Deep watering once or twice a week is usually best. Traffic should be kept off shady sites. The tender, shallow-rooted turf exhibits poor wear resistance and is easily damaged. Fungicide applications may be necessary to protect some grass species and varieties from diseases such as powdery mildew.

Turfgrass that is struggling in dense shade may be saved by trimming all the branches off a tree up to a height of 10 feet (3 meters) or higher. Removing

FIGURE 18-5
Removing the lower branches of a tree often results in improved turf quality.

the lower branches will allow a greater amount of sunlight to reach the turf (Figure 18-5). Selective pruning in the crown of a tree will open up the canopy and increase light penetration. A tree that is pruned properly will be healthier and have a more attractive appearance. Branches that are dead, diseased, crossing, or growing close together are logical choices for removal. When many branches must be trimmed to decrease light blockage, all of the limbs should not be cut off immediately. Gradual removal over a period of a few years is preferable to one severe pruning, which could harm the tree. Thinning should be done by someone with arboriculture training.

When leaves drop to the ground in autumn, prompt removal is essential (Figure 18-6). Leaves smother the grass and exclude light. Turf covered with leaves remains constantly damp, and this encourages disease problems. The amount of light available to the grass increases substantially once a tree loses all of its leaves. If the fallen leaves are collected immediately, these higher light intensities will reach the turf and stimulate a greater rate of photosynthesis. The grass will then be able to build up its food reserves before winter arrives. Mulching mowers may be used to chop up the leaves fine enough that the small pieces can be left in the lawn.

Practicing these cultural techniques enables turfgrass to be more adaptable to shaded conditions. Careful and expert management is necessary to grow good-quality turf beneath trees. On heavily shaded sites, however, even skillful

FIGURE 18-6

A large leaf removal machine. *(Courtesy of Jacobsen Division of Textron, Inc.)*

management may not be enough. The decline of properly maintained shade-tolerant grasses indicates that the shade is too dense for turf to survive. When light exclusion is this severe, there is no point in replanting grass unless the trees are cut down. Other alternatives are available. The area under the trees can be covered with materials such as wood chips, bark mulch, bricks, or marble chips. If plants are preferred, there are a number of vegetative ground covers that can grow at very low light intensities (Figure 18-7). These landscape plants can survive on extremely shady sites (Figure 18-8).

COMMON NAME	SCIENTIFIC NAME
Bugleweed or carpet bugle	*Ajuga reptans*
Wild ginger	*Asarum* spp.
Leather crassifolia	*Bergenia crassifolia*
Lily-of-the-valley	*Convallaria majalis*
Epimedium or bishop's hat	*Epimedium macrothum*
Wintercreeper	*Euonymus fortunei*
English ivy	*Hedera helix*
Plantain lily	*Hosta decorata*
Creeping lily	*Liriope spicata*
Moneywort or creeping jenny	*Lysimachia nummularia*
Creeping mahonia	*Mahonia repens*
Pachysandra or Japanese spurge	*Pachysandra terminalis*
Star jasmine	*Trachelospermum jasminoides*
Periwinkle or creeping myrtle	*Vinca minor*

FIGURE 18-7

Examples of vegetative ground covers that are adapted to the shade.

FIGURE 18-8
Vinca minor is a popular shade-tolerant ground cover.

FIGURE 18-9
Mosses are small, leafy plants which appear to be a mass of fine stems.

Moss and Algae

Moss and algae, when they invade a turf area, are usually found in the shade. These simple plants are unable to compete successfully with turfgrass if the growing conditions are satisfactory for turfgrass. Their presence is an indication of unfavorable environmental conditions and improper turf maintenance. Shade, poor drainage, overwatering, and soil acidity, infertility, and compaction all contribute to the decline of the turf and favor the development of moss and algae (Figure 18-9).

Any remedy that improves the health and competitiveness of the turfgrass helps to control moss and algae. Correct management of turf in the shade is very important. Low fertility is a common reason for their appearance. Adequate fertilization often solves the problem.

Hand-raking is the simplest method of ridding a turf of moss and algae. They can also be controlled with chemicals such as ferrous ammonium sulfate, hydrated lime, or copper sulfate. However, these unwanted plants will return unless the conditions that weakened the grass are corrected.

SOIL COMPACTION

A compacted soil is a soil in which the mineral particles have been pressed close together. Compaction is usually a result of excessive traffic. The particles are forced together because of the mechanical pressure exerted by foot traffic or the tires of machinery and vehicles. Compaction generally occurs in the top 2 to 3 inches (5 to 7.6 centimeters) of soil.

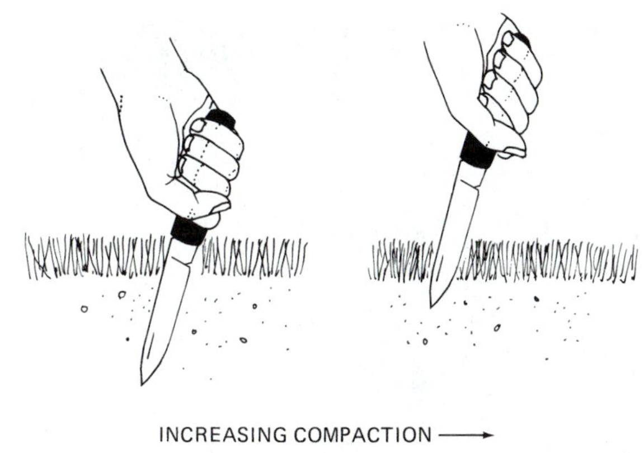

FIGURE 18-10
A knife blade can be pushed relatively easily into a noncompacted soil. As compaction increases, it becomes more difficult to force the blade into the soil.

An easy method of diagnosing compaction is to stick a knife into the soil. It is difficult to push the blade into a compacted soil (Figure 18-10). A soil sample can be examined. When the surface soil is compacted it looks and feels hard and dense. There are instruments available, called penetrometers, to measure compaction.

Soil compaction is a serious problem. As the mineral particles are crushed together, porosity decreases. There is a significant decrease in the large pore spaces—the macropores, and this decrease near the soil surface restricts the movement of air, water, fertilizer, lime, and pesticides into the soil. A soil with a good structure may be composed of 40 percent air on a volume basis. The air content of a severely compacted soil may be as little as 5 percent.

The result of this loss of pore space is a substantial decline in turf quality. Roots may be unable to penetrate the soil because the particles are tightly packed in a solid mass. Reduced aeration is also a problem because roots need oxygen to respire and produce energy. The barrier formed by the compacted layer can trap carbon dioxide and other gases in the soil. These gases are released by roots and soil organisms and may become toxic to grass roots as their concentrations increase.

A major problem is poor water infiltration and percolation. The soil turns into mud after rain or an irrigation. Puddles often form on the surface, and the turf becomes unusable. The soil drains slowly and remains extremely wet during periods of rainy weather. When the soil is wet it is more prone to further compaction and the turf is less wear resistant.

Studies have shown that sports injuries increase significantly when athletes play on compacted fields. Athletes are more likely to be hurt when they fall on a hard, compacted surface. Injuries are less common on a softer, noncompacted field that has a thick cover of turfgrass which serves as a cushion.

Factors Contributing to Compaction

The major reason for soil compaction is intense traffic. Compaction is usually associated with athletic fields, golf greens, and other heavily used recreational

FIGURE 18-11
A compacted path on a university lawn. *(Courtesy of James Bates, photographer)*

turf. Areas at the edges of driveways, highways, and sidewalks are often compacted. The problem occurs on lawns located at industrial parks, shopping centers, schools, colleges, and other complexes where there is a large amount of foot traffic (Figure 18-11). A home lawn seldom becomes seriously compacted unless the neighborhood children use it as a playground.

How many times people walk, run or drive equipment over a turf has the greatest effect on whether or not compaction develops. Other factors have an important influence as well. For example, severe compaction is much more likely on a fine-textured, clayey soil than on a coarser, sandy soil. The smaller, platelike clay particles can be crushed closer together. Clay soils also hold more water and drain slower than sandy soils. A film of water surrounding soil particles acts as a lubricant. When a wet soil is subjected to mechanical pressure the particles readily slide together.

A dense stand of turf helps to protect the soil from compaction. The leaves, stems, and thatch act like a cushion or shock absorber. Thin or bare spots are prone to compaction because a foot or tire exerts its mechanical pressure directly against the soil.

Preventing and Alleviating Compaction

The most effective method of preventing compaction is to reduce the amount of traffic on a turf area. Hedges, trees, flower beds, shrubs, fences, and stone walls can be used as barriers to keep people off grass. They also guide the flow of traffic in the desired direction (Figure 18-12). Correct placement of sidewalks and pathways is important. Their location should allow a pedestrian to walk directly and quickly to his or her destination. If walkways are positioned

FIGURE 18-12

Hedges create attractive barriers that prevent pedestrians from walking on the grass.

properly, people will be less tempted to take shortcuts across the grass. The walkways should be wide enough to accommodate groups of people.

Curbing stops drivers from parking on the grass. However, a curb can restrict the access of persons confined to wheelchairs. Signs that remind people to stay off the grass are helpful. Golf cart paths on golf courses are marked with signs to prevent carts from being driven on greens, tees, aprons, and collars.

On sports fields and other recreational areas it may be impossible to divert traffic to the extent that no compaction occurs. However, even some traffic reduction may decrease the severity of compaction. For example, the field that is used for football games on Saturdays should not be used at other times. During the week the team should scrimmage on a practice field. The two fields will recover more rapidly if they are not played on at all in the spring.

It is extremely important to keep traffic off turf when the soil is wet. Maintenance operations requiring the user of heavier equipment should be postponed until the soil is drier. The turfgrass manager should not irrigate a sports field immediately before a game.

Any vehicle or machine driven on turf should be equipped with special turf tires (Figure 18-13). These wide tires are designed to reduce compaction. The amount of pressure exerted by an object is determined by dividing its weight by the surface area that is in contact with the soil. As the amount of surface area in contact with the soil increases, the mechanical pressure exerted decreases. Wide turf tires distribute the weight over a larger area than regular tires; therefore, less compaction results.

FIGURE 18-13
Wide, turf-type tires are used to minimize compaction. (*Courtesy of Jacobsen Division of Textron, Inc.*)

Whenever possible, turf areas that receive intense traffic should be established on a sandy soil to minimize compaction potential. A coarse-textured soil that is 80 percent or more sand is recommended for heavy-use sites such as athletic fields, putting greens, tees, or turfed pathways. A 12-inch (0.3-meter) layer is ideal. If this extremely sandy media is used, organic material such as peat moss or sawdust is incorporated into the top 3 to 4 inches (7.6 to 10.2 centimeters) to increase the nutrient-holding ability. An irrigation system is required because the sand will drain rapidly. Frequent irrigation may be necessary. Replacing a clayey soil with sand is an excellent solution to the compaction problem, but it is very expensive.

Mechanical cultivation helps to alleviate compaction after it has occurred. Coring or core cultivation is a widely used maintenance practice. It is also referred to as aerating or aerifying. Hollow metal tubes called spoons or open tines are forced into the soil. When they are withdrawn cores or plugs of soil are removed, leaving holes (Figure 18-14). The distance between the holes and their diameter and depth depend upon the type of machine used. The holes are spaced 1 to 6 inches (2.5 to 15 centimeters) apart and are normally 0.25 to 0.75 inch (6.4 to 19.1 millimeters) in diameter. A machine with 0.75-inch diameter tines making holes 2 inches apart will remove over 5 percent of the surface soil in one pass. The deeper the holes, the more effective the core cultivation.

Removing plugs or cores relieves soil compaction. The holes allow better penetration of water, air, fertilizer, lime, and pesticides into the root zone. The result of this increased infiltration is deeper-rooted, healthier, more vigorous turfgrass. Toxic gases are able to escape from the soil through the openings

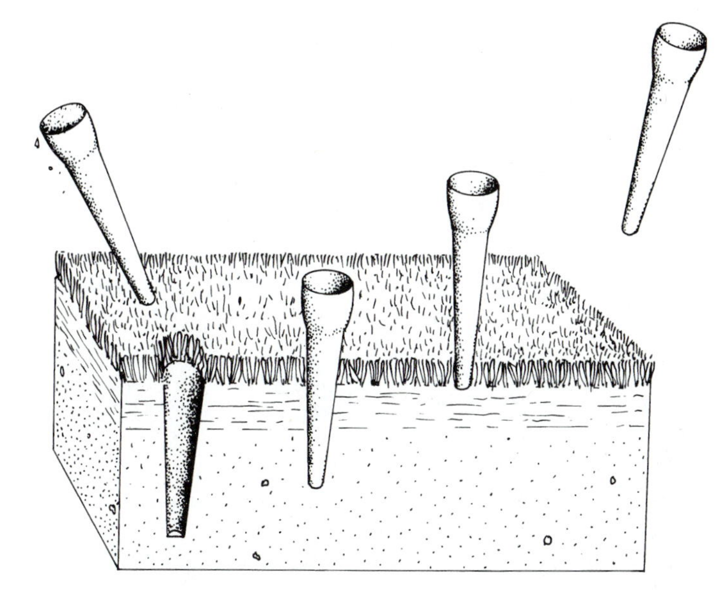

FIGURE 18-14 Core cultivation results in the removal of cores or plugs of soil.

made by the core cultivator. Improved surface drainage helps to dry out the soil and prevents the formation of puddles. There a number of benefits derived from core cultivation.

One type of cultivator has hollow tines mounted on a crankshaft (Figure 18-15). The tines move up and down vertically and allow deep penetration without damaging or disrupting the turf. These machines are popular for golf greens because they cause minimal disturbance to the putting surface. Vertical motion aerators have a relatively slow operating speed. Though most aerators penetrate only 2 to 4 inches (5.1 to 10.2 centimeters), some, called deep-tine

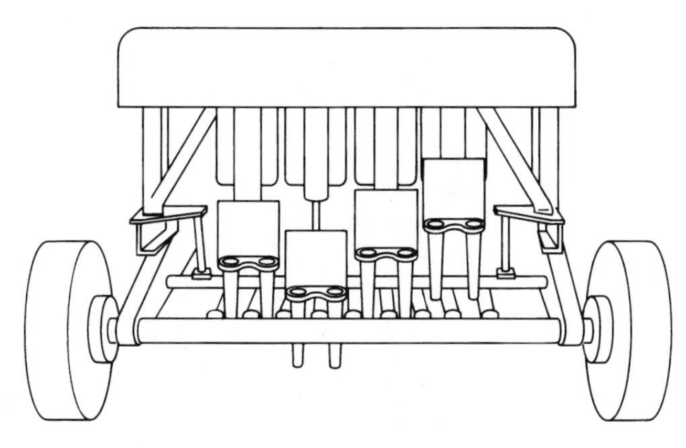

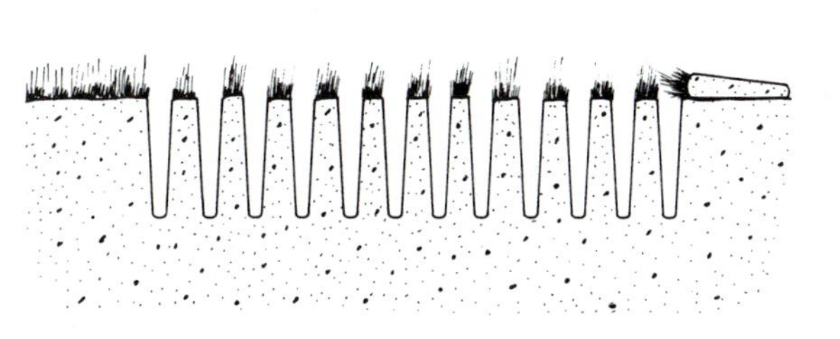

FIGURE 18-15 A vertical motion-type core cultivator.

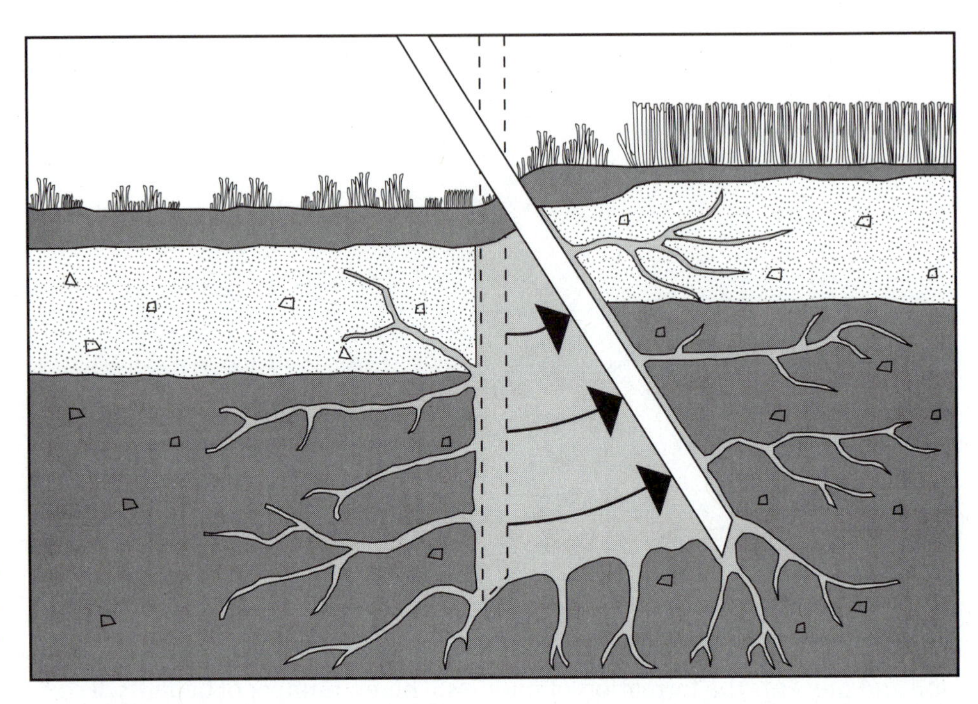

FIGURE 18-16

Improved aeration due to shattering effect of a solid tine.

aerifiers, can make holes 6 to 16 inches (15 to 41 centimeters) deep. Conventional aerators can create a compacted area, called a cultivation pan, which occurs just beneath the point of their tines' deepest penetration. A deep-tine aerifier can break this up and can also penetrate through different soil layers that may impede water movement and root development.

Deep-tine aerifiers use either solid or hollow tines. The solid tines usually penetrate deeper and are faster than hollow tines. There is also no cleanup because soil is not removed. Solid tines are also called shatter-core tines because the impact of the soil tine shatters or fractures the soil (Figure 18-16). Areas aerated with hollow tines, because soil has been removed, resist compaction better.

Deep-tine machines are significantly more expensive than conventional aerators. Rather than buying a machine, most turf managers contract a company to do the deep aeration for them.

A second type of core cultivator has open tines or spoons mounted on a drum or metal wheels (Figure 18-17). The drum or wheels turn in a circular motion and force the tines or spoons into the soil. These rotating units core cultivate turf more quickly than the vertical type and are preferred for aeration of larger areas. However, they cause more disruption to the turf surface and do not penetrate as deeply as the vertical motion cultivators. The penetration depth can be increased by placing weights on the aerator.

The plugs must be picked up when a closely mowed playing surface is aerated. These soils cores would interfere with putting and be an eyesore if left on a golf green. Traditionally the plugs are raked up or shoveled off the green.

FIGURE 18-17

Many aerators have tines or spoons mounted on disks or wheels. *(Courtesy of Jacobsen Division of Textron, Inc.)*

Core harvesters that attach to turf vehicles can be used to pick up the cores. Some aerators have the tines mounted on a hollow drum. The plugs are pushed into the drum and collected (Figure 18-18). When the drum is full it can be emptied on a compost pile.

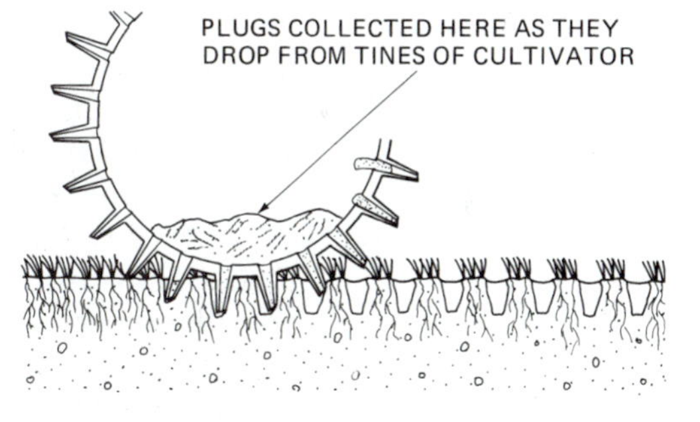

FIGURE 18-18A

A drum-type cultivator collects the plugs. *(Courtesy of OMC Lincoln)*

FIGURE 18-18B

The plugs removed by the cultivator are collected in the drum. *(Courtesy of OMC Lincoln)*

FIGURE 18-19
Breaking up soil cores with a mat. *(Courtesy of OMC Lincoln)*

The plugs do not need to be collected on higher-cut areas such as lawns or fairways. A metal drag mat or piece of chainlink fence can be pulled over the surface to help break up the soil cores (Figure 18-19).

Core cultivation should not be performed when the soil is too dry or wet. A dry soil is hard and difficult to penetrate. Damage to the turf is more likely if the soil is wet. Opening up the soil can cause root desiccation during hot, dry weather. Irrigation is necessary after core cultivation. Core cultivation is generally avoided in midsummer. The best time to perform this practice on warm season turf is in late spring or early summer. The ideal time on cool season turf is late summer or early fall.

A deep-drill aerifier is another approach. Drill bits bore holes up to 12 inches (30 centimeters) deep. The machine is slow, but there is very little surface disruption. It is primarily used on putting greens.

The newest innovation is high-pressure (5000 psi) water injection. Jets of water at speeds in excess of 600 mph make holes in the soil 6 to 16 inches (15 to 41 centimeters) deep (Figure 18-20). The holes are less than 0.125 inch (3.2 millimeters) wide. There is no surface disruption. High-pressure water injection can be performed during the summer.

Spiking and slicing are two other practices that help to relieve compaction. They provide temporary relief and are not as effective as core cultivation. However, they offer two advantages. First, plugs are not extracted, and no cleanup is necessary after the operation. Second, the turf is only minimally disturbed during these operations.

A spiker uses solid tines to punch shallow holes by forcing the soil downward and laterally. These holes are extremely compacted at the bottom and along the sides because the hole is made by crushing the soil instead of removing it. A slicer has thin, V-shaped knives that cut into the soil. Narrow slits 2 to 4 inches (5 to 10.2 centimeters) deep are sliced in the turf.

FIGURE 18-20

High-pressure water injection is a new method of aeration. *(Used with permission of The Toro Company. "Toro" is a registered trademark of The Toro Company, Minneapolis, Minnesota.)*

A subsurface cultivator can be used to reduce compaction. It has aerating blades with bulletlike tips which shatter the soil as the blades vibrate back and forth (Figure 18-21). These machines can cause serious surface disruption. The subsurface cultivator helps to break up a compaction layer which can develop at the bottom of holes made by a core cultivator or spiker.

Aeration is not a permanent solution to a compaction problem. The soil gradually expands and the holes fill in again. Heavily trafficked areas may require cultivation several times each year. Some athletic fields are aerified twice a month. The aerifier can be run across the turf in more than one direction if a soil is severely compacted. This produces a greater number of holes. It is com-

FIGURE 18-21

The subsurface cultivator blades shatter the soil 5 to 7 inches (12.7 to 17.8 centimeters) deep.

FIGURE 18-22
Power rollers. *(Courtesy of Brouwer Turf Equipment, Ltd.)*

mon for a golf course superintendent to use two or three different aerification techniques each year.

Rolling

Rolling is a maintenance practice used to push grass roots back into contact with the soil. Grass plants may be heaved up as the result of soil freezing and thawing in the winter. The spikes and cleats worn by athletes on sports fields contribute to uprooting. Grass will wilt and desiccate when its roots are removed from the soil. Raised plants are also likely to be damaged by mowers.

The weight of a roller presses down on the turf and forces roots into contact with the soil surface (Figure 18-22). The roots can then grow and reenter the soil. Water ballast rollers can be filled with water to increase their weight. A 20-gallon (76-liter) roller, for example, will hold approximately 170 pounds (77 kilograms) of water.

Rolling can cause compaction if a heavy roller is used on a wet soil. The turf manager should avoid rolling unless raised, uprooted sod is a problem. A light roller is usually sufficient to press plants into the soil. A soil that is too wet should not be rolled, but adequate moisture is necessary to enable roots to grow back into the soil quickly.

A roller is a valuable tool for smoothing and firming a seedbed. Tilling loosens up and expands a soil. The soil will eventually settle, but a light rolling will pack it down immediately. People can then walk on the seedbed without sinking in when they are seeding and mulching. Another light rolling after the seed is spread ensures good seed-soil contact, which results in better germination. Newly installed sod is often rolled.

FIGURE 18-23
The thatch layer is located above the soil. *(Courtesy of James Bates, photographer)*

THATCH

Thatch is a layer of partially decomposed and undecomposed plant tissue. It accumulates above the soil surface and is composed of dead and living roots and stems (tillers, rhizomes, and stolons). Thatch is brown in color and contains little or no soil (Figure 18-23). The mat, directly beneath the thatch, is a soil layer with organic matter mixed in it.

Thatch buildup occurs when the production of plant tissue is greater than the decomposition rate. Plant debris is primarily decomposed by soil microorganisms such as fungi, bacteria, and actinomycetes. Earthworms also feed on organic matter and help to break it down. Decomposed organic material is called humus and has many characteristics that are desirable for plant growth. Undecayed organic material, thatch, has many undesirable characteristics.

Lignin is a compound in plant tissue that makes cell walls strong, hard, and rigid. It is resistant to decomposition. Ligneous tissue is decomposed at a very slow rate by microorganisms. The parts of a grass plant that contain the greatest amounts of lignin are the major components of the thatch layer. Roots are highest in lignin content, but stems, rhizomes, stolons, and leaf sheaths also contain substantial quantities. Leaf tissue, which is relatively low in lignin, does not contribute significantly to thatch accumulation because it is decayed rapidly.

A thin layer of thatch is beneficial. It acts as a cushion and reduces sports injuries and compaction. The undecomposed organic matter serves as a mulch and protects the soil surface from drying. It insulates the crown of the plant from sudden temperature changes. Thatch can become quite dry, and this prevents weed seed germination.

For most turf areas a thatch layer less than 0.5 inch (1.3 centimeters) thick is desirable. Problems begin to develop when the thatch accumulation becomes greater than 0.5 inch. The layer should not exceed 0.3 inch (0.76 centimeter) on a putting green.

Problems Association with Thatch

A thick thatch provides a good environment for the survival and growth of some insect pests and disease organisms. Lawnmower wheels sink into thatch because it is soft and spongy. When the wheels sink the mower blade is lowered and scalping can occur.

Roots, rhizomes, stolons, and crowns are located in the thatch if the layer is thick. They are more likely to suffer cold, heat, and drought injury because the thatch is not as protective as soil. Shallow rooting is another consequence of heavy thatch accumulation. The roots become confined to the thatch layer. The turf is weakened and more susceptible to drought and other stresses because of the shallow root system.

Thatch is very porous and contains many macropores. It exhibits many of the same characteristics as sand. Water retention is poor because it drains rapidly. The layer is hydrophobic and tends to repel water. Once thatch dries out it is difficult to rewet. Dry thatch can significantly reduce the water infiltration rate. Thatch does not have any capillary activity, so water cannot move back into it from the soil below. However, when the thatch layer becomes wet, it remains wet, and water moves slowly down into the soil. Moisture levels in the thatch are often unsatisfactory. The layer is relatively infertile because of poor nutrient retention.

Thatch impedes the movement of pesticides into the soil. For example, some insecticides are tied up in the thatch. If they are absorbed, even a heavy watering after application may not move them through the layer. Grub control becomes more difficult when the downward movement of insecticides is significantly slowed by a thick layer of thatch.

The opposite problem occurs with preemergent herbicides. These pesticides form a chemical barrier at the soil surface which prevents the seedlings of crabgrass and other annual weeds from emerging. The herbicides kill these weeds by inhibiting cell division in their roots. The chemicals could also destroy the roots of the established, desirable turfgrasses. This does not normally happen because these herbicides are immobile in the soil and remain on the surface; they do not move down to where roots of mature, perennial grasses are located. Nevertheless, preemergent herbicides have a greater mobility in the thatch and will move around. Turf roots that are confined to a thick thatch layer can be injured.

Why Thatch Accumulates

Thatch accumulates when the rate of tissue production is greater than the rate of tissue decomposition. High-maintenance programs that stimulate rapid turfgrass growth can be partially responsible for this imbalance. The production of plant tissue is increased substantially when vigorous cultivars are fertilized and irrigated heavily on a regular basis. Excessive nitrogen fertilization is a major reason for thatch buildup.

Soil microorganisms decompose plant tissue. A large microorganism population is necessary to keep thatch under control because each individual microorganism is so small. These organisms are very prolific and have tremendous reproductive rates. If soil conditions are favorable for the development of these populations the thatch decomposition rate will be high. When soil conditions are unfavorable, organic matter will accumulate because of the lower populations.

Soil problems that inhibit microbial activity should be corrected. Many of these organisms are pH sensitive. Optimal development occurs at a soil pH in the 6.0 to 7.0 range. Populations drop off sharply when the pH is too acid or alkaline. Low-maintenance programs that do not provide enough lime and fertilizer contribute to thatch buildup. Soils deficient in nitrogen cannot support adequate microbial activity. Moderate fertility is best. There should be enough nitrogen applied to the turf to encourage microorganisms, but not so much that excessive plant growth is stimulated.

Compacted, poorly drained, overly wet soils lack oxygen, which is required by most of these microbes. Moisture is also necessary for microbial activity to occur. Decomposition stops when the thatch and soil are dry. Soil temperature is very important, but it is a factor that the turf manager cannot control. Optimal temperatures for thatch-decomposing microorganisms are in the 95°–100°F (35°–38°C) range. The warmer the temperature, the faster the decomposition rate.

Leaf tissue has little effect on thatch unless the clippings removed by the mower are quite long. Longer pieces of tissue do not break down as rapidly as short clippings. Frequent pesticide use may lead to thatchy turf. Soil insecticides that are lethal to earthworms will stop their decomposing activities. Fungi feed on organic matter and decay thatch. Fungicides used to control fungi that cause turf diseases also reduce populations of beneficial fungi.

Thatchy turf is often produced by turfgrasses that are vigorous growers, develop extensive root systems, and have higher concentrations of lignin in their tissue. Hybrid bermudagrass, St. Augustinegrass, zoysiagrass, and the fine fescues are very prone to thatching.

Controlling Thatch

A vertical mower removes thatch from a turf. It is also called an aeroblade, thatcher, verticutter, or dethatcher. The machine has a series of knives mounted on a horizontal shaft (Figure 18-24). As the shaft rotates at a high speed the blades slice into the thatch and rip it out of the turf (Figure 18-25). The blades are spaced 1 to 3 inches (2.5 to 7.6 centimeters) apart on the shaft.

Some vertical mowers can penetrate as deep as 2 to 3 inches (5.1 to 7.6 centimeters). The machines can be adjusted to cut into the turf to various depths. The proper depth setting depends upon the thickness of the thatch. The knives should penetrate to the bottom of the thatch layer, and some turf specialists recommend slicing into the surface soil beneath. Decomposition

FIGURE 18-24

Knives on a vertical mower. *(Courtesy of Jacobsen Division of Textron, Inc.)*

FIGURE 18-25

The vertical mower slices into the thatch layer.

may be encouraged if the blades pull soil up into the thatch. The slits in the soil surface may help to alleviate compaction. Vertical mowing is beneficial because new growth is stimulated when the stolons and rhizomes are severed. This can result in increased turf density and is the major reason why vertical mowing is performed on putting greens.

The frequency of dethatching operations is determined by the rate of thatch accumulation. When the thatch layer becomes thicker than 0.5 inch (0.3 inch) on greens, vertical mowing should be considered. Managers should periodically cut into the turf with a knife and check the thatch depth. Turf areas that are slow to build up thatch may require verticutting once every several years, or never at all. Areas that are prone to thatching may need to be dethatched once or even several times per year. The organic material that is removed should be raked up or vacuumed (Figure 18-26).

The best time to vertical-mow warm season grasses is in the late spring or early summer. Late summer or early fall is preferred for cool season grasses. The spinning knife blades injure the plants as they rip through the turf. This is less likely to be a problem if the grass experiences good growing whether after the operation. To recover quickly, turfgrass needs at least a few weeks of favorable growing conditions after it is verticut. Turfgrass injury is more extensive if the soil and thatch are too moist when vertical mowing is performed.

FIGURE 18-26

A turf vacuum. *(Used with permission of The Toro Company. "Toro" is a registered trademark of The Toro Company, Minneapolis, Minnesota.)*

There is some question about the effectiveness of vertical mowing as a thatch reduction technique. Relatively small percentages of the total amount of thatch are removed after vertical mowing in just one direction. Slicing the area in several directions to remove greater quantities of thatch can result in serious turfgrass injury. This is especially true if the grass is shallow rooted. Some studies have shown an overall decrease in turf quality after vertical mowing. The best approach is to try to limit the development of the thatch layer so thatch removal does not become necessary.

When the thatch layer is allowed to become excessively thick it may be uncontrollable. The only solution in an extreme situation is to remove the grass and thatch with a sod cutter. Soil must then be added to level the area before it is reestablished. The turf manager can avoid this drastic solution by keeping the thatch under control.

Many greensmowers have interchangeable vertical mower units. A greensmower can be converted to a vertical mower in half an hour by installing these units in place of the grass-cutting blades. The grass catchers can be used to collect the thatch. Another piece of equipment that removes thatch is a power rake. The flexible wire tines, similar to those on a leaf rake, are able to pull small amounts of thatch from the surface of the turf. A power rake is ineffective if the thatch layer is deeper than 0.5 inch (1.3 centimeters).

Topdressing is a very effective method of controlling thatch. A thin layer of soil is spread over a turf area and then dragged or brushed into the thatch. Adding topdressing to the thatch layer is very beneficial because the soil particles dilute the thatch and modify its adverse effects. It also creates a favorable environment for the development of microorganisms by improving water retention and other conditions. Topdressing results in increased microbial activity and increased thatch decomposition.

Topdressing is a more effective method of thatch control than vertical mowing. However, topdressing is expensive, and its use is limited primarily to putting greens. The soil is applied with a topdresser (Figure 18-27). This machine can be calibrated to apply topdressing material at various thicknesses. A metal drag mat or piece of chainlink fence can be attached to the rear of the topdresser. The drag mat works the soil into the thatch layer. The brush attachment on a greensmower can also be used to accomplish this. Another method is to brush the material in by hand with brooms. If the soil is dry, it can be applied with a fertilizer spreader.

The topdressing material should be identical to the topsoil beneath the thatch if the topsoil has favorable characteristics. The material commonly used contains 80 percent or more sand because most greens have a very sandy topsoil. Topdressing with pure sand has become popular because it results in a superior putting surface. The sand should be fine enough to be easily worked into the thatch, but not so fine that it can seal the surface and impede air and water movement.

The depth of the topdressing layer depends upon the frequency of topdressing and the thickness of the thatch. A typical program for greens is to

FIGURE 18-27
A topdresser. The inset shows the topdressing material being released from the hopper. *(Courtesy of OMC Lincoln)*

spread 0.03 to 0.06 inch (0.76 to 1.5 millimeters) of soil each application if the green is topdressed 12 or more times a season. When topdressing is performed 3 or 4 times annually, the rate is increased to 0.13 to 0.25 inch (3.2 to 6.4 millimeters) (Table 18-2). It is more effective to topdress at least 8 times a season, but frequency may be limited by budget and labor restrictions.

Where intensive topdressing programs are used, the thatch is covered with topdressing material every two to three weeks.

TABLE 18-2 Soil Volumes Required to Topdress 1,000 Ft² (93 M²) to Various Depths

Depth		Soil Volume		
in	mm	ft^3	yd^3	m^3
1/32 (0.03)	0.8	2.6	0.1	0.07
1/16 (0.06)	1.6	5.2	0.2	0.15
1/8 (0.125)	3.2	10.4	0.4	0.3
1/4 (0.25)	6.4	21	0.8	0.6
1/2 (0.5)	12.7	42	1.5	1.2

The golf course superintendent should not topdress more frequently than necessary. If the thatch layer is completely decomposed, the surface of the green becomes too firm. This hardness causes golf balls to bounce too much and not hold on the green.

Topdressing can be performed whenever the grass is actively growing. It often follows core cultivation. The sandy topdressing media will fill in the aeration holes if it is worked into the turf. These pockets of sand will increase air and water infiltration. This can be particularly helpful on native-soil greens, where drainage and aeration are inadequate. The plugs of undesirable soil should be removed from the surface of the green before topdressing. After a few years of incorporating sand into cultivation holes the characteristics of the growing media may improve significantly. However, if the topsoil has an undesirable texture, the best solution is to remove all of it and replace it with a sand that is appropriate for greens. This, of course, is expensive.

Selecting a topdressing material requires careful consideration. Choosing a topdressing source that is not physically compatible with the present media can lead to disaster. For example, placing a fine-textured layer over coarser particles will create a perched water table at the surface of the green. Reduced water movement and other serious problems will result.

Some people prefer straight sand topdressing, while others like to add peat or a small amount of soil to the sand. There are also great differences between the size and shape of sand particles. The turf manager must send samples of prospective topdressing materials to a soil testing laboratory to determine which one is most compatible with the growing media already present.

There is another benefit from topdressing besides thatch decomposition. When a green is matted after topdressing, the sand or soil is not spread equally over it. A greater amount of topdressing tends to be pulled into low spots, which reduces surface irregularities and levels the green. A smooth surface enables a golfer to putt accurately—the ball runs truer. Topdressing also occurs on athletic fields for leveling purposes. As much as 1 to 2 inches (2.5 to 5.1 centimeters) of soil or sand is spread and then drag-matted to smooth the surface. Large topdressing machines are manufactured for sports field use.

Core cultivation has a topdressing effect if the plugs are broken up with a mat or verticutter and the soil is mixed into the thatch layer. Working the soil particles from the cores into the thatch is a more effective method of thatch reduction than vertical mowing. One type of core cultivator grinds the plugs into smaller particles and spreads this soil as it aerates. Some thatch is also removed when the plugs are extracted. This amount is significant when the holes are large and spaced close together. The percentage of thatch removed increases even more if the site is core-cultivated in several directions. Core cultivation also produces soil conditions that are more favorable for thatch-decomposing organisms.

Lime will stimulate thatch decomposition if the pH of the thatch is too acid for microorganisms. Normally lime is applied to a soil only every few years. However, frequent, light lime applications may be most helpful because calcium and magnesium move through the thatch layer quickly. The pH of topdressing material should be in the 6.0 to 7.0 range.

Some companies sell thatch-decomposing microorganisms. Spreading a bag of these organisms on a turf area is supposed to result in thatch reduction. Studies have shown mixed results. Some products have no significant effect on the thatch decomposition rate, but a few have shown promise.

Research has shown that the use of wetting agents can improve water infiltration and movement in thatchy soils.

SELF-EVALUATION

1. A common turfgrass disease in the shade is ___________.
2. Some turfgrass species and varieties can adapt to ___________ shade.
3. As a general rule, the mower blade should be raised ___________ inch(es) when cutting grass growing in shaded areas.
4. ___________ or ___________ are often found growing on wet, shady sites.
5. Soil compaction is the result of intense ___________.
6. When soil is compacted there is a reduction in the number of ___________.
7. The machine that removes plugs of soil is called a ___________.
8. A ___________ is used to press uprooted grass plants back into contact with the soil.
9. A layer of undecomposed or partially decomposed organic matter is called ___________.
10. The compound in plant tissue that is very resistant to decay is ___________.
11. Microorganisms and ___________ help to decompose plant debris in the soil.
12. Optimal soil pH range for thatch decomposition is ___________.
13. Perhaps the major reason for thatch accumulation is overapplications of ___________.
14. The knife blades on a ___________ rip thatch out of the turf.
15. Spreading a thin layer of soil on a turf area is called ___________.
16. Discuss why thatch accumulates.
17. What techniques can be used to improve the quality of turf growing in the shade?
18. How would you construct an athletic complex to minimize compaction problems?

CHAPTER

19

Improving Unsatisfactory Turf

OBJECTIVES

After studying this chapter, the student should be able to

- Prepare and use a turf evaluation form
- Explain how a turfgrass improvement program is developed
- Discuss methods of turf renovation
- Describe the reestablishment and reconstruction of turf areas

INTRODUCTION

When something goes wrong, improving unsatisfactory turf becomes an important part of a turfgrass manager's job. If a turf area does not satisfy the expectations, needs, or standards of the turfgrass manager, it is of unsatisfactory quality. The level of quality that is desired varies from situation to situation. For example, a golf course superintendent may judge a putting green to be unsatisfactory if its quality is less than excellent, while a lawn that is only in fair condition is often acceptable to the average homeowner. A grounds superintendent who manages a professional sports stadium and has a large maintenance budget would be expected to produce an outstanding turf. However, a groundskeeper who maintains the athletic fields at a high school and has a limited budget would be satisfied with a turf of considerably lower quality.

The manager uses quality indicators such as density, uniformity, and color to evaluate the turf areas under his or her supervision. When the quality is judged to be unacceptable, the manager must determine why the turf has not attained a desirable quality. If the turf was satisfactory but has deteriorated, the reasons for the decline must be discovered. Attempts to improve turf quality are usually unsuccessful without an understanding of why the turfgrass is performing unsatisfactorily. Turfgrass recovery is unlikely if the problems are not corrected (Figure 19-1).

FIGURE 19-1

Grass does not survive at this location because of severe compaction resulting from foot traffic. Replanting the site is probably a wasted effort unless the traffic problem can be solved. *(Photo by Brian Yacur)*

Many lawn care companies have developed evaluation forms that help their employees identify the reasons for poor turfgrass performance. The forms are composed of a list of common turfgrass problems. The evaluator can use this type of diagnostic checklist as a guide when examining turf. After the survey is completed and the problems have been identified, the next step is to suggest corrective measures. These recommendations are written on the form. A written record of problems and solutions can be an invaluable aid to the turfgrass manger.

A SAMPLE TURFGRASS EVALUATION FORM

The survey shown in Figure 19-2 is a typical example of a turfgrass evaluation form. Noted in parentheses on this sample form are examples of recommendations that might be suggested by the evaluator. The form includes a listing of the major turfgrasses, weeds, insects, and diseases found throughout the United States. Many lawn care companies prefer to use a form that lists only the turfgrass species and pests common to the geographical area in which the survey is conducted.

TURFGRASS EVALUATION

TURF LOCATION: (identification of the turf surveyed—Tom Smith's lawn, the 9th fairway, Wiscasset Town Park, etc.)

SIZE: (the area of the turf measured in square feet, acres, square meters, or hectares)

ESTABLISHMENT DATE: (when the turfgrass was established)

DATE OF SURVEY:

SURVEY PERFORMED BY:

1. **DOMINANT TURFGRASSES**

   If possible, estimate the approximate percentage (%) of each species that is found in the turf area.

   _____ Kentucky bluegrass
   _____ Bentgrass
   _____ Fine fescue
   _____ Perennial ryegrass
   _____ *Poa trivialis* (rough bluegrass)
   _____ Tall fescue
   _____ Wheatgrass
   _____ Bermudagrass
   _____ Zoysiagrass
   _____ Carpetgrass
   _____ Bahiagrass
   _____ St. Augustinegrass
   _____ Centipedegrass
   _____ Buffalograss
   _____ Percentage (%) of turf area covered with desirable turfgrasses

2. **TURF DENSITY**

   _____ Dense _____ Medium _____ Thin

3. **SOIL TEST RESULTS**

   _____ pH _____ Salinity _____ Texture
   _____ Phosphorus _____ Potassium

   Recommendations: (lb of lime or sulfur needed per M,* etc.)

4. **FERTILIZER PROGRAM**

   Grade (analysis) of fertilizer use _____

   % of slowly available nitrogen in fertilizer _____

   Application dates and rates of nitrogen (lb N/M) applied each time _____

   Total lb of nitrogen (lb N/M) applied per year _____

   Recommendations: (apply more fertilizer, use a different analysis, change application dates, etc.)

5. **SHADE**

   _____ Heavy _____ Moderate _____ Full sun

   Recommendations: (prune trees, mow higher, rake up leaves more quickly, etc.)

*M is the abbreviation for 1,000 ft^2.

FIGURE 19-2
Sample turfgrass evaluation form.

6. THATCH

Is the thatch too thick? ______ Yes ______ No

Depth of the thatch layer ______

Recommendations: (core-cultivate, topdress, decrease nitrogen rates, etc.)

7. MOWING PROBLEMS

______ Cutting height too low

______ Cutting height too high

______ Grass not mowed frequently enough—too much leaf tissue removed at each mowing

______ Dull blade

Recommendations: (sharpen blade, cut grass to a higher height, mow more often, etc.)

8. SOIL COMPACTION

Degree of compaction

______ Severe ______ Moderate ______ Not compacted

Recommendations: (reduce traffic, core-cultivate, etc.)

9. WATER PROBLEMS

Soil drainage too slow ______ Poor water retention ______

Incorrect irrigation practices

______ Watering too frequently

______ Not watering often enough

______ Individual applications too light

______ Individual applications too heavy

______ Water source unsatisfactory

Recommendations: (water heavily once a week, apply more water when irrigating, install a tile drainage system, etc.)

10. CONTROLLABLE BROADLEAF WEEDS

______ Dandelion

______ Broadleaf plantain

______ Narrowleaf (buckhorn) plantain

______ Common chickweed

______ Mouse-ear (hairy) chickweed

______ Clover

______ Yellow rocket

______ Black medic

______ Chicory

______ Carpetweed

______ Curly dock

______ Others:

Recommendations: (herbicide to use, best time to apply, it shouldn't rain for 24 hours after application, etc.)

FIGURE 19-2
(continued)

11. DIFFICULT-TO-CONTROL BROADLEAF WEEDS

______ Knotweed
______ Ground ivy (creeping charlie)
______ Veronica (speedwell)
______ Red sorrel
______ Spurge
______ Oxalis
______ Mugwort
______ Violet
______ Oxeye daisy
______ Mallow
______ Pinappleweed
______ Wild garlic
______ Wild onion

______ Others:

Recommendations: (herbicide, rate, reapplication often necessary, etc.)

12. CONTROLLABLE GRASSY WEEDS OR SEDGES

______ Crabgrass
______ *Poa annua* (annual bluegrass)
______ Goosegrass
______ Foxtail
______ Barnyardgrass
______ Nutsedge

______ Others:

Recommendations: (herbicide, rate, date of application, don't disturb soil surface after application, etc.)

13. NONCONTROLLABLE PERENNIAL GRASSY WEEDS (weeds that cannot be controlled with selective herbicides)

______ Bermudagrass
______ Bentgrass
______ Tall fescue
______ Nimblewill
______ Quackgrass
______ Dallisgrass
______ Johnsongrass
______ Kikuyugrass
______ Bromegrass
______ Orchardgrass

______ Others:

______ % of turf area covered with noncontrollable grassy weeds

Recommendations: (spot treat with a nonselective herbicide, be sure not to apply to desirable turfgrasses, dig out clumps of weedy grasses, etc.)

14. INSECT PROBLEMS

______ Annual bluegrass weevil
______ Ant
______ Aphid
______ Armyworn
______ Billbug
______ Black turfgrass ataenius
______ Chinch bug
______ Cutworm
______ Ground pearl
______ Grub
______ Leafhopper
______ Mite
______ Mole cricket
______ Periodical cicada
______ Scale
______ Sod webworm
______ Wireworm

______ Others:

Recommendations: (insecticide, rate, when to apply, water in after application, wear rubber gloves, etc.)

FIGURE 19-2
(continued)

15. DISEASE PROBLEMS

_____ Anthracnose
_____ Brown patch
_____ Copper spot
_____ Dollar spot
_____ Fairy ring
_____ Gray leaf spot
_____ Helminthosporium leaf spot
_____ Necrotic ring spot
_____ Nematode
_____ Powdery Mildew
_____ Pythium blight
_____ Red thread
_____ Rust
_____ SAD
_____ Slime mold
_____ Smut
_____ Snow mold-gray
_____ Snow mold-pink
_____ Spring dead spot
_____ Summer patch
_____ Others:

Recommendations: (apply a fungicide, decrease nitrogen fertilization, reseed with resistant cultivars, etc.)

16. OTHER PROBLEMS (examples of other problems not already listed in the evaluation form such as shallow soil, winter injury, moss, pesticide injury, burn caused by dog urine, etc.)

Recommendations: (add topsoil, rake out moss, keep dogs off lawn, etc.)

17. SUMMARY OF PROBLEMS

18. SUMMARY OF RECOMMENDATIONS FOR IMPROVEMENT

FIGURE 19-2 *(continued)*

IMPROVING TURF QUALITY

The turfgrass specialist must answer three questions when developing a turf improvement program:

1. What are the reasons for the unsatisfactory condition of the turf?
2. What are the solutions to these problems?
3. How extensive is the damage to the turfgrass?

A careful examination of the turf area and an evaluation of the maintenance program should identify the reasons for unsatisfactory turfgrass quality. In many cases, the decline is the result of improper management. Either the maintenance practices are performed incorrectly or a necessary practice is not performed at all. For example, soil fertility levels or pH may be too low; therefore, the application of fertilizer or lime may be enough to restore turf quality. Cutting with a dull blade, removing too much leaf tissue at each mowing, or

FIGURE 19-3

Some turf problems can be solved easily. An application of a broadleaf herbicide and fertilizer will restore the quality of this lawn.

scalping are common problems that are corrected by improving the mowing program. Herbicides, fungicides, and insecticides will help to control weeds, diseases, and insects. If thatch or compaction is a problem, the addition of core cultivation or topdressing to the maintenance program may be the solution.

When the problems can be corrected by relatively simple changes in the maintenance program, the next step is to assess the extent of the damage to the turf. If the turf is only slightly unsatisfactory, acceptable quality may be restored in a short time by correcting the maintenance program. A minimally injured turf that is fairly dense and largely composed of desirable turfgrasses may recover in a few weeks or months (Figure 19-3). This is especially true when the turfgrasses present produce stolons or rhizomes that can spread quickly. Examples of such recuperative grasses are St. Augustinegrass, centipedegrass, bermudagrass, creeping bentgrass, and vigorous Kentucky bluegrass cultivars. Bare spots smaller than a dinner are soon filled in and covered. Broadleaf weeds and annual grasses that have invaded the weakened turf can be eradicated with herbicides.

An area that has suffered a more severe loss of quality will recover eventually if the reasons for its deterioration are remedied. However, when the turf is very thin and comprised of large or numerous bare spots and patches of weeds, a year or more may be required before complete recovery occurs. A turf that is primarily composed of nonspreading species such as perennial ryegrass, tall fescue, or fine fescue recuperates very slowly. The turfgrass manager usually does not want to wait a long time for a satisfactory quality to develop. He or she has other alternatives besides patience. When the turf will not be readily improved by routine maintenance practices, it can be repaired by introducing new seeds, sprigs, stolons, plugs, or sod to the area.

TOTAL REESTABLISHMENT

If the turf quality is poor, total reestablishment may be necessary. The entire area is replanted. This is the best solution when 50 percent or less of the area is covered with desirable turfgrasses, or large populations of noncontrollable perennial grassy weeds are present. This type of improvement should be considered when a thick thatch layer has developed or soil texture or depth is unsatisfactory. Sometimes the turfgrass species and cultivars are no longer adaptable to the site and should be replaced with newer, better selections. For example, older cultivars may have lost their ability to resist certain diseases.

Total reestablishment requires greater labor and is more costly than improving the existing turfgrasses. However, it should be seriously considered when the turf is so unsatisfactory that starting over with an entirely new stand of turfgrass seems more desirable than trying to rejuvenate the old stand.

Generally, the first step in reestablishment is to kill the turfgrass and weeds with a nonselective herbicide (Figure 19-4). Glyphosate (Roundup) is commonly used because the area can be replanted in a week or less after it is applied. This chemical, when it reaches the soil, becomes tightly bound to soil particles. This immediate inactivation allows the turf manager to replant soon after using the material without any injury to the new plants. Glyphosate will not be taken up by their roots. Other herbicides often require a longer waiting period because they remain active in the soil and retain their ability to kill plants for a month or more.

Fumigants can be used to eliminate weeds and pests in the soil.

The next step is to rototill, disk, or plow the entire area and prepare a new seedbed. When glyphosate is used, the area should not be tilled until a week after application. This herbicide is a systemic—it enters the leaves and is then translocated throughout the entire plant. This movement occurs in the vascu-

FIGURE 19-4

A nonselective herbicide is used to kill all of the plants growing in the turf area that is to be reestablished.

lar system. When cultivation practices are performed immediately after glyphosate is sprayed, the vascular tissue of these plants will be disrupted. If the transport system is damaged, the chemical cannot be translocated to all of the plant tissue. The surviving tissue may be able to produce new shoots. Many perennials, for example, can grow back if their roots, rhizomes, or stolons are not killed. Waiting a week before filling, however, ensures that glyphosate is transported to all parts of the plant.

In some cases, the old vegetation and thatch are removed with a sod cutter before working the soil. After the soil is prepared, the area is seeded or established vegetatively by installing sod or planting plugs or sprigs (pieces of stolons and rhizomes). The last step is to keep the site adequately watered until the turfgrass is established.

COMPLETE RENOVATION

Complete renovation is another method of replanting the entire site. It is cheaper and less time-consuming than the method of reestablishment described previously because it does not require working or preparing the soil. However, the results are not always as satisfactory.

The first step in complete renovation is to spray the site with glyphosate. Then, after waiting seven days, an overseeder is used to replant the areas. These machines slice narrow grooves in the turf and insert seeds into the slits (Figure 19-5). A sprig planter can also be used.

The overseeder, or disk seeder, has vertical mowing blades mounted on a horizontal shaft at the front of the machine. As the shaft rotates rapidly the knives cut narrow grooves in the vegetation and soil. Disks at the rear of the unit are in line with the knives at the front. A seed box is located above the disks. The seed drops down through metal tubes and falls into the slits that are held open by the disks (Figure 19-6). The overseeder places seed directly into the soil and ensures good soil–seed contact, which is necessary for germination.

FIGURE 19-5
A disk-type seeder which is used to renovate turf areas. *(Courtesy of Jacobsen Division of Textron, Inc.)*

FIGURE 19-6

A closeup view of a disk seeder or overseeder showing the seed box, tubes, and disks.

FIGURE 19-7

Seedlings emerging three weeks after overseeding a bare area.

These machines can be calibrated to seed at various rates. the grooves are usually 2–3 inches (5.1–7.6 centimeters) apart so that the new plants come up in lines or strips (Figure 19-7). Eventually the turfgrass spreads from the grooves and covers the entire area. Glyphosate takes two or three weeks to kill the old vegetation. The yellow turf gradually turns green as new plants emerge and fill in. This transition can be hastened by overseeding in several directions.

The thatch removed by the vertical mower blades should be raked up after replanting. The site is then watered regularly until the turfgrass is well established. Complete renovation is not usually successful if the site is heavily thatched. The thatch layer is often too dry for good seedling or sprig growth. To obtain water, seeds and sprigs need to be in contact with the soil.

RECONSTRUCTION

Reconstruction is a third method of reestablishment. It involves significantly modifying or rebuilding the soil. Reconstruction may be required when the turfgrass has performed unsatisfactorily because of poor soil texture or shallow soil depth. When serious soil problems exist, these conditions should be improved before replanting. If they are not, the new turfgrass may deteriorate rapidly.

Topsoil should be added when the present soil is too shallow to support a healthy, dense turfgrass population. The soil should be completely changed or at least modified if the texture is too clayey or sandy. Compacting and poor drainage are examples of problems associated with a soil that has a high clay content. A very sandy soil may not be able to supply the turfgrass with adequate amounts of water. Specific recommendations for rebuilding a soil are discussed in Chapter 6.

PARTIAL RENOVATION

In many cases, an unsatisfactory turf requires only partial replanting. This is the best solution when a large population of turfgrass plants has survived the decline, but the introduction of new seeds or sprigs is necessary to speed up recovery. If the majority of the area is covered with desirable grasses, there is usually no need to reestablish the entire site. It is inefficient to destroy the remaining turfgrass plants with herbicides or by tillage when the population is large enough to be worth saving.

Partial renovation is defined as reseeding or replanting a site without disturbing the existing turfgrass. Seeds, sprigs, or plugs are planted in the areas with sparse turf coverage, either by a machine or by hand. As the old turfgrass plants recuperate and spread, new plants also appear and help to thicken the turf. The result is increased density and improved quality. Partial renovation differs from complete renovation because the surviving turfgrass is not killed with glyphosate.

If broadleaf weeds are a problem the area should be treated with a broadleaf weed killer before replanting. The turf manager should then wait several weeks before seeding or sprigging. Broadleaf herbicides have little effect on mature grasses, but they can injure turfgrass seedlings. After approximately a month, the toxic residues in the soil are broken down or inactivated. Herbicide application is not necessary if only a small number of broadleaf weeds are present. Clumps of perennial grassy weeds can be spot treated with a nonselective herbicide.

The grass should be cut to a low height before renovation. A disk seeder or sprig planter is then used to plant seed or sprigs into the thin turf. The disk seeder is widely used for renovation because it places seeds directly into the soil. As previously explained, contact with the soil is required to ensure germination.

If a seeding machine is not available, the site should be core-cultivated five or six times or sliced with a vertical mower several times (Figure 19-8). Performing both practices is very desirable. Verticutting following core cultivation will cut up the plugs extracted from the soil. They can also be broken up with a drag mat. The holes made by coring and the grooves produced by vertical mowing will enable seeds or sprigs to be in contact with the soil. Seeding into core cultivation holes usually works well, and some turf managers prefer this technique to the use of a disk seeder.

The area is seeded at a heavy rate, often five times greater than the normal seeding rate. Seed is then raked or drag matted into the holes or grooves. It is best to insert sprigs into the holes or grooves by hand. After planting seed or sprigs, a light rolling helps to ensure good contact with the soil. Topdressing, if possible, is also recommended. The final step is to water frequently until the new turfgrass plants are well established.

The key to successful renovation is getting the seeds or sprigs into the soil. Small bare spots can be dug or raked up before planting. Spreading seed or sprigs on top of the thatch or soil and leaving them there is the only method that does not work.

FIGURE 19-8
A vertical mower.
(Courtesy of Jacobsen Division of Textron, Inc.)

Strip sodding is a method used with some warm season species such as zoysiagrass. Strips of the existing, weak turf are removed and replaced with pieces of desirable sod.

The best time to renovate or reestablish cool season turfgrasses is in the late summer or early autumn. Warm season grasses should be replanted in the spring or early summer.

SELF-EVALUATION

1. Spraying an area with a nonselective herbicide and then planting into the dead or dying vegetation with a renovating machine is called ___________.
2. Why is the turf manager able to seed or sprig shortly after applying glyphosate?
3. Why is reconstruction the most expensive method of reestablishing a turf area?
4. Explain how an overseeder or disk seeder works.
5. Why is it essential that seeds or sprigs be in contact with the soil?
6. When would partial renovation be preferable to total reestablishment?
7. What are the major reasons why a turf area declines?

CHAPTER

20

Golf Course Management

OBJECTIVES

After studying this chapter, the student should be able to

- Understand the basic design and layout features of a golf course
- Discuss the responsibilities of the superintendent and other golf course employees
- Describe the maintenance practices performed on greens, tees, fairways, sand traps (bunkers), and roughs
- Identify other miscellaneous work activities performed by golf course employees

INTRODUCTION

Golf course management is an important part of the turfgrass industry. The exact origins of golf are unclear, but the game is known to have been popular in Scotland as long ago as the early 1400s. Today, there are more than 14,000 golf courses and over 25 million golfers in the United States.

A golfer plays 18 holes in a round of golf. Consequently, golf courses consist of 18 holes (Figure 20-1). On smaller courses that have only 9 holes, the golfer completes a round by playing the 9 holes twice. Some large golf facilities have two or more 18-hole courses.

The area of an 18-hole golf course ranges from 120 to 200 acres (49 to 81 hectares). A golf hole is composed of four major areas—tee, fairway, rough and green (Figure 20-2). Play on each hole begins at the tee and ends when the ball is putted into a hole on the green. The tee and green are connected by the fairway, which is generally 125 to 450 yards (137 to 492 meters) long and 30 to 90 yards (33 to 94 meters) wide. The number of shots a golfer must hit to reach the green depends largely upon the length of the fairway (Table 20-1).

The rough is located at the sides of the fairway and usually receives consid-

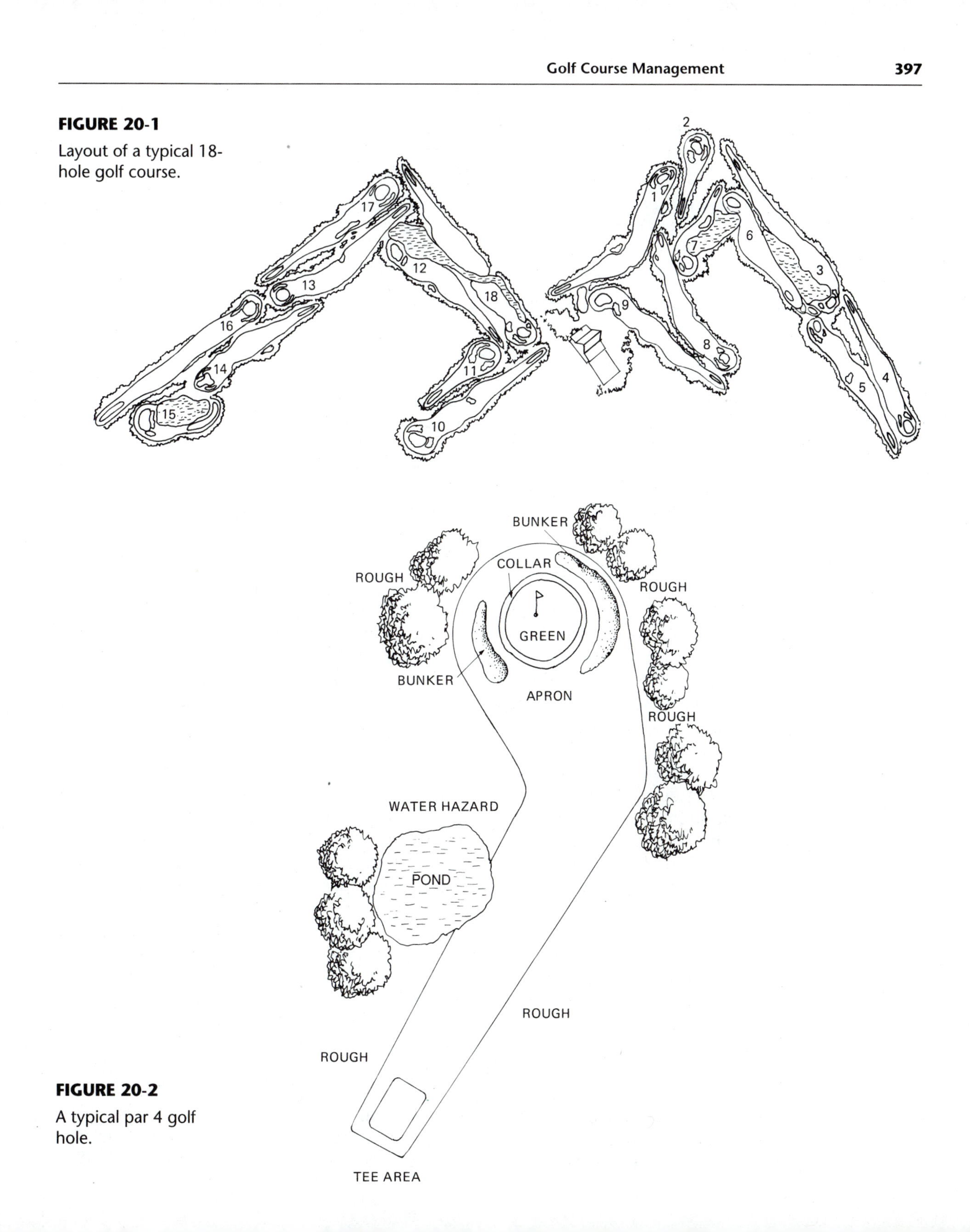

FIGURE 20-1

Layout of a typical 18-hole golf course.

FIGURE 20-2

A typical par 4 golf hole.

TABLE 20-1 Maximum and Minimum Par Ratings*

	Yardage (maximum and minimum)	
Par Rating	**Men**	**Women**
3	Up to 250	Up to 210
4	251–470	211–400
5	More than 470	401–575
6		More than 575

*Par is the number of strokes (shots) that are considered necessary for an expert player to complete a hole. A golfer is expected to need two putts on the green and one (par 3 hole), two (par 4 hole), or three (par 5 hole) strokes to reach the green. The number of strokes required to reach the green depends primarily upon the length of the hole. The United States Golf Association has developed the yardage guidelines given above for establishing par ratings.

erably less maintenance than the fairway. The purpose of the rough is to penalize golfers for inaccurate shots. It is more difficult to play out of the rough than it is to hit a shot from the fairway. This is because the rough is mowed higher than the fairway and has poorer playing surfaces. It may contain trees.

The strategic placement of various obstacles on a hole also increases the difficulty of the game. Bunkers or traps filled with sand, clumps of trees, and water hazards make the hole more challenging (Figure 20-3).

Most golf courses have a practice green where golfers can practice their putting. Some courses also have a driving range where golfers can hit practice drives and iron shots. Golf course equipment and supplies are stored in a maintenance building. Equipment maintenance and repair are performed in this building, and the golf course superintendent's office is located there as well.

The clubhouse varies in size and complexity from course to course. At some golf clubs it is a large facility and contains a restaurant, a lounge, locker rooms, showers, and meeting rooms. The pro shop is usually located in the clubhouse. At the pro shop the golf professional rents or sells golf equipment and collects greens fees—the money that is paid by golfers to play the course. Besides signing up to play, golfers can also rent carts and arrange for golf lessons at the pro shop.

The number of golf course workers employed by the club depends upon the size and the maintenance budget of the course. The golf course superintendent is the manager in charge of maintaining the golf course. His or her responsibilities include purchasing equipment and supplies, developing the maintenance program, and directing the activities of the other workers. The superintendent must possess both turfgrass and business management skills.

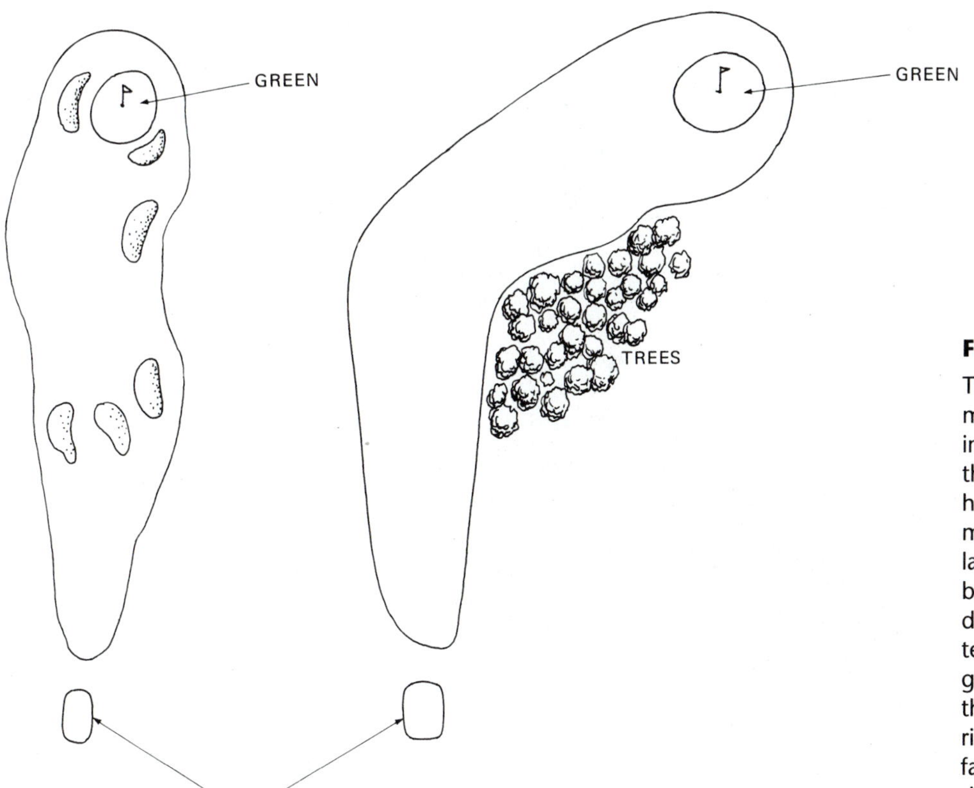

FIGURE 20-3

The hole on the left is more challenging and interesting because of the six bunkers. The hole on the right is more difficult because a large clump of trees has been left standing directly between the tee and the green. In golfing terminology, this hole is a "dogleg right" because the fairway bends to the right.

The assistant superintendent helps the superintendent implement the maintenance program and coordinate work activities. Foremen are responsible for the direct supervision of work crews out on the course. Many clubs employ a mechanic to maintain and repair the many pieces of equipment used on the golf course (Figure 20-4). There may also be an irrigation specialist who operates, maintains, and repairs the irrigation system. The majority of the employees, often called the green staff or greensworkers, perform the daily maintenance operations. Their major activity is to mow the greens, fairways, tees, and roughs. However, they do other jobs as well, such as irrigating or fertilizing. The greensworkers position provides valuable training for a person who plans to become a superintendent eventually.

Golf courses vary greatly in physical layout, design, size, and the amount of money available to operate the facility. On any course, however, golfers expect and demand a high level of turf quality. It is very difficult to play satisfying golf on unsatisfactory turf. For example, putting accuracy can be disrupted by minor imperfections on the surface of a green. Consequently, golf courses receive a higher level of maintenance than other types of turf areas. Obviously, a superintendent with a small maintenance budget and a few employees cannot produce the near-perfect turf found at some of this country's

FIGURE 20-4
Keeping the many pieces of equipment in operating condition is a full-time job on larger golf courses. *(Photo by Brian Yacur)*

outstanding golf clubs, but he or she is expected to maintain the course in as good a condition as possible.

Golf course owners and members are continually looking for ways to reduce maintenance expenses. However, any significant reduction in turf quality is unacceptable to most American golfers because they have become accustomed to superior playing conditions. Some methods of lowering maintenance costs without seriously affecting golfing quality have been developed. One example is contour mowing. The idea behind contour mowing is to decrease fairway acreage by simply mowing a smaller area at fairway height. This is accomplished by beginning the closely mowed fairway farther from the tee and by narrowing it in places (Figure 20-5). The result is an increase in the size of rough and "waste" areas, which are less expensive to maintain than fairways.

The remainder of this chapter is a discussion of the maintenance practices performed on golf courses. To keep a curse in satisfactory condition, many different tasks must be completed.

GREENS

Golfers judge the quality of a golf course primarily by the condition of its greens. Much of the actual play and many of the most important shots occur on the green. A superintendent restricted by a small maintenance budget will

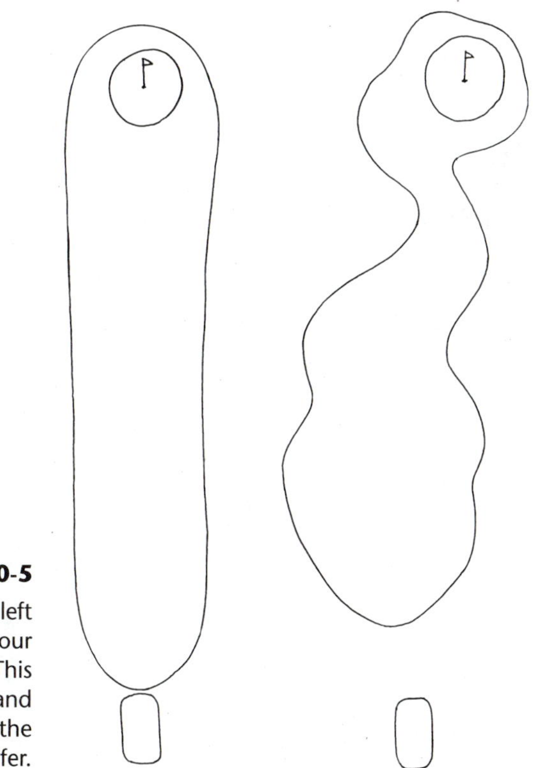

FIGURE 20-5

The size of the large fairway on the left can be reduced by using the contour mowing pattern shown on the right. This results in lower maintenance costs and provides a more challenging hole for the golfer.

attempt to keep the greens in excellent shape and give lower priority to the case of fairways and other turf areas.

A properly maintained green has a smooth, level putting surface, a dense turf cover, and a uniform and attractive appearance. Creeping bentgrass is generally the species used for greens in the cool season and transition zones, and it is popular in the cooler portions of the warm climatic region as well. Bermudagrass hybrids are the most common choices for warm season golf greens. These grasses have many outstanding characteristics, but the major reason for their selection is their ability to tolerate the extremely close mowing heights required for greens.

Greens are usually round or roughly circular and vary in size from 3,000 to 12,000 ft^2 (279 to 1,115 m^2). Most greens are in the 5,000 to 7,500 ft^2 (465 to 697 m^2) range. The total area of the 18 greens is usually 2 to 3 acres (0.8 to 1.2 hectares). The size of a green depends upon the difficulty of the approach shot. Normally, the longer par 3, 4, and 5 holes have larger greens than the shorter par 3, 4, and 5 holes.

The green is surrounded by the *collar,* a strip of grass 3 to 5 feet (0.9 to 1.5 meters) wide. It is closely mowed, enabling the golfer to putt from the collar onto the green. The area that lies immediately outside the collar is called the *apron* (Figure 20-6). A green may be ringed with sand bunkers.

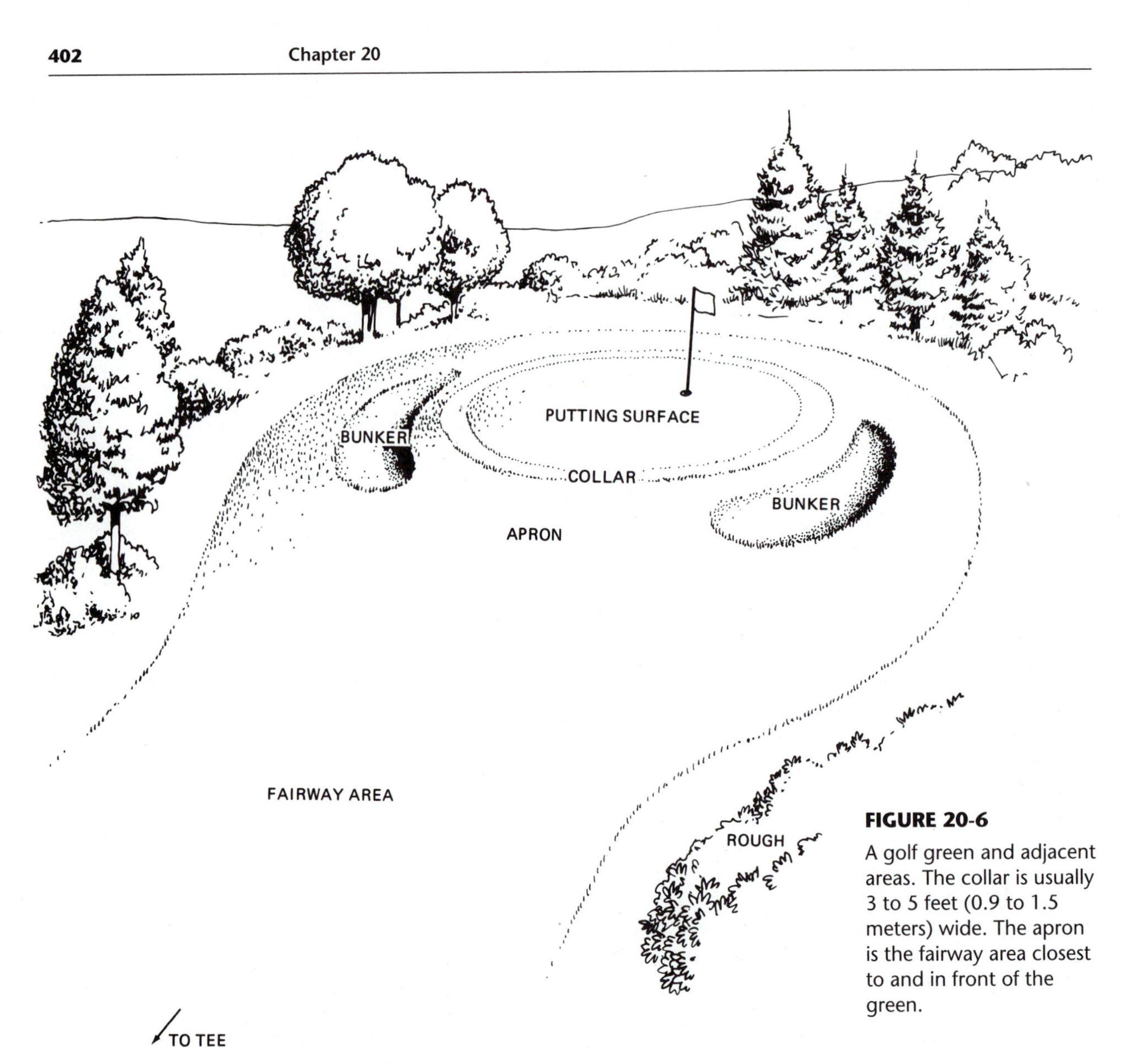

FIGURE 20-6
A golf green and adjacent areas. The collar is usually 3 to 5 feet (0.9 to 1.5 meters) wide. The apron is the fairway area closest to and in front of the green.

Construction

Golf course maintenance rather than construction is emphasized in this chapter. However, it is important to know how a green is built. Many greens are constructed on native soil, but it is preferable to use a sand base. Placing a deep layer of sand beneath a green allows excess water to drain rapidly from the surface. If puddles form on the green after a heavy rain or irrigation, play is interrupted. A green also receives a large amount of foot traffic. Compaction is less of a problem when a green is established on sand. Recommendations for the design of greens are available from the United States Golf Association (USGA) and other golf course construction specialists.

A method of greens construction developed by the USGA involves placing a 12-inch (0.3-meter) layer composed primarily of sand in the 0.25- to 1.0-millimeter-diameter range over a 4-inch (10-centimeter) layer of pea gravel or crushed stone. A 2-inch (5-centimeter) intermediate layer of coarse sand may be placed between the top layer of sand and gravel layers. The layer of gravel or coarse sand allows the formation of a *perched* water table. Water accumulates at the interface between the layers because it does not readily move from the finer-textured root zone sand into the underlying gravel or coarse-textured sand. The accumulated water can be absorbed by the roots of the grass plants. Eventually the water builds up to the point that it enters the coarser sand or gravel and rapidly drains away. A perched water table helps to compensate for the low water retention of the sand in the root zone.

A source of organic matter such as peat moss may be incorporated into the upper inches of sand to increase nutrient- and water-holding capacity. Uniform mixing is critical. Mixing should be performed off-site. A subsurface drainage system is installed beneath the sand and gravel layers. Properly spaced drainage lines surrounded by pea gravel ensure the rapid removal of excess water. The lateral lines should not be more than 15 feet (5 meters) apart. The cross-section of a typical sand-based green is illustrated in Figure 20-7. Constructing this type of green is considerably more expensive than simply seeding into the native soil, but the quality that results is highly satisfying to both the superintendent and golfers. It is extremely important that all sands, organic materials, and gravels considered for use be evaluated by a soil testing laboratory to ensure that they will perform satisfactorily.

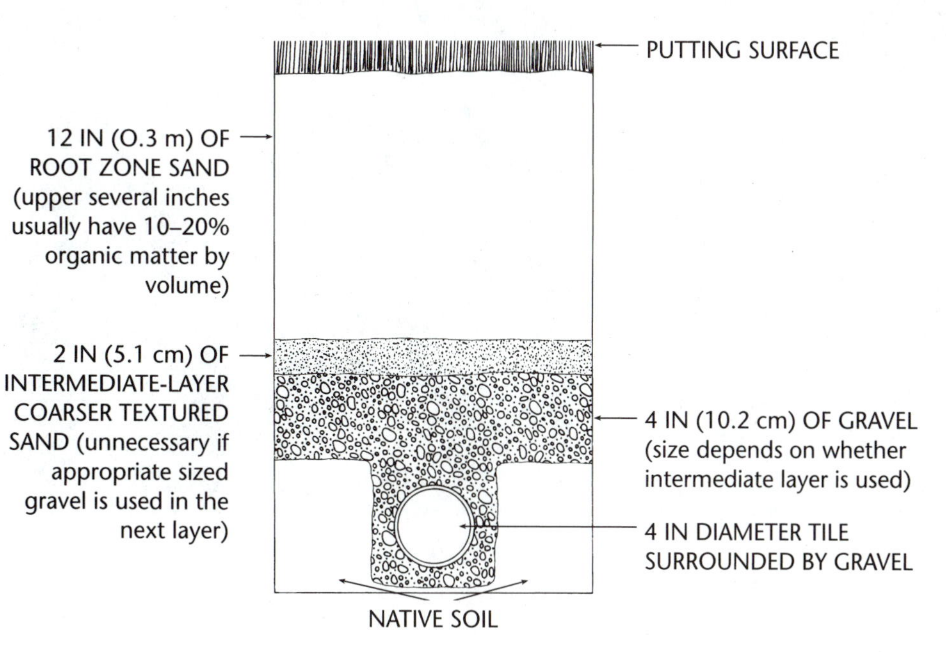

FIGURE 20-7

Cross-section of a USGA sand-based green.

Greens are usually elevated. Surface drainage is promoted by raising the green higher than the surrounding area. Elevating the green also increases its visibility to the golfers.

Maintenance

Mowing. Frequent mowing and an extremely close mowing height are necessary to maintain satisfactory putting conditions. Greens are normally cut daily, unless it is raining, to a height of 0.12 to 0.25 inch (3 to 6.4 millimeters). Mowing occurs early in the morning before many golfers are on the course.

Dew is removed from the green before mowing. The grass blades are normally covered with moisture early in the morning. Dew interferes with putting and encourages disease problems. The water can be removed by "whipping" the grass with a long, flexible bamboo or fiberglass pole. The pole is swept lightly over the surface, knocking the water from the leaves. Another method is to drag a long hose across the surface of a green. The lengths of hose can be pulled by hand or tied behind a turf vehicle. Turning on the irrigation system for a few minutes also results in dew removal. The heavy water droplets from the sprinklers wash the smaller dew droplets from the grass foliage.

Before mowing begins, the surface of the green should be checked care-

FIGURE 20-8

A walk-behind greensmower. *(Courtesy of Jacobsen Division of Textron, Inc.)*

FIGURE 20-9

A riding greensmower with brush attachment mounted on the front. *(Courtesy of Jacobsen Division of Textron, Inc.)*

fully for small stones or other objects that might be struck by the greensmower. Greens are cut with walk-behind or riding reel-type mowers. The walk-behind models have one cutting unit and usually a 21- to 22-inch (0.53- to 0.56-meter) cutting width (Figure 20-8). The riding greensmowers are often called triplex mowers because they have three cutting units (Figure 20-9), with a cutting width of 58 to 62 inches (1.5 to 1.6 meters). The use of triplex greensmowers results in reduced labor costs. For this reason, they are popular on golf courses with limited labor budgets. Greens can be cut faster with a triplex because of the greater cutting width provided by the three reel units. A riding mower is also more efficient because it can be driven from one green to another, whereas a walk-behind greensmower is normally transported on a trailer. However, the riding mower causes more wear and compaction than the walk-behind type. The walk-behind type produce the best quality.

A green is mowed by driving or walking the greensmower back and forth across it in straight paths. Turns are made off the green. The outermost few feet of the green are cut in a circular fashion (Figure 20-10). Ringing the outside of the green reduces the risk of scalping the collar by allowing the mower operator to raise or lower the reels farther away from the edge of the green.

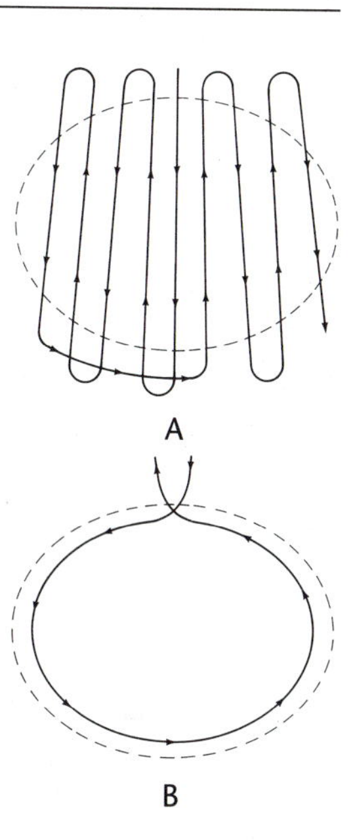

FIGURE 20-10

Recommended mowing pattern for a greensmower. The broken line represents the edge of the green. The circular swath (B) can be cut either first or last.

The mowing pattern should be varied to prevent the formation of grain. Four different directions should be used. Continually mowing the grass in the same direction causes it to lean in that direction. Grain can interfere with putting accuracy by deflecting a rolling ball from its true course. Brushing the green also helps to reduce grain formation by lifting leaves and stems. A brush or comb can be attached to the front of the greensmower. Groomers can be attached to the front of the reel. They have small knife blades that slice through stolons and stand-up leaves which are then cut by the reel blades and bedknife. This reduces grain and result, in a faster, truer putting surface.

Double-cutting is sometimes performed on days that a tournament is being played. Mowing the grass a second time at a right angle to the first direction of cut improves the uniformity of the putting surface and gives the green an attractive appearance.

Grass clippings are collected in baskets or catchers mounted on the mower. Clippings are dumped at a disposal site hidden from the view of golfers. At some courses the clippings are composted.

Irrigation. Frequent irrigation is required to ensure outstanding turf quality. Daily watering is common during the hotter, drier periods of the growing season. Frequent irrigation is necessary for several reasons. The very close mowing height results in shallow rooting and the plants have a reduced ability to retrieve water from the soil. Sand-based greens exhibit a limited water-holding capacity. For the grass to remain highly attractive, recuperative, and resistant to traffic, it must have a constant supply of water.

Irrigation occurs at night or early in the morning when the course is not in use. Heavy irrigation is usually inefficient because of the shallow-rootedness of the grass and low water retention of the sand. On a very hot, dry day the sprinklers may be turned on for a few minutes in the early afternoon. This practice is known as syringing. Syringing cools the plants and reduces water loss from the leaves.

Automatic, pop-up-type sprinkler systems are popular on golf courses because of their laborsaving features. Quick-coupler, manual-type systems are also used. They are less efficient than pop-ups because the sprinklers must be connected manually to water outlets before irrigating. Then, following irrigation, they must be removed. The least sophisticated method of irrigating involves connecting a hose to an outlet near the green and then moving the hose and attached sprinkler onto the green.

Cup Changing. The location of the hole on the green is changed several times per week. This is called cup changing because the hole contains a metal or plastic cup. The position of the cup is shifted regularly to prevent the grass around the hole from becoming badly worn. Moving the cup also adds variety and challenge to the game. The frequency of cup changing depends upon the intensity of traffic. It is a common practice to change the position of the cup every day. The new location should be relatively level and surrounded by dense, uniform turf, preferably 5 paces (approximately 15 feet), (4.6 meters) in from the edge of the collar and far enough from the previous location to allow the grass there to recover (Figure 20-11).

A cup cutter, which is similar to a post hold digger, is used to make the new hole. The cup is removed from the old hole with a cup puller or hook. The plug removed with the cup cutter to make the new hole is then placed in the old

FIGURE 20-11
Several factors must be considered before choosing the new cup location. *(Courtesy of Standard Golf Company)*

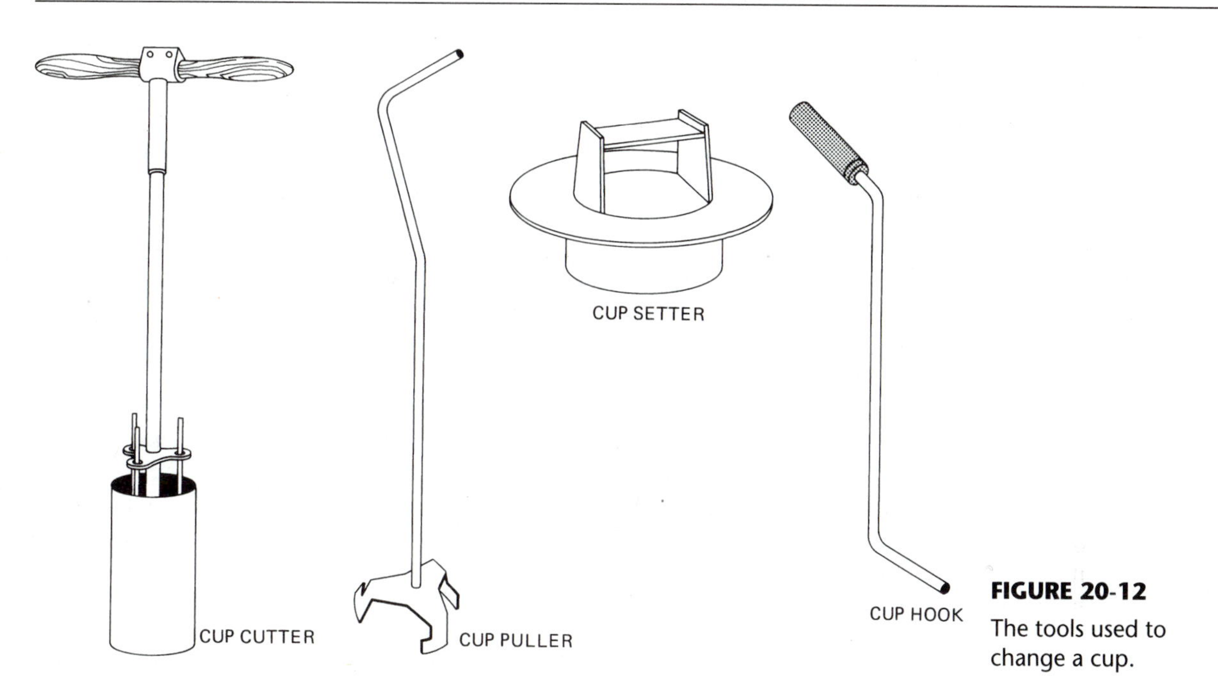

FIGURE 20-12 The tools used to change a cup.

hole to fill it. It is important that the plug be level with the surface of the green or bumpiness will result. The cup is set in the new hole with a cup setter (Figure 20-12). Stepping on the setter forces the cup into the hole and sets it at the correct depth. The top of the cup should be 1 inch (2.5 centimeters) beneath the putting surface. The plug inserted in the old hole is subject to drying out because its roots are severed by the cup cutter. Some greensworkers carry a watering can with them and soak the plug if the green is not to be irrigated soon after cup changing.

Fertilization. Greens are fertilized regularly to keep the grass attractive and vigorous. Fertilization rates have been moderated somewhat in recent years in an attempt to decrease disease occurrence and to increase the speed of the putting surface. The exact amount of fertilizer required depends upon the turfgrass species and cultivar present, the soil mixture, weather conditions, type of fertilizer used, and other variables.

The amount of phosphorus and potassium applied is based on soil test results. Complete fertilizers containing nitrogen, phosphorus, and potassium are necessary in a greens fertilization program; however, some applications often consist of nitrogen alone. It is important that the amount of nitrogen applied does not greatly exceed the amount of potassium. On a yearly basis, the number of pounds of nitrogen put down should not usually be more than double the potassium (K_2O) rate.

The frequency of fertilizer applications depends upon the same factors that affect monthly requirements. Applications may occur once every few weeks during the growing season, but longer intervals are common when slowly available nitrogen carriers are used. Nitrogen fertility should be reduced on bentgrass greens during the midsummer heat stress period and when the grass is hardening off (becoming cold tolerant) before winter dormancy. Bermudagrass greens should receive reduced rates when growth slows in the fall.

Soil test results will indicate whether micronutrients should be applied. Iron deficiencies are sometimes a problem on sand-based greens. Iron sulfate or chelated iron carriers are used to restore a healthy green color to the grass when an iron deficiency occurs. Lime or sulfur is applied to raise or lower the soil pH when adjustments are necessary. Greens soils are normally maintained in the 5.5 to 6.5 pH range.

Sulfur can be applied to lower soil pH, but on some greens the element contributes to the problem called black layer. Sulfur combines with hydrogen under anaerobic (without oxygen) conditions and forms hydrogen sulfide. Hydrogen sulfide is toxic to turf and reacts in the soil with metals such as iron to produce a layer of black gluelike, insoluble compounds. This black layer is usually only a few inches below the surface and is impenetrable by water, air, and roots.

Black layer is most common on sand greens that drain poorly. This can be the result of improper construction, layering resulting from different topdressing materials, compaction, or overwatering. The impact of a black layer can be reduced by deep core cultivation, cutting back on irrigation, using wetting agents to improve water movement, and increasing the height of cut to stimulate rooting.

Aeration. Core cultivation is performed at least once a year, and other types of aerification often occur several times (Figure 20-13). Besides benefiting the grass, core cultivation also improves golfing quality by helping to alleviate compaction. If a green becomes compacted, shots hit onto it will bounce too much when the ball strikes the hard surface. The green does not "hold" shots properly.

Vertical Mowing. Vertical mowing is performed to reduce grain and to stimulate the grasses to spread and increase density (Figure 20-14).

Topdressing. Topdressing is an essential part of a greens maintenance program. Mixing topdressing material into the thatch layer dilutes the thatch and increases the rate of decomposition. If this layer of underdecomposed organic matter is allowed to become too thick (more than 0.3 inch, or 0.76 centimeter), turf quality deteriorates. Playing quality is also adversely affected because the surface of a thatchy green is too soft. Another advantage of topdressing is its leveling effect. Accumulations of the sand or soil mix in low spots on the green result in a smoother surface.

Frequent, light topdressing applications are preferable. Many superintendents topdress their greens every three to four weeks. It is important that the topdressing material not form layers that impede water movement. Proper

FIGURE 20-13

Core cultivation. *(Courtesy of OMC Lincoln)*

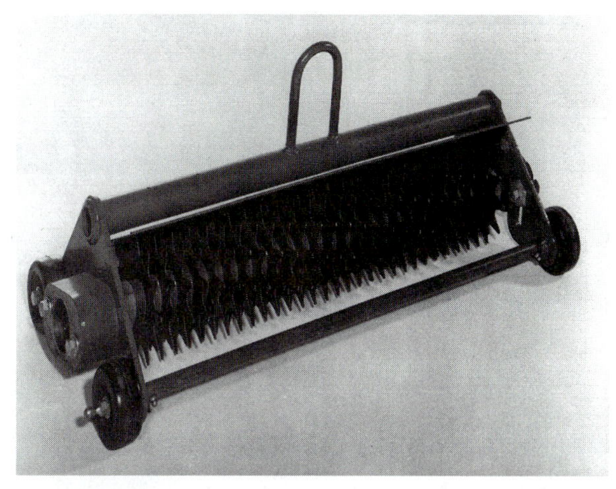

FIGURE 20-14

A greensmower vertical mowing unit. *(Used with permission of The Toro Company. "Toro" is a registered trademark of The Toro Company, Minneapolis, Minnesota.)*

particle size selection is critical. Sand is widely used because many greens are constructed on sand and it can improve native soils if worked into core cultivation holes.

Pest Control. Pest control is critical. A number of disease pathogens and insects can be extremely damaging to greens. Even relatively minor injury can temporarily ruin a green's putting quality and aesthetic appearance. The appropriate pesticide should be sprayed or spread immediately when disease or insect injury first becomes noticeable.

The grass on a green is predisposed to disease problems because of constant traffic and maintenance practices such as daily, close mowing. Because of this susceptibility, a preventive fungicide program is performed on most golf courses. The greens may be treated with a fungicide every week or two during periods when weather conditions favor disease development (Figure 20-15). Nematodes can be a threat to greens in the southern United States, and nematicide treatments may be necessary.

Weed Control. With the exception of *Poa annua* (annual bluegrass), weeds are not usually a major problem on putting greens. Weed competition is often

FIGURE 20-15
Fungicide applications are often necessary when hot, humid weather occurs. *(Used with permission of The Toro Company. "Toro" is a registered trademark of The Toro Company, Minneapolis, Minnesota.)*

minimal because of the close mowing height and high turf density. Some low-growing broadleaf weeds such as chickweed and white clover and some annual grasses such as crabgrass and goosegrass can become established and persist on a green. Nutsedge can be a serious problem on bermudagrass greens.

Extensive use of herbicides on greens should be avoided. Phytotoxicity and greater disease occurrence can result. When a small weed population exists on a green, it is often best to pull them by hand. Bentgrass is sensitive to a number of herbicides, so the selection of chemicals that can be used safely is limited. A larger number of preemergence and postemergence herbicides can be used safely on bermudagrass greens.

Annual bluegrass is the most serious weed problem on greens because of its ability to thrive and produce seeds at extremely low mowing heights. It also adapts well to the high level of maintenance performed on a green. Annual bluegrass control is discussed in detail in Chapter 14. Keeping the bentgrass or bermudagrass dense and vigorous is the best method of minimizing annual bluegrass problems.

The area of a green must be determined before it is treated with a pesticide. Most greens have a round shape, but they are seldom perfectly circular. How to determine their area in square feet is shown in Figure 20-16.

Repairing Ball Marks. When a shot lands on the green, the ball has backspin and may dig into the turf. These indentations are called ball marks. One of the greensworker's responsibilities is to repair ball marks. This is accomplished by

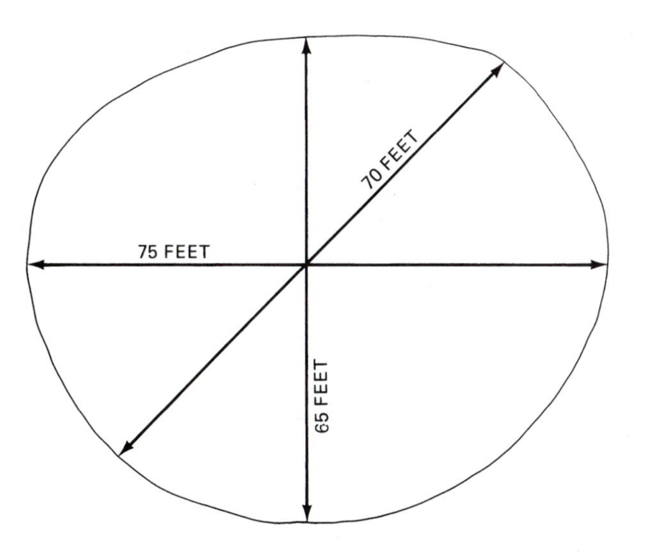

1. Find the average diameter by measuring the distance across the green in several directions.

   75 ft + 70 ft + 65 ft = 210 ft

   $\frac{210}{3}$ = 70 ft (average diameter)

2. Divide the average diameter by 2 to find the average radius.

   $\frac{70 \text{ ft}}{2}$ = 35 ft (average radius)

3. Multiply the radius by itself.

   35 ft × 35 ft = 1,225

4. Multiply this number by 3.14 to determine the area.

   1.225 ft × 3.14 = 3,846.5 ft^2

FIGURE 20-16
Determining the area of a green that is roughly circular.

inserting a knife blade, tee, or a special ball mark repair tool into the turf around the edges of the ball mark. The instrument is used to stretch the turf back over the bruised area and to lift the soil and grass upward (Figure 20-17). Once the compressed grass and soil are raised slightly above the putting surface, the turf is gently tamped down with the hand or foot until it is level with the surface. Failure to repair ball marks results in many small depressions on a green, which make it difficult to putt accurately.

Other Maintenance Requirements. Occasionally grass may be seriously damaged because of vandalism, accidents, pest attacks, or other problems. The superintendent cannot usually afford to wait for the grass to recover if putting quality is severely disrupted. Whenever possible, the injured turf should be replaced immediately. Many golf courses have a sod nursery where replacement grass for greens and tees is grown. The dead turf is removed from the

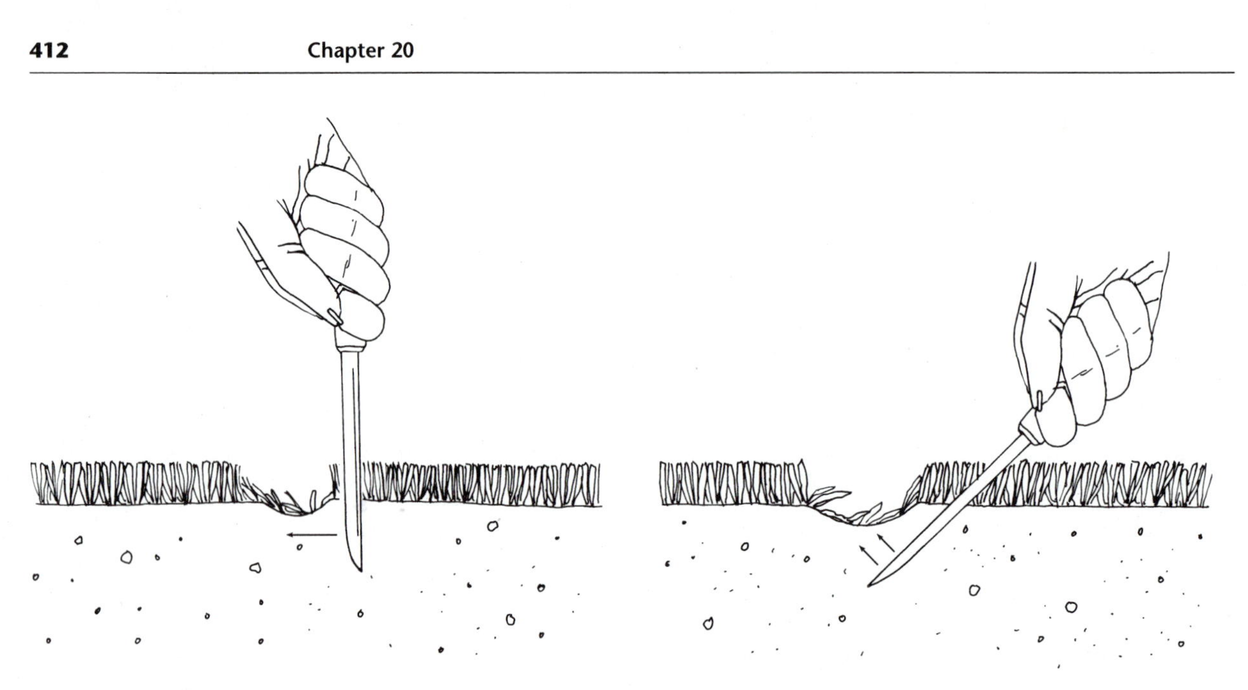

FIGURE 20-17

Ball marks are repaired by stretching the turf back over the injured area *(left)* and raising the compressed soil and turf *(right)*.

green with a knife or sod cutter, and the spots are repaired by installing pieces of healthy sod.

On some courses putting green speed is checked periodically. Putting green speed is a measure of how hard or firmly a ball must be hit with a putter to move it a certain distance. Mowing practices have the greatest influence on green speed. The closer the mowing height, the "faster" the green. Speed increases as the mowing height is lowered because less leaf tissue remains to resist the movement of the ball. Nitrogen fertilization, topdressing, vertical mowing, grooming, and core cultivation also have an effect.

To ensure uniform putting, all greens on a golf course should exhibit a similar speed. Uniformity can be checked by using an instrument called a Stimpmeter or "speedstick," which was developed by the United States Golf Association (USGA). A ball is placed in a groove on the Stimpmeter, and then one end of the instrument is slowly raised above the surface of the green until the ball begins to roll (Figure 20-18). The speed is expressed by the distance that the ball rolls across the green's surface. Three balls are rolled in at least two directions to determine the average speed. The USGA recommends that green speed measure at least 8 feet, 6 inches (2.6 meters) for tournament play (Table 20-2).

There has been a trend toward fast green speeds at many golf courses. This is partly because professional golfers prefer fast greens and because people who view the tournaments on television want the greens at their courses to also "stimp" at very fast speeds. This is accomplished by lowering cutting height. In some cases the grass is seriously weakened by the very close mowing necessary to attain fast greens.

Rolling is a technique that helps the superintendent increase green speed without having to cut the grass too short (Figure 20-19). Rolling the greens just

FIGURE 20-18
Three balls are rolled in one direction and then are rolled back from where they stopped toward the starting point. The average distance is then calculated. *(Photo by Brian Yacur)*

once in the morning increases the green speed by 10 to 20 percent for the rest of the day by reducing surface resistance to ball roll. Though the pressure per square inch exerted by rollers is low, they should be used only on sand greens. It is also best not to overdo the practice. Most superintendents who roll their greens only do it a few times per week.

Greens in colder regions are susceptible to winter injury. This is particularly true if they contain significant populations of *Poa annua*.

Desiccation can occur when greens are not protected by a snow cover. On windy, exposed sites grass leaves dry out and roots can't replace the water because the soil is frozen. This can be prevented by using synthetic covers, which

TABLE 20-2 Comparison of Green Speeds Developed by the United States Golf Association*

	Average Length of Roll			
Relative Green Speed	Regular Play		Tournament Play	
Fast	8.5 (feet)	2.6 (meters)	10.5 (feet)	3.2 (meters)
Medium-fast	7.5	2.3	9.5	2.9
Medium	6.5	2.0	8.5	2.6
Medium-slow	5.5	1.7	7.5	2.3
Slow	4.5	1.4	6.5	2.0

*Reduce the length of roll by 6 inches (15 centimeters) for bermudagrass greens.

FIGURE 20-19

A greens roller. *(Used with permission of The Toro Company. "Toro" is a registered trademark of The Toro Company, Minneapolis, Minnesota.)*

are often made out of pieces of polyethylene, polyester, or polypropylene that are woven or bonded together. Covers also warm the green which help to protect less tolerant grasses from low-temperature injury.

The timing of cover removal in the spring is very important. When covers are removed too early, the grass can be damaged by a sudden drop in the temperature. However, if covers are removed too late, the grass will be too tall. Cutting it to putting height necessitates scalping the grass, which weakens the turf. There is often a brief period in the spring when the superintendent has to remove the covers during the day and replace them at night. In the South greens may be covered any night that the air temperature goes below 32°F (0°C) and then uncovered the next day when it warms up.

Ice can be a major problem on greens. If a layer of ice is on the surface for more than 60 days, there can be a buildup of toxic gasses in the soil, which will kill *Poa annua*. However, the greatest problem with ice is crown hydration. As ice melts, the crown of the plant and the area around it becomes saturated. If this water refreezes, the ice crystals formed injure the cells in the crown. *Poa annua* is especially susceptible.

Good surface and subsurface drainage promotes a rapid removal of water and reduces crown hydration injury. Some superintendents remove ice from their greens. Aerifiers with solid tires can be used to break up the ice. Any snow or ice removal project must be performed with great care to avoid damaging the green. Darkening agents such as Milorganite fertilizer or charcoal can be broadcast on the green to help melt the ice.

To minimize winter injury, it is important to avoid nitrogen fertilization late in the fall before the grass is dormant. Potassium levels should be adequate. Traffic must be kept off turf when it is frozen or frosted. Thawing soils are very easily compacted.

In some sections of the southern United States bermudagrass becomes dormant in the winter. This presents a problem to superintendents because golfers continue to play throughout the winter. The problem is solved by overseeding the greens with cold-hardy cool season species. Such a temporary winter turf helps to protect the dormant permanent grass from traffic injury and improves both the playability and appearance of the green. When warmer weather returns in late winter or early spring the cool season grass dies out and the bermudagrass begins to grow again. The procedure for establishing a temporary winter turf is discussed in detail in Chapter 9.

COLLARS AND APRONS

The collar around a green is generally 3 feet (0.9 meter) wide, but greater widths are also common. Collar areas are maintained the same as greens, with one exception—collars are cut higher. The mowing height is usually from 0.4 to 0.75 inch (1.0 to 1.9 centimeters). The apron outside the collar receives less intensive culture. Its maintenance is similar to that performed on fairways.

FAIRWAYS

Fairways compose the largest maintained area on a golf course. Their total area is normally from 30 to 60 acres (12 to 24 hectares). The level of fairway maintenance varies greatly, but it is always less than that performed on greens. A superintendent with a sizable budget and ample labor force may be expected to keep the fairways in excellent condition. A low-maintenance program may be acceptable at a less affluent club.

Fine-quality fairways are irrigated whenever necessary, kept free of weeds by herbicide applications, core cultivated or sliced regularly, and fertilized several times a year. Cool season grasses require 2 to 4 pounds (0.9 to 1.8 kilograms) of nitrogen per 1,000 ft^2 (93 m^2) annually. Warm season fairway grasses usually need a larger amount of fertilizer. At the other extreme, a superintendent restricted by a small budget may fertilize once a year, apply herbicides only when weed problems become severe, and may not be able to water at all because of the lack of a fairway irrigation system.

Fairway mowing height ranges from 0.5 to 1.25 inches (1.3 to 3.2 centimeters). Playing quality is best when the grass is cut to 0.5 to 0.75 inch (1.3 to 1.9 centimeters). The mowing height used depends upon the grass species and cultivars that compose the fairways. In the warm season zone hybrid bermudagrass is the major fairway species. It can tolerate close mowing to 0.5 inch (1.3 centimeters), or less. Kentucky bluegrass and perennial ryegrass are widely used on fairways in the cool season zone. Some Kentucky bluegrass cultivars can tolerate a cutting height as low as 0.75 inch (1.9 centimeters) if they are irrigated regularly. However, many Kentucky bluegrass cultivars are

FIGURE 20-20

A large fairway mower. *(Courtesy of Jacobsen Division of Textron, Inc.)*

FIGURE 20-21

A lightweight fairway mower. *(Courtesy of Jacobsen Division of Textron, Inc.)*

unable to maintain a satisfactory density when mowed shorter than 1 inch (2.5 centimeters). Creeping bentgrass produces a high-quality fairway turf that can be cut lower than 0.75 inch (1.3 centimeters), but it is expensive to maintain. There has been a trend toward creeping bentgrass fairways because of their superior characteristics. Zoysiagrass and cold-hardy bermudagrass cultivars are popular in the transition zone. Common bermudagrass, buffalograss, fairway wheatgrass, and blue grama may be used on unirrigated fairways in semiarid regions.

Some fairways are usually cut with large reel-type gang mowers that have five, seven, or nine cutting units. The reel units are either mounted together on a frame and pulled behind a tractor or individually attached directly to the mowing tractor (Figure 20-20). Nine-gang mowers have a cutting width of approximately 20 feet (6.1 meters) and enables the operator to cut a fairway quickly. The trend today is to cut fairways with lightweight, very maneuverable three-reeled mowers (Figure 20-21). The lightweight mowers cause less compaction and less wear and tear on the turf than the larger mowers. They have baskets so clippings can be removed. Though they are slower than the large units, the use of light weight mowers results in better-quality fairways.

Fairways are mowed lengthwise, traveling back and forth between the tee and green. Periodically they should also be cross-mowed from side to side. With the large mower wide, slow turns are necessary to prevent the wheels from digging into the turf. The path of the tractor wheels should be staggered each time the fairway is cut to avoid excessive turf wear. Driving the tractor too fast causes the mower units to bounce and results in strips of higher-cut grass and an uneven surface.

Pest control is important. Annual bluegrass can be a serious problem on fairways. If large areas of annual bluegrass die out during hot, dry periods, playability is significantly decreased. Overseeding with competitive, desirable turfgrasses is helpful when chemical control is not effective. Excessive irrigation tends to encourage *Poa annua* infestations. Many superintendents have to live with annual bluegrass and try to prevent it from succumbing to heat and moisture stress by various management techniques.

TEES

Tees are usually more highly maintained than fairways areas but receive less care than greens. However, on courses with a limited budget tees may be treated the same as fairways. At some of the finer courses, the maintenance programs for tees and greens may be virtually identical.

The golfer hits his or her first shot on a hole from the tee. The ball is normally placed on top of a small wooden peg called a tee, but on shorter holes the ball may be hit directly off the turf. The grass on a tee is constantly damaged because golfers' clubs often strike the turf during their swing (Figure 20-22). The small pieces of turf that are ripped up by the club are called divots. Consequently, tees require repair work as well as the normal turfgrass maintenance practices.

Tee markers such as plastic balls are used to designate the location on the tee from which golfers are to hit their shots. To prevent the turf in that area from becoming excessively worn, the markers are relocated regularly a few feet forward or backward. Three different sets of markers are commonly placed on a tee. Women tee off behind red markers, which are located at the front of the tee. Men tee off behind white markers, which are in the middle of the tee. The

FIGURE 20-22
Severe wear on a tee.

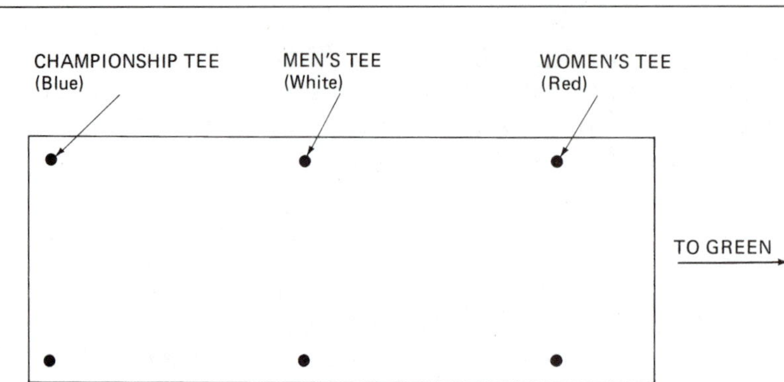

FIGURE 20-23
Tee marker locations.

blue championship or professional markers are placed at the back of the tee, farthest from the green (Figure 20-23). Courses may also have separate tees for women which are closer to the greens than the men's tees.

Large tees take more time to maintain than small tees, but are preferable because the larger area allows a greater number of tee marker locations. For this reason excessive wear is less common on a big tee. Small tees are practical only on courses that do not receive very much play. Tees are normally elevated a few feet above the fairway. This allows the golfer standing on the tee to have a better view of the hole and promotes surface drainage. A 1- to 2-percent slope on the surface of the tee also encourages water movement. Tees with a sandy root zone exhibit the best internal drainage and lowest rate of compaction.

Hybrid bermudagrass is widely used on tees in the warm season zone. Kentucky bluegrass and creeping bentgrass are the two best choices in the cool season zone. Tees in the transition zone are often comprised of zoysiagrass or bermudagrass. The preferred mowing height is 0.5 inch (1.3 centimeters) or less, but Kentucky bluegrass may have to be cut at 0.75 inch (1.9 centimeters).

Tees should be fertilized following a program similar to that for greens, and regular irrigation is required. Core cultivation, vertical mowing, and topdressing are necessary. Topdressing is very valuable for tees because it helps to fill in damaged spots where divots were removed. Seed may be mixed with the topdressing material. Divots should be replaced if the piece of turf is intact. Major repairs are accomplished by overseeding or resodding.

ROUGHS

Roughs receive relatively minimum maintenance compared to the turf areas previously discussed in this chapter. Their purpose is to penalize golfers who land in them by forcing them to hit their next shot from an inferior playing surface (Figure 20-24). Usually a primary rough is maintained adjacent to the fairway. This swath is also called a close or intermediate rough, or semirough. The golfer who hits a ball into this area is not penalized too severely because the grass is cut to 1.0 to 3 inches (2.5 to 7.6 centimeters). The primary rough is maintained similar to the fairway with the exception of the higher mowing

FIGURE 20-24

The presence of trees increases the difficulty of hitting a shot from the rough.

height. The secondary rough or far rough is more distant from the fairway, and it is very difficult to hit a good shot out of this area. The secondary rough usually is not irrigated or fertilized. It is cut at 2 to 5 inches (5.1 to 12.7 centimeters) or is not mowed. On some courses a primary rough is not provided.

SAND BUNKERS

A sand bunker is a hollowed-out depression that is filled with sand. Golfers often call these areas traps. Most bunkers are positioned near a green, but they can be located anywhere on the fairway (Figure 20-25). Bunkers surrounding a

FIGURE 20-25

Bunkers surrounding a green.

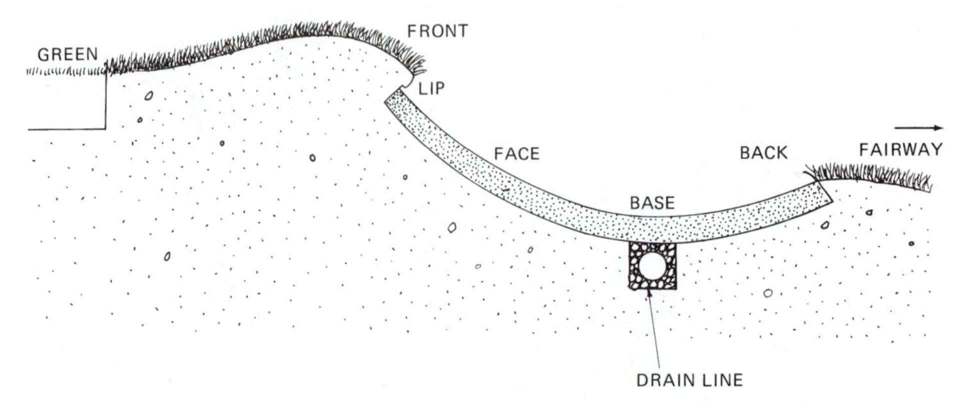

FIGURE 20-26

A side view of a bunker. The slope of the bunker nearest to the green is called the face. The upper part of the face that is covered with sod is the lip.

green are usually deeper and have a higher "lip" than fairway bunkers (Figure 20-26). The bunker penalizes golfers for inaccurate shots, and a golfer must be skilled to hit a good shot out of one.

Bunkers require a great deal of maintenance. Their edges must be kept neat and well defined. This is accomplished with a hand or power edger; plant growth regulators may also be used. The turf right next to the bunker usually must be mowed with a walk-behind rotary mower because riding mowers cannot be driven close enough to the trap to cut all the grass. Grass that spreads into the bunker is chemically controlled, removed with a hoe, or pulled by hand.

Sand bunkers should be raked daily to smooth out footprints and level the playing surface. This job can be performed with a power rake— a motorized, three-wheeled vehicle that has a rake attached behind (Figure 20-27). Hand-raking usually produces the best surface, but takes more time than raking with a machine.

FIGURE 20-27

A power rake used to smooth the sand in bunkers. *(Used with permission of The Toro Company. "Toro" is a registered trademark of The Toro Company, Minneapolis, Minnesota.)*

The layer of sand in the bunkers should be 4 to 6 inches (10 to 15 centimeters) deep except on steep faces where the thickness is less. Every time a golfer hits a shot from the bunker, sand is blasted out of it. Some sand is lost from bunkers because of water and wind erosion. Periodically sand must be added to the bunkers to compensate for these losses. The preferred sand is medium textured (0.25 to 0.50 millimeter in diameter) with sharp, angular edges. This type of sand shifts less under the weight of the golfer.

If bunker sand becomes contaminated with silt or clay, it must be replaced. These finer soil particles cause the surface to crust and can create drainage problems. Some superintendents clean their old sand by washing it rather than bringing in new sand.

OTHER ACTIVITIES

A number of other tasks are performed by golf course employees. Trees on the course are pruned, fertilized, and protected from diseases and insects. Flower beds are common around the clubhouse and must be maintained. Ball washers located next to tees are regularly filled with a water and soap solution. Golf cart paths must be kept in good condition. Aquatic weeds growing on ponds and water hazards are controlled with chemicals. The sod nursery where replacement turf for greens and tees is grown requires a high level of maintenance. Equipment repair and maintenance are major responsibilities.

SELF-EVALUATION

1. Which areas on a golf course are mowed with reel-type mowers?
2. Name the two turfgrass species that are commonly grown on greens.
3. Why are greens often prone to disease problems?
4. Why are greens and fairways usually irrigated at night?
5. List the three tools needed to change a cup.
6. A perched water table helps to compensate for the low water retention exhibited by ___________.
7. The depression resulting from a shot landing on the surface of a green is called a ___________.
8. The common name of *Poa annua* is ___________.
9. The fairway species that cannot usually tolerate an 0.5-inch (1.3-centimeter) mowing height is ___________.
10. Sand bunkers must be ___________ daily.
11. ___________ are mowed higher than other turf areas.
12. ___________ must be relocated before excessive divot removal occurs.
13. The level of maintenance performed on fairways is greatly influenced by the size of the ___________.
14. The micronutrient most likely to be deficient in the root zone of a green is ___________.
15. Develop a maintenance program for greens.
16. Discuss various methods of increasing green speed.

CHAPTER

21

Lawns and Other Turf Areas

OBJECTIVES

After studying this chapter, the student should be able to

- Discuss the establishment and maintenance of lawns, athletic fields, and other types of turf areas
- Describe the role of professional lawn care service companies in the turf industry
- Describe how seeds and sod are produced

INTRODUCTION

This chapter describes the care of lawns, athletic fields, and all the other types of turf areas that were not discussed in Chapter 20. It summarizes many of the maintenance practices mentioned earlier in the text and describes yearly management programs. Methods of producing sod and seeds are also explained.

LAWNS

Millions of lawns surround private residences throughout the United States and Canada. Turf quality varies greatly from one lawn to another. Some homeowners are indifferent to the appearance of their lawns and may do little more than mow the grass occasionally. Other homeowners, however, desire a beautiful lawn and spend large amounts of time and money striving for perfection. A recent trend among homeowners is the hiring of professional lawn care services to apply fertilizer, herbicides, insecticides, and fungicides.

The same variation in turf quality is found among lawns at schools, churches, cemeteries, apartment buildings, condominiums, government facilities, industrial parks, business complexes, and shopping centers. These lawn areas are usually cared for by the maintenance staff of the facility or by professional lawn service companies.

Most people prefer a nice-looking lawn, but many would not be willing to spend a great deal of money to obtain a near perfect lawn. Achieving and maintaining a good-quality lawn does not require unreasonably large expenditures, but it does require intelligent management. An attractive, health lawn is attainable if a few basic errors are avoided. Common management errors that frustrate the attempts of nonprofessionals to maintain a satisfactory lawn include planting grass species or cultivars that are not adaptable, mowing with a dull blade, cutting grass too short and too infrequently, applying too little or too much fertilizer, fertilizing at the wrong time of the year, poor watering practices such as frequent shallow irrigation, and ineffective use of herbicides.

Proper turfgrass selection for lawn establishment and correct maintenance programs differ throughout the country. Establishment and management practices were discussed in previous chapters in this text. Additional information can be obtained from the local office of the Cooperative Extension Service or from turfgrass specialists at the agricultural college in each state.

Selection of Turfgrass Species

In the cool season grass zones Kentucky bluegrass is the most widely used lawn species. Many excellent cultivars are available. Fine fescues and improved, turf-type perennial ryegrass cultivars are often mixed with Kentucky bluegrass. Common bermudagrass is the most popular choice for the warm season zones. Improved, hybrid bermudagrasses produce high-quality lawns. St. Augustinegrass, centipedegrass, and zoysiagrass are commonly used in areas where they are adaptable (Figure 21-1). In transition zones several species, including common bermudagrass, zoysiagrass, tall fescue, and Kentucky bluegrass, are planted on lawns.

Establishment is the critical step in developing a satisfactory lawn. If unsuitable, nonadaptable species or cultivars are selected for planting, the lawn will be a continual disappointment despite skillful postplanting management. Serious, thoughtful consideration must be given to the choice of turfgrasses before establishing a lawn. Soil improvement is important if the soil conditions present are unsatisfactory for good turfgrass growth.

Mowing

Lawns composed of bermudagrass or zoysiagrass are normally mowed to a height of 0.5 to 2 inches (1.3 to 5.0 centimeters), depending upon the cultivar. The best mowing height for St. Augustinegrass depends upon the cultivar that is grown, but most are cut to 2 to 3 inches (5.1 to 7.6 centimeters). Lawns comprised of Kentucky bluegrass, fine fescues, and perennial ryegrass are mowed at 1-1/2 to 3 inches (3.8 to 7.6 centimeters), with the higher heights preferable during the summer stress period. Tall fescue should not be cut

FIGURE 21-1
A warm-season-zone lawn. *(Courtesy of H. D. Hudson Manufacturing Co.)*

shorter than 2 inches (5.1 centimeters) unless a turf-type cultivar that tolerates close mowing is used.

Reel-type mowers generally produce a superior mowing quality, but rotary mowers are very popular because of their greater convenience (Figure 21-2). Rotary mowers can be satisfactory if the blade is kept well sharpened. Lawns should be cut regularly to avoid serious violations of the "one-third" rule. Clippings should be left on lawns whenever possible.

Fertilization

Soil test results, the nutritional requirements of the lawn grasses, and the desired level of turf quality determine the specific fertilizer program. Hybrid bermudagrass is a heavy feeder and may require as much as 1 pound (0.45 kilogram) of nitrogen per 1,000 ft^2 (93 m^2) per growing month. Fine fescues, however, should not receive more than 2 pounds (0.9 kilogram) of nitrogen per 1,000 ft^2 (93 m^2) per year. Cool season lawns receive from 2 to 5 pounds (0.9 to 2.3 kilograms) of nitrogen per 1,000 ft^2 (93 m^2) per year. The higher rates in the range are applied to the best quality lawns. In order of priority, application times for cool season lawns are late summer or early fall, late fall, spring, and early summer.

FIGURE 21-2
Reel mowers produce the best mowing quality. *(Used with permission of The Toro Company. "Toro" is a registered trademark of The Toro Company, Minneapolis, Minnesota.)*

The length of the growing season also has a major influence on a lawn fertilization program. Warm season lawns often receive more fertilizer than cool season lawns because of the greater number of growing months in the southern United States. Applications occur throughout the spring, summer, and fall, and even during the winter in southern areas where grass grows year-round. Four to six applications or more may be necessary.

The increased cost of turfgrass fertilizer in recent years has caused some people to fertilize less frequently. Despite the expense, fertilizer is still the best investment for anyone who wants a good lawn. Insufficient fertilization results in a dramatic decrease in turf quality. The turfgrass manager should use the appropriate spreader setting recommended on the bag. In most cases this will result in the application of approximately 1 pound (0.45 kilogram) of nitrogen per 1,000 ft^2 (93 m^2). If a quickly available, fast-release form of nitrogen is to be applied, lighter rates should be used.

Every two or three years soil tests should be performed to determine fertility levels and pH values.

Other Maintenance Practices

Better-quality lawns require supplemental irrigation during the hotter, drier periods of the growing season.

Core cultivation or vertical mowing is sometimes performed when the thatch layer is thicker than 0.5 to 0.75 inch (1.3 to 1.9 centimeters) (Figure 21-3). Dethatching may be necessary once a year on lawns composed of thatch-prone species such as hybrid bermudagrass and St. Augustinegrass. Cool season lawns usually require less frequent thatch control. Core cultivation is also necessary when compaction is a problem. It can be very beneficial for compacted, thatchy lawns. The best time to vertical-mow or core-culti-

FIGURE 21-3
Thatch control is an important part of a lawn maintenance program.

vate warm season lawns is late spring or early summer; late summer or early fall is recommended for cool season lawns.

Pest Control

Preemergence herbicides should be applied before annual grassy weeds such as crabgrass emerge in the spring. Lawns should be treated with postemergence herbicides when broadleaf weeds become a nuisance. Lawn managers must watch for insect injury and be prepared to apply an insecticide before significant damage occurs. In southern areas nematodes can be a threat to turfgrass. Fungicides have not been widely used on lawns, but their use is recommended if a disease develops into a serious problem.

Professional Lawn Care Services

The tremendous growth of the professional lawn care service industry began in the 1970s and continues today. Many homeowners pay professional lawn care specialists to maintain their lawns because they are dissatisfied with their own amateur efforts or because they do not enjoy lawn work. Others may not have the time to care for their lawns properly. Individuals or businesses with large lawn areas often find it more economical to hire lawn services rather than employ a maintenance staff and purchase turf equipment.

Lawn care companies vary greatly in size, performance, and the services they provide. The larger companies often specialize in fertilizer and pesticide applications. Their employees fertilize lawns several times a year, apply preemergence and postemergence herbicides, and treat the grass with insecticides and fungicides when necessary. These materials can be mixed together in a tanker truck and sprayed on the lawn (Figure 21-4). This is an efficient, quick method of application and results in reduced costs for the customer. In many

FIGURE 21-4
A lawn specialist applying fertilizer and herbicide. *(Courtesy of ChemLawn Corp.)*

cases the expense to customers is less than if they treated the lawn themselves. Some lawn care companies apply only granular materials. This is partially because some homeowners are uncomfortable with spray applications.

The lawn care specialists consult with the lawn owner if any turfgrass problems occur and also periodically test the soil. They will make recommendations concerning lawn establishment, mowing, irrigating and dethatching. However, at present most major lawn care companies do not actually perform all of these practices.

Smaller lawn care businesses often provide a wider range of services, including mowing and thatch control. However, the technical expertise of their personnel may not be as extensive as that of the specialists working for larger lawn companies, which provide their employees with excellent training programs and support staff. The best way to judge the performance of a lawn care service is to inspect lawns it maintains.

PARKS

Parks often contain large lawn areas. Consequently, large equipment is generally needed to maintain the turf (Figure 21-5). Areas that are heavily used by the public are kept in good condition if the maintenance budget is adequate. The expense of maintaining large expanses of turfgrass usually necessitates a medium level of maintenance consisting of regular mowing, a modest fertilization program, herbicide applications when weeds become a noticeable problem, and occasional watering if an irrigation system is present. Playing fields in the park will require core cultivation to correct compaction and higher rates of fertilizer to allow the grass to recover from hard use. Limited-use areas are basically ignored except for several mowings a year to a height of 2 to 4 inches (5.1 to 10.2 centimeters).

FIGURE 21-5

Parks may have large expanses of turf. *(Courtesy of Jacobsen Division of Textron, Inc.)*

CEMETERIES

As with other turf areas, the quality of the grass in a cemetery is primarily determined by the size of the maintenance budget. The lawns should be in as good a condition as possible in late May because of the large number of visitors on Memorial Day weekend. The major problem for cemetery lawn managers is mowing. Much of the mowing has to be accomplished with small walk-behind mowers because the grave markers are close together (Figure 21-6). Extensive trimming with "weed-eater"-type, nylon-cord trimmers is necessary around the gravestones. Plant growth regulators are used at cemeteries to suppress plant growth and reduce maintenance costs.

FIGURE 21-6

Mowing is a problem in cemeteries.

FIGURE 21-7
A typical utility turf.
(Courtesy of Gulf States Paper Corp.)

UTILITY TURF

The primary function of a utility turf is to prevent soil erosion. These areas receive little use, and appearance is not very important. Utility turfs are grown along roadsides, airport runways, river and pond embankments, on ski slopes, and at parks, fairgrounds, and other locations where grass plants are needed to provide soil stabilization (Figure 21-7). Because aesthetic quality is not a major concern, utility turf generally receives a minimum level of maintenance.

The grass species planted on utility areas must be relatively self-sufficient because of the lack of maintenance. Drought tolerance and the ability to perform adequately at low fertility levels and to survive infrequent mowing are important considerations when selecting species and cultivars. Cheaper seed mixtures are commonly used because the areas that are to be established are often quite large. Common Kentucky bluegrass, perennial ryegrass, fine fescues, common bermudagrass, bahiagrass, tall fescue, buffalograss, wheatgrasses, and blue gramagrass are popular choices in their respective zones of adaptation. Other types of ground covers such as crownvetch may also be mixed with the grasses.

Seeding is often accomplished with a hydroseeder, which applies seed, fertilizer, and wood fiber mulch in a liquid slurry. Mulching the seedbed is essential because irrigation generally is not possible. Sod is occasionally installed on steep slopes, but its use is limited because of the greater expense.

Fertilization is minimal. Some utility turf areas are not fertilized after establishment; others may receive one application per year or every few years. Mowing is also infrequent. Some utility areas are never mowed, but most are cut at least twice per year. The grass is seldom mowed more than four or five times each year. The mowing height is generally 3 to 4 inches (7.6 to 10.2 centimeters) or higher. Growth regulators are sometimes used to retard the growth of the grass and decrease the need for mowing.

Irrigation is extremely uncommon. Pesticide applications are also unlikely. A few sites, however, may be sprayed with herbicides if broadleaf weeds start to predominate.

Of all the different types of turf, utility areas require the least maintenance.

ATHLETIC FIELDS

Football, soccer, baseball, lacrosse, and field hockey fields are found in communities throughout the United States. It can be very difficult to maintain satisfactory turf quality on these fields because of the damaging effects of the sports activities played on their surfaces. Compaction and severe turfgrass wear are common problems on athletic fields (Figure 21-8). These problems can be significantly reduced by proper construction and careful selection of turfgrass species and cultivars.

The key to building a good athletic field is providing a good root zone. As mentioned in the section on golf green construction in Chapter 20 a layer of medium-textured sand at least 12 inches (0.3 meter) deep is most satisfactory. The pure sand base allows rapid drainage and resists compaction. A drainage line surrounded by gravel beneath the sand is essential (Figure 21-9). An irriga-

FIGURE 21-8

Compaction and wear result in thin turf or even bare soil.

FIGURE 21-9

An athletic field should have a subsurface drainage system. (*Courtesy of Charles Machine Works*)

tion system is also necessary because of the low water retention of the sand. Incorporating organic matter into the upper inches of sand can help to increase water- and nutrient-holding capacity.

Another method of constructing athletic fields was developed at Purdue University. At those locations where this method is used, excellent turf quality has resulted The method is called Prescription Athletic Turf (PAT) and provides a very efficient water management system. Impermeable plastic sheeting is placed beneath a layer of sand. Drain tubes are placed directly above the plastic. The plastic barrier conserves water by preventing it from moving deep into the soil. If excess water is present because of a heavy rain, suction pumps are used to remove it from the soil. The water is pulled into the drain tubes and removed through outlets.

The PAT system provides the athletic field manager with a large degree of control over soil moisture levels. Removing excess water before a game or practice results in less compaction, reduced turf injury, and better playing conditions. If dry weather occurs, irrigation water that percolates through the root zone can be conserved and redistributed. PURR-WICK is a similar system that is constructed beneath golf greens.

The superior quality of an athletic turf established on a deep layer of sand cannot be overemphasized. Even with careful management, fine-textured soils are often unsatisfactory, especially if the field is used frequently. They can be adequate if the field's use is limited and a drainage system is installed. When a sports turf is established on native fine-textured soil, the field should be crowned to encourage surface runoff. The center of the field is often 12 to 18 inches (0.3 to 0.46 meter) higher than the sidelines. Extensive core cultivation and slicing are necessary to control compaction, (Figure 21-10).

FIGURE 21-10

Core-cultivating an athletic field. *(Used with permission of The Toro Company. "Toro" is a registered trademark of The Toro Company, Minneapolis, Minnesota.)*

Selection of Turfgrass Species

Species and cultivars selected for athletic turfs should be wear tolerant and recuperative. Wear-resistant, vigorous Kentucky bluegrass cultivars are the best choices for cool season locations. Turf-type perennial ryegrasses are frequently mixed with the Kentucky bluegrass. Perennial ryegrass wears well, but its recuperative ability is not good because of a lack of rhizomes and stolons. Bermudagrass is the preferred species for warm season athletic turf. Zoysiagrass is also used. Tall fescue exhibits excellent wear tolerance and is planted in the transition zone, but its inability to persist at close mowing heights can be a problem.

Maintenance Practices

Core cultivation to alleviate compaction is essential. This maintenance practice was discussed in detail in Chapter 18. The need for aeration is determined by the severity of traffic on the field and the texture of the soil beneath it. Heavily used fields established on finer-textured soils are most prone to compaction and require extensive core cultivation, which is commonly performed in the spring if the field is used for fall sports such as football or soccer. Holding practice sessions on a different field also reduces compaction.

Grass that experiences severe wear and is seriously injured should be replaced. This can be accomplished by seeding, spot-sodding, plugging, stolonizing, or sprigging. If sod is used, it must have soil that is very similar to the soil on the field. Major repairs are performed during the off-season so that the new grass can become well established before it is used.

Overseeding has become a popular method of renovating athletic fields. Kentucky bluegrass, perennial ryegrass, tall fescue, and common bermudagrass seeds are planted into the turf with an overseeding machine. This is a nondisruptive and relatively inexpensive way to improve turfgrass density. Seeds of quick-germinating species such as perennial ryegrass may be overseeded into a field immediately after a game. Seed can also be drag-matted into core cultivation holes. A less sophisticated technique is to broadcast seed before a game and rely on players' cleats to work the seeds into the soil. Perennial ryegrass is often overseeded into southern fields for winter play.

A recent innovation is the placement of mesh material in the surface soil. The pieces interlock and absorb and reduce some of the force caused by the impact of athletes' feet on the surface of the field. This results in less compaction. Also, turfgrass is less likely to be ripped out of the field because its roots intertwine with the mesh material, anchoring the plants more securely. Nylon netting is another material sometimes used to help hold the sod together.

Kentucky bluegrass and perennial ryegrass are usually cut to a height of 1-1/2 inches (3.8 centimeters) during the playing season. Improved, hybrid bermudagrasses are generally cut to 0.5 to 1 inch (1.3 to 2.5 centimeters), and common bermudagrass is cut to 1 to 1.5 inches (2.5 to 3.8 centimeters).

The field should be well fertilized to produce healthy, recuperative grass plants and a dense cover (Figure 21-11). A fertilization a few weeks before the playing season is especially important.

FIGURE 21-11

Fertilizer applications help sport turf to recover from injury. *(Courtesy of Vicon Farm Machinery, Inc.)*

Irrigation should be provided to keep the grass dense and vigorous. Sand fields require a conscientious irrigation program because of the low water retention of the root zone media (Figure 21-12). Water should not be applied the day before a game, to ensure that the surface soil is not overly wet. If rain is expected soon before a game, the field can be covered with a tarp. Irrigation immediately after a game is strongly recommended to hasten recovery of the grass. Some clay fields become very hard if they are not irrigated properly.

FIGURE 21-12

An unirrigated sand field.

FIGURE 21-13

Lining a football field. *(Courtesy of Fox Valley Marking Systems, Inc.)*

Athletic fields are generally lined with white spray paint or lime which are applied with a piece of equipment called a marker or a striper (Figure 21-13).

Any pests that threaten turf density or vigor should be controlled. Weeds are a common problem when excessive wear results in spots where the soil is bare.

Sports such as football that continue late into the fall present special problems because the grass loses vigor at the onset of colder temperatures. In some locations bermudagrass becomes dormant before the end of the playing season. Clear, perforated plastic can be used to cover the field to raise temperatures and reduce freezing. At a few stadiums electric heating cables are installed beneath the field. Perennial ryegrass may be overseeded into southern fields for a late fall, winter turf.

Contact sports inevitably result in injuries. Studies have shown that the injury rate is significantly higher on athletic fields that are in poor condition. Players who fall down on a hard, compacted surface are obviously more susceptible to injury. Sand fields are generally safest because it is easier to maintain a dense turf cover and a noncompacted playing surface. On fine-textured soils extensive core cultivation has been shown to reduce the likelihood of serious injury.

Artificial Turf

Artificial turf made of nylon fibers was introduced in the 1960s and installed at a number of major sports stadiums in the United States (Figure 21-14). This synthetic turf was initially popular because it provided a permanent, uniform playing surface and was relatively maintenance-free compared to natural grass. Beneath the nylon grass is a foam pad which cushions the impact of a fall. A layer of asphalt is under the pad. Vacuuming and water removal machines are the primary maintenance equipment. The cost of installation in the 1960s was $250,000 to $500,000, so its use was limited to more affluent athletic facilities.

FIGURE 21-14
Artificial turf.

Enthusiasm for artificial turf has waned in recent years. The nylon material is a petroleum derivative, and its cost rose sharply as soil prices increased in the 1970s. Some other problems became apparent. Injuries were more prevalent on artificial turf, and surface temperatures were extremely high during hot weather. Moreover, the synthetic material was not as permanent as originally thought. It could be damaged by oil leaking from equipment, cigarette burns, and other accidents; in some cases the color faded. The cost of cleaning and repairing the turf was considerably higher than anticipated.

The trend has been away from artificial turf and back to natural grass. Much of the dissatisfaction with grass athletic fields that was common in the 1960s has been overcome by the introduction of improved construction techniques, more adaptable turfgrass cultivars, and better maintenance programs. A survey of professional football players showed that 80 percent preferred natural grass. Artificial turf is necessary at covered stadiums and may still be the ideal choice for fields that receive such extensive use that grass cannot be satisfactorily maintained.

SOD FARMS

Sod production has increased steadily in recent years because of the popularity of "instant" turf. A deep soil and good water supply are necessary for sod production. Organic, muck soils are best because they are lighter than mineral soils. Sod grown on muck soils does not have to be as well knit and can be harvested in a shorter time. Well-drained sandy loam, silt loam, or loam soils are the most satisfactory mineral soils for sod production. Sod farms must be located near their prospective customers because sod should be installed within 48 hours of harvest (Figure 21-15).

FIGURE 21-15
Freshly harvested sod on the way to a garden center. *(Courtesy of Brouwer Equipment, Ltd.)*

The shorter the period between planting and harvesting, the greater the potential profit. However, harvesting cannot occur until the sod is strong enough to hold together well when handled. Species and cultivars that produce aggressive rhizomes or stolons are used because they develop sod strength more quickly. Species that have limited spreading ability such as fine fescue knit the sod together slowly and are less desirable.

Sod production generally takes several months to two years, depending on the grass species and the length of the growing season. The process can be speeded up by placing netting in the seedbed. The netting helps to reinforce sod strength during harvesting and installation. Growing sod over plastic also hastens the development of sod strength.

The turfgrasses selected should exhibit high quality. Besides having a vigorous spreading ability, they should also be attractive, disease resistant, and adaptable to local conditions. People who are willing to pay the extra expense of establishment by sodding expect excellent quality. High levels of maintenance are necessary to produce quality sod in the most profitable length of time. Adequate fertilization, constant irrigation, and correct mowing practices are essential. Pesticides must be used to keep the sod totally free of weeds and insect or disease injury.

FIGURE 21-16
A sod harvester. *(Courtesy of Brouwer Equipment, Ltd.)*

The sod is cut with large harvesting machines (Figure 21-16). Some of the machines are quite sophisticated and cut, roll, and stack the sod on pallets in one operation. Typically, sod pieces are 12 to 24 inches (0.3 to 0.6 meters) wide and 4 to 6 feet (1.2 to 1.8 meters) long. Strips 4 feet (1.2 meters) by 45 feet (13.7 meters) are harvested when the sod will be installed by unrolling it from a bar on the back of a tractor. Sod should be cut as thin as possible to minimize soil loss, to make the sod lighter and easier to handle, and to encourage rapid rooting.

SEED PRODUCTION

Most cool season grass seed is produced in Washington, Oregon, Idaho, and western Canada (Figure 21-17). These regions have a climate that is ideal for seed production. The weather is dry in the summer when the seed is harvested, fall rains stimulate regrowth, and mild winters result in a lengthy growing season. Many warm season grass seeds are grown in irrigated areas of Arizona.

Establishment occurs in either the spring or fall. Seed is planted at very light rates in rows 16 inches (0.38 meter) apart. The field must be kept weed-free. To help accomplish this, the rows of seed are covered with a band of activated charcoal and then sprayed with a preemergence herbicide. The activated charcoal absorbs the herbicide near the seed and protects the seedlings from injury. An older method of establishment is to till the field in the fall, let it lie fallow over the winter, then treat with herbicides, and finally plant in the spring. Vigorous postplanting weed control is necessary to ensure that only seeds of the planted cultivar are harvested.

FIGURE 21-17

A seed field in Oregon. *(Courtesy of the O. M. Scott and Sons Company)*

Harvesting begins in late June or early July, when the upper stems are mowed and windrowed. After drying in the field for several days the seed is picked up and threshed by large combines (Figure 21-18). The seed is then shipped to the seed companies and is tested, cleaned, and packaged. For some species, burning the fields after harvesting results in significantly greater seed production the following year. Burning destroys the straw that is left on the fields from harvesting; reduces insect, disease, weed, and thatch problems; and increases tillering. Because of environmental concerns, field burning has been restricted in recent years. Growers either chop the straw fine and leave it on the field or vacuum it up. A seed field is kept in production for five years or less.

FIGURE 21-18

Seed harvesting. *(Courtesy of International Seeds, Inc.)*

FIGURE 21-19
Seed-cleaning machinery. *(Courtesy of Art Wick, Lesco Products)*

Cleaning operations are as important as field production. Specialized equipment, including vibrating screens, shakers, and air jets, is used to remove chaff, soil, weed seeds, and other impurities (Figure 21-19).

SELF-EVALUATION

1. Explain why southern lawns may require more fertilizer than northern lawns.
2. Homeowners should select species and cultivars for their lawns that are ___________ to local conditions.
3. ___________ is a major problem in cemeteries.
4. ___________ turf receives the lowest level of maintenance.
5. ___________ athletic fields resist compaction and are well drained.
6. ___________ results in less injuries because the practice helps to alleviate compaction.
7. An athletic field should not be ___________ right before a game.
8. Sod cannot be harvested until sufficient sod ___________ develops.
9. ___________ control is essential in a seed production field.
10. Bermudagrass and St. Augustinegrass lawns may require ___________ once a year.
11. Develop a yearly maintenance program for lawns in your area.
12. Discuss sod production.

CHAPTER 22

Business Management Practices

OBJECTIVES

After studying this chapter, the student should be able to

- Discuss the importance of business management skills to the turfgrass manager
- Discuss personnel management practices that encourage good manager–employee relations
- Describe the various types of records that are kept by the turfgrass manager
- Explain how to plan work schedules and prepare a budget
- Discuss other business-oriented activities such as estimating job costs and using computers

INTRODUCTION

The previous chapters provided a detailed description of turfgrass establishment and maintenance practices. Familiarity with these practices is the key to producing satisfactory turf. However, the ability to grow grass does not guarantee a successful turfgrass management program. A good program is the result of both maintenance expertise and business management skills. A person who cannot effectively supervise workers or accurately determine the amount of money needed for the maintenance budget will be an inefficient turfgrass manager, no matter how extensive his or her knowledge of turf care. A turfgrass management textbook would be incomplete without a discussion of business management practices.

All turfgrass managers become involved with basic business activities such as personnel management, planning work schedules, recordkeeping, budget preparation, and purchasing equipment and supplies. This chapter is primarily concerned with these topics because they are relevant to managers throughout the turfgrass industry. People who own or manage a turfgrass business such as a lawn care service often require more sophisticated business skills and may be responsible for product markups, job cost estimation, and writing maintenance specifications. A brief discussion of these tasks is also included.

Failure to develop business management skills can severely limit a turfgrass professional's opportunities for career advancement. The more financially and professionally rewarding a managerial position, the greater is the need for these skills. For example, it is not unusual for golf course managers at larger facilities to spend the majority of their time working at business-oriented activities. Managers may enjoy riding around on a mower or pushing a fertilizer spreader, but it is inefficient for them to spend much time performing basic tasks that can be accomplished just as satisfactorily by lower-paid, nonmanagement employees.

There is such a strong correlation between business skills and job success that turfgrass students should be encouraged to enroll in as many business management courses as their schedules will allow. People already working in the turfgrass industry usually have the opportunity to take business courses at local colleges during the evening or at other convenient times. Many colleges offer continuing education programs designed to meet the educational needs of people who have full-time jobs. A number of excellent business management textbooks and videos are available for the turf manager's reference library.

PERSONNEL MANAGEMENT

The manager is responsible for planning and coordinating the turfgrass maintenance program. However, most or all of the actual maintenance practices are usually performed by employees under the manager's supervision. The success of the program greatly depends on the quality of their work (Figure 22-1). It is difficult to implement a successful program without well-trained, conscientious workers who derive satisfaction from their jobs. The presence of good employees, in turn, is often the result of good leadership. A good personnel manager has the ability to hire, develop, and retain good workers. Poor leadership can cause employees to become dissatisfied with their jobs and indifferent about the quality of their work. Such negative attitudes undermine the turfgrass management program and are a threat to the manager's job security.

The turfgrass manager should keep certain rules in mind when dealing with employees. They are valid whether the manager directs the activities of one person or one hundred. The principles of effective personnel management are simply common sense—a manager should treat the people he supervises the same way he would like to be treated by his own supervisor.

Successful manager–employee relationships are based on mutual respect.

FIGURE 22-1

The manager's program is implemented by the employees. How successful the program is depends largely upon their efforts. *(Used with permission of The Toro Company. "Toro" is a registered trademark of The Toro Company, Minneapolis, Minnesota.)*

To be respected by her employees, the manager must be hardworking and knowledgeable about turf care. Equally important, the manager must show that she respects the people she supervises by being considerate of their feelings, concerned about their problems, supportive of their efforts, and interested in their professional development. Workers should be able to view their supervisor as both boss and friend.

Positive reinforcement is an essential management technique. The manager should congratulate employees for a job well done and pass on compliments about their work received from customers, golfers, and other people. Words of praise encourage workers to continue to do a good job. The employee who consistently does fine work should be given a promotion or a pay raise. The most effective method of motivating workers is to provide monetary rewards for outstanding job performance. A lack of financial incentive discourages achievement and forces energetic employees to look elsewhere for jobs. When an employee does not deserve a raise, the manager should explain to him why his job performance did not merit additional pay and how he can improve the quality of his work in the future.

Criticism should always be constructive. When a worker makes a mistake, the supervisor should politely and clearly explain what was done wrong, why it is wrong, and describe the correct way to perform the task. Reprimands should occur in private. Criticizing an employee in front of fellow workers causes pointless humiliation.

Many employee mistakes can be avoided if the manager gives explicit instructions and questions the worker to be certain that she understands how to accomplish the assigned task. Vague, unclear directions often result in confusion and unnecessary errors. Because turfgrass managers are well acquainted with maintenance practices, they may feel that a job is easy and need not be

FIGURE 22-2
Injury to a practice green caused by excessively deep vertical mowing. The worker did not understand the manager's instructions.

explained in detail. They can forget that an employee may not be equally familiar with these practices. Thorough explanations are especially important when the worker is new or has not performed the assigned task before (Figure 22-2).

Training

Employee training is an important part of the manager's job because her employees are often not educated in turfgrass maintenance. They must learn the majority of their turf care skills on the job from their supervisor. This training benefits both the manager and the employees. Workers become more capable and require less supervision. Their greater knowledge results in more skillful turfgrass maintenance. Employee job satisfaction increases. People enjoy learning because it makes their jobs more interesting. Workers who receive in-depth instruction realize that the manager cares about their professional development. A highly trained employee is also a valuable asset because he may be able to suggest ways to improve the maintenance program.

A training program, either informal or formal, is extremely beneficial and very cost efficient. Turf managers should not feel that they are too "busy" to spend the time it would take to train their employees thoroughly in maintenance skills. Some supervisors conduct training sessions on rainy days or during the winter when outside work activities are limited by the weather. To be a good teacher the manager must clearly explain the concept or practice she is discussing, encourage questions, and be patient when employees do not immediately understand her explanations (Figure 22-3).

The manager should not have to do all of the teaching. New employees can receive their initial training from more experienced workers. Employees should attend educational sessions at turfgrass conferences and seminars sponsored by

FIGURE 22-3
Teaching is an important part of a manager's job.

professional organizations, the Cooperative Extension Service, or equipment and supply companies. Workers should be encouraged to read turfgrass journals, magazines, and books and view training videos. If a turfgrass course is taught at a local college, the management can offer to pay the expenses of any employee who wishes to attend the class.

Pesticide training is especially important. Numerous federal and state regulations require that workers exposed to hazardous chemicals receive pesticide training from their employer.

The manager's style of leadership should be relaxed rather than dictatorial. Looser supervision demonstrates confidence in the employees' reliability and promotes independence and resourcefulness. Opinions and advice should be solicited from workers. People like to be consulted and often respond with valuable suggestions.

Hiring

Hiring is an important managerial responsibility. To hire people who will be good employees, the manager should be a skillful interviewer. During the interview the manager tries to determine the prospective employee's potential strengths and limitations. The interviewer should pose thoughtful questions designed to measure the job applicant's turfgrass knowledge, common sense, and ability to get along with other people. The turf manager should also remember that good candidates, because of nervousness, do not always give good interviews. It is usually wise to speak personally to each of the people listed as references on an application form or résumé.

Safety

The manager should be extremely concerned with employee safety. The physical well-being of workers is of the highest priority in a turfgrass management program. When employees perform hazardous duties they must be required to wear appropriate protective clothing such as hard hats, safety glasses, steel-tipped boots, gloves, or respirators (Figure 22-4). Emergency telephone numbers should be posted in a conspicuous location. Fire extinguishers and first aid supplies must be checked regularly. Employees should be taught how to administer first aid treatment in case of an emergency. Safe operation of equipment must be emphasized. Accidents are costly and demoralizing. Most accidents can be prevented or their severity greatly reduced if the supervisor trains employees to be careful but also to be prepared for emergencies.

Communications

All of the techniques used to develop good manager-employee relations rely on one important skill—the ability to communicate. The manager must clearly communicate her desires and concerns to her employees. They, in turn, must be

FIGURE 22-4

The mower operator is wearing a hard hat and steel-tipped boots. *(Used with permission of The Toro Company. "Toro" is a registered trademark of The Toro Company, Minneapolis, Minnesota.)*

given the opportunity to state their opinions to the supervisor. When people fail to communicate effectively, misunderstandings can occur, which may lead to serious personnel problems. Some successful managers promote good communications by meeting regularly with their employees. These group discussions enable the supervisors to disseminate information to the entire work force and give employees the chance to express their views and air grievances.

The turfgrass manager also has to deal regularly with his own supervisors, golfers, customers, athletic coaches, or other people who use the turf. An open line of communications is vital in these relationships as well. For example, animosities can develop between golf course superintendents and golfers when communications break down. This is unfortunate because the superintendent and the golfers have the same goal—fine quality turf and good playing conditions. However, when the golfer does not understand the superintendent's program and problems, and the superintendent is not aware of the golfer's concerns, friction is inevitable.

Two methods of improving communications used by golf course superintendents illustrate how to prevent many misunderstandings. One method is to distribute a questionnaire to members of the club. Golfers are asked whether the greens, fairways, tees, rough areas, and bunkers are in excellent, good, fair, or poor condition. They are also asked to rate the friendliness of the staff and the appearance of the clubhouse, and to supply other information relevant to the superintendent's responsibilities. A space should be left at the bottom of the questionnaire for suggestions. The responses allow the superintendent to evaluate her management program and determine what changes are necessary.

The superintendent can communicate his reasons for certain decisions and actions to golfers by writing a weekly or monthly report. These reports are sent to the green committee (the club members who supervise the superintendent) and can be posted in the locker room or pro shop where all of the members can read them. These reports help to educate golfers to accept maintenance practices and can be used to explain problems that the superintendent encounters. An example of such a report is shown in Figure 22-5.

Superintendent's Report—3/18/94

1. To all of you frustrated "early birds"—the course is closed for play until 9:00 a.m. at this time of year because there is still frost on the grass in the early morning. When grass is "frosted" it is easily damaged by foot and cart traffic. The weight of feet and cart tires can rupture the plant cells and may kill the grass. We're tempted to open the course earlier in the morning to accommodate early risers but we decided not to because the injury that occurs now might result in poor playing conditions for the rest of the season. As soon as the weather warms up a bit the course will be open for play 7:00 a.m.
2. We're quite pleased about the minimum amount of snow mold injury that occurred over the winter. Last fall we tried a new fungicide which seems to have worked very well. We're fortunate because a number of courses in the area were hit hard by snow mold.
3. Many of you who responded to our questionnaire last year commented that the 8th tee was in bad shape. We plan to solve the problem this spring. The tee is too small to move the tee markers any significant distance from their previous location. This results in severe wear to the grass because golfers always have to tee off in the same small area. Within the next month we will enlarge the 8th tee and this will allow us to spread the wear over a bigger area and reduce divot injury. During the reconstruction a temporary tee will be established at the beginning of the 8th fairway.
4. There will be the sound of chain saws this week near the 17th green. The large elm behind the green finally succumbed to Dutch elm disease. We're taking it down immediately because I'm worried that the dead tree might fall onto the green. Anyone who needs firewood and has the energy to split elm should contact the Green Committee because we'd like to give the logs away. Don Metz, the owner of Shadyside Nursery, has donated a young maple tree which we will plant where the elm stood.
5. Some golfers have mentioned that we should plant more flowers around the clubhouse and build another practice green. There are good suggestions. There will be more flowers this year. Unfortunately, we can't construct another green this year. The estimated cost of a new practice green is approximately $12,000. There is not enough money in the budget this year to pay for it. The Green Committee is considering the project for next year.
6. Lastly, I'm proud to announce that we have a new Assistant Superintendent—Holly Kornahrens. Holly will be starting work on April 1. She comes to us with a strong background in turfgrass management. Holly worked for several summers at Indian Hill Country Club and has just recently received her degree in golf course management from Cobleskill College. I know that Holly will be a very valuable addition to our staff. Please feel free to introduce yourself when you see her working around the course.

FIGURE 22-5
A superintendent's report.

Work Schedules

Another important personnel management duty is the preparation of work schedule. A good manager, before she arrives at work in the morning, has already decided what she wants to accomplish that day and has planned work activities for each of her employees. Studies of various businesses by efficiency experts have shown that 10 to 50 percent of work time is often wasted. Most of this lost work time is a result not of employee laziness but of poor supervision. Inadequate instructions and planning can result in gross inefficiency. When the manager comes to work in the morning without a clear idea of the day's work schedule, employees end up standing around waiting for directions and valuable time is wasted. The manager is then forced to make hasty, spur-of-the-moment decisions that may not be properly thought out.

It is wise to prepare in advance a tentative schedule for the next week's activities. This is a very efficient method of work scheduling, although the schedule may have to be modified to adjust to rainy weather, employee illness, equipment breakdown, and other unanticipated problems. A firm schedule of the next day's activities can be planned the afternoon or evening before each work day.

The work schedule can be communicated verbally to workers or can be posted in a written form. Some managers inform employees of their next day's work activities the afternoon before; others wait until the morning of the scheduled day to announce assignments. Advance notice can be advantageous because workers often like to know what they will be doing the following day, and it allows them to prepare for their assigned activities.

Flexibility is important. For example, the manager should maintain a list of alternative activities that can be performed when bad weather disrupts the planned work schedule. On rainy days the manager can consult his list of inside jobs, quickly assign them, and keep his workers busy accomplishing constructive tasks. Typical inside activities include equipment maintenance and repair, interior work on the maintenance building, training sessions, and taking inventory.

The type and amount of work scheduled for each employee depends on the length and difficulty of the job and the abilities of the employee. Tasks such as spraying pesticides that require higher levels of skill should be assigned to the better-trained, more capable workers. Simpler jobs such as mowing can be assigned to the less experienced employees. The manager must be able to roughly estimate how long it should take to complete the activity so that he can schedule the correct amount of time for it. Time estimation tables are especially helpful to inexperienced supervisors (Table 22-1). Management studies have repeatedly shown that scheduling a heavy work load for employees is best. Not only is more accomplished, but workers are also happiest when busy. A light schedule is less efficient and can cause employees to become bored with their work. However, continually pushing workers too hard by assigning inadequate periods of time to complete tasks can result in employee frustration and hurried, inferior work.

For reasons of efficiency, employees should spend the majority of their time working at the tasks at which they are best. However, occasionally the manager should change the workers' assignments so that they have the opportunity to perform new or different activities. Some alternation of work activities is necessary to broaden the employees' skills and to relieve the boredom that can result from performing the same jobs every day. It is also beneficial for

TABLE 22-1 Time Estimation for Mowing Operations*

Speed		Width of Cut**							
Miles per Hour	per Minute	Feet 6″	12″	18″	24″	36″	48″	60″	72″
.25	22	91	46	31	23	16	12	10	8
.5	44	46	23	16	12	8	6	5	4
1.0	88	23	12	8	6	4	3	3	2
1.25	110	19	10	7	5	4	3	2	66
1.50	132	16	8	6	4	3	2	66	55
1.75	154	13	7	5	4	3	71	57	48
2.0	176	12	6	4	3	2	62	50	42
2.5	220	10	5	3	3	66	50	40	33
3.0	264	8	4	3	83	55	42	33	28
3.5	308	7	4	95	71	48	36	29	24
4.0	352	6	3	83	62	42	31	25	21
4.5	396	6	110	74	55	37	28	22	19
5.0	440	198	99	66	50	33	25	20	17
5.5	484	180	90	60	45	30	23	18	15
6.0	528	165	83	55	42	28	21	17	14
8.0	704	124	62	42	31	21	16	13	11

Some average speeds

Slow walk, pushing mower . . 150′/min.
Slow walk . . . . 200′/min.
Fast pushing mower . . . . 250′/min.
Modest riding yard mower . . . 275′/min.
Fast, brisk walk . . . . 300′/min.
Good riding yard mower . . . 300–350′/min.
Tractor-towed riding mower . . . . . 6–7 mph

*The times required to mow a turf area are given in minutes per 1,000 ft^2 (above the thick black line) and minutes per acre (below the line). The information was compiled by the Professional Grounds Management Society.

**These figures are for effective width of cut, that is—width of blade less operational overlap averages = 4" to 10".

the managers to have people who are experienced enough to "fill in" when a coworker is ill or on vacation.

An example of typical weekly and daily work schedules is shown in Figure 22-6.

Activities for next week (August 23–27)

Top priority

1. Repair end zone on football field that was damaged by vandals—use sod
2. Weed flower beds near administration building (complaints received)
3. Routine mowing, irrigation
4. Fertilize lawn areas and athletic fields
5. Core-cultivate (aerate) softball field
6. Seed area (1,200 ft^2) dug up by telephone company
7. Trim around buildings with weedeater
8. Cut down dead maple by library
9. Overseed lawn at president's residence
10. On Friday, do activities 1, 2, 3, 4 on rainy-day list if not already completed

Rainy-day activities

1. Take truck to Bill's garage for inspection (sticker expires Aug. 31)
2. Adjust carburetor on turf vehicle
3. Check sprayer (need it for broadleaf weed control week of Aug. 30–Sept. 3)
4. Meet with employees for brief refresher training session on pesticide safety and to discuss ways to improve appearance of campus (1 hour)
5. Pick up peat moss in Albany
6. Mower maintenance (sharpen blades, lubricate, change oil, replace belts on Thompson riding mower)
7. Update supply inventory
8. Paint employees' lunch room

Monday, August 23

Joe B.	8–12	Cut sod, repair football field end zone
	1–5	Fertilize athletic fields
Sally K.	8–10	Work with Joe on sod
	10–12	Mow athletic fields
	1–5	Continue mowing
Al M.	8–10	Weed flower beds by administration building
	10–12	Fertilize lawns
	1–5	Continue fertilizing
Heather S.	8–10	Work with Al on flower beds
	10–12	Mow lawn areas
	1–5	Continue mowing

note – 1/2 of mowing done
Sod job looks good
All athletic fields are fertilized
[The daily work schedule may also be prepared without assigning approximate completion times, as shown in the following schedule.]

Tuesday, August 24

Joe B.	1. Irrigate athletic fields
	2. Seed area dug up by phone company
	3. Core-cultivate softball field
Sally K.	1. Seed area dug up by phone company (Joe will help)
	2. Mow
Al M.	1. Finish fertilizing
	2. Trim around buildings and fences with weedeater
Heather S.	Mow

FIGURE 22-6 Examples of work schedules prepared by a college grounds supervisor.

TABLE 22-2 Typical Labor-Hours Record Sheet

	Week of 8/4–8/10									Week of: 8/11–8/17								
								Total									Total	
Employee	S	M	T	W	T	F	S	Reg.	Overtime	S	M	T	W	T	F	S	Reg.	Overtime
Joe B.	0	H*	8	8	8	8	0	40	0	0	8	8	9	8	8	2	40	3
Sally K.	0	H	8	8	8	8	0	40	0	0	8	8	8	S	8	0	40	0
Al M.	0	H	8	8	V	V	0	40	0	0	V	8	8	8	8	0	40	0
Heather S.	0	H	8	8	8	11	0	40	3	0	8	8	8	8	V	0	40	0

*H = Holiday, S = Sick day, V = Vacation day. The employees are paid for holidays, sick days, and vacation days.

RECORD KEEPING

Recordkeeping is a tedious but essential duty. Many turf managers consider paper work the least enjoyable part of their job. However, a successful management program is impossible without good records. The type and quantity of records kept by the manager are largely determined by the specific needs of the manager and federal and state regulations. The kinds of records to be discussed are commonly used in most segments of the turfgrass industry.

Labor Hours

The supervisor is responsible for keeping a record of the number of hours worked by her employees. This information is used to compute the amount of wages earned by each employee and to keep track of sick days and vacation time. A simple labor-hours timecard is shown in Table 22-2.

Time-Use Records

Some managers compile records of the amount of time employees spend performing each activity to which they are assigned (Figure 22-7). These time-use records are kept by the supervisor or employee and can be used to help evaluate labor efficiency. For example, the manager may discover from time-use records that mowing operations are inefficient because an unacceptably large amount of time is spent cutting grass. He can then take corrective actions such as applying growth suppressants to certain areas, encouraging employees to mow faster, mowing unimportant areas less frequently, or buying larger equipment that will speed up mowing operations. Time-use records enable the manager to become aware of inefficiency problems and to justify his corrective actions to his supervisors.

DATE	MAINTENANCE ACTIVITIES
8/23	Joe B. 4hr. sodding football field – 4hr fertilizing fields
"	Sally K. 2hr. sodding football field – 6hr mowing fields
"	Al. M. 2hr. hand weeding flower beds – 6hr fertilizing lawns
"	Heather S. 2hr. hand weeding flower beds – 6hr mowing lawns
8/24	Joe B. 1½ hr. irrigating fields – 3hr seeding – 3½hr core cultivating
"	Sally K. 4½hr. seeding – 3½hr mowing lawns
"	Al M. 5hr. fertilizing lawns – 3hr trimming around buildings
"	Heather S. 8hr. mowing lawns

Turf areas and commonly performed jobs may be coded for easier recordkeeping. For example, if A = lawns and 1 = mowing, the entry "Heather S. 8A1" means that she spent eight hours mowing lawns.

FIGURE 22-7
A sample daily time-use record.

By reviewing records of past work performances the manager can estimate how long a job should take to complete and how much it will cost. An evaluation of time-use information allows the manager to know who performed each activity so that he can rotate work assignments, to keep track of when maintenance practices occurred, and to substantiate budget requests. Managers may also use the records as documentation of the accomplishments of the workers under their supervision. The time-use records are generally summarized on a monthly or yearly basis (Table 22-3).

TABLE 22-3 Summary of Time-Use Records for Maintenance of the 19 Greens and Collars and 40 Greens Sand Bunkers at Hidden Valley Country Club, 1994

Activity	Number of Times Performed per Year	Total Labor-Hours per Year	Average Hours Spent Each Time Activity Performed on All Greens
Mowing	142	811	5.71
Irrigation (regular)	79	162	2.05
Irrigation (syringing)	22	33	1.50
Changing cups	92	94	1.02
Repairing ball marks	100	123	1.23
Fertilizer application	8	42	5.25
Fungicide application	11	69	6.27
Insecticide application	2	15	7.50
Herbicide application	1	10	10.00
Lime application	1	19	19.00
Core cultivation	2	255	127.50
Spiking	12	64	5.33
Vertical mowing	7	39	5.57
Topdressing	8	332	41.50
Overseeding	1	8	8.00
Checking greens	62	65	1.05
Miscellaneous (leaf removal, repairs, etc.)		72	
Raking bunkers	211	703	3.33
		Total = 2,916	

An analysis of the data in Table 22-3 can provide the golf course superintendent with valuable information. For example, if she is looking for ways to reduce labor costs, an examination of the time-use record suggests several possibilities, even though the overall greens maintenance program is relatively efficient. Presently one greensworker spends the equivalent of three full work weeks (123 hours) repairing ball marks. Educating golfers to repair their own ball marks could significantly decrease the amount of time the greensworker must devote to this activity. Core cultivation, while performed only twice a year, is excessively labor intensive (255 hours). The practice is too valuable to discontinue, but it can be made less time-consuming by purchasing a core harvester, which attaches to a turf vehicle and collects the cores or plugs. Using this piece of equipment reduces the number of hours required for core cultivation because presently much time is spent cleaning up the greens after aeration.

The summary also shows that almost every day of the season more than three hours is spent raking sand bunkers. The superintendent might consider removing some of the bunkers or simply assign a smaller amount of time for their maintenance. Leaving an inexpensive rake by each bunker and encouraging golfers to smooth the sand after they hit out of a trap would also help to decrease labor expenses.

Inventory

The manager must keep a record of supplies and equipment on hand. This itemized list of materials in stock is called the inventory. Keeping updated inventory records enables the manager to know what supplies and equipment are presently available and what needs to be ordered for future use. Taking inventory is also necessary because these items owned by the business are assets and their value must be periodically calculated for tax purposes and to help determine the financial condition of the business. To keep the record current, inventory that is used up or sold or loses its value should be deleted from the list.

A materials inventory record is shown in Table 22-4. In this example, the quantity of material used is subtracted from the inventory list on the date of use and the remaining amount is recorded. Another method, which may be preferred if small quantities of an item are used frequently, is to simply record the remaining amounts of material at the end of the month rather than change the inventory list each time a few ounces or pounds of a product are used.

An equipment inventory record is shown in Table 22-5. Life expectancy is estimated from past experience and by use-life information available from the manufacturer. The number of hours per year that the equipment is used, how well it is maintained, and the skill of the operator have a major effect on life expectancy. The equipment manual should be studied thoroughly by all personnel who will be using the machine.

Equipment Maintenance Records

All equipment requires preventative maintenance to ensure that it is in proper operating condition. Preventative maintenance lengthens the useful life span of machinery and results in reduced equipment repair and replacement costs. The manager, mechanic, or equipment operator must keep records of when maintenance services are performed on a machine and how many hours the machine is in use. The equipment manual contains a preventative maintenance schedule based on the number of hours of equipment operation. The person responsible for equipment maintenance should rigidly adhere to this schedule.

The list of maintenance services necessary after 50, 100, 300, and 600 hours of equipment use shown in Figure 22-8 is fairly typical and will help the reader to understand the importance of equipment maintenance records. However, the manager should consult the manual for each piece of equipment under her supervision to determine exact maintenance schedule specifications. The oil level must always be checked before a machine is used.

TABLE 22-4 Materials Inventory Record

Materials	Dates of Use and Remaining Quantities										Total Use
25-10-10	1/1	5/20	5/21	6/11	8/28	8/29	9/3	11/1	11/2	12/31	
(50-lb bags)	180	125	100	200	120	40	37	137	80	80	300
15-15-15	1/1	5/3	9/5	9/9	11/6					12/31	
(80-lb bags)	100	85	60	20	100					100	80
Insecticide	1/1	6/4	8/17							12/31	
(5-gal cans)	5	2 1/2	2							2	3
Fungicide	1/1									12/31	
(25-lb bags)	7									7	0
Preemergent herbicide	1/1	4/29	4/30							12/31	
(36-lb bags)	62	36	9							9	53
Broadleaf herbicide	1/1	5/18	5/19	5/20	9/8					12/31	
(5-gal container)	4	3 1/2	2 3/4	2	1 1/2					1 1/2	2 1/2
Roundup	1/1	6/1	6/9	7/4	8/27	9/8	9/10			12/31	
(1-gal containers)	2	1 7/8	1 3/4	1 1/2	1 1/3	1	1/2			1/2	1 1/2
Lime	1/1									12/31	
(tons)	0									0	0
Peat moss	1/1	5/3	5/17	5/28	6/14	8/26	9/5	9/9		12/31	
(6-ft^3 bales)	50	28	23	13	2	27	13	1		1	74
Topsoil	1/1	4/26	5/2	9/1						12/31	
(yd^3)	0	80	40	10						10	70
Lawn seed mix	1/1	5/4	5/19	5/22	8/15	9/6	9/8	11/8		12/31	
(50-lb bags)	6	3 1/2	3	2	5	4	2 1/2	2		2	7
Perennial ryegrass	1/1	3/24	4/27	8/15	9/17	10/16				12/31	
(50-lb bags)	0	3	2	1 1/4	1	2				2	2
Kentucky bluegrass	1/1	4/29	5/3	8/15	9/17	10/16				12/31	
(50-lb bags)	2 3/4	2 1/2	2 1/4	1 1/2	1/2	2 1/2				2 1/2	2 1/4

Note: When quantities increase it means that products ordered were received on that date.

Other Types of Records

Many other types of records may be kept by the turfgrass manager. It is necessary to record the dates on which pesticides are applied, the product and the rates used, and the target pest. Fertilizer application dates and rates are usually recorded as well. Some managers keep detailed records of a wide assortment of information, such as the sizes of the various areas they maintain, monthly precipitation, accidents that occur to employees, and evaluations of employee performance which can be used to decide who deserves promotion or pay raises. The manager may be required to keep track of all expenditures.

TABLE 22-5 Equipment Inventory and Status Record for 1994

Equipment	Serial Number	Purchase Date	Original Cost	Equipment Status	Life Expectancy (years/hours)	Projected Replacement Year
Knoop riding mower	1786230	7/22/92	8,900	Fair	3/3,000	'95
Bruetsch riding mower	002176	3/5/93	8,100	Poor	3/3,000	'96
Bates 7-gang mower	205-A-13	11/19/89	20,351	Good	7/5,000	'96
Acme walk-behind mower	086127	7/6/91	616	Poor	3/1,000	'94
Hyatt walk-behind mower	097123	5/3/92	699	Good	3/1,000	'95
Helme walk-behind mower	K-5139	8/1/93	832	Excellent	3/1,000	'96
Bayzon diesel tractor	6B-980-526	4/12/86	17,118	Good	12/5,000	'98
Testa turf vehicle	16708-B	10/8/90	9,375	Fair	4/1,000	'94
Crosby pickup truck	RN-622-2	3/29/85	10,187	Poor	10/2,500	'95
Turf King power sprayer	8365-5	10/8/89	2,800	Good	7/3,000	'96

50 hours

- Check tire pressure and water level in battery
- Clean air filter (wash or replace if necessary)
- Remove corrosion from battery terminal connections
- Check oil level in transmission and hydraulic system
- Grease lubrication points

100 hours

- Clean engine
- Drain oil and refill engine crankcase
- Lubricate clutch and throttle linkage
- Tighten loose screws and nuts

300 hours (or every six months)

- Inspect and clean spark plugs
- Check ignition point gap
- Change hydraulic oil filters
- Check fan and drive belts

600 hours (or yearly)

- Change air cleaner and spark plugs
- Touch up with paint

FIGURE 22-8

A generalized preventative maintenance schedule. The list gives examples of maintenance practices that are commonly performed after 50, 100, 300, and 600 hours of equipment operating time.

Very impressed with Supergreen variety planted near park entrance—excellent appearance—stayed green during drought—no disease problems yet

Need another rotary mower for trimming—perhaps buy two more if money becomes available

Received several complaints about low branches at picnic area—must prune trees

Joe B. was handling granular insecticide without gloves—mention importance of safe pesticide handling to employees at next group discussion

Debbie K. suggested I check with Holmes Fertilizer Co.—they have 20-10-10 on sale this month

Read in August Turf Association Bulletin about a new growth regulator—should reread article and talk to sales representative about the product

FIGURE 22-9
Typical entries in the journal or diary of a grounds supervisor at a park.

Many outstanding turf managers maintain a journal or diary in which they record miscellaneous information they consider to be important (Figure 22-9). A busy supervisor can easily forget ideas that may help to improve his management program unless they are written down for future reference. Mistakes commonly made by workers may be recorded so that the manager will know what should be discussed at training sessions. The manager may write down mistakes that he makes so that he can remember to avoid them in the future. Ideas obtained from reading turfgrass magazines, innovations learned at educational meetings, suggestions from other managers, and criticism received can be entered in the diary. The supervisor may record his opinion of a turfgrass cultivar that he has planted, the daily environmental conditions, his observations concerning the effectiveness of a pesticide, or his thoughts about ways to improve employee morale. Any information worth remembering can be noted in the journal.

Photographs or videos are an excellent way to provide a permanent record of the manager's work. They can be used to show proof of the manager's skill when her performance is being evaluated by her employers or when she is applying for a new job. "Before" and "after" shots illustrating problems that the manager has solved and the resulting improvements serve as excellent evidence of her accomplishments. Many successful turf managers use camcorders to make training videos or to document problems and improvements.

BUDGET PREPARATION

Many turfgrass managers are required to prepare an annual budget. The budget proposal is an estimate of how much money is needed to finance the turfgrass management program for the coming year. Some managers are required to develop a highly accurate estimate of projected expenditures, while others are asked to provide only a rough estimate. Considerable time and care are necessary to prepare a precise estimate of expenditures.

TABLE 22-6 Summary of Previous Year's Expenditures

Month	J	F	M	A	M	J	J	A	S	O	N	D	Total	%
Labor	3,906	4,016	4,381	6,719	7,162	7,942	8,016	7,185	6,231	5,709	3,972	3,881	69,120	68
Supplies	604	1,437	5,002	3,071	1,663	389	1,000	195	2,816	795	1,215	2,333	20,520	20
Equipment (purchase and repair)	62	14	0	6,137	117	38	54	136	2,796	0	16	79	9,449	9
Miscellaneous (utilities, phone, etc.)	417	436	206	175	57	54	43	67	39	106	201	295	2,096	2
													$101,185 yearly total	

Generally 65 to 80 percent of the total budget is consumed by labor expenses—wages, salaries, payroll taxes, employee insurance, and benefits such as pensions and paid holidays and vacations. Supplies, materials, and equipment purchase and repair usually account for 15 to 30 percent of yearly expenditures. Miscellaneous expenses such as telephone and utility costs range from 1 to 3 percent.

The previous year's budget and record of actual expenditures are the most valuable source of information for the manager when he is preparing a budget proposal (Table 22-6). The effect of inflation must be considered when the budget proposal is based on past costs. A larger amount of money has to be requested to compensate for price increases, and employees should receive higher wages to offset increased cost-of-living expenses. Sales representatives should be able to provide the manager with accurate predictions of product price changes for the coming year.

The manager should examine the materials inventory record before determining how much money must be included for supplies in the budget proposal. Some quantities of the items needed during the next year may already be on hand (Figure 22-10). To determine the amounts of fertilizers and pesticides required, the manager must know the size of the areas under his supervision. Measuring wheels are used by many people in the turfgrass industry (Figure 22-11).

Checking the equipment inventory and replacement schedule enables the manager to be aware of what new equipment should be purchased during the coming budget year. Any expected increases in expenditures because of proposed changes in the management program must be included in the budget request. Methods of decreasing some expenses can be discovered by examining time-use records.

Managers who can clearly explain the necessity of all budget requests are more likely to receive adequate financial support than managers who are not prepared to show how each item in the proposal is essential to the maintenance program. It is especially important to justify any major changes in the

Fertilizer to be purchased is a 25-8-12, which comes in 80-lb bags.
Cost per bag is $25.00.

Area to be fertilized is 20 acres.

Application time	Rate	Number of bags needed
Spring	20 lb N/Acre	20
Late summer	40 lb N/A	40
Late fall	40 lb N/A	40
		100 total

We have 15 bags on hand

100 – 15 = 85 needed

$$
\begin{array}{r} \$25 \\ \times\ 85 \\ \hline \$2{,}125 \end{array} = \text{fertilizer cost for year}
$$

FIGURE 22-10

A manager's worksheet showing how the fertilizer budget is prepared.

budget that result in requests for large sums of money. The manager should substantiate the cost of new activities or the purchase of new equipment by explaining how the expenditure will improve the program and perhaps save money in the long run. A sample justification appears in Figure 22-12.

When a manager is strictly held to her budget and not allowed to exceed it, she must be careful to include every possible anticipated expenditure in the

FIGURE 22-11

A measuring wheel is used to quickly and accurately determine the dimensions of an area.

I believe that the request for $15,400 to purchase a Yelle 84" Model 19 riding mower is well justified. Our present Lamco riding mower has a 42" cutting width and is in poor shape. Last year the breakdown time for the mower totaled over 27 days. This was a great inconvenience, seriously disrupted the maintenance schedule, and resulted in numerous complaints about the height of the grass.

The cost of parts and labor to repair the mower and get it in decent running condition is approximately $3,200. It does not seem worthwhile to spend this much money to temporarily repair a machine that is seven years old (the manufacturer estimates its normal life expectancy to be five years).

The Yelle 84" Model 19 has twice the cutting width that our present mower has and can mow an area twice as fast. Reducing mowing time by 50% will result in labor savings of over $800 per year. This figure was arrived at by the following calculation:

15 minutes (time saved per acre) × 10 acres (total mowing area) × 30 (number of times mowed per year) = 70 hours saved per year

70 hours × $12.00 (hourly labor cost) = $840

Purchasing a new mower will solve several problems—time wasted repairing the old mower, inability to mow when necessary, and complaints about tall grass. If we buy a new Yelle that has double the cutting width of our present mower, we can achieve significant time and labor savings.

FIGURE 22-12
A sample budget request justification.

proposal. Some managers purposely overorder supplies in years when financial support is generous and stockpile the excess to ensure that adequate amounts are on hand during periods when money is less available. Pesticides, however, are generally ordered only when they will be needed in the near future because of the limited shelf life of some of these chemicals and the safety problems associated with storing hazardous materials.

OTHER BUSINESS SKILLS

Managers of lawn care services and other turfgrass maintenance companies often have to estimate the cost of a job and submit a bid to potential customers. Generally managers are expected to commit themselves to a relatively firm price. Accurate cost estimation is essential because the manager will not be awarded the job if his bid is too high, and if it is too low he may get the job and lose money on it.

A number of factors must be considered when developing a cost estimate for a yearly maintenance job. The manager must first determine which maintenance services (mowing, fertilization, weed control, irrigation, etc.) the prospective customer wants performed. The next step is to measure the square footage of the areas to be maintained. Then the manager must estimate the number of times each maintenance activity will be performed annually and

TABLE 22-7 Sample Maintenance Estimating Table*

Maintenance Operation	Minutes per 1,000 ft²	
Mowing		
19-inch walk-behind power mower	5	
21-inch walk-behind power mower	4	
48-inch riding mower	36	Minutes per acre
72-inch riding mower	24	Minutes per acre
Fertilizing		
broadcast (rotary) spreader	3	
Weed control		
drop (gravity) spreader	15	
3-gal hand pump sprayer	15	
power sprayer with 15-inch boom	8	
power sprayer with 30-inch boom	4	
Vertical mow or power rake	10	
Cleanup		
leaf rake	25	
power sweeper or vacuum	5	
trimming with string trimmer	10	

*This information was developed for lawn care professionals by the Professional Grounds Management Society. *(Courtesy Professional Grounds Management Society)*

how many minutes are required to complete the activity per 1,000 ft^2 (93 m^2) each time it is performed. This information can be obtained from the manager's own records or by consulting maintenance estimating tables (Table 22-7). An estimate of the total time per year for each activity is arrived at by multiplying the number of minutes the activity takes per 1,000 ft^2 by the number of 1,000 ft^2 to be maintained by the number of times the service is performed annually (Figure 22-13).

Once the total number of hours per year required for each maintenance activity is calculated, the manager then determines labor and machinery costs per hour for each of these services. Labor costs reflect more than simply the worker's hourly wage. Other expenses such as medical insurance, worker's compensation, taxes, paid vacation and holidays, sick days, pensions, unemployment insurance, and nonproductive work time must be included in the hourly labor cost (Figure 22-14). If machinery is used to perform the activity, its hourly cost is estimated in a similar manner by considering factors such as the original purchase price, life expectancy, and maintenance, repair, fuel, in-

Number of minutes to fertilize 1,000 ft^2 = 3 minutes
Number of 1,000 ft^2 to be maintained = 10
Number of times service is performed annually = 4
Total time per year for fertilization =
3 minutes × 10 (no. of 1,000 ft^2) ×
4 applications = 120 minutes =

$$\frac{120}{60} = 2 \text{ hours}$$

FIGURE 22-13

Estimating the amount of time it will take to fertilize a 10,000-ft^2 lawn on a yearly basis.

surance, storage, and licensing expenses. The manager can then determine how much it will cost to do the job by multiplying hourly costs times the total number of hours required to complete the activities. This estimated cost figure is increased by an appropriate percentage to ensure an adequate profit.

The manager also has to determine the amount of materials that will be

COMPUTING PRODUCTIVE WORK COSTS*

Days Per Year		= 365 Days
Less – 10 Holidays		
12 Vacation Days		
6 Sick Days Used		
2 Misc.		
30 Days Paid Leave	= 30	
52 Weekends × 2 Days Each	= 104	
Total Nonworking Days	= 134	= – 134 Days
Total Work Days		= 231 Days
Average Productive Hours/Day = 6 Hours × 231 Days Discounting time for paperwork, meetings, travel, prep time, breaks, put away, etc.		= 1386 Hours
Employee Wage	= $16,000.00/Year (7.70/Hour)	
Plus 15% Benefits	= $2,400.00	
Cost/Productive Hour	= $18,400.00/Year ÷ 1386 Hours	= $13.28/Hour
Add 10% for overhead (Low)		= 1.33/Hour
⟶ Total Cost Per Hour of Productive Time		= 14.61/Hour
Just to Break Even—You Must Charge		= 14.61/Hour
If You Want 20% Profit Add		2.92/Hour
	For a Total of	$17.53/Hour

*Do not use these figures—develop your own.

FIGURE 22-14

Labor cost calculations. *(Courtesy of Professional Grounds Management Society)*

used. The quantity of materials needed for regular maintenance (fertilizer, herbicides, seeds, etc.) can be predicted fairly easily by knowing the size of the area and application rates and frequency. Insecticide and fungicide use is more difficult to predict because the manager cannot be certain whether insect or disease problems will occur. Some turf care businesses include preventative applications of insecticides and fungicides in their normal maintenance program, while others inform customers that they will be charged extra if insect or disease control becomes necessary.

The manager adds a percentage onto the cost of materials used on the job. This is called a markup. Charging more for the product than the manager has paid to purchase it offsets business expenses such as freight, handling, and storage costs and allows a profit. Professional turf care businesses buy materials at wholesale prices, which are lower than the retail prices the general public pays for the same products. The manager often charges maintenance customers the higher retail price, which is the same price the customer would pay if he bought the materials at a garden center or hardware store.

The retail price is the wholesale price plus the markup. The markup is usually from 25 to 50 percent based on the retail price. The percentage of the markup is determined by dividing the difference between the wholesale and retail price by the retail price. When a product wholesales for $10.00 and retails for $15.00, the markup is 33 1/3 percent ($5.00/$15.00). Pricing techniques vary. Some turf care businesses, for example, ignore the retail price and simply charge double the wholesale cost for the material.

The price that the manager quotes a potential customer for a maintenance job is based on estimated labor, machinery, and material costs. The cost figures are then increased by a large enough percentage to guarantee a sufficient profit. Determining labor and machinery costs and appropriate hourly charges and product markups can be a complicated process. Larger businesses use accountants or bookkeepers to prepare this information.

An alternative to cost estimation is the time plus materials method. The manager tells the customer the hourly charges for labor and machinery and that the normal retail rate will be charged for any materials used, but no figure is given for the total cost of the job. The manager presents the bill after the job is completed. This is the safest pricing technique because the manager knows exactly how many hours were worked and how much material was used when the bill is prepared. Time plus materials is an especially good method for inexperienced managers who doubt their ability to develop an accurate estimate. However, this method is not always satisfactory to the customer because she does not know what she will be charged until the job is finished.

To avoid misunderstandings between the turf care service and the customer, the manager should prepare a maintenance agreement which specifies the activities to be performed. Whether the agreement is a formal contract or a nonbinding specifications statement, it should clearly explain the nature and frequency of services provided (Figure 22-15).

Customer name:

Address:

The services provided will be performed in a professional, expert manner to ensure that the lawn is maintained in a superior condition. This lawn cared for by the Quality Lawn Company shall receive no less than the following:

A. Mowing shall occur from April 1 to November 1. When the lawn is actively growing it shall be cut no less than once every seven days. During periods of slow growth less frequent mowing will be required. If clippings are long, they will be collected. Clippings short enough not to cause an aesthetic problem will be left on the lawn to allow valuable nutrients to be returned to the soil. Mowing height shall be 2 inches in sunny areas and 2 1/2 inches in the shade.

B. The lawn will be irrigated whenever necessary to keep the grass throughout the growing season. Irrigation frequency will depend on temperature and rainfall. During drier, hotter periods the lawn will be watered heavily either once or twice a week.

C. Fertilization will be sufficient to maintain the lawn in a healthy, attractive condition. Turf areas shall receive 4 pounds of actual nitrogen per 1,000 ft^2 as well as ample amounts of phosphorus and potassium. Fertilizer will be applied in the spring, early summer, late summer, and late fall. A top-quality fertilizer specially formulated to meet the nutritional requirements of lawns in this area will be used.

D. Herbicides will be applied at the correct time to ensure excellent weed control. A preemergent herbicide will be applied in the spring to prevent crabgrass. A broadleaf weed killer will be applied in the late spring and again in the early fall if necessary to control dandelions, plantain, chickweed, and other broadleaf weeds. A totally weed-free lawn cannot be guaranteed because a few weed species are not killed by chemicals.

E. All edges of the lawn will be edged at least once a month to maintain a clean appearance.

F. In the fall leaves will be raked up.

G. In the late summer or early fall the lawn will be dethatched with a verticutting machine. Removing the thatch helps to revigorate the lawn and improves the health of the grass though the area will look a bit unattractive for a brief period after this necessary practice is performed.

H. The lawn will be "cleaned up" before each mowing. Undesirable debris such as twigs, branches, trash, stones, and paper will be picked up and removed from the area.

I. Maintenance activities other than those previously stated in this agreement may be performed if requested by the customer. However, customers will be charged an additional fee for any service performed that is not listed in sections A–H of this agreement. Examples of these "extra" services include aerating, liming, seeding, sodding, tree or shrub pruning and fertilization, and insect and disease control. Our lawn specialists will constantly check for insects and disease during their regular maintenance visits.

IMPORTANT: IF AREAS OF THE LAWN SUDDENLY BEGIN TO DIE, CALL OUR CUSTOMER SERVICE NUMBER—597-8300—IMMEDIATELY. A LAWN SPECIALIST WILL COME WITHIN 24 HOURS AFTER THE CALL AND DIAGNOSE THE PROBLEM. IF THE PROBLEM IS CAUSED BY INSECTS OR DISEASE, CHEMICAL CONTROL WILL BE APPLIED AT A REASONABLE COST.

Dates of maintenance services—April 1, 1994 to November 1, 1994.

Signed ____________________ Signed ____________________

FIGURE 22-15 A maintenance specifications agreement.

COMPUTERS

Computers have had a major impact on business management practices in the United States. The development of personal computers has had a revolutionary effect on management programs (Figure 22-16).

Many turfgrass managers perform virtually all of their recordkeeping with a computer. They use computers to keep track of budgets and expenses, inventory and purchase orders, personnel and labor records, equipment maintenance information, pesticide use, and maintenance activities. Software is available for many diverse applications. Lawn care companies use computer programs that schedule work activities, estimate the time and cost for maintenance procedures, and set up routes so that employees will only have to travel short distances between customers.

Turf managers can also buy programs that calibrate a sprayer or operate an automatic irrigation system. Word processing software is used on a daily basis.

Traditionally, turfgrass industry personnel have kept themselves updated by attending educational meetings and reading trade magazines. Today this is changing. Modems and communications software enable data transmission between computers. A modem converts computer signals into a form that can be transmitted over telephone lines. A computer user can go online (on a phone line) and obtain information from a database, or communicate with other turf professionals by typing a message and sending it to an electronic bulletin board.

The Turfgrass Information File (TGIF) is a turfgrass information database located at Michigan State University. TGIF contains thousands of records. This

FIGURE 22-16

A personal computer. *(Photo by Brian Yacur)*

information comes from periodicals, research and technical reports, dissertations, trade and professional magazines, and books. The turf manager can go online from her office and find information on bunker renovation, fairway mowing, disease control, organic fertilizers, the effect of different mowing heights, what varieties are best in the shade, and thousands of other topics. The manager can then download the information into her computer and print the information at leisure.

TurfByte is an electronic bulletin board system for golf course superintendents. Bulletin boards are an easy way for people with computers and modems to exchange information. A superintendent can call up TurfByte and type in a question about how to overseed a fairway or control a disease. Other superintendents read the message and reply if they have suggestions or comments. Turf managers can have electronic conversations with their colleagues throughout the United States and Canada without leaving their office. The most attractive feature of a bulletin board is that users can join in at their convenience because these conversations take place over an extended period of time.

Many states have computerized agricultural information systems which contain turfgrass information. The Internet is a worldwide network that allows access to these and many other resources.

SELF-EVALUATION

1. Studies show that the majority of wasted work time is caused by __________ __________.
2. __________ __________ should be planned before the manager arrives at work in the morning.
3. __________ records show how much time employees spend completing their assigned activities.
4. A list of materials presently in stock is called the __________.
5. Always check the __________ level before using a piece of equipment.
6. __________ expenses account for 65 to 80 percent of the annual budget.
7. The manager should be prepared to __________ all budget requests.
8. The __________ __________ __________ method of preparing a bill is a safer pricing technique than estimating the cost of a job before hand.
9. What is the percentage markup when a product that has a wholesale price of $25.00 is sold at a retail price of $50.00? __________
10. Is the following statement true or false? The only information the manager needs to know to calculate hourly labor costs is the hourly wages paid to her employees. __________
11. Develop a budget for the maintenance of the athletic fields at your school.
12. Estimate how long it should take to mow the athletic fields at your school with the equipment used by the grounds staff. Determine if the equipment is efficient.
13. Discuss the qualities a turf manager needs to be a successful leader.

CHAPTER

23

Turfgrass Calculations

OBJECTIVES

After studying this chapter, the student should be able to

- Understand why math skills are important to a turfgrass manager
- Describe the types of math problems a turf specialist encounters
- Solve typical math problems that are common in the turfgrass industry

INTRODUCTION

The turfgrass manager is confronted with mathematical problems every day. Fortunately, these problems are relatively simple, and their solution requires only basic arithmetic skills and common sense. Accuracy is essential. If, for instance, a mistake is made when calculating the number of ounces of pesticide to add to a spray tank, the results can be disastrous.

Many people feel uncomfortable about working with numbers. Turf specialists, however, cannot avoid calculations, so they must develop the necessary math skills. This is accomplished by doing practice problems, enrolling in a math course, or by studying an introductory math textbook. Using a calculator is very helpful.

When solving problems, answers should always be double-checked. Putting decimal points in the wrong place is a common cause of serious errors. The following calculations are examples of typical math problems encountered by turf managers. Some have been discussed in previous chapters. *The capital letter M is the abbreviation for 1,000 ft*^{2}.

PROBLEM 1

The turf manager wants to apply a 10-5-5 fertilizer at the rate of 1 pound of nitrogen per 1,000 ft^2. How many pounds of the fertilizer should be applied per 1,000 ft^2?

There are several methods that can be used to solve any of these problems. This problem will be solved by three different methods to illustrate some of the mathematical approaches available.

Solution

Ask the mathematical question: 10 percent of what amount of fertilizer contains 1 pound of nitrogen (N)? Ten percent is the same as 10/100, 1/10, or .1/1. When multiplying or dividing by 10 percent, it is written as .10.

$$.10x = 1 \text{ lb N per M}$$

$$x = \frac{1 \text{ lb}}{.10} = \frac{10}{1} = 10 \text{ lb fertilizer}$$

Ten pounds of a 10-5-5 fertilizer should be applied per 1,000 ft^2 to supply 1 pound of nitrogen per M.

The problem can also be solved by setting up a proportion:

$$\frac{10}{100} = \frac{1 \text{ lb N}}{x}$$

where

$$\frac{10}{100} = $$ 10%, 1 lb is the amount of nutrient needed, and

x is the total amount of fertilizer that should be applied. To solve for x, cross-multiply:

$$10x = 100 \text{ lb}$$

$$x = \frac{100}{10}$$

$$x = 10 \text{ lb fertilizer/M}$$

A short method used by agriculturists is to divide the number of pounds of nutrient to be applied by the percentage of that nutrient in the fertilizer.

$$\frac{1 \text{ lb N}}{.10} = \frac{10}{1} = 10 \text{ lb fertilizer}$$

PROBLEM 2

Using the information obtained in the previous problem, determine how much it will cost to fertilize an area that is 200 feet long and 90 feet wide at the rate of 1 pound N per M. Each bag of 10-5-5 fertilizer weighs 50 pounds and costs $10.00.

Solution

First, calculate how many square feet are in the area to be fertilized:

$$200 \text{ ft} \times 90 \text{ ft} = 18 \text{ M}$$

The amount of fertilizer needed per M was determined in problem 1 to be 10 pounds.

$$\frac{10}{100} = \frac{1 \text{ lb N}}{x}$$

$$10x = 100 \text{ lb N}$$

$$x = 10 \text{ lb/M}$$

How much fertilizer is required to cover 18,000 ft^2? There are 18 M areas in 18,000 ft^2.

$$10 \text{ lb/M} \times 18 \text{ M} = 180 \text{ lb fertilizer needed}$$

How many 50-pound bags is this? Divide the number of pounds of fertilizer by the weight per bag:

$$\frac{180 \text{ lb}}{50 \text{ lb}} = 3.6 \text{ bags}$$

Suppliers will sell only whole bags, not part of a bag, so four bags must be purchased. To find the cost:

$10.00 (cost per bag) × 4 = $40.00/4 bags

The actual cost to fertilize 18,000 ft^2 is:

$10.00 × 3.6 bags = $36.00

PROBLEM 3

It is stated on a 36-pound bag of 28-8-12 fertilizer that the bag covers 10,000 ft^2. The turf manager must determine what rate of N/ per M is being applied if the spreader setting recommended on the bag is used.

Solution

The first question to be answered is, How many pounds of nitrogen does the bag contain?

$$\begin{array}{rl} 36 \text{ lb} & \text{(wt. fertilizer in bag)} \\ \times\ .26 & \text{(\% N in fertilizer)} \\ \hline 9.36 & \text{(lb N in bag)} \end{array}$$

We can round off 9.36 to 9.4 pounds. Now, if the bag is spread over 10,000 ft^2, how many pounds of N per M will be applied?

$$\frac{9.4 \text{ lb N}}{10} = .94 \text{ lb N/M}$$

(no. of M areas)

PROBLEM 4

Before seedbed preparation the turf specialist sends a soil sample to a laboratory for analysis. The soil test results indicate that 2 pounds of phosphorus (P_2O_5) should be incorporated per 1,000 ft^2. He has triple superphosphate (0-46-0) on hand. The site contains 12,470 ft^2. How much superphosphate is needed?

Solution

$$\frac{46}{100} = \frac{2 \text{ lb } P_2O_5}{x}$$

$$46x = 200$$

$$x = \frac{200}{46}$$

$$x = 4.3$$

4.3 pounds of superphosphate per 1,000 ft^2 supply 2 pounds of P_2O_5 per M. To fertilize 12,470 ft^2, it would take 54 pounds of triple superphosphate.

$$12.47 \text{ M} \times 4.3 \text{ (lb/M)} = 53.6 \text{ (54) lb}$$

PROBLEM 5

An extension bulletin recommends applying Milorganite (6-2-0) at the rate of 20 pounds per 1,000 ft^2. How much nitrogen will be applied per M?

Solution

$$\begin{array}{rl} 20 \text{ lb} & \text{(fertilizer applied/M)} \\ \times \ .06 & \text{(\% N in fertilizer)} \\ \hline 1.2 & \text{(lb N/M)} \end{array}$$

PROBLEM 6

On the label of a gallon of 20-10-10 liquid fertilizer, it states that the gallon should be applied to 5,000 ft^2. The label also states that the gallon of liquid fertilizer weighs 9.6 pounds. How much N per M will be applied if the recommendation is followed?

Solution

First determine the number of pounds of nitrogen in the gallon:

$$\begin{array}{rl} 9.6 \text{ lb} & \text{(wt. gallon)} \\ \times \ .20 & \text{(\% N)} \\ \hline 1.92 & \text{(lb N)} \end{array}$$

To calculate the rate of N per M divide the pounds of nitrogen applied by the size of the area fertilized:

$$\frac{1.92 \text{ lb N}}{5 \text{ M}} = .38 \ (.4) \text{ lb N/M}$$

PROBLEM 7

A fertilizer label states:

24% total nitrogen
6% water-insoluble nitrogen

What percentage of the total amount of nitrogen is in the slow-release form?

Solution

Divide the percentage of water-insoluble nitrogen by the total percentage of nitrogen in the fertilizer.

$$\frac{.06}{.24} = .25 \ (25\%) \text{ slow-release N}$$

PROBLEM 8

The turf manager is deciding which of two fertilizers is the better buy. One, a 16-8-8, comes in a 44-pound bag and sells for $14.95. The other, a 25-12-12, costs $15.95 for a 35-pound bag. Each fertilizer contains approximately 30 percent slow-release nitrogen.

Cost comparisons between fertilizers are valid only when they have approximately the same ratio and percentage of slow-release nitrogen. In this case, they are similar in both regards and can be compared.

Solution

This problem is solved by determining the cost per pound of nutrients. The first step is to calculate how many pounds of nutrients are in each bag. Add up the percentages of N, P_2O_5, and K_2O and multiply this number times the weight of the bag.

16-8-8	25-12-12
.16	.25
.08	.12
.08	.12
.32	.49
44 lb × .32 = 14.08 lb/nutrients	35 lb × .49 = 17.15 lb/nutrients

The final step is to divide the cost of each bag by the pounds of nutrients in it. This calculation determines the price per pound. The numbers can be rounded off.

$$\frac{\$14.95}{14.1 \text{ lb}} = \$1.06 \qquad \frac{\$15.95}{17.2 \text{ lb}} = \$.93$$

The 25-12-12 at $.93 per pound of nutrients is the better buy, if the two fertilizers are similar in quality.

Some turf people consider only the pounds of nitrogen in the bag when they perform fertilizer cost analysis. Phosphorus and potassium are ignored, and the cost comparison is based on the price per pound of nitrogen alone. Using this method, the 25-12-12 is again shown to be a more economical fertilizer source than the 16-8-8:

16-8-8	25-12-12
44 lb × .16 = 7.04 lb N	35 lb × .25 = 8.75 lb N

$$\frac{\$14.95}{7.04 \text{ lb}} = \$2.12/\text{lb N} \qquad \frac{\$15.95}{8.75 \text{ lb}} = \$1.82/\text{lb N}$$

PROBLEM 9

A sandy loam soil has a pH of 5.2, and the turf specialist wants to raise the pH to 6.5. He consults a liming recommendation chart and finds that approximately 50 pounds of dolomitic limestone per 1,000 ft^2 will increase the pH of a sandy loam soil by 1 pH unit. How much dolomitic lime is required per M?

Solution

To determine how many units the soil pH needs to be raised, subtract the present pH from the desired pH:

$6.5 - 5.2 = 1.3$

Then multiply this number times the pounds of lime required to increase the soil pH by one unit:

$50 \text{ lb} \times 1.3 = 65 \text{ lb}$

Add 65 pounds of dolomitic lime per 1,000 ft^2 to correct the soil pH.

PROBLEM 10

An extension agent recommends applying 100 pounds of limestone per 1,000 ft^2 to raise the pH of a clayey soil 1 unit. The present pH is 5.8; the desired pH is 6.5. The area contains 63,000 ft^2. How much limestone is needed?

Solution

$6.5 - 5.8 = .7$

$100 \text{ lb} \times .7 = 70$ lb limestone/M

70 lb per M $\times$ 63 M = 4,410 lb limestone required.

PROBLEM 11

A sports field (300 ft × 150 ft) is to be topdressed with .25 inches of sand. How much sand is needed to place a 1/4-inch layer of topdressing on the field?

Solution

The first step is to determine the cubic feet of sand required. The next step is to convert cubic feet to cubic yards. Cubic measurement is calculated by multiplying length times width times height.

Length × width = 300 ft × 150 ft = 45,000 ft^2

The third dimension, height, is .25 inches. This number must be converted to feet because the other two measurements are in feet. There are 12 inches in a foot—.25 in is equal to 1/4 in. The height is 1/48 feet (1/4 × 1/12).

$$\text{number ft}^3\ (1 \times w \times h) = 45{,}000\ \text{ft}^2 \times \frac{1}{48} = \frac{45{,}000}{48} = 937.5\ (938)\ \text{ft}^3$$

A cubic yard contains 27 ft^3 (3 ft × 3 ft × 3 ft). To calculate the number of cubic yards, divide 938 ft^3 by 27 ft^3.

$$\frac{938}{27} = 34.7\ (35)\ \text{yd}^3$$

Some trucking companies base their charge on weight rather than volume. One cubic yard of sand weighs approximately 1.5 tons, so in this case 53 tons (35 yd^3 × 1.5 tons) would be ordered.

PROBLEM 12

A golf course superintendent plans to build a 2,000 ft^2 tee. To elevate the tee and establish a desirable root zone, he decides to place a 9-inch layer of sand over the existing soil. What will be the total cost for the sand if the price per cubic yard is $15.00?

Solution

Nine inches is the same as .75 feet or 3/4 feet. The cubic feet of sand needed is:

$$2{,}000\ \text{ft}^2 \times 3/4\ (.75)\ \text{ft} = 1{,}500\ \text{ft}^3$$

$$\frac{1{,}500\ \text{ft}^3}{27\ \text{ft}^3} = 55.6\ \text{yd}^3$$

55.6 can be rounded off to 56 cubic yards.

$$\begin{array}{rrl} & \$\ 15.00 & (\text{cost/yd}^3) \\ \times & 56 & \text{yd}^3 \\ \hline & \$840.00 & \end{array}$$

The sand will cost $840.00.

PROBLEM 13

A turf specialist intends to improve soil structure by incorporating peat moss. The instructions on the 4 ft^3 bale recommend using 1 bale per 200 ft^2. The price of a bale is $10.00. The site is 60 feet long and 50 feet wide. How many bales of peat will be needed and what will be the cost?

Solution

First, the square area is calculated:

$$60 \text{ ft} \times 50 \text{ ft} = 3{,}000 \text{ ft}^2$$

Next, a proportion can be set up:

$$\frac{1 \text{ bale}}{200 \text{ ft}^2} = \frac{x \text{ bales}}{3{,}000 \text{ ft}^2}$$

Cross-multiply to solve for x—the number of bales needed to improve 3,000 ft^2:

$$200x = 3{,}000$$

$$x = \frac{3{,}000}{200}$$

$$x = 15 \text{ bales}$$

The cost of 15 bales is $10.00 × 15 = $150.00.

PROBLEM 14

A growth retardant costs $200.00 a gallon. The recommended application rate is 1.5 pints of growth regulator per acre (43,560 ft^2). What is the cost of the chemical per acre?

Solution

There are 8 pints in a gallon.

$$\frac{8 \text{ p}}{\$200 \text{ (gal. price)}} = \frac{1.5 \text{ p}}{x \text{ (price of 1.5 p)}}$$

$$8x = 300$$

$$x = \frac{300}{8} = \$37.50$$

PROBLEM 15

A gallon of herbicide costs $90.00. The recommended application rate is 2 ounces per 1,000 ft^2. How much does it cost to treat 1,000 ft^2?

Solution

There are 128 ounces in a gallon.

$$\frac{128\text{ oz}}{\$90} = \frac{2\text{ oz}}{x}$$

$$128x = 180$$

$$x = \$1.41$$

PROBLEM 16

A 5-pound container of fungicide costs \$80.00. The material is a wettable powder, and the recommended application rate is 3 ounce per 1,000 ft^2. How much will it cost to spray 8,500 ft^2?

Solution

There are 16 ounces in a pound. A 5-pound bag contains 80 ounces (5 lb × 16 oz/lb). The next step is to determine how many ounces are needed to spray 8,500 ft^2.

$$3\text{ oz/M} \times 8.5\text{ M} = 25.5\text{ oz}$$

The answer can then be calculated by using the proportion:

$$\frac{80\text{ oz}}{\$80} = \frac{25.5\text{ oz}}{x}$$

$$80x = 2{,}040$$

$$x = \$25.50$$

An alternative method is to calculate the price per ounce:

$$\frac{\$80}{80\text{ oz}} = \$1.00/\text{oz}$$

Then multiply the cost per ounce times the number of ounces needed:

$$\$1.00 \times 25.5\text{ oz} = \$25.50$$

PROBLEM 17

A turf manager has a choice between two granular herbicides. Herbicide A costs \$33.50 per bag and each bag treats 18,000 ft^2. A bag of herbicide B costs \$25.95 and covers 7,500 ft^2. Which one is more economical?

Solution

The problem is solved by determining how much it will cost to treat 1,000 ft^2 with each product. This is accomplished by dividing the number of 1,000 ft^2 that each bag covers by the cost of the bag.

Herbicide A $\frac{\$33.50}{18\text{ M}} = \$1.86/\text{M}$

Herbicide B $\frac{\$25.95}{7.5\text{ M}} = \$3.46/\text{M}$

PROBLEM 18

The recommended application rate for a granular insecticide is 5 pounds per 1,000 ft^2. It comes in 25-pound bags. How many bags would it take to treat an acre?

Solution

There are 43,560 ft^2 in an acre. This number can be rounded off to 44,000 ft^2. The recommended application rate is 5 pound per M. There are 44 1,000 ft^2 areas in an acre. The following calculation determines the pounds needed:

$5\text{ lb} \times 44\text{ M} = 220\text{ lb insecticide}$

Or a proportion can be used:

$$\frac{5\text{ lb}}{1\text{ M}} = \frac{x}{44\text{ M}}$$

$$1x = 220$$

$$x = 220\text{ lb}$$

How many bags is this? Divide the amount of insecticide required by the weight of a bag:

$$\frac{220\text{ lb}}{25\text{ lb}} = 8.8\text{ bags}$$

Nine bags must be purchased

PROBLEM 19

The recommended pesticide application rate may be expressed in pounds of active ingredient (a.i.) per acre. If a 5 percent granular (5G) insecticide is to be applied at the rate of 5.5 pounds of active ingredient per acre, how many pounds of the granular material should be applied per acre?

Solution

The answer can be determined by dividing the pounds of active ingredient needed per acre by the percentage of active ingredient in the granular pesticide:

$$\frac{\text{lb a.i./acre}}{\text{\% a.i. in pesticide}} = \frac{5.5\text{ lb}}{.05} = 110\text{ lb}$$

A proportion can also be set up, using 5/100 to represent 5 percent:

$$\frac{5}{100} = \frac{5.5\text{ lb}}{x}$$

$$5x = 550\text{ lb}$$

$$x = 110\text{ lb}$$

PROBLEM 20

An emulsifiable concentrate (E or EC) formulation is to be used to treat a 0.5-acre turf area. The recommended application rate is 4 pounds of active ingredient per acre. The formulation contains 2 pounds of active ingredient per gallon (stated as 2E or 2EC on the label). How much pesticide is needed to spray the area?

Solution

$$\frac{\text{lb a.i. recommended per acre}}{\text{lb a.i./gal}} \times \text{area (acres)} = \text{gal pesticide to be applied}$$

$$\frac{4\text{ lb}}{2\text{ lb}} \times \frac{1}{2}\text{ acre} = 1\text{ gal}$$

PROBLEM 21

A golf course superintendent plans to seed a new green with creeping bentgrass. The seeding rate is 1 pound per M. The green is not perfectly round, so the diameters differ. To determine the average diameter of this irregular circle, the superintendent measures the width at several points. The diameter measurements are 72 feet, 67 feet, 69 feet, 68 feet, and 74 feet. How much seed is necessary to establish the green?

Solution

First, the area of the green must be calculated. The area of a circle = πr^2. The quantity $\pi = 3.14$; r is the radius, which is half of the diameter.

To find the radius, the average diameter must be determined. This is accomplished by adding up the different diameters and dividing the total by the number of measurements taken.

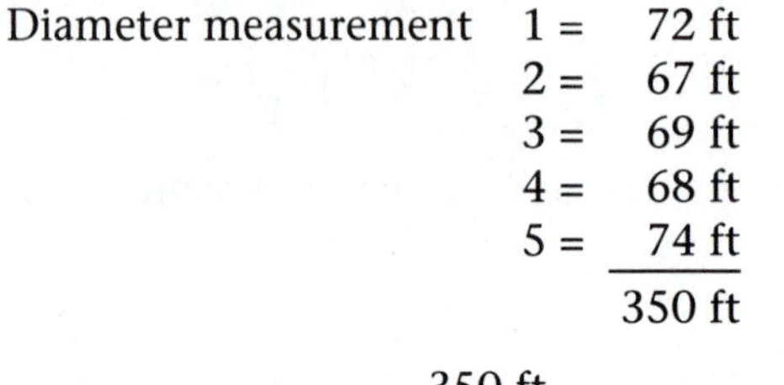

Diameter measurement		
	1 =	72 ft
	2 =	67 ft
	3 =	69 ft
	4 =	68 ft
	5 =	74 ft
		350 ft

$$\text{Average diameter} = \frac{350 \text{ ft}}{5} = 70 \text{ ft}$$

Radius = 1/2 diameter = 70 ft/2 = 35 ft

Area = $\pi r^2 = 3.14 \times 35 \text{ ft} \times 35 \text{ ft} = 3{,}846.5 \text{ ft}^2$

We can round off 3,846.5 to 4,000 ft^2. How much seed is needed?

1 lb (rate per m) $\times$ 4 M = 4 lb

$$\text{or } \frac{1 \text{ lb}}{1 \text{ M}} = \frac{x}{4 \text{ M}}$$

$$1x = 4$$

$$x = 4 \text{ lb}$$

PROBLEM 22

A lawn is to be established vegetatively. The sprigging rate is 4 bushels of stolons per 1,000 ft^2. The backyard measures 90 feet by 150 feet, the front yard is 90 feet by 50 feet. How many bushels will it take to sprig the lawn site?

Solution

Calculate the total area to be established.

Backyard:	90 ft $\times$ 150 ft	=	13,500 ft^2
Front yard:	90 ft $\times$ 50 ft	=	4,500 ft^2
			18,000 ft^2

Then, the rate per M is multiplied times the number of 1,000 ft^2:

4 bushels × 18 M = 72 bushels

or

$$\frac{4 \text{ bushels}}{1 \text{ M}} = \frac{x}{18 \text{ M}}$$

$$1x = 72$$

$$x = 72 \text{ bushels}$$

PROBLEM 23

A recreational field at a park is to be seeded. The park manager can buy an athletic field seed mix in 4-pound boxes that cost $8.00 each. The same mix is also sold in 50-pound bags for $85.00. Which would be cheaper?

Solution

The problem is solved by calculating the price per pound of seed in each case and then comparing them. Divide the cost per box and bag by the weight of each.

$$\text{Box} = \frac{\$8.00}{4 \text{ lb}} = \$2.00/\text{lb}$$

$$\text{Bag} = \frac{\$85.00}{50 \text{ lb}} = \$1.70/\text{lb}$$

PROBLEM 24

Using the answers obtained in Problem 23, determine how much money the park manager would save by purchasing bags instead of the smaller boxes. The recreational area contains 40,000 ft^2. The seeding rate is 3 pounds per 1,000 ft^2.

Solution

The first step is to calculate how much seed is needed:

3 lb/M × 40 M = 120 lb

The price per pound of seed in the bag is $.30 less than the price per pound in the box ($2.00 – 1.70 = $.30). Multiply the difference in price per pound times the number of pounds needed:

$.30 × 120 lb = $36.00

The amount saved by purchasing the bags of seed is $36.00.

PROBLEM 25

A southern lawn is to be overseeded with ryegrass. The overseeding rate is 15 pounds of seed per 1,000 ft^2. The ryegrass seed costs $1.00 a pound. How much will it cost to overseed 20,000 ft^2?

Solution

$15 \text{ lb} \times 20 \text{ M} = 300 \text{ lb seed needed}$

$300 \text{ lb} \times 1.00 \text{ (price/lb)} = \$300.00$

The cost of the seed is $300.00.

PROBLEM 26

A golf course superintendent is preparing a report on irrigation costs for the green committee. She needs to know what percentage of the maintenance budget was spent on water. Her records for the past year show that the water bill was $18,600. The total maintenance budget was $372,000.

Solution

Percentage is calculated by dividing the water cost by the total budget figure.

$$\frac{\$18{,}600}{\$372{,}000} = .05 = 5\%$$

PROBLEM 27

Professional turf managers are able to buy turf supplies at wholesale prices. The general public has to pay higher retail prices. For example, the wholesale price for a bag of fertilizer is $10.00, and the retail price is $15.00. What percentage is the retail markup?

Solution

Determine how much less the wholesale price is:

$\$15.00 - 10.00 = \$5.00$

Then divide the difference in price by the retail cost:

$$\frac{\$5.00}{\$15.00} = .333 = 33\text{-}1/3\% \text{ markup}$$

PROBLEM 28

When preparing a bid or a bill, the turf manager often adds 50 to 100 percent or more onto the wholesale cost of the materials used. The manager pays the wholesale price and is able to make a profit by charging the customer a higher price for the supplies.

A homeowner wants 5,000 ft^2 of sod installed. The turf manager pays $.14 a square foot for the sod. What will the manager charge the homeowner for the sod if he chooses to add 50 percent to the price he paid for it?

Solution

Step 1 is to calculate the cost of the sod to the turf manager:

$$5\text{ M} \times \$.14 = \$700.00$$

Step 2 is to add 50 percent to the amount that the turf manager pays for the sod:

$$\$700 \times .50 = \$350.00$$

$$\$700.00 + \$350.00 = \$1,050.00$$

The amount the homeowner is billed for the sod is $1,050.00.

PROBLEM 29

A person in the lawn care business has fertilized and mowed a lawn. The wholesale cost of the fertilizer used is $30.00. It took her two hours to fertilize and mow. How much should her bill be?

Solution

There are several ways to calculate a bill. One method is to add 50 percent to the price of any materials used and also charge an hourly cost for the labor.

$$\begin{aligned} \$30.00 + 50\% &= \$45.00 \\ 2\text{ hr} \times \$35.00\text{ (hourly cost)} &= \underline{\$70.00} \\ &\quad \$115.00 \end{aligned}$$

The bill is $115.00.

Other methods are used as well. Material expenses may be doubled or even tripled. The labor charge is usually increased if there is little or no money made on the markup of materials. The expense of using the equipment and the time it takes to get to the job site should also be considered.

PROBLEM 30

Turfgrass problems are solved in exactly the same manner when metric units are used. For example, a certain pesticide is applied at the rate of 60 milliliters (ml) per 100 square meters (m^2). The site is 64 meters long and 55 meters wide. How much pesticide is needed to treat this area?

Solution

$$64 \text{ m} \times 55 \text{ m} = 3{,}520 \text{ m}^2$$

$$\frac{60 \text{ ml}}{100 \text{ m}^2} = \frac{x}{3{,}520 \text{ m}^2}$$

$$x = 2{,}112 \text{ ml}$$

$$1{,}000 \text{ ml} = 1 \text{ l}$$

$$2{,}112 \text{ ml} = 2.112\ (2.1) \text{ l}$$

PROBLEM 31

Conversions to and from metric measures are necessary. For example, it states in a journal that 97 kilograms (kg) of nitrogen per hectare (ha) is a recommended fertilizer rate. How many pounds of nitrogen per 1,000 ft^2 would this be equivalent to?

Solution

A metric conversion table shows that 1 hectare equals 2.47 acres and 1 kilogram equals 2.2 pounds.

$$1 \text{ ha} = 2.47 \times 43{,}560 \text{ (ft}^2\text{/acre)} = 107{,}593.2 \text{ ft}^2$$

$$97 \text{ kg} = 97 \times 2.2 \text{ lb} = 213.4 \text{ lb}$$

$$\frac{213.4 \text{ lb N}}{107.6 \text{ M}} = 1.98 \text{ lb N/M}$$

SELF-EVALUATION

1. How many pounds of 20-10-10 fertilizer are necessary to fertilize 18,000 ft^2 at the rate of 2 pounds N/M?__________
2. An 80-pound bag of 15-15-15 fertilizer costs $10.95. The recommended rate is 1 pound N/M. How many bags are needed to fertilize an area 100 feet long and 150 feet wide? How much will it cost?__________
3. A fertilizer salesperson says that a 37-pound bag of 16-8-12 covers 5,000 ft^2. What is the rate of N/M?__________
4. An area containing 40,000 ft^2 needs 1 pound of P_2O_5 per 1,000 ft^2. How much 0-20-0 is required?__________
5. A turf manager applies 6 pounds of 33-5-5 per 1,000 ft^2. What rate of N per M has been applied?__________
6. A fertilizer label states:

   15% total nitrogen

   5% water-insoluble nitrogen

   What percentage of the total amount of nitrogen is in the slow-release form?__________
7. How much sand is required to spread 1/4 inch of topdressing over an area measuring 75 feet by 108 feet?__________
8. Topsoil costs $13.00 per cubic yard. How much will it cost to place a layer of topsoil 6 inches deep on a site containing 6,800 ft^2?__________
9. A gallon of fungicide costs $155.00. The recommended application rate is 1.5 ounces per 1,000 ft^2. How much will it cost to treat 36,000 ft^2?__________
10. A herbicide is applied at the rate of 3 ounces per 1,000 ft^2. How many gallons are needed to spray 10 acres?__________
11. What is the area of a green that has a diameter of 83 feet?__________
12. How many bushels of stolons are needed to establish an area 39 ft × 77 ft if 5 bushels are used per 1,000 ft^2?__________
13. A granular insecticide is applied at the rate of 7 pounds per 1,000 ft^2. It comes in 35-pound bags. How many bags are needed to treat 55,000 ft^2?__________
14. A turf manager spent 6 hours maintaining a lawn. During this period she spread 2 bags of fertilizer for which she paid $29.00 each. She also sprayed a quart of herbicide that cost her $27.00. If she charges $30.00 an hour for her labor and adds 50 percent to the price of the materials, what will her bill be?__________
15. A wettable powder formulation which contains 50 percent active ingredient (50 WP) is to be applied at the rate of 3 pounds a.i. per acre. What amount of the material is needed to treat 15,000 ft^2?__________

APPENDIX

A

Conversion Tables

Abbreviations

a	=	are
cm	=	centimeter
g	=	gram
ha	=	hectare
kg	=	kilogram
km	=	kilometer
l	=	liter
m	=	meter
ml	=	milliliter
mm	=	millimeter
MT	=	metric ton

Metric to Customary			Customary to Metric		
VOLUME					
1 cm^3	=	0.061 in^3	1 in^3	=	16.39 cm^3
1 ml	=	0.034 fl oz	1 fl oz	=	29.57 ml
1 m^3	=	35.31 ft^3	1 ft^3	=	0.028 m^3
1 l	=	2.12 pt	1 pt	=	0.473 l
	=	1.06 qt	1 qt	=	0.945 l
	=	0.264 gal	1 gal	=	3.785 l
WEIGHT					
1 g	=	0.053 oz	1 oz	=	28.35 g
1 kg	=	2.205 lb	1 lb	=	453.6 g
1 MT	=	1.102 ton	1 ton	=	0.097 MT
LENGTH					
1 cm	=	0.394 in	1 in	=	2.54 cm
	=	0.033 ft	1 ft	=	30.48 cm
1 m	=	39.37 in		=	0.304 m
	=	3.28 ft	1 yd	=	0.914 m
1 km	=	0.621 mi	1 mi	=	1.609 km
	=	1,093.6 yd		=	1,609.3 m
AREA					
1 cm^2	=	0.155 in^2	1 in^2	=	6.45 cm^2
1 m^2	=	10.76 ft^2	1 ft^2	=	0.0929 m^2
	=	1.196 yd^2		=	929 cm^2
1 a	=	1.076 ft^2	1 yd^2	=	0.836 m^2
	=	119.6 yd^2	1,000 ft^2	=	92.9 m^2
1 ha	=	2.47 acre	1 acre	=	0.405 ha
			1 mi^2	=	2.59 km^2
RATE					
1 kg/ha	=	0.89 lb/acre	1 lb/acre	=	1.12 kg/ha
1 kg/a	=	2.05 lb/1,000 ft^2	1 lb/1,000 ft^2	=	0.488 kg/a
1 g/cm^2	=	2.048 lb/ft^2	1 oz/gal	=	7.8 ml/l
1 g/cm^3	=	62.4 lb/ft^3	1 pt/20 gal	=	1.65 ml/l
1 kg/cm^2	=	14.223 lb/in^2	1 mph	=	1.61 km/hr
1 km/hr	=	0.621 mph			

APPENDIX

B

Sprayer Calibration

The pressure created by a pump causes the water/pesticide mixture to be sprayed out through the nozzles. The amount of liquid applied per 1,000 ft^2 or acre by a sprayer depends upon factors such as nozzle size, operating pressure, and the speed at which the applicator walks while spraying or at which the tractor or turf vehicle pulls the sprayer.

A nozzle tip has an orifice (hole) through which the liquid is sprayed. A nozzle with tip number 8005 has a spray angle of 80 degrees and applies 0.5 gallons per minute at 40 psi. A 730462 tip has a spray angle of 73 degrees and applies 0.462 gallons per minute at 40 psi. It is easy to quickly change the nozzle tips on a sprayer boom from one size to another. Nozzle tips are made out of brass, stainless steel, hardened stainless steel, and a ceramic material.

The greater the pressure, the greater the flow rate through the orifice. To double the application rate, the pressure must be increased four times. As pressure is increased, droplet size decreases. Smaller droplets are more likely to drift off target, so pressures over 60 psi are usually avoided when spraying turf.

If an applicator were spraying at 40 psi and needed to double the rate, he would have to increase the pressure to 160 psi. Many turf sprayers cannot produce pressures this high, and even if they could, drift would be a serious problem. However, the capacity of an 8005 nozzle tip changes from 45 to 78 ounces per minute when the pressure is increased from 20 psi to 60 psi, so rate changes are possible by manipulating pressure as long as the pressure is kept in the proper range.

Increasing the speed at which the nozzles pass over the turf results in a decreased application rate; decreasing the speed increases the rate. An 8005 tip at 40 psi with 20 inch-nozzle spacing sprays 74 gallons per acre (gpa) at 2 miles per hour (mph), 50 gpa at 3 mph, 37 gpa at 4 mph, and 30 gpa at 5 mph. Adjusting the speed until the correct application rate is achieved is another method of calibrating a sprayer. However, the applicator may have a speed that she prefers to use and may not want to vary from it by much. Speeds less than 3 mph are avoided unless the area is difficult to spray because it is rough, small, or has many obstacles. Faster speeds (4 to 5 mph) are often preferred for fairways and larger, open areas.

There are several methods that can be used for calibration. The one described next is good for inexperienced applicators.

The recommended application rate for a certain pesticide is 2 ounces of pesticide in 1 gallon of water per 1,000 ft^2. The sprayer that will be used has nozzles mounted on a small boom and sprays a swath 5 feet wide. It is pulled by a small garden tractor.

The first step is to put a can under each nozzle and collect the amount of water that is sprayed out in 60 seconds. The total volume of all the water collected is measured. If the total output from all of the nozzles was 1.2 gallons in 60 seconds, then the desired amount—1 gallon of water—would be applied in 50 seconds.

$$\frac{1.2 \text{ gallons}}{60 \text{ seconds}} = \frac{1 \text{ gallon (the recommended rate)}}{x \text{ (number of seconds to apply 1 gallon)}}$$

$$1.2\,x = 60$$

$$x = 50 \text{ seconds}$$

The applicator now knows that the sprayer will apply the amount of liquid that is recommended per 1,000 ft^2 in 50 seconds (as long as the same pressure is used). She also knows that she must spray the 1,000 ft^2 in 50 seconds to apply the right amount of liquid.

The next step is to measure and mark a distance of 200 feet. Because the boom has a spray width of 5 feet, driving the tractor a distance of 200 feet results in spraying 1,000 ft^2 (5 ft × 200 ft = 1,000 ft^2). The applicator drives the distance and uses his watch to time the tractor. He varies the speed until he drives the 200 feet in 50 seconds. When the distance is finally covered in the correct amount of time he marks the throttle setting and notes the gear used.

The applicator is now able to apply the pesticide at the recommended rate by traveling at the correct speed.

The applicator should check the amount of water collected in each can to see if each nozzle is applying the same amount of water. If one can has less, the nozzle may be partially plugged and should be cleaned with compressed air or a soft brush.

The amount of spray material that is applied can also be determined by filling the tank with water and then spraying it over a measured area such as 1,000 ft^2 or an acre. The applicator then refills the tank to determine the amount used. If 20 gallons are added to refill the tank after spraying an acre, the application rate is 20 gallons per acre.

Many sprayers have clear plastic tubes, called sight tubes, which are quite useful in sprayer calibration. Behind the tube on the tank various gallonages are marked, usually at 20-gallon increments. The applicator can tell how much water is in the tank by looking at the height of the water in the sight tube. The following problem illustrates how a sight tube is used.

A pesticide label states that 1.5 quarts of pesticide should be applied in 50 to 150 gallons of water per acre. The boom sprays a width of 15 feet, and the spacing between the 8008 nozzle tip is 10 inches. The applicator fills the tank until the sight tube measures 100 gallons, sets the pressure at 30 psi, and then drives at 4 mph until the water drops to the next increment, which is 80 gallons.

She then measures how many feet she traveled to spray 20 gallons. The total is 580 feet. Therefore, she covered 8700 ft^2 (15 ft width × 580 ft length).

$$\frac{20 \text{ gallons}}{8700 \text{ ft}^2} = \frac{x \text{ gallons}}{43{,}560 \text{ ft}^2 \text{ (acre)}}$$

$$8700\,x = 871{,}200$$

$$x = 100 \text{ gallons per acre}$$

The application rate at 4 mph is 100 gallons per acre. Thus the applicator adds 1.5 quarts of pesticide to 100 gallons of water.

The recommended range is 50 to 150 gallons per acre. But what if the applicator chooses 50 gallons per acre rather than 100 because he can spray twice as large an area per tank without having to stop to mix and fill his tank as many times as he would at the 100-gallon rate? To cut the rate in half (100 gallons to 50 gallons), he has to double his speed and drive at 8 mph. This is too fast. A more likely solution is to switch to 8004 nozzle tips, each of which sprays only half as much as an 8008 tip. If 8004 tips are used, 1.5 quarts of pesticide will be added per 50 gallons of water.

Calibration kits are available and can be very helpful. The TeeJet Calibration Calculator is used by many turf managers. Given the nozzle spacing, ounces sprayed by a nozzle in a minute (which is measured by collecting the liquid), and speed (mph) it will calculate the gallons per acre or per 1000 ft^2 that will be applied. The calculator will solve for any one of the four variables if the other three are known and entered into it.

Monitors which measure ground speed and flow rate and give a digital readout of the spray rate on a small screen are popular. Some will help to control the spray rate by changing the flow rate.

Tables are available which show how many gallons per acre or gallons per 1,000 ft^2 will be applied based on nozzle size, pressure, nozzle spacing (usually 10 to 20 inches), and speed. This abbreviated table is an example:

		20-Inch Nozzle Spacing (GPA)			
Nozzle Tip	**Liquid Pressure (PSI)**	**2 MPH**	**3 MPH**	**4 MPH**	**5 MPH**
8005	20	52	35	26	21
	30	64	43	32	26
	40	74	50	37	30
	60	91	60	45	36

If the applicator wants to apply 50 gallons per acre with 8005 tips, she selects a speed of 3 mph and a pressure of 40 psi. If she wants to go 3 mph and apply 60 gallons per acre, she increases the pressure to 60 psi.

The speed of many tractors and turf vehicles can be set by looking at a table which tells how many miles per hour the machine will travel in high or low range, in different gears, and at various rpms. The applicator may find that in low range, 2nd gear, and a 2,300 engine rpms the tractor should go 3 mph. To go 3 mph, then, the applicator can shift into low range, 2nd gear, and lock the engine rpms at 2,300 using a governor or throttle control.

It is a good idea to check vehicle speed by timing how long it takes to travel 100 or 200 feet.

	Time Required to Travel in Seconds a Distance of:	
Speed in MPH	**100 feet**	**200 feet**
2.0	34.1	68.2
2.5	27.3	54.5
3.0	22.7	45.5
3.5	19.5	39
4.0	17	34.1
4.5	15.2	30.3
5.0	13.6	27.3

If the vehicle is correctly set to travel at 3 mph, it should go 200 feet in 45.5 seconds.

It has been determined that the applicator will apply 50 gallons per acre with 8005 nozzles spaced 20 inches apart if he travels at 3 mph and the pressure is 40 psi. To double-check this the applicator can spray out 20 gallons using the sight tube. If everything is operating properly, she should travel 830 feet to spray out 20 gallons:

$$\frac{\text{20 gallons}}{\text{50 gallons per acre}} = \frac{\text{x feet}}{\text{43,560 ft}^2\text{ (acre)}} = \text{17,424 ft}^2$$

The 50 gallons per acre rate is equivalent to spraying 20 gallons per 17,424 ft^2. If the boom has a 21-foot spray width, the distance that should be traveled is 830 feet:

$$\frac{\text{17,424 ft}^2}{\text{21 feet (width)}} = \text{829.7 (830) feet}$$

The application rate of a spray gun is affected by nozzle size, pressure, and walking speed. Nozzles commonly spray at one of the following rates: 1, 2, 2.5, 3, or 4 gallons per minute. A lower pressure such as 20 psi is normally used.

The best way to determine gallons per minute is to spray into a bucket for 30 to 60 seconds and measure the amount collected.

If the applicator wants to apply 1 gallon per 1,000 ft^2, and she collects 2 gallons in 60 seconds, she should cover 1,000 ft^2 in 30 seconds. The applicator must then determine the spray width with the gun. If the distance she sprays when she swings her arm from side to side is 10 feet, she should adjust her walking speed so she travels 100 feet (10 ft × 100 ft = 1,000 ft^2) in 30 seconds.

When pesticide is mixed with water in a spray tank, the weight of a gallon of the mixture may be more or less than the weight of a gallon of water alone. This can have an effect on the spray and should be considered when the change is significant.

After using a sprayer the inside and outside of the tank should be rinsed three times with clean water. Clean water should be sprayed through the nozzles until the water coming out is perfectly clear. A small amount of liquid detergent can be added to the water. Ammonia or activated charcoal can be mixed with the water if the pesticide residues are difficult to remove.

To avoid double applications or missing an area, the applicator can add a special dye to the spray mixture to mark the turf that has been treated. The color pattern is temporary and disappears in a short time.

APPENDIX C

Spreader Calibration

DROP (GRAVITY) SPREADER

Method A

Make a catch tray or pan and fasten it beneath the spreader. Heavy cardboard, wood, tin, plastic, or a piece of gutter (eaves trough) can be used. Then operate the spreader over a 500-ft^2 area. If the spreader application width is 4 feet, the spreader is pushed 125 feet (4 ft × 125 ft = 500 ft^2). Weigh the collected material and multiply by 2 to determine the application rate per 1,000 ft^2. Adjust the rate setting on the spreader and continue the process until the correct amount of material is applied. Smaller areas can be used, but calibration accuracy is better with larger areas.

Method B

Place a piece of plastic on the ground and operate the spreader on it. If, for example, the measured distance on the sheet is 12.5 feet long and the spreader application width is 4 feet, the spreader is run over the plastic 10 times to cover 500 ft^2 (4 ft × 12.5 ft × 10 = 500 ft^2). The material is swept up, then weighed, and the weight multiplied by 2 to determine the application rate per 1,000 ft^2.

ROTARY (CENTRIFUGAL) SPREADER

Weigh the fertilizer before putting it in the hopper. Operate the spreader over a known area and then weigh the amount of material remaining in the hopper. If, for example, the area covered was 1,000 ft^2, the initial weight of the material in the hopper was 10 pounds, and the final weight after application is 4 pounds, the application rate is 6 pounds of material per 1,000 ft^2 (10 lb – 4 lb = 6 lb). Overlap is usually necessary to ensure a uniform distribution and must be considered when determining application rate.

APPENDIX

D

Identification of Grasses By Vegetative Characteristics

The first step in identification is to determine whether the leaves are folded in the bud (Figures 1 and 2), or rolled in the bud (Figure 3). This is accomplished by examining the youngest, unexpanded leaf at the top of the stem. If the leaf appears round, it is rolled. If the bud leaf is more flat than round, it is folded. Other important structures used for identification purposes, such as the ligule (Figures 4 and 5), collar (Figures 6 and 7), and auricles (Figures 8 and 9) are found at the collar area where the leaf sheath and blade join (Figures 10 to 13). A hand lens is helpful when examining plant parts.

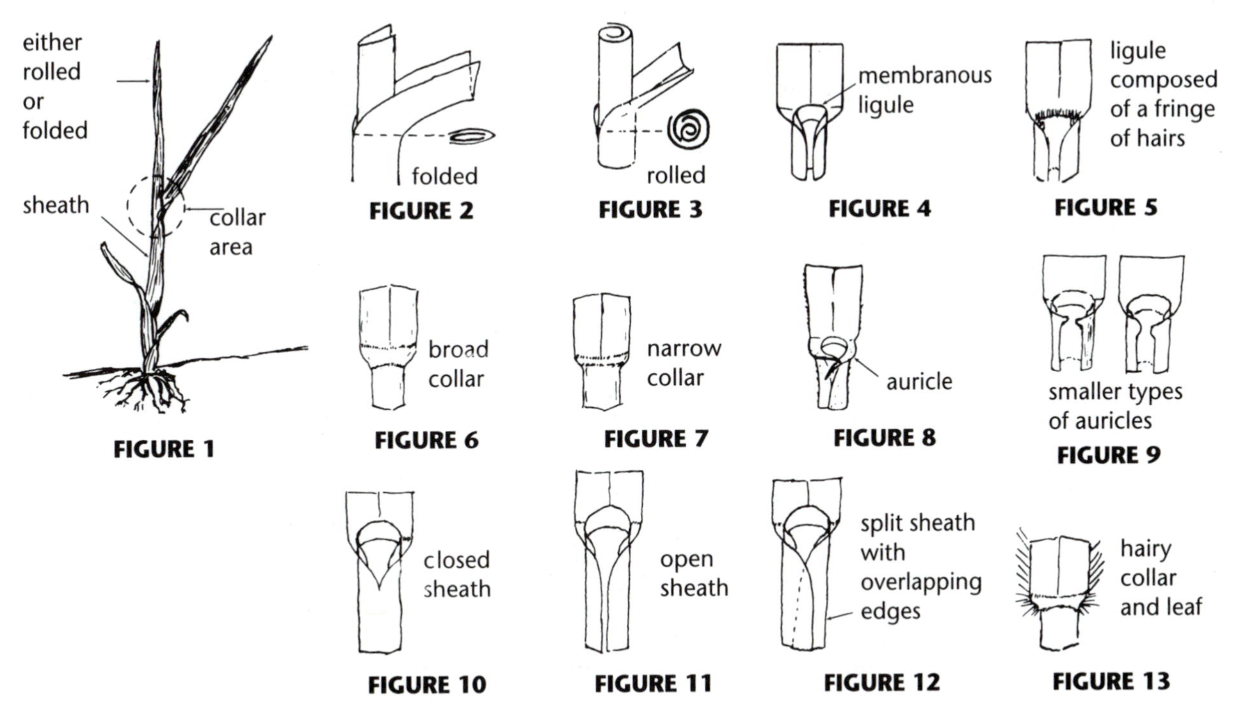

Description	*Grass Species*
I. Leaves folded in the bud	
A. Auricles present, small to long and clawlike; ligule a short membrane; veins of blades prominent; blades shiny on underside, dull on upper surface, 2 to 5 mm wide; bunch-type growth	Perennial ryegrass
B. Auricles absent	
1. Creeping stolons present	
a. Blades narrowed to form short stalk at base of blade; sheaths greatly compressed	
(1) Ligule a fringe of very short hairs; blade 4 to 10 mm wide, blunt and rounded at tips; collar smooth without hairs	St. Augustinegrass
(2) Ligule membranous with short hairs at top; collar has hairs; blade 3 to 5 mm wide, edges hairy toward base	Centipede grass
b. Blades not constricted at base; sheaths compressed	
(1) Ligule a fringe of hairs	
(*a*) Collar broad, hairy; blades and sheaths very hairy; blades 4 to 5 mm wide, V-shaped; stolons and rhizomes present	Kikuyugrass
(*b*) Collar narrow, slightly hairy; sheaths and blades smooth or slightly hairy	
i. Blades 1.5 to 4 mm wide; stolons and rhizomes present; blade tip tapers to a point	Bermudagrass
ii. Blades 4 to 10 mm wide, blunt and rounded at tip; rhizomes absent; nodes hairy	Carpetgrass
(2) Ligule very short, membranous; blades 4 to 8 mm wide, slightly hairy toward the base; stolons and rhizomes short and thick	Bahiagrass (leaves may also be rolled in the bud)
2. Stolons absent	
a. Blades narrow, curled, bristlelike, prominent veins on upper surface	
(1) Rhizomes present; culms reddish at base; ligule membranous and medium-short; blades smooth	Red fescue
(2) Rhizomes absent	
(*a*) Leaves bluish-green, 0.5 to 1.5 mm wide; culms green or pinkish at base; ligule membranous and very short; sheath split	Sheep fescue
(*b*) Leaves bright green, 1 to 2.5 mm wide; culms red at base; sheath closed almost to the top; very similar to red fescue except for absence of rhizomes	Chewings fescue
b. Blades flat to V-shaped; veins inconspicuous	
(1) Blades with boat-shaped tip; transparent lines on either side of midvein (may have to hold up to light to see this)	
(*a*) Rhizomes normally absent	
i. Blades usually light green; smooth sheath whitish at base; membranous ligule long and pointed; seed heads often present	Annual bluegrass
ii. Thin stolons present; sheaths have rough, bumpy feel; membranous ligule long and pointed	Rough bluegrass
(*b*) Rhizomes present	
i. Blade tapers to boat-shaped tip; sheaths strongly compressed; membranous ligule medium-short	Canada bluegrass

Description	*Grass Species*
ii. Blade does not taper to point, entire leaf has similar width; sheaths not strongly compressed; very short membranous ligule	Kentucky bluegrass
(2) Blades without boat-shaped tip; no transparent lines on either side of midvein; prostrate growth habit (stems lie on ground)	Goosegrass
II. Leaves rolled in bud	
A. Auricles present	
1. Sheaths reddish at base; blades shiny on underside	
a. Edges of blades smooth, leaf width 3 to 7 mm; long, clawlike auricles	Annual ryegrass
b. Edges of blades rough; short auricles	
(1) Auricles usually without hairs	Meadow fescue
(2) Auricles and collars have few short hairs, auricles very small	Tall fescue
2. Sheaths not reddish at base; underside of leaves not shiny	
a. Strong rhizomes present; long clasping auricles	Quackgrass
b. Rhizomes absent; long clawlike auricles	Crested wheatgrass
B. Auricles absent	
1. Sheaths round	
a. Collar somewhat hairy	
(1) Sheaths not hairy	
(*a*) Strong stolons present	
i. Collar with long hairs; ligule a fringe of hairs; blades 2 to 5 mm wide; rhizomes	Japanese lawngrass *(Zoysia japonica)*
ii. Collar sparsely hairy; blades 2 to 3 mm wide; shape similar to *Z. japonica*	Manilagrass *(Zoysia matrella)*
(*b*) Stolons not present; weak rhizomes; collar has long hairs; leaf width 1 to 2 mm	Blue grama
(2) Sheaths hairy; no rhizomes	Downy bromegrass
b. Collar not hairy	
(1) Ligule a fringe of hairs; collar broad	
(*a*) Rhizomes present; blade smooth	Mascarenegrass *(Zoysia tenuifolia)*
(*b*) Rhizomes absent, stolons present; blade hairy, 1 to 3 mm wide, grayish-green	Buffalograss
(2) Ligule membranous; collar narrow	
(*a*) Sheath closed almost to top; blade 8 to 12 mm wide, smooth	Smooth bromegrass
(*b*) Sheaths split with overlapping edges; ligule pointed; blades with prominent veins on upper surface	
i. Stolons absent or weak	
Blades 3 to 7 mm wide; long ligule	Redtop
Blades 1 to 3 mm wide; ligule medium-short	Colonial bentgrass
ii. Strong stolons	
Blades 1 mm wide; ligule medium-short	Velvet bentgrass
Blades 2 to 3 mm wide; ligule medium length; long stolons	Creeping bentgrass
2. Sheaths compressed (flattened); ligule very short, membranous; blades 4 to 8 mm wide; stolons and rhizomes short and thick	Bahiagrass

APPENDIX

E

Seed Identification

Seeds are not drawn to scale.

Creeping bentgrass

Colonial bentgrass

Kentucky bluegrass

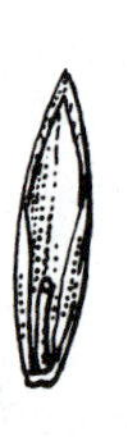

Rough bluegrass

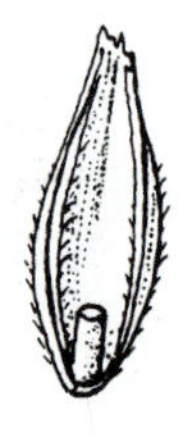

Annual bluegrass

Tall fescue

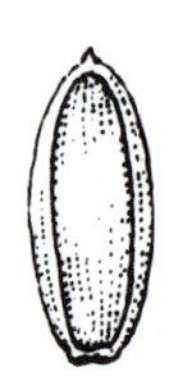

Carpetgrass

Centipedegrass

Bermudagrass

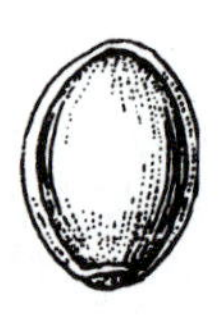

Bahiagrass

Blue gramagrass

Buffalograss (female)

Perennial ryegrass

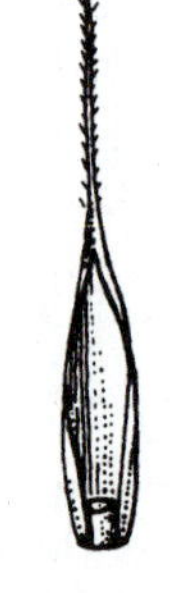

Annual ryegrass

Glossary

Absorption. The entrance of a chemical into a plant or microorganism.

Acid soil. Soil with a pH below 7.0.

Active ingredient (ai). The actual toxic material present in a pesticide formulation.

Adaptation. The ability to adjust to and perform well under certain environmental conditions.

Adsorption. The binding of a chemical to the surface of soil particles or plant parts.

Aeration or aerification. The practice of improving the movement of air, water, and nutrients into the soil by making holes or slits in it. Specialized machines are commonly used to remove plugs of soil or to cut slices in the soil.

Algae. Microscopic plants that lack true stems, leaves, and roots but contain chlorophyll. Algae are occasionally found growing on the soil surface in poorly drained, infertile, or shaded turf areas.

Alkali soil. An alkaline soil having a high sodium content that is characteristic of arid climates.

Alkaline soil. Soil with a pH higher than 7.0.

Annual, summer. A plant that starts from seed, usually in the spring, and completes its life cycle in one growing season.

Annual, winter. A plant that starts from seed in the fall, lives through the winter, and dies the following spring or summer.

Apron. The fairway area immediately in front of a putting green. It is often called the approach by golfers.

Arid regions. Regions where the soil is usually very dry because of the lack of rainfall.

Artificial turf. A synthetic covering composed of nylon fibers that resembles grass. It has been used primarily on sports fields.

Auricles. Clawlike appendages that project from the collar where the leaf sheath and blade join.

Ball mark. The injury caused to turf by the impact of a golf ball when it lands on the green.

Bedknife. The stationary bottom blade on a reel mower. As the reel turns, its blades push the grass against the bedknife which does the actual cutting.

Bench setting. The height at which a rotary blade or bedknife is set above a firm, level surface.

Biennial. A plant that completes its life cycle in two years.

Blade. The expanded or flattened part of a grass leaf located above the sheath.

Blend, seed. A combination of seeds of two or more cultivars of a single turfgrass species.

Blight. A serious disease which causes the rapid death of plant tissue.

Broadleaf. A dicot plant.

Brushing. Lifting stems and leaves with a brush before mowing.

Bud. Cells which may develop into a stem, leaf, or flower.

Bunch-type growth. Plant development by tillering without the production of rhizomes or stolons. Grasses that spread by tillers alone are also referred to as clump grasses.

Bunker. A depression, normally filled with sand, which serves as a hazard or obstacle on a golf course. Golfers often refer to them as sand traps.

Calcium carbonate equivalent. The number of pounds of calcium carbonate ($CaCO_3$) required to neutralize the acidity caused by a ton of fertilizer. This information is stated on the fertilizer bag.

Capillary action. The upward movement of water held in small pores in the soil.

Carbohydrate. Chemical compounds such as sugars, starches, and cellulose that are composed of carbon, hydrogen, and oxygen. Simple carbohydrates are manufactured by plant photosynthesis.

Cation. An ion carrying a positive charge of electricity. Examples of common soil cations are calcium (Ca^{++}), magnesium (Mg^{++}), and potassium (K^{+}).

Cation exchange capacity. The total quantity of exchangeable cations that a soil can absorb. The higher a soil's cation exchange capacity, the greater its ability to hold and store nutrients.

Certified seed. Seed that is guaranteed to be genetically pure or true to type by a certifying agency, which conducts inspections in the production field and after the seed is harvested.

Chlorosis. The yellowing of normally green plant tissue because of the lack of chlorophyll.

Cold water insoluble nitrogen (CWIN). Forms of fertilizer nitrogen that are insoluble in cold water (25°C or 77°F).

Cold water soluble nitrogen (CWSN). Forms of fertilizer nitrogen that are soluble in cold water (25°C or 77°F).

Collar. A narrow band where the leaf blade and sheath join.

Collar. A strip of grass, usually 3 feet (0.9 meter) wide, that surrounds a putting green. It is normally mowed closer than the fairway but higher than the putting surface.

Colorant. A dye or paintlike material which is sprayed on discolored turf to give it an attractive green appearance.

Combing. Lifting stems and leaves with a metal comb so that they will be cut by the mower.

Compaction. An unfavorable soil condition caused by mechanical forces such as heavy traffic. The soil particles are pressed closely together, and pore space is eliminated from the soil.

Compatible chemicals. Chemicals that can be mixed together without losing their desired effects.

Complete fertilizer. A fertilizer product that contains nitrogen, phosphorus, and potassium.

Cool season turfgrass. A turfgrass species that grows best when temperatures are in the 60°–75°F (15°–24°C) range.

Core cultivation. A method of soil cultivation which results in the removal of soil cores (plugs).

Creeping growth habit. The ability to spread horizontally, exhibited by grasses which produce stolons and/or rhizomes.

Crown. The major meristematic area of a grass plant from which all growth is initiated. It is located at the base of the plant near the soil surface.

Culm. The stem of a grass plant.

Cultipacker seeder. A mechanical seeder which plants seeds in a prepared seedbed and firms the soil around the seeds. It is used to seed large areas.

Cultivar. A variety of plant that differs from others in the same species and retains these distinguishing features when reproduced.

Cultivation. Working the soil without destroying the turf by methods such as coring, spiking, and slicing.

Cup. The "hole" on a golf course putting green. It is 4.25 inches (10.8 centimeters) in diameter. The plastic or metal lining placed inside the hole should be sunk at least 1 inch (2.5 centimeters) below the putting green surface.

Cutting height. Also called the mowing height, it is the actual height above the soil surface at which the grass is cut.

Density. See **Shoot density.**

Desiccation. Loss of moisture from a plant because of hot, dry weather or chemicals.

Dethatch. To reduce an excessive thatch accumulation, usually with a machine such as a vertical mower.

Dicot, dicotyledon. A type of plant that has two cotyledons (see leaves) present when the seedling emerges from the soil. Dicots are often referred to as broadleaf plants.

Disease. An abnormal, unhealthy condition caused by pathogens such as fungi, bacteria, and nematodes which results in injury to the grass plant.

Divot. A piece of turf torn up by the golf club when the golfer strikes the ball. Golfers may also refer to the injured area where the turf was removed as the divot.

Dormancy. A resting stage during which a plant's growth rate is greatly reduced because of extended drought, heat, or cold stress.

Dormant seeding. Planting seed during late fall or early winter after temperatures become too cold for seed germination to occur until the following spring.

Drift. Movement of pesticides by air currents away from the intended target area.

Drought resistance. The broad term for the ability of a grass plant to avoid or tolerate desiccation. Drought avoidance occurs when a plant reduces water loss or increases water uptake. This is accomplished by adaptations such as reduced leaf area, thicker cuticles, deeper rooting, and the extensive development of root hairs. Drought tolerance is the ability of a grass plant to tolerate desiccation by mechanisms such as decreasing cell size or becoming dormant. Despite these distinctions, in the turf industry the term *drought tolerance* has traditionally meant any type of tolerance or avoidance. Because this practice is still widespread, no distinction is made between drought tolerance and resistance in this book.

Embryo. The part of a seed that develops into a young plant.

Endosperm. The part of a seed that contains stored food.

Environmental conditions. All external conditions that influence the development of a living organism.

Evapotranspiration. The combined loss of water from a given area because of evaporation from the soil surface and transpiration from plants.

Extravaginal. Growth that occurs when stems penetrate through the basal leaf sheath and grow outside of the leaf sheath. Rhizomes and stolons are examples of extravaginal growth.

Fertigation. The application of fertilizer through an irrigation system.

Fertilizer analysis. The percentage by weight of nitrogen, available phosphoric acid (P_2O_5), and soluble potash (K_2O) found in a fertilizer.

Fertilizer grade. The guaranteed minimum analysis of a fertilizer. A 10-5-5 fertilizer contains 10% nitrogen, 5% available phosphoric acid (P_2O_5), and 5% soluble potash (K_2O).

Fertilizer ratio. A comparison of the amunts of nitrogen, phosphorus, and potassium in a fertilizer. A 10-5-5 fertilizer has a 2:1:1 ratio or two parts of nitrogen to one part available phosphoric acid (P_2O_5) and one part soluble potash (K_2O).

Field capacity. The moisture content of the soil two or three days after the soil has been saturated by rain or irrigation. The soil is at field capacity after the free or gravitational water has drained away.

Flail mower. A type of mower that has a series of individual cutting blades that hang loose from a shaft and are extended by centrifugal force when the shaft rotates at high speeds. It is also called a hammerknife mower.

Foliar burn. Dehydration of leaf or stem tissue due to contact with a high concentration of fertilizer or pesticide.

Footprinting, frost. Foot-shaped discolored areas caused by walking on live, frosted turf.

Footprinting, wilt. Foot-shaped impressions that temporarily appear after walking on a turf experiencing moisture stress. This type of footprinting occurs when leaves are wilting and lack sufficient turgor to spring back up immediately after being stepped on.

Formulation. The form in which a pesticide is offered for sale. Examples are emulsifiable concentrates, granulars, wettable powders, and flowables.

Fungicide. A chemical used to control a disease caused by a fungus.

Fungus. A lower form of plant life that lacks chlorophyll and is therefore unable to produce its own food. Most turfgrass diseases are caused by fungi that feed on grass plants.

Germination. The sprouting of a seed is the technical definition, but many turf managers use the term to mean the first visible appearance of a seedling when it emerges from the soil.

Grade. To establish elevations and contours before planting.

Grain. The horizontal orientation of grass stems and leaves in one direction.

Granules. Solid particles such as clay, ground corn cobs, or vermiculite that serve as a carrier for pesticides.

Gravitational water. Water which moves through the soil under the influence of gravity.

Grub. The larva (immature stage) of various beetle species.

Herbicide. A chemical used to control weeds, also called a weed killer.

Hot water insoluble nitrogen (HWIN). Forms of fertilizer nitrogen that are insoluble in hot water (100°C or 212°F).

Humus. The well-decomposed, more or less stable portion of soil organic matter.

Hybrid. The offspring of genetically dissimilar parents.

Hydroseeding. Mixing seed with water and spraying it on the planting site. The mixture may also contain fertilizer and mulch. This method of seeding is also referred to as hydraulic seeding.

Hyphae. The threadlike filaments that form the mycelium (vegetative part) of a fungus.

Infection. The invasion and establishment of a disease-causing microorganism (pathogen) within a plant.

Infestation. The presence of insect pests, disease-causing organisms, and weeds in a turf area.

Infiltration rate. The maximum rate at which water can enter a soil.

Inflorescence. The flowering part of a plant where seed develops.

Internode. The portion of the stem between nodes.

Intravaginal. Growth that occurs when stems grow upward within an enclosing leaf sheath. A tiller is an example of intravaginal growth.

Lapping. Turning the reel of a mower backward while applying grinding compound between the reel and bedknife. This practice, also known as backlapping, results in a proper fit or "clearance" between the reel blades and bedknife.

Larva. The immature stage of insects that undergo complete metamorphosis. Examples are caterpillars, grubs, and maggots.

Lateral growth. The sideward or horizontal (rather than upward) growth of a plant.

Layering. The development of layers of different-textured materials in the root zone. Layering can negatively affect water movement, aeration, and rooting.

LD_{50}. The amount of pesticide required to kill (lethal dose) 50% of a given test animal population. LD_{50} is usually expressed as milligrams of pesticide per kilogram of body weight (mg/kg).

Leaching. The removal of materials dissolved in the soil solution caused by the movement of water down through the soil. Usually refers to the loss of nutrients from the root zone.

Lignin. A component of plant tissue, especially roots, that is difficult for microorganisms to decompose.

Ligule. An upright, membranous or hairy projection located where the leaf sheath and blade join.

Lime. Various calcium and magnesium compounds used to decrease soil acidity (raise soil pH).

M. The abbreviation for 1,000 ft^2.

Mat. A layer composed of thatch intermixed with soil.

Matting. A method of breaking up soil cores or working in topdressing by dragging a steel mat over the surface of the turf area.

Media. The soil or material in which a plant's roots grow.

Membranous. Thin, soft, often transparent or translucent.

Metamorphosis. The transformation or change in appearance that occurs when an immature insect becomes an adult.

Microclimate, microenvironment. The environmental conditions in the immediate vicinity of a plant.

Micronutrient. A nutrient that plants need only in extremely small amounts to grow and complete in their life cycle.

Microorganism. A tiny living organism such as a bacterium or fungus.

Midvein. The main vein in the center of a leaf, also called the midrib.

Mixture, seed. A combination of the seeds of two or more turfgrass species.

Monocot, monocotyledon. A type of plant that has only one seed cotyledon (seed leaf) present when the seedling emerges from the soil. Grasses are monocots.

Monostand. A turf area composed of only one cultivar.

Moss. A low-growing, small primitive plant that may be found in shaded, wet, or poorly drained turf areas.

Mulch. A material such as straw or peat moss that is spread on a newly planted site to protect the soil and preserve moisture.

Mulch blower. A machine that shreds bales of straw or hay and blows the mulching material onto a newly planted site.

Mycelium. The mass of branching, threadlike filaments that compose the vegetative part of a fungus.

Necrosis. The death of plant tissue.

Nematode. Small (often microscopic), unsegmented roundworms with threadlike bodies. Many nematode species attack plant roots.

Nitrogen activity index (AI). The percentage of cold water insoluble nitrogen that is soluble in hot water, used to estimate the potential solubility of ureaformaldehyde compounds. A UF carrier should have an AI of at least 40% to ensure sufficient nitrogen availability the first several months after application.

Node. The joint of a grass stem; an enlarged area where buds and leaves occur.

Nonselective herbicide. A herbicide that kills many buds and leaves occur.

Noxious. Any plant which has been defined by state law to be a particularly undesirable or harmful weed.

Nymph. The immature stage of insect species that undergo simple metamorphosis (chinch bugs, scales, aphids, grasshoppers, etc.). Unlike larvae, nymphs are similar in appearance to the adult stage.

Organic matter. Material resulting from the growth of living organisms. The term usually refers to animal and plant residues that are in the soil.

Organic soil. A soil which contains a high percentage (>20%) of organic matter, such as a muck soil.

Overseed. Seeding into an existing turf.

Parent material. The rock material or organic matter from which a soil develops.

Pathogen. A disease-causing organism.

Percolation. The downward movement of water through the soil.

Perennial. A plant that has a life span of more than two years.

Permeability. The ease with which plant roots, water, air, or other gases penetrate or move through the soil.

Persistent. A pesticide that retains its toxicity for a relatively long time; also called residual.

pH, soil. The measurement of the acidity or alkalinity of a soil. A pH of 7.0 is neutral; below 7.0 is acidic; above 7.0 is alkaline.

Phloem. The food-conducting tissue of the vascular system.

Photosynthesis. The process by which chlorophyll-containing cells in green plants use the energy from sunlight to convert carbon dioxide and water into carbohydrates.

Phytotoxic. Injurious to plants. The term usually refers to chemicals that can be toxic to plants.

Plug. A small piece of sod used for planting grasses vegetatively.

Plugging. A method of vegetatively establishing turfgrass by planting small pieces of sod in the soil.

Polystand. A turf area composed of two or more cultivars and/or species.

Pore space. The volume of the soil not occupied by solid particles. Pore spaces are filled with either air or water.

Postemergence herbicide. A chemical applied to control weeds after they have emerged from the soil.

Preemergence herbicide. A chemical applied to control weeds before their emergence from the soil.

Pure live seed (PLS). The percentage of seed that is pure and viable (capable of germinating).

Purity, seed. The percentage by weight of the seed of each cultivar in a seed blend or mixture.

Reconstruction. Significantly modifying or rebuilding the soil before replanting a turf area.

Recuperative potential. The ability of a plant to recover from injury; often refers to how rapidly the plant can spread and fill in injured spots.

Reel mower. A mower that has a rotating reel which cuts against a stationary bedknife.

Reestablishment. Replanting an entire area after destroying the old turf and tilling the soil. It does not involve reconstructing the soil.

Release rate. The rate of nutrient release, primarily nitrogen. Water-soluble fertilizers are said to be fast-release or quickly available sources, whereas fertilizers that are insoluble are called slow-release or slowly available sources.

Renovation, complete. Replanting into a turf that has been killed with a nonselective herbicide without tilling the soil.

Renovation, partial. Replacing into a living turf without disturbing the existing turfgrass.

Residual. A characteristic of a pesticide that is effective or persists for more than a short time. The term also refers to the long-lasting response of a slow-release fertilizer.

Resistance. The ability of a plant to overcome, completely or partially, the effect of a pathogen, or the ability of a pest to survive being treated with a pesticide.

Respiration. The process by which plants release energy from carbohydrates.

Rhizome. A spreading stem that grows underground and produces new shoots and roots at the nodes.

Root hairs. Numerous tiny hairlike projections on the outside of roots. They absorb water and nutrients from the soil.

Root zone. The area of the soil where plant roots develop and grow.

Rotary mower. A mower which cuts because of the impact of a rapidly rotating blade.

Rough. The hazard area of a golf course surrounding the fairway and green which is mowed higher than the fairway and has less desirable playing surface.

Saline soil. A nonsodic soil containing high enough levels of soluble salts to interfere with plant growth.

Saline-sodic soil. A soil containing sufficient amounts of soluble salts and exchangeable sodium to interfere with plant growth.

Salt tolerance. The ability of a plant to withstand the potentially injurious effect of soluble salts in the soil or saltwater.

Saprophyte. An organism that lives on dead or decaying organic matter.

Saturate. To fill all of the pore spaces between soil particles with water.

Scalp. To mow a turf so close that an excessive quantity of green leaves is removed. After a scalping the turf has a brown, stubby appearance.

Selective herbicide. A herbicide that kills one type of plant but does not seriously affect other types of plants.

Semiarid turfgrass. Turfgrasses adapted to dry, low rainfall regions. They can be grown without irrigation and include species such as buffalograss and blue gramagrass.

Sheath. The tubular, nonexpanded lower portion of a grass blade which surrounds the stem.

Shoot. A stem and its attached leaves.

Shoot density. The relative number of shoots per unit area.

Sign. Any part of a pathogen that can be seen.

Sod. Pieces or strips of live grass, including leaves, crowns, stolons, rhizomes, roots, and adhering soil, which are used for vegetative planting.

Sod strength. The ability of sod to resist tearing during handling and installation.

Sodic soil. A soil that contains sufficient sodium to interfere with plant growth.

Soil. The top layer of the earth's surface, which is the natural medium for plant growth.

Soil probe. A soil sampling tool also referred to as a core sampler.

Soluble. Capable of being dissolved in a liquid.

Solution. A mixture in which a substance is dissolved in a liquid. The substance in solution remains dissolved indefinitely and does not settle out.

Species. A group of organisms with common characteristics and capable of interbreeding to produce fertile offspring which are like the parents.

Spiking. A cultivation practice in which solid tines or spikes penetrate the soil. Holes are produced, but no plugs of soil are removed.

Spore. The reproductive structure produced by a fungus, comparable in function to a seed.

Spot treatment. Treating a small area infested with a pest rather than applying the pesticide to the entire turf area.

Spreader. A mechanical device that uniformly distributes fertilizer, seed, and granular pesticides over a turf area. The two major types are drop (or gravity) spreaders and rotary spreaders. The second type is also called a broadcast, centrifugal, or cyclone spreader.

Sprig. A stolon, rhizome, or tiller used for propagating certain grasses vegetatively.

Sprigging. Vegetative establishment by planting sprigs (stolons, rhizomes, or tillers) in furrows or small holes.

Stand. A group of plants growing together in one area.

Stem. The part of a plant that has nodes and internodes and which supports a leaf or flower.

Sticker. A material added to a spray mixture to improve its ability to adhere to a plant surface.

Stimpmeter. A device used to measure the speed and uniformity of putting greens.

Stolon. A spreading stem that grows along the surface of the ground and produces new shoots and roots at the nodes.

Stolonizing. Vegetative establishment by broadcasting stolons over the planting site. The stolons are covered by topdressing or pressrolling.

Stomate. One of the tiny openings in the surface of a leaf or stem through which water vapor and gases pass.

Structure, soil. The arrangement or combination of soil particles in a soil.

Succulent. Tender, soft, full of juice or sap.

Surfactant. A material which reduces surface tension between two unlike materials such as oil and water. Surfactants are spray additives which can improve spreading, wetting, sticking, and emulsifying characteristics.

Susceptible. A plant characteristics meaning that the plant lacks the ability to resist disease.

Suspension. A mixture in which a substance is temporarily dispersed in a liquid. Agitation may be required to prevent the substance from settling out.

Symptom. The external or internal changes in a plant that indicate the presence of a disease.

Syringing. A light watering lasting no longer than a few minutes.

Systemic. A chemical that is absorbed into treated plants through leaves or by the roots. Once inside the plant, the chemical moves through the vascular tissue to various parts of the plant.

Tee. The area from which the first shot on each golf hole is hit.

Tensionmeter. An instrument that measures the tension at which water is held in the soil. Tensionmeter readings can be used to determine irrigation needs.

Texture, leaf. Principally, the width of the leaf.

Texture, soil. The relative proportion of sand, silt, and clay in a soil.

Thatch. The accumulation of undecomposed or partially decomposed organic material located above the soil surface. The thatch layer is primarily composed of dead and living root and stem tissue.

Tillage. Preparation of the soil for planting by loosening it mechanically with a rotary tiller, plow, disk, or similar device.

Tiller. A stem that develops from the crown of the parent plant and grows upward within the enclosing leaf sheath of the parent plant.

Tillering. The production of tillers. Tillering results in a greater shoot density.

Tolerance. The ability of a plant to tolerate the effect of a pathogen or environmental stress.

Topdressing. Spreading a thin layer of soil mix over a turf area and working it into the turf to stimulate thatch decomposition and to smooth the surface.

Trace element. A micronutrient.

Traffic. The movement of people, equipment, and vehicles on a turf area.

Transition zone. The zone between the warm season and cool season zones, where both warm season and cool season grasses can be grown but where the climate is not optimal for either.

Translocate. The movement of materials from one part of a plant to another part.

Transpiration. The loss of water vapor to the atmosphere through open stomata in the leaves.

Turf. A covering of mowed vegetation, usually a turfgrass, growing intimately with an upper soil stratum of intermingled roots and stems.

Turfgrass. A special or cultivar of grass, usually of spreading habit, which is maintained as a mowed turf.

Turfgrass quality. A visual assessment of the uniformity, texture, density, color, and growth habit of a turf.

Turgidity. A condition in which plant tissue is swollen or expanded because the cells are filled with water.

Variety. Another term for cultivar.

Vascular system. The tissue that transports materials such as water and nutrients within the plant.

Vertical mowing. Cutting slices in the turf with a machine that has blades mounted on a vertically rotating shaft. Vertical mowing is an important method of thatch reduction.

Verticutting. See **Vertical mowing.**

Volatilization. The transformation of a solid material into a gas, usually referring to a chemical that evaporates at normal air temperatures.

Warm season turfgrass. A turfgrass species that grows best when temperatures are in the 80°–95°F (27°–35C) range.

Wear. The injurious effects of traffic on turf, other than the problems caused by soil compaction.

Weed. A plant growing where it is not wanted.

Wetting agent. A material that is added to a spray mixture to cause better spreading of the liquid on plant surfaces, or that is applied to a soil to improve wettability.

Winter overseeding. Seeding cool season grasses into a permanent warm season turf to provide green, actively growing turf during colder periods when the warm season grasses are brown and dormant.

Winterkill. Any injury to turfgrass plants that occurs during the winter period.

Xylem. The water-conducting tissue of the vascular system.

Bibliography

BOOKS

Baker, J.R. *Insects and Other Pests Associated with Turf.* North Carolina Extension Service, 1986.

Balogh, J.C., and W.J. Walker, Eds. *Golf Course Management and Construction: Environmental Issues.* Lewis Publishers, 1992.

Beard, J.B. *Turfgrass: Science and Culture.* Prentice-Hall, Inc. 1972.

Beard, J.B. *Turfgrass Management: The Golf Course.* Burgess Publishing Company, 1981.

Cockerham, S.T. *Turfgrass Sod Production.* ANR Publications, 1988.

Converse, J. *Proturf Guide to the Identification of Dicot Turf Weeds.* O.M. Scott & Sons, 1974.

———. *Proturf Guide to the Identification of Grasses.* O.M. Scott & Sons, 1973.

Decker, H. and J. Decker. *Turf Management Handbook: A Guide to Professional Lawn Care.* Prentice-Hall, Inc., 1988.

Knoop, W. *The Complete Guide to Texas Lawn Care.* Texas Gardener Press, 1986.

Leslie, A.R., Ed. *Handbook of Integrated Pest Management for Turf and Ornamentals.* Lewis Publishers, 1994.

Niemczyk, H. *Destructive Turf Insects.* HDN Book Sales, 1981.

Shetlar, D.J., P.R. Heller, and P.D. Irish. *Turfgrass Insect and Mite Manual.* Pennsylvania Turf Council, 1983.

Shurtleff, M. *Controlling Turfgrass Pests.* Prentice-Hall, Inc., 1987.

Smiley, R.W., P.H. Dernoeden, and B.B. Clarke. *Compendium of Turfgrass Diseases.* APS Press, 1992.

Smith, J.D., N. Jackson, and A.R. Woolhouse. *Fungal Diseases of Amenity Turf Grasses.* Routledge, Chapman, and Hall, 1989.

Sprague, H.B. *Turf Management Handbook.* Interstate Printers and Publishers, Inc., 1994.

Tashiro, H. *Turfgrass Insects of the United States and Canada.* Cornell University Press, 1987.

Waddington, D.V., R.N. Carrow, and R.C. Shearman, Eds. *Turfgrass*. American Society of Agronomy, 1992.

Watkins, J.A. *Turf Irrigation Manual: The Complete Guide to Turf and Landscape Sprinkler Systems*. Telsco Industries, 1987.

TRADE MAGAZINES

Golf Course Management. Published by the Golf Course Superintendents Association of America, Lawrence, Kansas.

Greenmaster. Published by the Canadian Golf Course Superintendents Association, Toronto, Ontario, Canada.

Green Section Record. Published by the United States Golf Association, Far Hills, New Jersey.

Grounds Maintenance. Published by Intertec Publications, Overland Park, Kansas.

Journal of Environmental Turfgrass. Published by the American Sod Producers Association, Rolling Meadows, Illinois.

Landscape Management. Published by ADVANSTAR Communications, Cleveland, Ohio.

Lawn and Landscape Maintenance. Published by Gie, Inc., Cleveland, Ohio.

Park and Grounds Management. Published by Madisen Publishing Division, Appleton, Wisconsin.

Southern Golf. Published by Brantwood Publications, Clearwater, Florida.

SportsTURF. Published by Adams Publishing Co., Cathedral City, California.

Index